D0084668

Introduction to Industrial and Systems Engineering

PRENTICE HALL INTERNATIONAL SERIES
IN INDUSTRIAL AND SYSTEMS ENGINEERING

W. J. Fabrycky and J. H. Mize, Editors

ALEXANDER *The Practice and Management of Industrial Ergonomics*
AMOS AND SARCHET *Management for Engineers*
AMRINE, RITCHEY, MOODIE, AND KMEC *Manufacturing Organization and Management*, 6/E
ASFAHL *Industrial Safety and Health Management*, 2/E
BABCOCK *Managing Engineering and Technology*
BADIRU *Expert Systems Applications in Engineering and Manufacturing*
BANKS AND CARSON *Discrete-Event System Simulation*
BLANCHARD *Logistics Engineering and Management*, 4/E
BLANCHARD AND FABRYCKY *Systems Engineering and Analysis*, 2/E
BUSSEY AND ESCHENBACH *The Economic Analysis of Industrial Projects*, 2/E
BUZACOTT AND SHANTHIKUMAR *Stochastic Models of Manufacturing Systems*
CANADA AND SULLIVAN *Economic and Multi-Attribute Evaluation of Advanced Manufacturing Systems*
CHANG AND WYSK *An Introduction to Automated Process Planning Systems*
CHANG, WYSK, AND WANG *Computer Aided Manufacturing*
CLYMER *Systems Analysis Using Simulation and Markov Models*
ELSAYED AND BOUCHER *Analysis and Control of Production Systems*
FABRYCKY AND BLANCHARD *Life-Cycle Cost and Economic Analysis*
FABRYCKY AND THUESEN *Economic Decision Analysis*, 2/E
FRANCIS, MCGINNIS AND WHITE *Facility Layout and Location: An Analytical Approach*, 2/E
GIBSON *Modern Management of the High-Technology Enterprise*
HALL *Queuing Methods: For Services and Manufacturing*
HAMMER *Occupational Safety Management and Engineering*, 4/E
HUTCHINSON *An Integrated Approach to Logistics Management*
IGNIZIO *Linear Programming in Single- and Multiple-Objective Systems*
KUSIAK *Intelligent Manufacturing Systems*
OSTWALD *Engineering Cost Estimating*, 3/E
PULAT *Fundamentals of Industrial Ergonomics*
TAHA *Simulation Modeling and SIMNET*
THUESEN AND FABRYCKY *Engineering Economy*, 8/E
TURNER, MIZE, CASE, AND NAZEMETZ *Introduction to Industrial and Systems Engineering*, 3/E
WOLFF *Stochastic Modeling and the Theory of Queues*

Introduction to Industrial and Systems Engineering

THIRD EDITION

WAYNE C. TURNER
JOE H. MIZE
KENNETH E. CASE
JOHN W. NAZEMETZ

Oklahoma State University

PRENTICE HALL, Englewood Cliffs, N.J. 07632

Library of Congress Cataloging-in-Publication Data

Introduction to industrial and systems engineering / Wayne C. Turner
. . . [et al.].—3rd ed.
p. cm.—(Prentice-Hall international series in industrial and systems engineering)
Rev. ed. of: Introduction to industrial and systems engineering / Wayne C. Turner. 2nd ed. ©1987.
Includes index.
ISBN 0-13-481789-3
1. Industrial engineering. 2. Systems engineering. I. Turner, Wayne C. II. Turner, Wayne C., Introduction to industrial and systems engineering. III. Series.
T56.T86 1993
670—dc20 91-48050
CIP

Acquisitions editor: Marcia Horton
Production editor: Irwin Zucker
Prepress buyer: Linda Behrens
Manufacturing buyer: David Dickey
Supplements editor: Alice Dworkin
Editorial assistant: Dolores Mars

A Simon & Schuster Company
Englewood Cliffs, New Jersey 07632

Printed in the United States of America

10 9 8 7 6 5 4 3 2 1

ISBN 0-13-481789-3

Prentice-Hall International (UK) Limited, *London*
Prentice-Hall of Australia Pty. Limited, *Sydney*
Prentice-Hall Canada Inc., *Toronto*
Prentice-Hall Hispanoamericana, S.A, *Mexico*
Prentice-Hall of India Private Limited, *New Delhi*
Prentice-Hall of Japan, Inc., *Tokyo*
Simon & Schuster Asia Pte. Ltd., *Singapore*
Editora Prentice-Hall do Brasil, Ltda., *Rio de Janeiro*

Contents

PREFACE **xvii**

Part 1 History and Perspective of Industrial Engineering **1**

1 HISTORY OF ENGINEERING AND DEVELOPMENT OF INDUSTRIAL ENGINEERING **1**

1.1. Introduction 1
1.2. Early Developments 2
1.3. The Modern Era 3
1.4. The Engineering Process 5
1.5. Engineering as a Profession 6
1.6. Professional Ethics 8
1.7. Professional Licensing 8
1.8. Engineering Education and ABET Accreditation 10
1.9. Chronology of Industrial Engineering 12

1.10. Industrial Engineering Organizations 15
1.11. Definition of Industrial Engineering 18
1.12. Industrial Engineering Education 18
1.13. Impact of Related Developments 19
1.13.1 Impact of Operations Research, 19
1.13.2 Impact of Digital Computers, 20
1.13.3 Emergence of Service Industries, 21
1.14. Relationship to Other Engineering Disciplines 21
1.15. Challenges of the Future 23

2 INDUSTRIAL AND SYSTEMS ENGINEERING 25

2.1. Introduction 25
2.2. Industrial and Systems Engineering Design 26
2.2.1 Human Activity System, 26
2.2.2 Management Control System, 26
2.3. Typical I.&S.E. Activities 28
2.3.1 Production Operations, 28
2.3.2 Management Systems, 30
2.3.3 Corporate Services, 31
2.4. Relationship to Total Organization 33
2.5. Internal Organization of the I.&S.E. Function 36
2.6. Effectiveness Measures for the I.&S.E. Function 36
2.7. The Nature of "Systems" 36
2.7.1 Definitions, 37
2.7.2 System Classifications, 38
2.8. Feedback Control in Systems 38

Part 2 Industrial and Systems Engineering Methodologies 43

3 MANUFACTURING ENGINEERING 43

3.1. Introduction 43
3.2. Product–Production Design Interaction 44
3.3. Process Engineering 45
3.3.1 Defining Product Structure and Specifications, 45
3.3.2 Assessing Manufacturability, 48
3.3.3 Determining Processes Capable of Producing the Part, 48
3.3.4 Evaluating the Cost of Each Process, 50

3.3.5 Determining the Sequence of Operations, 55
3.3.6 Documenting the Process, 55

3.4. Industrial Processes 56
3.4.1 Refining and Alloying, 56
3.4.2 Casting, 58
3.4.3 Metal Forming, 59
3.4.4 Metal Cutting, 60
3.4.5 Welding, 64
3.4.6 Assembly, 67
3.4.7 Finishing, 67

3.5. Ancillary Functions 68
3.5.1 Tool, Jig, and Fixture Design, 68
3.5.2 Cost Estimating, 70
3.5.3 Maintenance Systems Design, 71
3.5.4 Packaging Systems, 72

3.6. Example 73

3.7. Computer Applications 73

4 FACILITIES LOCATION AND LAYOUT 80

4.1. Introduction to Facilities Location 80

4.2. Considerations 81

4.3. Analytical Techniques 83
4.3.1 Transportation Method of Linear Programming, 84
4.3.2 Multiple Objectives, 88
4.3.3 Mathematical Programming (Optional), 89
4.3.4 Public-Sector Location Problems, 96

4.4. Introduction to Facilities Layout 99

4.5. General Considerations 102

4.6. Systematic Layout Planning 102

4.7. Computerized Layout Planning (Optional) 113

4.8. Impact of Computers 118

5 MATERIAL HANDLING, DISTRIBUTION, AND ROUTING 125

5.1. Introduction 125

5.2. Material Handling 126
5.2.1 Equipment Concepts, 127
5.2.2 Principles of Material Handling, 131
5.2.3 Quantitative Techniques, 134

5.3. Distribution 137
5.3.1 *Warehouse Location, 137*
5.3.2 *Operations Management—Routing, 137*
5.3.3 *Routing in the Public Sector, 145*

6 WORK DESIGN AND ORGANIZATIONAL PERFORMANCE—WORK MEASUREMENT 151

6.1. Introduction 151
6.2. Methods Improvement 154
6.2.1 *Flow Process Charts, 155*
6.2.2 *Left-Hand–Right-Hand Charts, 156*
6.2.3 *Other Charts, 160*
6.2.4 *Principles of Motion Economy, 163*
6.2.5 *Human Engineering, 164*
6.3. Work Measurement 165
6.3.1 *Direct Time Study, 167*
6.3.2 *Time Study Standard Data, 171*
6.3.3 *Predetermined Times, 173*
6.3.4 *Predetermined Time Standard Data, 174*
6.3.5 *Work Sampling, 174*
6.4. Organizational System Performance Measurement 175
6.4.1 *Productivity Measurement Basics, 175*
6.4.2 *Normative Productivity Measurement Model, 176*
6.4.3 *Multifactor Productivity Measurement Model, 178*
6.5. Computers and Work Measurement and Design 178

7 OPERATIONS PLANNING AND CONTROL 183

7.1. Introduction 183
7.2. Overview of Operations Planning and Control 184
7.2.1 *Demand Forecasting (I), 184*
7.2.2 *Operations Planning (II), 186*
7.2.3 *Inventory Planning and Control (III), 187*
7.2.4 *Operations Scheduling (IV), 188*
7.2.5 *Dispatching and Progress Control (V), 188*
7.2.6 *Interfaces, 189*
7.2.7 *Integrating the Functions, 190*
7.3. Techniques for Demand Forecasting 190
7.3.1 *Moving Average, 190*
7.3.2 *Exponentially Weighted Moving Average, 191*
7.3.3 *Regression Analysis, 192*

7.4. Techniques for Operations Planning 194
7.5. Techniques for Inventory Planning and Control 197
7.6. Techniques for Operations Scheduling 200
7.6.1 Purpose of Operations Scheduling, 201
7.7. Dispatching and Progress Control 203
7.8. MRP Systems 203
7.9. Just-in-Time Manufacturing 207

8 QUALITY CONTROL 212

8.1. Introduction 212
8.2. A Bit of History 213
8.3. The Malcolm Baldrige National Quality Award 213
8.4. Deming's Thoughts on Continuous Improvement 216
8.5. Juran's Contributions to Quality Thought 217
8.6. Tools for On-line vs. Off-line Quality Control 218
8.7. Quality Function Deployment 218
8.8. Quality Cost Systems 220
8.9. Benchmarking 221
8.10. Tools of Statistical Process Control 223
8.10.1 Flowchart, 223
8.10.2 Cause-and-Effect Diagram, 226
8.10.3 Data Collection Form, 227
8.10.4 Pareto Analysis, 228
8.10.5 Histogram, 230
8.10.6 Scatter Plot, 232
8.10.7 Designed Experimentation, 233
8.11. Background on Control Charts 234
8.12. Control Charts for Variables 236
8.13. Sensitivity Checks for Control Charts 241
8.14. Process Capability Analysis 241
8.15. Control Charts for Attributes 243
8.15.1 The p *Control Chart, 243*
8.15.2 The c *Control Chart, 244*

9 FINANCIAL COMPENSATION 252

9.1. Introduction 252
9.2. Job Analysis 253

9.3. Job Evaluation 254
9.3.1 Ranking Method of Job Evaluation, 256
9.3.2 Classification or Grade Description, 256
9.3.3 Factor Comparison, 257
9.3.4 Point Rating, 259

9.4. Wage Surveys 262

9.5. Wage Payment 266
9.5.1 Daywork, 267
9.5.2 Measured Daywork, 267
9.5.3 Piecework Incentive, 267
9.5.4 Standard Hour, 268
9.5.5 Group Plans, 269

10 CAD/CAM, ROBOTICS, AND AUTOMATION 275

10.1. The Second Industrial Revolution 275
10.1.1 A Brief History of Manufacturing, 275
10.1.2 Impact of Computers and Electronics, 277
10.1.3 Other Recent Developments, 277
10.1.4 The Factory of the Future, 278

10.2. Computer-Aided Design 279
10.2.1 Computers in Product Design, 279
10.2.2 Computers in Process Design, 280
10.2.3 Computers in Electronics Design, 281

10.3. Computer-Aided Manufacturing 282
10.3.1 Computer-Aided Process Planning, 282
10.3.2 Numerical Control, 283
10.3.3 The Concepts of Group Technology, 285
10.3.4 Automated Storage, Retrieval, and Handling, 286
10.3.5 Computer-Aided Testing and Inspection, 288
10.3.6 Computer-Aided Factory Management, 290
10.3.7 The Concepts of Flexible Manufacturing Systems, 290

10.4. Robotics 291
10.4.1 Definition and Basic Concepts, 293
10.4.2 Physical and Technical Aspects of Robots, 293
10.4.3 Robotic Applications, 295

10.5. Automation 296

10.6. The Promise of CIM 297

10.7. Opportunities for I.E.'s 298

11 HUMAN FACTORS 301

11.1. Perspective 301
11.2. Physiological Aspects of Human Performance 302
11.3. Psychological Aspects of Human Activities 304
11.4. Human Interface with the World of Work 305
11.4.1 Human Interface with the Work Environment, 306
11.4.2 Human Interface with Machines, 307
11.4.3 Human Interface with Information/Communication Systems, 308
11.4.4 Human Interface with Organizational/Supervisory Structure, 308
11.4.5 Human Interface with Robots and Intelligent Machines, 309

12 RESOURCE MANAGEMENT 312

12.1. Introduction 312
12.1.1 Energy Management, 312
12.1.2 Water Management, 313
12.1.3 Hazardous Material Management, 313
12.1.4 This Chapter, 314
12.2. Energy Management 314
12.2.1 Why Bother?, 315
12.2.2 Why Industrial Engineering?, 315
12.2.3 Required Ingredients, 315
12.2.4 Understanding Rate Schedules, 316
12.2.5 Alternate Rate Schedules, 319
12.2.6 Energy Management Opportunities, 319
12.3. Water Management 322
12.4. Hazardous Material Management 324
12.4.1 Government Regulations, 324
12.4.2 The Role of Industrial Engineering, 325
12.5. Summary 327

13 FINANCIAL MANAGEMENT AND ENGINEERING ECONOMY 329

13.1. Introduction 329
13.2. Accounting 330
13.3. Cost Accounting 333

13.4. Engineering Economy 336

13.5. Interest Factors 337

13.5.1 Single-Payment Compound Amount Factor, 338

13.5.2 Other Interest Factors, 339

13.5.3 Examples, 341

13.6. Back to Gadgets—Present Worth Calculations 343

13.7. Impact of the Computer on Accounting and Engineering Economy 345

14 DETERMINISTIC OPERATIONS RESEARCH 349

14.1. Introduction—Definition 349

14.2. Similarity to Industrial Engineering 350

14.3. Nature of Operations Research 351

14.3.1 Economic Order Quantity, 351

14.3.2 Plant Location, 351

14.3.3 Job Evaluation, 352

14.3.4 Quality Control, 352

14.3.5 Others, 352

14.4. Categorization of Operations Research 352

14.4.1 Deterministic Approach, 353

14.4.2 Probabilistic Approach, 353

14.5. Deterministic Operations Research 354

14.6. Mathematical Programming 354

14.7. Unconstrained Optimization 355

14.8. Linear Programming 362

14.8.1 Assignment Problem, 367

14.8.2 Transportation Problem, 369

14.9. Other Techniques 372

14.9.1 Nonlinear Programming, 372

14.9.2 Integer Programming, 373

14.9.3 Zero–One Programming, 373

14.9.4 Quadratic Programming, 373

14.9.5 Geometric Programming, 373

14.9.6 Other Programming, 373

14.10. Impact of Computers 373

15 PROBABILISTIC MODELS 378

15.1. Introduction 378

15.2. Queueing Theory 378

15.2.1 Queueing System Structure, 379
15.2.2 Queueing Notation, 380
15.2.3 Single-Service Channel, 381

15.3. Inventory Control 384
15.3.1 Single-Period Model—No Setup Cost, 384
15.3.2 Lot Size–Reorder Point Models, 387
15.3.3 Periodic Review Models, 388

15.4. Markov Chains 389
15.4.1 Regular Markov Chains, 390
15.4.2 Absorbing Markov Chains, 392

15.5. Impact of Statistics and Computers 393

16 SIMULATION 396

16.1. Introduction 396

16.2. Simulation Examples 397

16.3. Random Number Generation 403

16.4. Time-Flow Mechanism 407

16.5. Simulation Languages 407

17 PROJECT MANAGEMENT 411

17.1. Introduction 411

17.2. Project Planning Networks 413

17.3. Critical Path Method 415
17.3.1 Forward Pass, 417
17.3.2 Backward Pass, 417
17.3.3 Total Activity Slack, 419
17.3.4 Critical Path, 419

17.4. Program Evaluation and Review Technique 419

17.5. Time–Cost Trade-offs 424

17.6. Resource Leveling 429

Part 3 Integrated Systems Design 432

18 SYSTEMS CONCEPTS 432

18.1. Introduction 432

18.2. Introduction to Systems Thinking 433
18.2.1 Origin of Systems Thinking, 433
18.2.2 Hierarchical Nature of Systems, 434

18.3. Definitions and Terminology 436

18.4. Systems Engineering 440

18.4.1 Systems Analysis and Design, 440
18.4.2 The Systems Design Process, 440

18.5. System Representation 442

18.5.1 Block Diagrams, 442
18.5.2 Transfer Functions, 443

19 MANAGEMENT SYSTEMS DESIGN 448

19.1. Introduction and Perspective 448

19.2. A Systems View of an Organization 449

19.2.1 Gaining a Perspective, 449
19.2.2 Finding a Starting Point, 450
19.2.3 Universal Outcome Goals, 452
19.2.4 Determining Goals and Objectives, 453
19.2.5 A Unified Framework, 454

19.3. Organization Design 456

19.3.1 Specification of Objectives, 458
19.3.2 Determination of Functions, 459
19.3.3 Grouping the Functions, 459
19.3.4 Functional Objectives, 460
19.3.5 Job Descriptions, 460
19.3.6 Management Controls, 460
19.3.7 Organization Design Is Continuous and Dynamic, 461
19.3.8 Organization Structures, 461
19.3.9 Coordination within the Organization, 462
19.3.10 Keeping the Design Current, 463

19.4. Providing Management Controls 463

19.5. The Organization Life Cycle 464

19.5.1 Life Cycle Stages of an Organization, 464
19.5.2 Organizational Renewal and Redesign, 465
19.5.3 The Learning Organization, 466

20 COMPUTERS AND INFORMATION SYSTEMS 469

20.1. Perspective 469

20.2. Basic Concepts of Information Systems 470

20.3. The Process of Designing Information Systems 472

20.3.1 Feasibility Study, 472
20.3.2 Systems Analysis, 472
20.3.3 General Systems Design, 474
20.3.4 Systems Evaluation and Justification, 474

20.3.5 Detail Systems Design, 475
20.3.6 Systems Implementation, 475
20.3.7 Systems Operation and Maintenance, 476

20.4. Data-Base Management Systems 476

20.5. Data Communications Networks 479

21 PERSONNEL MANAGEMENT 483

21.1. Introduction 483

21.2. Selection, Testing, and Placement 484

21.3. Performance Appraisal, Training, Education, and Promotions 486

21.4. Job Analysis and Description 490

21.5. Labor Relations 490

21.6. Safety Programs 492

21.7. Benefits and Services 493

21.8. Motivation, Supervision, and Communications 494
20.8.1 Motivation, 494
20.8.2 Supervision, 496
20.8.3 Communications, 497

21.9. Engineering Management 498

Appendices 501

A PROBABILITY AND STATISTICS 501

A.1. Introduction 501

A.2. Basic Probability Theory 502
A.2.1 Sample Space, 502
A.2.2 Events, 502
A.2.3 Probability of an Event, 503
A.2.4 Rules of Operation, 503
A.2.5 Combinations, 504

A.3. Random Variables 505

A.4. Estimating Probabilities 506

A.5. Some Important Probability Distributions 508
A.5.1 Discrete Distribution Properties, 509
A.5.2 Binomial Distribution, 509
A.5.3 Poisson Distribution, 511
A.5.4 Uniform Distribution, 513
A.5.5 Continuous Distribution Properties, 514

A.5.6 Normal Distribution, 514
A.5.7 Exponential Distribution, 516
A.5.8 Rectangular Distribution, 516
A.5.9 Distribution Summary, 517

A.6. Expected Values and Variability 518
A.6.1 Mean, 518
A.6.2 Variance, 518

A.7. Populations and Samples 520
A.7.1 Population, 520
A.7.2 Sample, 521
A.7.3 Sample Statistics, 521
A.7.4 Distribution of Sample Means, 522

A.8. Central Limit Theorem 523

B TABLES 527

Table B.1. Poisson Distribution—Cumulative 528
Table B.2. Normal Distribution—Cumulative 531

INDEX 533

Preface

This book provides an introduction to industrial and systems engineering. It is especially designed for use as a text in an introduction to industrial engineering course. The purpose is to define industrial and systems engineering, describe its place in the business world, and give a broad picture of the functional areas with some solution techniques. The book is also useful to anyone desiring an overview of industrial engineering.

This book is not a detailed text for any of the individual techniques presented, but it does show what an industrial and systems engineer is capable of doing in a wide variety of organizations. Special attention is given to describing situations in which the tools or techniques may be applied. Instead of taking the classical "technique looking for a problem approach," the problems are first described and then the technique(s) applicable to the problems is (are) discussed.

The book is divided into three parts. In Part 1 the history of engineering in general and industrial engineering specifically is given in an attempt to show the range and growth of the discipline's objectives. Then, a modern definition and discussion of industrial and systems engineering are given. They define the purpose and objectives of the discipline and show areas where it is applicable. Part 1 discusses the place of industrial and systems engineering in an organization and how to manage and control the function. Finally, Part 1 introduces the concepts of elementary systems theory and feedback.

Part 2 constitutes the largest portion of the book, with a chapter devoted to each of the major methodologies of the discipline of industrial and systems engineering. For each general area of industrial and systems engineering, a typical problem is presented to provide a concrete framework, and then the tools and techniques appropriate for that situation are developed. The purpose in this approach is to emphasize the proper use of the various techniques. Since modern computing methods have had a significant effect on industrial and systems engineering, almost all chapters in Part 2 discuss computerization of the techniques. Included in Part 2 are some of the newer tools, such as CAD/CAM, robotics, and resource management, as well as tools that have been around for many years. The relationship of industrial engineering to such areas as operations research and ergonomics is emphasized.

Part 3, Integrated Systems Design, is intended to show how the I.E. must bring together all the detailed pieces into an integrated system. Elementary concepts from systems engineering are used as a vehicle for portraying the complex interactions among system components. A chapter is included on computers and information systems because of their critical importance in the design of integrated systems.

Those familiar with the second edition of this book will notice that we have added a new chapter on simulation. All chapters have been thoroughly updated, with some being completely revised. Also, other chapters have been combined and rearranged for a more effective organization of topics.

We are very grateful to our many colleagues who, having taught from the first and second editions of this book, provided many helpful suggestions for the third edition. We have incorporated most of their suggestions, and we feel that the book is stronger because of them. We would also like to thank the following reviewers: Avinash Waikar, Louisiana State University; Roger Berger, Iowa State University; Paul McCright, Kansas State University; Sabah U. Randhawa, Oregon State University; Chris Styliandis, North Dakota State University; John R. English, Texas A & M University; Robert L. Williams, Ohio University; Jill A. Swift, University of Miami; and Timothy J. Greene, Virginia Polytechnic University. We also wish to acknowledge the useful comments offered by many students who have studied using this text.

In using the book at Oklahoma State University for many years, we have found that the course is greatly enhanced through the use of a workbook containing laboratory exercises, which give the students hands-on experience in applying the concepts covered in the text. This workbook is now available from Prentice Hall. We wish to acknowledge the help of our colleagues Pat Koelling and Jim Shamblin in preparing the workbook.

W. C. Turner
J. H. Mize
K. E. Case
J. W. Nazemetz

PART 1 HISTORY AND PERSPECTIVE OF INDUSTRIAL ENGINEERING

CHAPTER 1

History of Engineering and Development of Industrial Engineering

1.1. Introduction

Industrial engineering is emerging as one of a small number of "glamor" professions that will be counted on for solving complex problems in the highly technological world of the future. We are seeing rapid advances in the use of automation and other forms of high technology in factories and production systems around the world.

Designing a "factory of the future" is a challenging and complex task—one that requires a knowledge of fundamental sciences, engineering sciences, behavioral sciences, computer and information sciences, economics, and a broad array of topics concerning the basic principles of production systems.

Industrial engineering (I.E.) curricula are designed to prepare students to meet the challenges of the future. Many I.E. graduates will, indeed, design modern manufacturing facilities. Others will elect to design health-care delivery systems, transportation systems, or other systems that provide services rather than manufactured products.

The demand for I.E.s is strong, and growing each year. In fact, the demand for I.E.s greatly exceeds the supply. This demand/supply imbalance is greater for

I.E. than for any other engineering or science discipline and is projected to exist for many years in the future.[1]

The purpose of this book is to provide an introduction and overview of the exciting field of industrial engineering. Our intent is to provide a framework, so that the student may have a basic frame of reference as he or she encounters in-depth courses in each of the specialty areas within I.E.

1.2. Early Developments

How did the two words "industrial" and "engineering" become combined to form the label "industrial engineering"? What is the relationship of industrial engineering to other engineering disciplines, to business administration, to the social sciences?

To understand the role of industrial engineering in today's complex world, it is helpful to learn the historical developments that were instrumental in the evolution of I.E. There are many ways to write a history of engineering. The treatment in this chapter is brief because our interest is in reviewing the highlights of engineering development, particularly those leading to industrial engineering as a specialty. More complete histories are available in the references.[2–4]

Engineering and science have developed in a parallel, complementary fashion, although not always at the same pace. Whereas science is concerned with the quest for basic knowledge, engineering is concerned with the application of scientific knowledge to the solution of problems and to the quest for a "better life." Obviously, knowledge cannot be applied until it is discovered, and once discovered, it will soon be put to use. In its efforts to solve problems, engineering provides feedback to science in areas where new knowledge is needed. Thus, science and engineering work hand in hand.

Although "science" and "engineering" each have distinguishing characteristics and are regarded as different disciplines, in some cases a "scientist" and an "engineer" might be the same person. This was especially true in earlier times when there were very few means of communicating basic knowledge. The person who discovered the knowledge also put it to use.

A complete chronology of early engineering practice is beyond the scope of this text. The interested reader will find an excellent treatise in *The Ancient Engineers* by de Camp (footnote 2), in which numerous early developments are described and illustrated.

---

[1] *Science and Engineering Education for the 1980's and Beyond*, prepared by the National Science Foundation and the Department of Education, Oct. 1980.

[2] L. S. de Camp, *The Ancient Engineers* (Cambridge, Mass.: MIT Press, 1963).

[3] J. D. Kemper, *The Engineer and His Profession*, 2nd ed. (New York: Holt, Rinehart and Winston, 1975).

[4] R. J. Smith, *Engineering as a Career*, 3rd ed. (New York: McGraw-Hill Book Company, 1969).

We naturally think of such outstanding accomplishments as the pyramids, the Great Wall of China, the Roman construction projects, and so forth, when we think of early engineering accomplishments. Each of these involved an impressive application of fundamental knowledge.

Just as fundamental, however, were accomplishments that are not as well known. The inclined plane, the bow, the wheel, the corkscrew, the waterwheel, the sail, the simple lever, and many, many other developments were very instrumental in the engineer's efforts to provide a better life.

Almost all engineering developments prior to 1800 had to do with physical phenomena: overcoming friction, lifting, storing, hauling, constructing, fastening. Later developments were concerned with chemical and molecular phenomena: electricity, properties of materials, thermal processes, combustion, and other chemical processes.

Fundamental to almost all engineering developments were the advances made in mathematics. Procedures for accurately measuring distances, angles, weights, and time were necessary for almost all early engineering accomplishments. As these procedures were refined, greater accomplishments were realized.

Another very important contribution of mathematics was the ability to represent reality in abstract terms. A mathematical model of a complex system can be manipulated such that relationships between variables in the system can be understood. The simple relationship commonly called the *Pythagorean theorem* is such an example. This theorem states that the hypothenuse of a right triangle is expressed as the square root of the sum of the squares of the adjacent sides. The use of abstract models representing complex physical systems is a fundamental tool of engineers.

As a final comment on early developments, let us discuss an early development that *did not* occur. The missing early development is related to the behavioral sciences. The understanding of human behavior has lagged considerably behind developments in the mathematical, physical, and chemical sciences. This is important to industrial engineers because the systems designed by I.E.s involve people as basic components. The lack of progress in behavioral science has hindered the industrial engineer in his efforts to design optimal systems involving people.

1.3. The Modern Era

We shall arbitrarily define the *modern* era of engineering as beginning in 1750, even though there were many important developments between 1400 and 1750. We choose 1750 as the beginning of modern engineering for two reasons:

(1) Engineering schools appeared in France in the eighteenth century.
(2) The term *civil engineer* was first used in 1750.

Principles of early engineering were first taught in military academies and were concerned primarily with road and bridge construction and with fortifications. This portion of academic training was referred to as *military engineering*. When some of the same principles were applied to nonmilitary endeavors, it was only natural to refer to these as *civilian engineering*, or simply *civil engineering*.

Interrelated advancements in the fields of physics and mathematics laid the groundwork for practical applications of mechanical principles. An important advancement was the development of a practical steam engine that could perform useful work. Once such an engine was available (approximately 1700), many mechanical devices were developed that could be driven by the engine. These efforts culminated in the emergence of *mechanical engineering* as a distinct branch in the early nineteenth century.

Another example of such an advancement was the fundamental work conducted in the latter part of the eighteenth century on electricity and magnetism. Although early scientists had known about magnetism and static electricity, an understanding of these phenomena did not commence until Benjamin Franklin's famous kite-flying experiment in 1752. The next half-century saw the laying of the foundation of electrical science, primarily by German and French scientists.

The first significant application of electrical science was the development of the telegraph by Samuel Morse (approximately 1840). Thomas Edison's invention of the carbon-filament lamp (approximately 1880) led to widespread use of electricity for lighting purposes. This, in turn, spurred very rapid developments in the generation, transmission, and utilization of electrical energy for a variety of labor-saving purposes. Engineers who chose to specialize in this activity were naturally labeled *electrical engineers*.

Along with the developments in mechanical and electrical technology were accompanying developments in the understanding of substances and their properties. The science of chemistry is concerned with understanding the nature of *matter* and in learning how to produce desirable changes in materials. Fuels were needed for the new internal combustion engines being developed. Lubricants were needed for the rapidly growing array of mechanical devices. Protective coatings were needed for houses, metal products, ships, and so forth. Dyes were needed in the manufacture of a wide variety of consumer products. Somewhat later, synthetic materials were needed to perform certain functions that could not be performed as well or at all by natural materials. This field of engineering endeavor naturally became known as *chemical engineering*.

As industrial organizations emerged to capitalize on the rapidly developing array of technological innovations, the size and complexity of manufacturing units increased dramatically. *Mass production* was made possible through two important concepts:

- Interchangeability of parts.
- Specialization of labor.

Through mass production the unit cost of consumer products was reduced dramatically.

The groundwork was now laid for a dramatic shift in the lifestyles and cultures of industrialized nations. Within half a century the United States and other developing countries changed from largely rural, agricultural economies and societies to urban, industrialized economies and societies. The suddenness of this change is probably the cause of many of today's pressing problems, for example, pollution and crowding.

During the early part of this movement it was recognized that business and management practices that had worked well for small shops and farms simply were inadequate for large, complex manufacturing organizations. The need for better management systems led to the development of what is now called *industrial engineering*. This development will be traced more completely in later sections of this chapter.

These five major engineering disciplines (civil, chemical, electrical, industrial, and mechanical) were the branches of engineering that emerged prior to the time of World War I. These developments were part of the industrial revolution that was occurring worldwide, and the beginning of the *technological revolution* that is still occurring.

Developments following World War II led to other engineering disciplines, such as *nuclear engineering, electronic engineering, aeronautical engineering*, and even *computer engineering*. The Space Age led to *astronautical engineering*. Recent concerns with the environment have led to *environmental engineering* and *bioengineering*. These newer engineering disciplines are often considered specialties within one or more of the ''big five'' disciplines of civil, chemical, electrical, industrial, and mechanical engineering.

1.4. The Engineering Process

We have sketched the development of several engineering branches, and we all have a general understanding about what particular engineers actually do. For example, we know that mechanical engineers are involved with developing better automotive engines and that electrical engineers are concerned with developing better television sets.

Are there characteristics that distinguish engineers from other professionals? If so, what are these characteristics? Let us explore some of the general activities in which all engineers are involved.

- Engineers *solve problems*, but so do mathematicians.
- Engineers *analyze*, but so do statisticians and economists.
- Engineers *design systems*. Do others?

Certainly, architects can legitimately claim that they design systems, if one can call a building a system. To do a complete job of building design requires a knowledge of materials and forces. Thus, it can be argued that only an *architectural engineer* is truly qualified to design a building. Architects who do not have a basic engineering background must rely on engineers for much of their design work.

Within the spirit of the framework being developed, therefore, we can say that *the distinguishing characteristic of engineering is that it is concerned with the design of systems*.

What is design? Certainly, design involves a considerable amount of creativity. Is design then simply an art form that one learns through experience? Or are there design principles that can be learned and applied?

Synthesis is a term whose meaning is almost the same as *design*. Perhaps we can develop an understanding of design by discussing the two terms synthesis and analysis.

Analysis is concerned with resolving something into its basic elements; synthesis is concerned with combining elements into a whole. When freshmen engineering students take statics, they learn much about analysis. For example, they are given a beam having certain properties (size, material, loading points, and so forth) and they are asked to analyze the beam in terms of its loading characteristics. The analysis may show, for example, that the beam will support a maximum of 100 tons. There is only one answer to problems involving analysis.

Now, let us consider a different question: Suppose you are asked to design a beam that will support 100 tons. There are a large variety of beams that will support 100 tons; thus, there is not necessarily one best answer to the design problem.

Typically, analysis is concerned with existing systems. Synthesis is usually concerned with a new or improved system.

The engineering process involves both synthesis and analysis. Figure 1.1 is an illustration of the engineering process. Note that this process is used by all engineers: civil, chemical, electrical, industrial, mechanical, and others. The manner in which industrial engineers engage in this process will be discussed in a later section of this chapter.

The dual arrows in Figure 1.1 between the two blocks representing analysis and synthesis are intended to portray the iterative nature of the engineering process. One stage of analysis provides sufficient insight to postulate a particular design. This design, in turn, provides additional information that changes the basis of the analysis. This iterative process is repeated until an acceptable solution, system, or method is derived. (The word "acceptable" is used instead of the word "optimal" because a true optimal design is rarely achieved, and then only within the limitations of the assumptions under which the engineer is working.)

1.5. Engineering as a Profession

Although there can never be general agreement on this matter, the vocations that are commonly referred to as *professions* are medicine, teaching, architecture, law, ministry, and engineering. Smith[5] points out that these vocations have four common characteristics:

---

[5]Smith, p. 10.

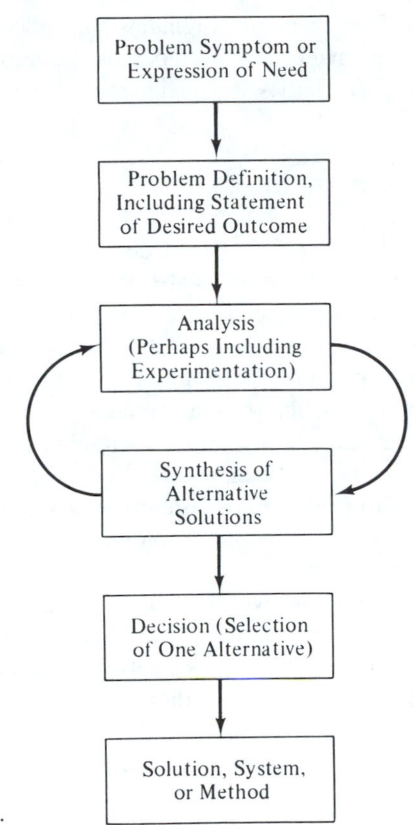

Fig. 1.1. Basic engineering process.

(1) Associated with a profession is a significant body of special knowledge.
(2) Preparation for a profession includes an internship-like training period following the formal education.
(3) The standards of a profession, including a code of ethics, are maintained at a high level through a self-policing system of controls over those practicing the profession.
(4) Each member of a profession recognizes his responsibilities to society over and above responsibilities to his client or to other members of the profession.

Other definitions of a profession go further and state that its members must engage in *continued study* and that its prime purpose must be rendering of a *public service*. Although these matters are subject to interpretation, they are evidence that professionalism is indeed a serious matter and that much is expected of people who claim to be professionals.

Chapter 4 of Kemper's book (footnote 3) is entitled ''Is Engineering Really a Profession?'' and is highly recommended reading. Kemper concludes that engineering is indeed a profession. He cautions, however, that '' . . . just because engineering is a profession it does not necessarily follow that every engineer is a professional. Every individual remains charged with the responsibility to meet the criteria on a personal basis.'' Kemper also points out that the definition of a professional in the Taft–Hartley Act includes engineers.

1.6. Professional Ethics

Engineers are frequently involved in decisions that have a profound effect on society. The design of particular devices almost always involves the safety of the user. The design of processes frequently affects the environment. The design and location of a factory affect the community and its citizens. The design of a management system greatly affects the individuals working for the organization—their comfort, their sense of worth, their financial status, and so on.

Engineers have been unjustly accused of creating almost all the current problems in our society, for example, pollution and crowding. The truth is that engineers have worked diligently and persistently to provide society with the things the social scientists claimed were needed. Pollution, crowding, and unsightly freeways are simply the byproducts.

Engineers do, however, have a tremendous responsibility to protect the public welfare. They must continually engage in trade-offs between costs and factors affecting public welfare, such as safety. An automobile can be made almost completely safe for its occupants, but the automobile would cost approximately $100,000 and would weigh approximately 14 tons. Even if someone were willing to pay this much for an automobile, there are other considerations: the safety of people in lighter cars traveling on the same roadways and the fuel consumption of a 14-ton vehicle.

Engineers are also frequently caught in a controversy between their employer or client and the public. Companies must produce items for sale at competitive prices, even though a ``cheap'' design may mean an unsafe or unreliable product.

The engineering profession enjoys a very favorable reputation regarding its adherence to professional ethics. Fortunately, early engineers were perceptive and conscious of the public good. The National Society of Professional Engineers (see Section 1.8) publishes a set of Canons of Ethics for engineers. They do not answer every question or resolve every controversy, but they do provide a good foundation from which one may extrapolate to cover almost any situation. A copy of the Canon of Ethics may be obtained by writing to your State Engineering Registration Board.

1.7. Professional Licensing

How would you like to be treated by a medical doctor who had not been examined and approved by your state's licensing board? How would you like to be represented in a court case by a lawyer who had not proven his knowledge of the law by passing the bar exam?

If we agree that doctors and lawyers should be required to be examined before being admitted to practice, let us examine the same general question about the engineering profession. Should bridges be designed by people who have not proven their knowledge of civil engineering? Should airplanes be designed by people who have not proven their knowledge of aerodynamics?

Professional licensing of engineers has not been as rigidly pursued as with the professions of law and medicine. Indeed, many states require barbers to be licensed, but not engineers! One of the reasons for this is that almost all engineers work for large companies, but doctors and lawyers (and even barbers) are usually self-employed.

All states do have provisions for licensing engineers, but this is not required as a condition for practicing engineering, except in certain situations, such as consulting firms. But as companies find themselves more subject to liability suits over product safety and environmental violations, it is expected that almost all engineers will eventually be subject to professional licensing.

Licensing of engineers is typically referred to as *professional registration*. Whereas all 50 states, the District of Columbia, and 4 U.S. territories now have engineering registration laws, there are significant differences in the requirements for professional registration. An engineer can become registered in some states by simply applying and demonstrating that he has performed engineering work for a certain length of time. Other states require extensive examinations on fundamental engineering principles.

Some uniformity appears to be emerging, although complete uniformity may never be realized. The emerging pattern seems to be as follows: Engineering students take the Engineer in Training (E.I.T.) exam during their last year in college; the students graduate, take jobs and practice the engineering profession for a specified period of time; they then take an examination as a final step in the registration process.

A current major issue concerns continuing certification of an engineer's ability. Some states are beginning to consider requiring engineers, doctors, and lawyers to renew their licenses periodically. There is considerable debate whether license renewal should involve reexamination on current engineering material or should simply certify that a certain minimum amount of continuing education and professional development has been achieved.

In view of the current state of uncertainty regarding professional registration, a natural question is "Why should I bother?" Kemper[6] states six compelling reasons why every engineer should become registered as quickly as possible after graduation:

(1) Registration will most likely become more important in the future, and will probably be required in some states.
(2) No young engineer can foresee the total future course of his career: he may believe that he will always be working in areas not requiring registration, but

[6]Kemper, pp. 245–246.

many unexpected events may occur which would require registration in order to practice the engineering profession.

(3) A court of law generally will not recognize an individual as an engineer unless he is registered, and therefore will not accept an unregistered engineer's testimony as an expert witness.

(4) More and more states require that an engineer be registered in order to engage in a growing array of activities. If an unregistered person engages in practice required by law to be performed by registered engineers, that person may be penalized and, at the very least, the courts will not aid him in collecting his fee. Many states now prohibit the word ''engineering'' from appearing in the name of a firm or in its advertisement if it has no registered engineers in responsible positions.

(5) An increasing number of companies believe it is desirable for members of their engineering management group to be registered; therefore, registration could be an aid in promotion.

(6) As the passage of written examinations becomes a more universal requirement, registration will likely be increasingly regarded by employers as an indicator of technical competence.

1.8. Engineering Education and ABET Accreditation

The complete education of an engineer occurs in at least three definable stages:

(1) *Preparatory*—High school courses in mathematics and science.
(2) *University*—A formal engineering curriculum.
(3) *Continuing*—Lifelong learning through professional practice and professional development.

Students who are college sophomores in 1993 began their preparatory work in 1988 and will not reach their full growth as practicing engineers until after the year 2000. Engineering education, in its broadest sense, is indeed a long process.

Engineering education has progressed through several stages of evolution. Prior to World War II engineering education was concerned with the art and practice of engineering principles. Engineering students spent long hours learning to operate lathes, drill presses, molding machines, foundries, and so on. They learned to wind motors and to build radio sets. There was a considerable amount of ''hands-on'' experience involved in the educational process.

The so-called ''Grinter Report''[7] in 1955 instigated significant changes in engineering education. Much of the ''art'' of engineering was removed from the engineering curricula and was replaced by a much greater emphasis on the basic science underlying engineering. The great achievements in space, communications,

[7] ''Report on Evaluation of Engineering Education,'' American Society for Engineering Education, June 15, 1955.

and other sophisticated systems can be attributed at least in part to the new emphasis on the science content of engineering curricula.

Critics of the aforementioned trend claim that in many cases engineering curricula became almost indistinguishable from curricula in mathematics and physics. The greater emphasis on analysis resulted in a diminished emphasis on engineering design. Consequently, current trends are toward a more balanced approach to engineering education.

Strong cases have been made[8] for extending the length of engineering curricula to include a year of graduate study leading to the master's degree. Entry into almost all professions (law, medicine, teaching) requires completion of a significant amount of post-baccalaureate education. Proponents of expanded engineering curricula argue that five years of formal training is a minimum time in which to acquire the tools needed for practicing engineering.

Traditionally, graduate education in engineering has led to a "master of science" degree. The emphasis on science education often includes a research effort and a formal thesis. Almost all engineers who receive a master's degree, however, do not go into research or teaching. It appears that there is a need for a path at the master's level that emphasizes *engineering practice* rather than research. Such an approach is often called a *professional engineering program*[9] and includes a period of professional practice in the real world prior to completion of degree requirements.

The Accreditation Board for Engineering and Technology (ABET) is the official agency in the United States for examining and accrediting engineering curricula. The purpose of ABET accreditation is to assure the public and employers of engineering graduates that certain minimum standards have been met.

Students often ask, "Why should I have to take English, differential equations, or thermodynamics?" Many engineering students are particularly negative toward courses in humanities and the social sciences.

Indeed, why *do* engineering students have to take the particular courses specified on their curriculum sheets? Several reasons can be stated:

- There should certainly be some degree of commonality among the many branches that are called "engineering."
- Companies hiring engineering graduates need some assurance that certain minimum content has been provided. This is especially needed when companies are hiring engineering graduates from several universities in different parts of the country.
- Much of the content of an engineering curriculum is intended as a general foundation from which an engineer may pursue a particular career path. Few engineering students know specifically what they will be doing in the future.
- Courses in humanities and social sciences are included to provide the engineer with an awareness of social values and concerns of nontechnical people in our

[8]M. R. Lohmann, "Professional Schools for Engineers," *Engineering Education*, 60 (1970), p. 954.

[9]Lohmann, p. 954.

society. Also, problems are not just technical in nature; there are always social, economic, and perhaps political factors that must be considered.

The ABET requirements for the curriculum content of a four-year engineering program are as follows:

- The equivalent of approximately two and one-half years of study in the area of mathematics, science, and engineering. The course work should include at least one year of an appropriate combination of mathematics and basic sciences, one year of engineering sciences, and one-half year of engineering design.
- The equivalent of one-half year as the minimum content in the area of the humanities and social sciences.

For those institutions which elect to prepare graduates for entry into the profession at the advanced level, ABET normally requires that students' curricular work (1) satisfy ABET engineering criteria at the basic level for the program being evaluated, and (2) have the equivalent of one additional year of study beyond that required for a basic level program. This additional year of study must include at least two-thirds year of advanced mathematics, basic sciences, engineering science and engineering design. The additional year of study must include a considerable amount of material and treatment at an advanced level not normally associated with the basic level.

It may appear that the ABET is attempting to force a high degree of standardization with the above described criteria. Actually, the ABET encourages experimentation and custom-tailored programs to fit local conditions. The criteria provide considerable flexibility while assuring the inclusion of critical essential elements.

1.9. Chronology of Industrial Engineering

The preceding sections are intended to portray the broad context in which the field of industrial engineering exists. Although there is considerable commonality among the different branches of engineering, each branch has distinguishing characteristics that are important to recognize. The remainder of this text is concerned with industrial engineering.

Industrial engineering emerged as a profession as a result of the industrial revolution and the accompanying need for technically trained people who could plan, organize, and direct the operations of large complex systems. The need to increase efficiency and effectiveness of operations was also an original stimulus for the emergence of industrial engineering. Some early developments are explored, in order to understand the general setting in which industrial engineering was born. For more details, see the excellent work of Emerson and Naehring.[10]

---

[10] Howard P. Emerson and Douglas C. E. Naehring, *Origins of Industrial Engineering* (Atlanta: IE&M Press, 1988).

Charles Babbage visited factories in England and the United States in the early 1800's and began a systematic record of the details involved in many factory operations. For example, he observed that the manufacture of straight pins involved seven distinct operations. He carefully measured the cost of performing each operation as well as the time per operation required to manufacture a pound of pins. Babbage presented this information in a table, and thus demonstrated that money could be saved by using women and children to perform the lower-skilled operations. The higher-skilled, higher-paid men need only perform those operations requiring the higher skill levels. Babbage published a book containing his findings, entitled *On The Economy of Machinery and Manufactures* (1832). In addition to Babbage's concept of division of labor, the book contained new ideas on organizing and very advanced (for the time) concepts of harmonious labor relations. Significantly, Babbage restricted his work to that of observing and did not attempt to improve the work methods or to reduce the operation times.

The concept of *interchangeable manufacture* was a key development leading to the modern system of mass production. This concept was to produce parts so accurately that a specific part of a particular unit of a product could be interchanged with the same part from another unit of the product, with no degradation of performance in either unit of the product. Eli Whitney received a government contract to manufacture muskets using this method. Another of his contributions was the design and construction of new machines that could be operated by laborers having only a minimal amount of training. Through the successful application of these two concepts, Whitney created the first mass production system.

In the period around 1880 industrial operations were conducted in a much different manner from today. There was very little planning and organizing, as such. A first-line supervisor was given verbal instructions on the work to be done and a crew of (usually) poorly trained workers. The supervisor was expected to work his men as hard as he could. Any improved efficiency in work methods usually came from the worker himself in his effort to find an easier way to get his work done. There was virtually no attention given to overall coordination of a factory or process.

Frederick W. Taylor is credited with recognizing the potential improvements to be gained from analyzing the work content of a job and designing the job for maximum efficiency. Taylor's original contribution, constituting the beginning of industrial engineering, was his three-phase method of improving efficiency: Analyze and improve the *method* of performing work, reduce the times required, and set standards for what the times should be. Taylor's methods brought about significant and rapid increases in productivity. Later developments stemming from Taylor's work led to improvements in the overall planning and scheduling of an entire production process.

Frank B. Gilbreth extended Taylor's work considerably. Gilbreth's primary contribution was the identification, analysis, and measurement of fundamental motions involved in performing work. By classifying motions as "reach," "grasp," "transport," and so on, and by using motion pictures of workers performing their tasks, Gilbreth was able to measure the average time to perform each basic motion under varying conditions. This permitted, for the first time, jobs to be *designed* and the time required to perform the job known before the fact. This was a fundamental step in the development of industrial engineering as a profession based on "science" rather than "art."

Dr. Lillian Gilbreth, wife of Frank, is credited with bringing to the engineering profession a concern for human welfare and human relations. Having received a doctoral degree in psychology from Brown University, Dr. Gilbreth became a full partner with her husband in developing the foundational concepts of industrial engineering. During Dr. Gilbreth's long life (1878–1972), she witnessed and contributed to the birth, growth, and maturation of the I.E. profession. She became known as the "first lady of engineering" and the "first ambassador of management." She received many, many honors and awards from professional organizations, universities, and governments from around the world. She was the first woman to be elected to the National Academy of Engineering.

Another early pioneer in industrial engineering was Henry L. Gantt, who devised the so-called *Gantt chart*. The Gantt chart was a significant contribution in that it provided a systematic graphical procedure for preplanning and scheduling work activities, reviewing progress, and updating the schedule. Gantt charts are still in widespread use today.

W. A. Shewhart developed the fundamental principles of statistical quality control in 1924. This was another important development in providing a scientific base to industrial engineering practice.

Many other industrial engineering pioneers contributed to the early development of the profession. During the 1920s and 1930s much fundamental work was done on economic aspects of managerial decisions, inventory problems, incentive plans, factory layout problems, material handling problems, and principles of organization. Although these pioneers are too numerous to mention in this brief chronology, more complete historical accounts are available elsewhere.[11–14]

The significance of the early development of industrial engineering is far greater than most people realize, and in fact, ranks among the greatest achievements of all time. In 1968, Peter Drucker[15] described the importance of "scientific management," which we today would call industrial engineering:

> Scientific management (we today would probably call it "systematic work study," and eliminate thereby a good many misunderstandings the term has caused) has proved

[11] Ross W. Hammond, in H. B. Maynard (ed.), *Industrial Engineering Handbook*, 3rd ed. (New York: McGraw-Hill Book Company, 1971), pp. 1–3 through 1–17.

[12] John A. Ritchey (ed.), *Classics in Industrial Engineering* (Delphi, Ind.: Prairie Publishing Company, 1964).

[13] W. R. Spriegel, and C. E. Myers (eds.), *The Writings of the Gilbreths* (Homewood, Ill.: Richard D. Irwin, Inc., 1953).

[14] L. Urwick (ed.), *The Golden Book of Management* (London: Newman Neame Ltd., 1956).

[15] Peter Drucker, *The Age of Discontinuity* (New York: Harper & Row, Publishers, 1968), p. 271.

> to be the most effective idea of this century. It is the only basic American idea that has had worldwide acceptance and impact. Whenever it has been applied, it has raised the productivity and with it the earnings of the manual worker, and especially of the laborer, while greatly reducing his physical efforts and his hours of work. It has probably multiplied the laborer's productivity by a factor of one hundred.

Shown in Figure 1.2 are a number of significant events and developments that have occurred in the evolution of industrial engineering. The position of each event relative to the time axis is intended only to show the approximate time at which that event occurred. In many cases, such as *time studies*, the event on the chart merely indicates its beginning; time studies are still a fundamental tool of industrial engineering.

At the top of the chart are shown four overlapping periods of emphasis. The labels are somewhat arbitrary. The period from approximately 1900 through the mid-1930s is generally referred to as *scientific management*. The next period, labeled *industrial engineering*, begins in the late 1920s and is shown extending to the present time. The period marked *operations research* had a great influence on industrial engineering practice and is shown beginning in the mid-1940s and extending somewhat past the mid-1970s. The fourth period, *industrial and systems engineering*, is shown beginning around 1970 and extending indefinitely into the future.

Notice that each period blends into the succeeding period. In this context, a particular period never really "ends." Instead, the cumulative influences of successive periods simply result in the attainment of the particular characteristics of following periods. The influences of Taylor and other pioneers certainly continue to the present time and are reflected in industrial engineering practices.

The authors believe that the profession has evolved to the point that a significant new direction has been emerging for several years. We envision this new orientation as not only building on the previous foundations of industrial engineering and operations research, but also including appropriate concepts from feedback control theory, computer science, behavioral theory, *systems engineering*, and cybernetics.

Some of the most exciting and productive years for our profession are just ahead. The opportunities for service to humankind have never been greater.

1.10. Industrial Engineering Organizations

Much can be learned about any profession by tracing the organizations which members of the profession form and/or join. The American Society of Mechanical Engineers provided the first forum for a discussion of the works of the early pioneers, particularly Taylor and his associates. Then, in 1912 the Society to Promote the Science of Management was formed. The name was changed in 1916 to the Taylor Society.

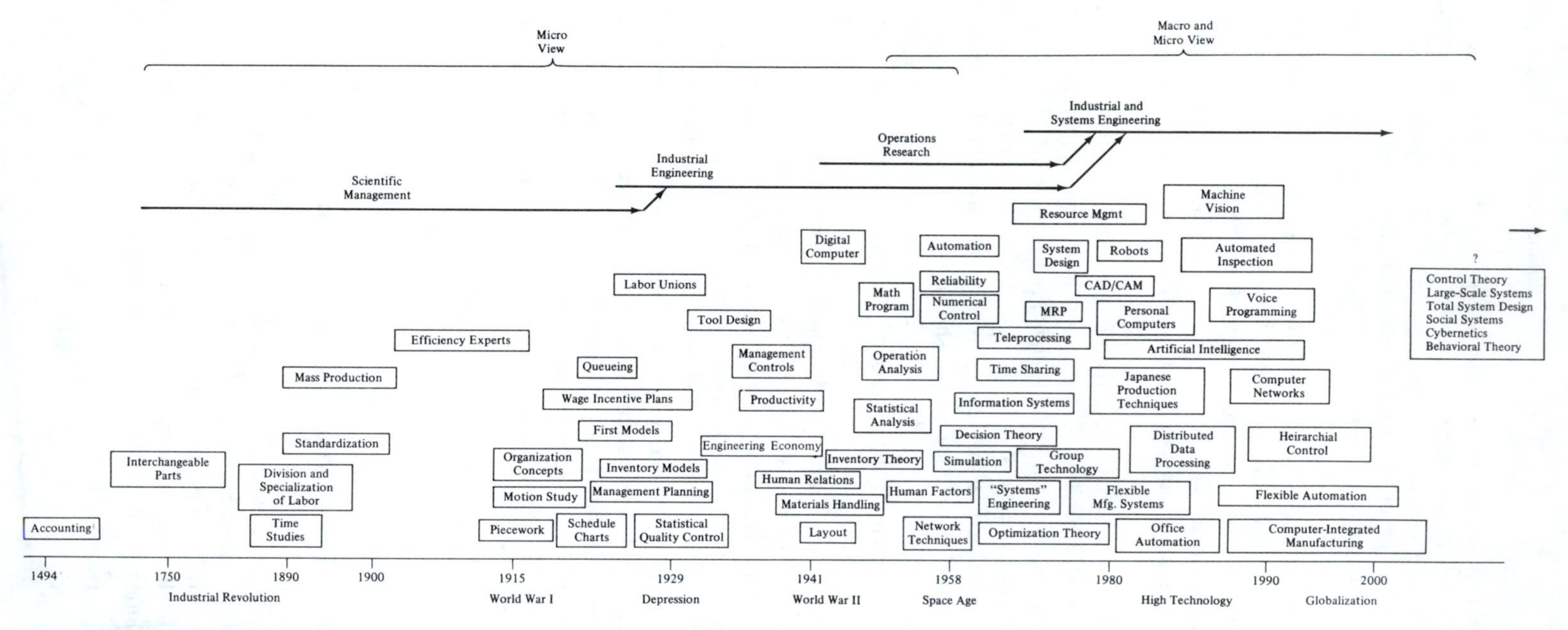

Fig 1.2. Chronology of significant events and developments in the evolution of industrial and systems engineering.

The Society of Industrial Engineers was formed prior to 1920. The American Management Association was formed in 1922, and many industrial engineers were active in this organization. In 1936 the Taylor Society and the Society of Industrial Engineers were combined to form the Society for the Advancement of Management.

The American Institute of Industrial Engineers was founded in 1948. The AIIE provided, for the first time, a professional organization devoted exclusively to the interests and development of the industrial engineering profession. Previously, I.E.s were associated with organizations whose main interests were in management or another branch of engineering.

The AIIE was an instant success in all respects. Within one year student chapters were formed at 11 major universities. The official publication of the AIIE, *The Journal of Industrial Engineering*, also made its initial appearance the following year, the first issue being published in June, 1949. In 1969 the *Journal* was divided into two publications. *Industrial Engineering* is published monthly and is devoted primarily to industrial engineering practice. *IIE Transactions* is published quarterly and is devoted primarily to research and new developments within the profession.

In 1981, through a vote of its membership, the Institute changed its corporate name from AIIE to IIE. By dropping the word "American," the Institute officially recognized the international nature of its activities. The IIE has members in more than 80 countries around the world.

Many practicing industrial engineers belong to other organizations that are related to the I.E. field. Some of these are:

- Operations Research Society of America.
- The Institute for Management Sciences.
- Association for Computing Machinery.
- American Society for Quality Control.
- Society for Decision Sciences.
- American Production and Inventory Control Society.
- Society of American Value Engineers.
- American Association of Cost Engineers.
- Society of Manufacturing Engineers.
- Robot Society of America.

It is significant that in 1970 the AIIE conducted a vote of its membership on a proposal to change the name of the institute to Institute of Industrial and Systems Engineers. Although this proposal was narrowly defeated (46% for; 54% opposed),[16] it indicates that a sizable portion of IIE membership view their professional activities as concerned with broad aspects of total system design. This concept will be developed further in Chapter 2.

The IIE is the technical society for all industrial engineers, beginning with university students who are majoring in industrial engineering. A very important part of the overall organization are the university chapters. Students participating

[16]Frank Cotton, "On the Institute Name," *Industrial Engineering*, 2 (1970), 49.

in these chapters receive the IIE publications and are considered an integral part of the overall organization.

Several nationwide contests among students are conducted each year, such as the Outstanding Technical Paper Award. The winners of the several contests are recognized and honored at the annual conference of the IIE.

By beginning participation in his or her technical society as a student, a person is more likely to continue participation in professional activities after graduation.

1.11. Definition of Industrial Engineering

The following formal definition of industrial engineering has been adopted by the IIE:

> Industrial Engineering is concerned with the design, improvement, and installation of integrated systems of people, materials, information, equipment, and energy. It draws upon specialized knowledge and skill in the mathematical, physical, and social sciences together with the principles and methods of engineering analysis and design to specify, predict, and evaluate the results to be obtained from such systems.

As used in this context, the term *industrial* is intended to be interpreted in the most general way. Although the term *industrial* is often associated with manufacturing organizations, here it is intended to apply to *any* organization. The basic principles of industrial engineering are being applied widely in agriculture, hospitals, banks, government organizations, and so forth.

Read the definition again. Think about any large factory that you have seen in which thousands of workers, hundreds of machines, a large variety of materials, and millions of dollars must be combined in the most productive, cost-effective manner. Think about a large city that also requires thousands of workers, hundreds of vehicles and other machinery, materials, and millions of dollars in order to deliver services required by the public. Imagine how much more effectively the city could be run if the principles of industrial engineering were applied.

Consider again the right-hand side of Figure 1.2. The current definition of industrial engineering needs little or no modification to be a suitable definition for "industrial and systems engineering" in which the overall system design is emphasized.

1.12. Industrial Engineering Education

Topics that later evolved into industrial engineering subjects were initially taught as special courses in mechanical engineering departments. The first separate departments of industrial engineering were established at Pennsylvania State University and at Syracuse University in 1908.[17] (The program at Syracuse was short-lived, but it was reestablished in 1925.) An I.E. option in mechanical engineering was established at Purdue University in 1911. A rather complete history of industrial engineering academic programs may be found in Chapter 5 of *Origins of Industrial Engineering*.[18]

The practice of having an I.E. option within a mechanical engineering department was the predominant pattern until the end of World War II. Following World War II, separate I.E. departments were established in colleges and universities throughout the country.

There was very little graduate level work in industrial engineering prior to World War II. Once separate departments were established, master's level programs began to appear.

By 1990 over 150 universities were offering curricula in industrial engineering, of which 92 were accredited by the Accreditation Board for Engineering and Technology.[19] A large number of the accredited programs also offer the master's degree.

Graduate programs in industrial engineering leading to the doctoral degree are recent developments. Prior to 1960 fewer than 100 total doctoral degrees in industrial engineering had been granted. By 1990 approximately 175 students received the doctoral degree each year.[20] Many doctoral degree holders go into teaching, but more and more are going into positions in industry and government.

1.13. Impact of Related Developments

The evolution of the industrial and systems engineering profession has been affected significantly by a number of related developments. We discuss several of these in this section.

1.13.1 Impact of Operations Research

The development of industrial engineering has been greatly influenced by the impact of an analysis approach called *operations research*. This approach originated in England and the United States during World War II and was aimed at solving difficult war-related problems through the use of science, mathematics, behavioral science, probability theory, and statistics. The approach enjoyed rather good success.

Following World War II the concepts of operations research were extended

---

[17] Hammond, pp. 1–15.

[18] Emerson and Naehring, pp. 43–58.

[19] 1990 Annual Report, Accreditation Board for Engineering and Technology, New York, Sept. 1990, pp. 69–70.

[20] Records of the Council of Industrial Engineering Academic Department Heads.

to problems in industry and commerce. A large number of mathematicians and scientists began devoting attention to a wide variety of operational problems. This resulted in considerable interaction between industrial engineers and members of other scientific disciplines. The infusion of new ideas and new approaches to problem solving had a dramatic impact on industrial engineering education and practice.

Industrial engineering departments at many universities began offering course work and complete options (usually at the graduate level) in operations research. In fact, some academic I.E. departments have changed their names to Industrial Engineering and Operations Research.

As it was originally conceived, the operations research approach was as follows: A specific problem was identified; specialists from appropriate fields were formed into an interdisciplinary task force to develop a solution; appropriate scientific methods and principles (usually involving mathematical models) were brought to bear on the problem; consideration was given to the interaction of the various components in the system being studied; the ''best'' solution was decided upon and presented to management. Once the work of the task force was completed, the task force was disbanded.

An approach such as this essentially precludes the establishment of operations research as a separate discipline. One person simply cannot have expertise in all the scientific areas that might be needed for problem solving.

As currently used, the term operations research connotes a set of quantitative methods that are applicable to a wide range of managerial and operational problems. Consequently, very few people consider themselves as operations researchers. Many industrial engineers do consider operations research as being their primary interest.

1.13.2 Impact of Digital Computers

Another development that has had a significant impact on the I.E. profession is the digital computer. Digital computers permit the rapid and accurate handling of vast quantities of data, thereby permitting the I.E. to design systems for effectively managing and controlling large, complex operations.

Many of the methods of operations research, discussed in the preceding section, require extensive calculations. The digital computer permits the I.E. to exercise a wide variety of optimization techniques to assist decision makers to better allocate scarce resources.

The digital computer also permits the I.E. to construct computer simulation models of manufacturing facilities and the like in order to evaluate the effectiveness of alternative facility configurations, different management policies, and other management considerations. Computer simulation is emerging as the most widely used I.E. technique.

A recent development that is having a profound (although still uncertain) impact on industrial engineering is computer-aided design (CAD) and computer-aided manufacturing (CAM). The I.E. is now designing facilities and work spaces directly on a computer workstation. The computer is also being used to automatically generate process plans, bills of material, tool release orders, work schedules, operator instructions, and so on. During production, computers are used to control the cutting path of machine tools, send instructions to robots and other devices, record production progress, dynamically reschedule work centers, and automatically generate reports to all levels of management.

The development and widespread utilization of personal computers is having a dramatic impact on the practice of industrial engineering. I.E.s use PCs extensively to perform various analyses, to execute mathematical models, to plan and manage complex projects, to create data bases, and to implement a wide range of decision support tools.

Computer networks are being developed for companies that will permit all computers (even those made by different companies) to "talk to each other." In this mode, managers, supervisors, and machine operators may gain access to vast amounts of information that is stored in different computer systems, but accessible through the network.

The I.E.s role in the computer environment described above is to perform the overall system design, which effectively links together the various components of the system. Essentially every I.E. tool and technique discussed in the remainder of this book will be affected by developments in computers and communications.

1.13.3 Emergence of Service Industries

In the early days of the industrial engineering profession, I.E. practice was applied almost exclusively in manufacturing organizations. After World War II there was a growing awareness that the principles and techniques of I.E. were also applicable in nonmanufacturing environments.

One of the first service industries to utilize industrial engineering on a broad scale was the health-care industry. Many hospitals and clinics employ I.E.s to improve their operations, eliminate waste, control inventories, schedule activities, and for a wide variety of other functions.

A more recent area to utilize I.E.s is that of government agencies. This is now occurring at the national, state, and local levels. Thousands of I.E.s are employed by government organizations to increase efficiency, reduce paperwork, design computerized management control systems, implement project management techniques, monitor the quality and reliability of vendor-supplied purchases, and for many other functions.

Many other service industries are utilizing industrial engineers to improve their operations. The demand for I.E.s in all types of industry is far greater than the supply.

1.14. Relationship to Other Engineering Disciplines

As discussed in Section 1.3, the industrial engineering profession evolved from other engineering disciplines. Figure 1.3 illustrates the general nature of this evolution as well as the inputs to industrial engineering from certain nonengineering disciplines.

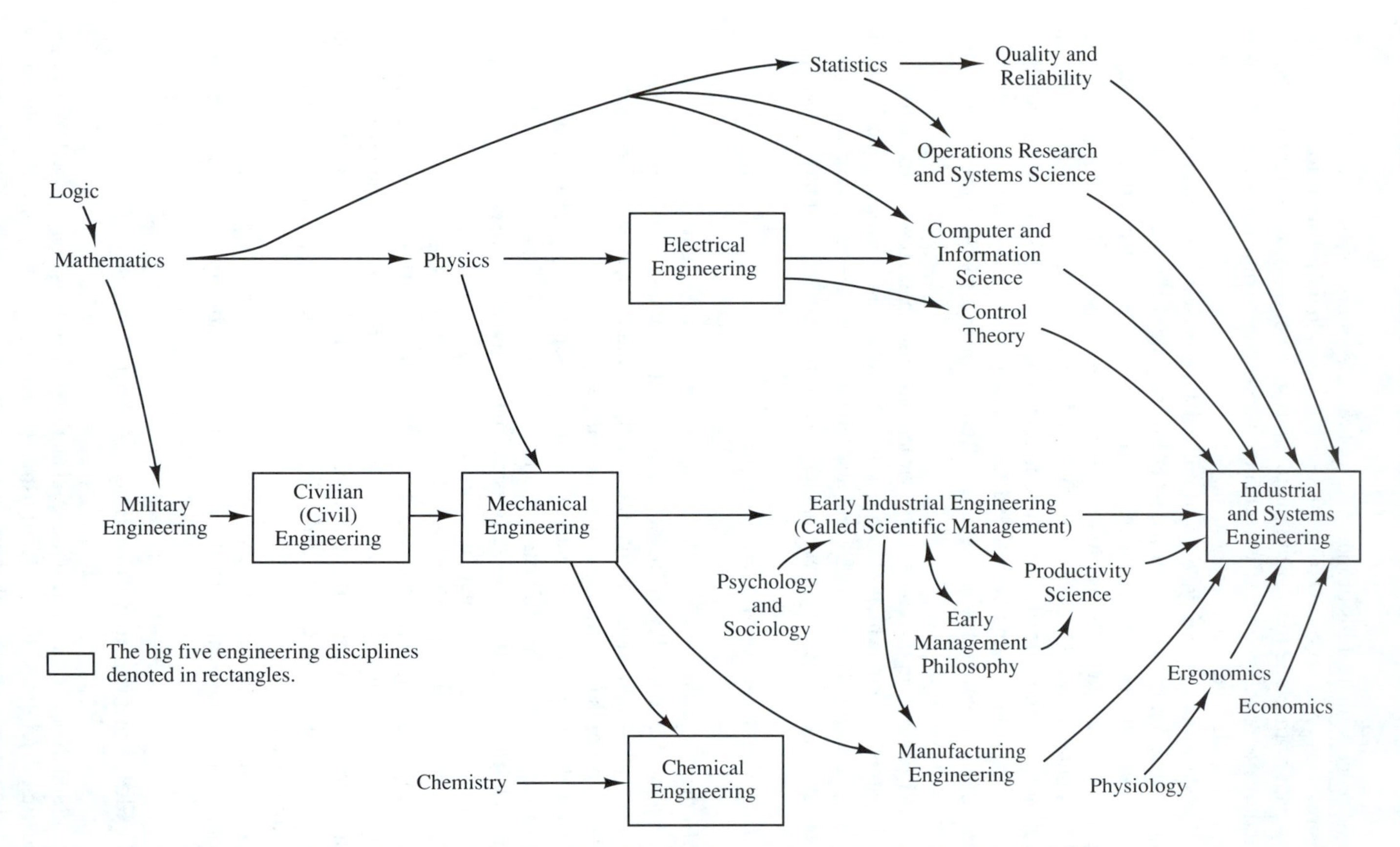

Fig. 1.3. Relationship of industrial and systems engineering to other engineering and scientific disciplines.

1.15. Challenges of the Future

We have briefly reviewed the role of the engineer in the technological developments that have led to the complex world in which we live. The remarkable achievements of the engineering profession have, ironically, contributed to some of the problems now facing human society. In the petroleum industry, engineers have designed highly efficient processes and systems for locating, extracting, processing, and distributing petroleum products. Increased efficiency led to increased usage. Increased usage led to increased profits, which led to even greater efficiency. This process could continue indefinitely if there were unlimited amounts of oil to be found. The oil shortages of the 1970s have made the world painfully aware that petroleum reserves are indeed limited.

Similar examples could be cited for other natural resources. The world has a finite amount of nonrenewable resources. One of the major challenges for future engineers is to learn to accomplish the engineer's mission in recognition of these constraints.

Another major challenge facing the engineering profession is to design systems and processes that are compatible with our natural environment. Dumping wastes into a river or allowing harmful gases to escape into the atmosphere are no longer permissible design strategies.

A major challenge that future engineers will encounter is that of designing products that are safe and reliable. More and more companies are being held accountable for faulty products that result in injury or harm to the buyer or to the public. Some states are now enacting legislation requiring that the design of all products affecting the public be approved by a registered professional engineer.

Through the application of sound industrial engineering practices, American industry became the strongest industry in the world during the two decades following World War II. Through the application of technological advancements and the never-ending search for improved work methods, American industry attained the highest level of productivity in the world. American products were regarded in the world market as being of the highest quality and reliability.

Productivity is the ratio of outputs to inputs. If factory A produces a 19-inch television set using 13.6 labor hours, and factory B produces an equivalent 19-inch television set using only 6.8 labor hours, then factory B has attained twice the productivity level of that attained by factory A. This concept will be developed further in Chapter 6.

American industry continues to lead the world in labor productivity, but other countries are rapidly closing the gap. Japan, France, West Germany, and other European countries are far outpacing the United States in the annual rate of productivity increase. Since labor costs in many of these countries are still below those in the United States, products manufactured in these countries can be sold on the world market at prices considerably below those of U.S. products. Consequently, the United States has lost large shares of the world market in many industries. This has had a negative impact on our balance of payments, inflation, and national debt.

Another major problem facing U.S. industry is the declining quality of American products relative to those made in many other countries. This has resulted, in part, from the aging equipment and processes in our factories. American industry, for a variety of reasons, has not made adequate investments in new technology, new equipment, and new manufacturing concepts.

There are encouraging signs in the early 1990s that American industry is beginning to take the steps necessary to regain a competitive position in world markets. An increasing number of companies are investing in automation, robots, and computer-controlled manufacturing processes. There is also an increasing emphasis on improved quality. The critical importance of a well-educated work force is also being recognized.

The business magazines are now saying that a "second industrial revolution" is under way in American industry. The engineering profession, particularly industrial engineering, will have a major role to play in this movement. We will look in greater detail at some of the aspects of this movement in Chapter 10.

As legislative leaders grapple with the many problems facing humankind, a realization will soon emerge that their solutions require an input from technically competent professionals. For Congress to enact a technologically infeasible national energy policy would be a disservice to the nation. *Perhaps the greatest challenge facing the engineering profession is to become involved in political issues and to provide assistance to lawmakers in the design of social systems.*

Despite the many remarkable achievements of engineers in the past, their greatest contributions to society will occur in the future. They will be expected to design systems that optimally utilize available resources for the satisfaction of human needs. Thus, the basic mission of the engineer remains essentially the same as always. Additional constraints, however, such as resource limitations, environmental concerns, and political issues, must be considered as an integral part of the engineering process.

DISCUSSION QUESTIONS

1. Using the references cited in this chapter, plus others that you may find in your library, develop a detailed chronology of the development of fundamental I.E. concepts during the period 1880–1930.
2. Describe how the work of Frederick Taylor was related to that of Frank Gilbreth, and vice versa.
3. Develop a convincing argument that industrial engineering is a "true" engineering discipline, analogous to the disciplines of civil, chemical, electrical, and mechanical engineering.
4. Find case studies that illustrate the applicability of industrial engineering principles and methodologies to nonmanufacturing systems, such as banks, hospitals, and government functions.

CHAPTER 2

Industrial and Systems Engineering

2.1. Introduction

We proposed in Section 1.4 that the principal activity engaged in by engineers that distinguishes them from other professionals is that of the *design of systems*. Basic courses in engineering science that are taken by all engineering students provide a general foundation upon which design principles in particular engineering curricula are later built.

Mechanical engineers design systems that are largely mechanical in nature, such as engines, machines of all types, and transportation vehicles. Electrical engineers design the electrical components of systems, such as integrated circuits and control mechanisms. If M.E.s design mechanical systems, and if E.E.s design electrical systems, can the analogy be extended to claim that I.E.s design industrial systems? *Indeed they do*! Provided, that is, that we define *industrial* to include all types of organizations, not just manufacturing.

2.2. Industrial and Systems Engineering Design

Just what do we mean when we say that industrial and systems engineers *design* industrial systems?[1]

Industrial and systems engineers (I.&S.E.s) design systems at two levels. The first level is what Blair and Whitston[2] have called *human activity systems* and is concerned with the physical workplace at which human activity occurs. The second level is called *management control systems* and is concerned with procedures for planning, measuring, and controlling all activities within the organization.

2.2.1 Human Activity System

The human activity system within an organization consists of the following elements that are designed or prescribed by I.&S.E.s:

- The manufacturing process itself (or the processing procedures of a service organization).
- Materials and all other resources utilized in the production process.
- Machines and equipment.
- Methods by which workers perform tasks.
- Layout of facilities and specification of material flow.
- Material handling equipment and procedures.
- Workplace design.
- Storage space size and location.
- Data recording procedures for management reporting.
- Procedures for maintenance and housekeeping.
- Safety procedures.

2.2.2 Management Control System

The management control system of an organization consists of the following elements that are designed or prescribed by I.&S.E.s:

- Management planning system.
- Forecasting procedures.
- Budgeting and economic analyses.
- Wage and salary plans
- Incentive plans and other employee relations systems.
- Recruiting, training, and placement of employees.
- Materials requirement planning.

---

[1] At this point, the reader is strongly urged to review the engineering process discussed in Section 1.4 and the definition of industrial engineering presented in Section 1.11.

[2] R. N. Blair and C. W. Whitston, *Elements of Industrial Systems Engineering* (Englewood Cliffs, N.J.: Prentice Hall, 1971).

- Inventory control procedures.
- Production scheduling.
- Dispatching.
- Progress and status reporting.
- Corrective action procedures.
- Overall information system.
- Quality control system.
- Cost control and reduction.
- Resource allocation.
- Organization design.
- Decision support systems.

Although the elements just described are expressed in manufacturing terminology, the framework is applicable to *any* system.

How well the I.&S.E. performs the above design tasks determines to a very great extent how efficiently and effectively his or her organization will operate. The difficulty involved in performing these design tasks is enormous. Many years of professional practice following a formal educational program in industrial and systems engineering are required before an engineer is capable of understanding how to perform various design tasks and, more important, how these tasks should all be related for an optimal design for the total system.

Chapters 3 through 21 present a quick introduction and overview of some of the major tools that must be mastered by the industrial and systems engineer in order to adequately perform his or her professional activities.

Truthfully speaking, almost all human activity systems are not really *designed*; they simply *evolve*. Few I.&S.E.s really engage in overall system design. Almost all are concerned with *improving* a very small piece of an existing system.

Compared to other engineering disciplines, the design methodology for industrial and systems engineering is relatively primitive and nonrigorous. Much effort needs to be directed toward the development of better design procedures. Even though it can truthfully be argued that the systems with which I.&S.E.s deal are much more complex than purely mechanical or electrical systems, we still have not made satisfactory progress in developing design methodology.

The *teaching* of other engineering disciplines is made relatively easy because the instructor is able to have the students work with an actual example of a physical system. The mechanical engineering instructor can have the students experiment with different fuel mixtures and burn these mixtures in an engine in a laboratory. The electrical engineering instructor can have the students design a circuit in a laboratory based upon fundamental knowledge presented in a lecture.

A professor of industrial and systems engineering cannot bring a production system into a laboratory. It is too big, too complex. Furthermore, it involves human beings who are not only unpredictable and difficult to measure, but who would also be unwilling to be part of a laboratory experiment.

The I.&S.E. student must learn the fundamental nature of the systems he or she is studying by experimenting with *models* of the systems rather than the actual systems. During the 1940s and 1950s I.E. students typically constructed *scale models* of production systems during their final semester. These scale models, with accompanying engineering drawings, represented the students' "solutions" to particular design problems.

Today, *symbolic models* and *conceptual models* are used rather than physical scale models. Often, these are converted to *computer models* to simulate real-world systems.

Unfortunately, symbolic and conceptual models are very difficult to comprehend. Also, the way these models are typically used in I.&S.E. education does not involve the student in the *design process*. The basic structure of the model is presented by the faculty member and the student merely manipulates independent variables. This exercise is certainly worthwhile, because it provides the student insight into the behavior of a system in response to manipulation of the associated decision variables. The weakness is that the student engages in very little *design activity*.

Obviously, much development work is urgently needed in better methods for teaching design principles to I.&S.E. students. The authors believe that this will receive major attention in the near future.

2.3. Typical I.&S.E. Activities

Different companies expect their I.&S.E.s to perform different kinds of activities. Traditionally, I.E. functions have been concentrated at the *operations* level of a firm. Other firms, recognizing the broad-based skills of their I.E.s, have expanded their activities to include the design of *management systems*. In recent years, as the I.E. role has added a systems flavor, I.&S.E.s are expected to engage in activities at the *corporate* level. To gain an appreciation for the broad spectrum of activities in which an I.&S.E. might be engaged, we present in outline form a list of activities grouped according to the three categories just discussed: production operations, management systems, and corporate services.

2.3.1 Production Operations

A. Related to the *product* or *service*:
   1. Analyze a proposed product or service.
      - Determine whether it would be profitable, at various production volumes.
      - Is it compatible with the existing product line?
      - Assess the manufacturability of the design, as prepared by the engineering design department.
      - Determine the best (most cost-effective) material to utilize.
   2. Constantly attempt to improve existing products or services.
      - Analyze field data on product usage.
      - Coordinate with the engineering design department on design changes.

3. Perform analyses relating to distribution of the product or delivery of the service.

B. Related to the *process* of manufacturing the product or producing the service:
1. Determine the best process and method of manufacture.
   - Apply classification and coding to each component.
   - Apply principles of group technology when appropriate.
   - Apply process planning algorithms to determine production sequence; determine alternate routings.
   - Design and specify appropriate tooling and fixturing.
2. Select equipment, determine degree of automation, use of robots, and so on.
3. Balance assembly lines.
4. Determine the best material flow and material handling procedures and systems.

C. Related to the *facilities*:
1. Determine the best layout of equipment.
2. Determine the appropriate storage facilities for raw materials, work in process, and finished goods inventory.
3. Determine appropriate preventive maintenance systems and procedures.
4. Provide for appropriate inspection and test facilities.
5. Provide adequate utilities for the operation.
6. Provide for security and emergency services.

D. Related to *work methods* and *standards*:
1. Perform work measurement studies; establish time standards; update as required.
2. Perform methods of improvement studies.
3. Perform value engineering analyses, eliminating cost and waste to the maximum extent possible.

E. Related to *production planning and control*:
1. Forecast the level of activity. (How many units will be sold?)
2. Analyze the capacity and resource constraints.
3. Perform operations planning:
   - Facilitate any necessary facility rearrangement.
   - Carry out make-or-buy decisions.
   - Plan production rates; rebalance assembly lines.
   - Construct master production plans.
4. Perform inventory analysis:
   - Determine current levels of raw materials inventory, work in process, and finished goods inventory.
   - Perform multilevel inventory analyses.
   - Determine appropriate reorder levels, reorder quantities, and safety stock levels.
   - Maintain vendor performance measures on quality, lead time, dependability, and so on.
   - Utilize principles of ''just in time,'' as appropriate.

5. Perform materials requirement planning (MRP).
6. Perform operations scheduling:
   - Allocate resources.
   - Schedule component production.
   - Schedule assembly operations.
   - Design procedures for schedule review and update.
7. Design the quality control system and inspection procedures.
8. Simultaneous smoothing of production, inventory, and work force.
9. Design systems and procedures for shop floor control:
   - Progress and status reporting.
   - Shortage reports.
   - Quality and rework reports.
   - Corrective action procedures.
   - Cost accumulation.
   - Labor utilization reports.
   - Reports on productivity, efficiency, and effectiveness.

2.3.2 Management Systems

A. Related to *information systems*:
1. Determine management information requirements:
   - Identify the decisions that are made by managers at all levels; specify timing of each decision.
   - Determine the specific data/information needed for each decision.
   - Identify the sources of each data element.
   - Determine the preferred form of data and the appropriate media for its transmission.
2. Design the *data base* to support the information system:
   - Specify *input formats* from data sources.
   - Specify the specific *files* required.
   - Within each file, specify the *records* it will contain.
   - For each record, specify the specific data elements (fields) that collectively make up the record.
   - Design the interfaces between the various data files.
3. Design the *management reports* that will be produced:
   - By management level.
   - By time period.
   - Provide for interactive inquiry.
4. Perform *data analyses*, as required.
5. Provide *feedback* to all levels of the organization.
6. Develop and implement *decision support systems*.
7. Analyze the requirements for *data communications* and *computer networks*.

B. Related to *financial and cost systems*:
   1. Design a budgeting system.
   2. Perform a variety of engineering economy studies.
   3. Design, implement, and track cost-reduction programs.
   4. Design procedures for systematically updating standard costs.
   5. Design systems for generating cost estimates for various purposes.
   6. Develop procedures for tracking and reporting cost data for management decision making.

C. Related to *personnel*:
   1. Design procedures for employee testing, selection, and placement.
   2. Design training and education programs for personnel at all levels in the firm.
   3. Design and install job evaluation and wage incentive programs.
      - Individual incentives.
      - Group incentives.
      - Profit-sharing programs.
      - Nonfinancial incentives.
   4. Design effective labor relations programs and procedures.
   5. Apply the principles of *ergonomics* and *human engineering* to the design of jobs, workplaces, and the total work environment.
   6. Develop effective programs of job enhancement.
   7. Coordinate the activities of quality circle groups.
   8. Design, implement, and monitor effective safety programs.

2.3.3 Corporate Services

A. Relative to *comprehensive planning*:
   1. Design, implement, and monitor a multilevel planning system:
      - Specify mission of organization.
      - Identify key results areas.
      - Specify long-term goals.
      - Determine short-term objectives.
      - Design system for tracking actual results, comparing to plan, and determining corrective action.
   2. Assist corporate management in performing strategic planning.
   3. Assist corporate management in rationalizing the firm's strategy in the international arena.
   4. Perform enterprise modeling:
      - Develop a high-level "business model," in which the major data flows are mapped between the major corporate functions.
      - Employ structured modeling methods to develop a hierarchical breakdown structure of the enterprise functions, subfunctions, and so on.
      - Use the enterprise model to measure business response cycles, determine critical pressure points, and identify strategic opportunities.

5. Perform systems integration activities:
   - Identify key functional interfaces.
   - Determine interdependencies between functions.
   - Rationalize the physical processes; simplify to the extent possible.
   - Rationalize the information interchanges.
   - Implement standard data exchange protocols.
6. Perform capacity analyses.
7. Participate in decisions relative to plant expansion and new plant siting.
8. Provide project management services:
   - Project definition and planning.
   - Work breakdown structures.
   - Network analysis.
   - Project tracking and follow-up.
9. Assist in implementing the concepts of total quality management throughout the organization.
10. Provide leadership in resource management:
    - Provide diagnostic services regarding utilization of energy, water, and other resources.
    - Suggest effective means of providing cogeneration capability within the company.
    - Devise effective systems for the management of hazardous wastes, scrap, and other by-products.
    - Provide leadership in constantly seeking means of reducing adverse environmental impacts of the company's products and the processes used in making those products.

B. Relative to *policies and procedures*:
1. Perform studies relative to organizational analysis and design.
2. Perform analyses of various functional groupings, recommend improvements to top management.
3. Develop and maintain policy manuals.
4. Develop and maintain current procedures relative to all management practices and systems.

C. Relative to *performance measurement*:
1. Design meaningful performance measures for the key results areas of each organizational unit.
2. Identify the "critical success factors" or measures of merit for each unit.
3. Develop methods and systems for analyzing operating data of all units and interpreting the results.
4. Specify corrective action procedures.
5. Design reports for all levels of management.

D. Relative to *analysis*:
1. Analyze systems and construct models:
   - State explicitly the problem being studied.
   - Determine the appropriate solution method.
   - Apply fundamental solution methodologies.
   - Recognize all assumptions pertaining to model and the solution method.

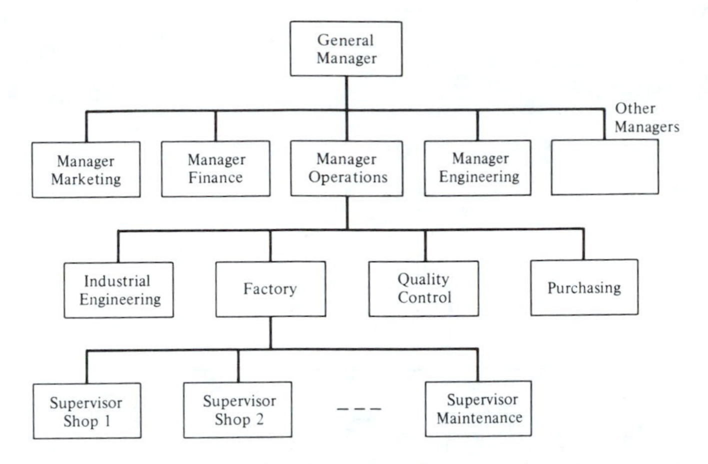

Fig. 2.1. Typical organizational structure for a manufacturing operation.

- Interpret the results of the solution in terms of the original problem statement and the underlying assumptions.

2. Perform simulation studies as appropriate.
3. Perform operations research studies as appropriate.
4. Perform statistical analyses.
5. Recognize the dynamic nature of the system being studied and include this feature in proposed solutions.
6. Apply the concepts of artificial intelligence and expert systems, as appropriate.
7. Conduct designed experiments on appropriate portions of the organization, in an attempt to continuously improve the overall performance of the organization.

One person cannot possibly perform all of the foregoing activities for an organization. I.&S.E. education programs, however, are designed to provide the fundamental principles involved in many of these activities. An individual engineer will specialize in a subset of the foregoing activities. The industrial engineering department will coordinate the individual efforts to address the spectrum of activities outlined.

2.4. Relationship to Total Organization

The various activities listed in the preceding section are not always performed by the industrial engineering department of a company. There are great differences in the scope of activities performed by industrial engineers in different companies. There are also great differences among companies as to where the industrial engineering function fits into the total organizational structure.

Typically, the industrial engineering department is a staff function that reports to the line manager who is in charge of operations. Figure 2.1 illustrates such an organizational arrangement. It is, however, not unusual for the industrial engineering department to be at the "manager" level shown in Figure 2.1.

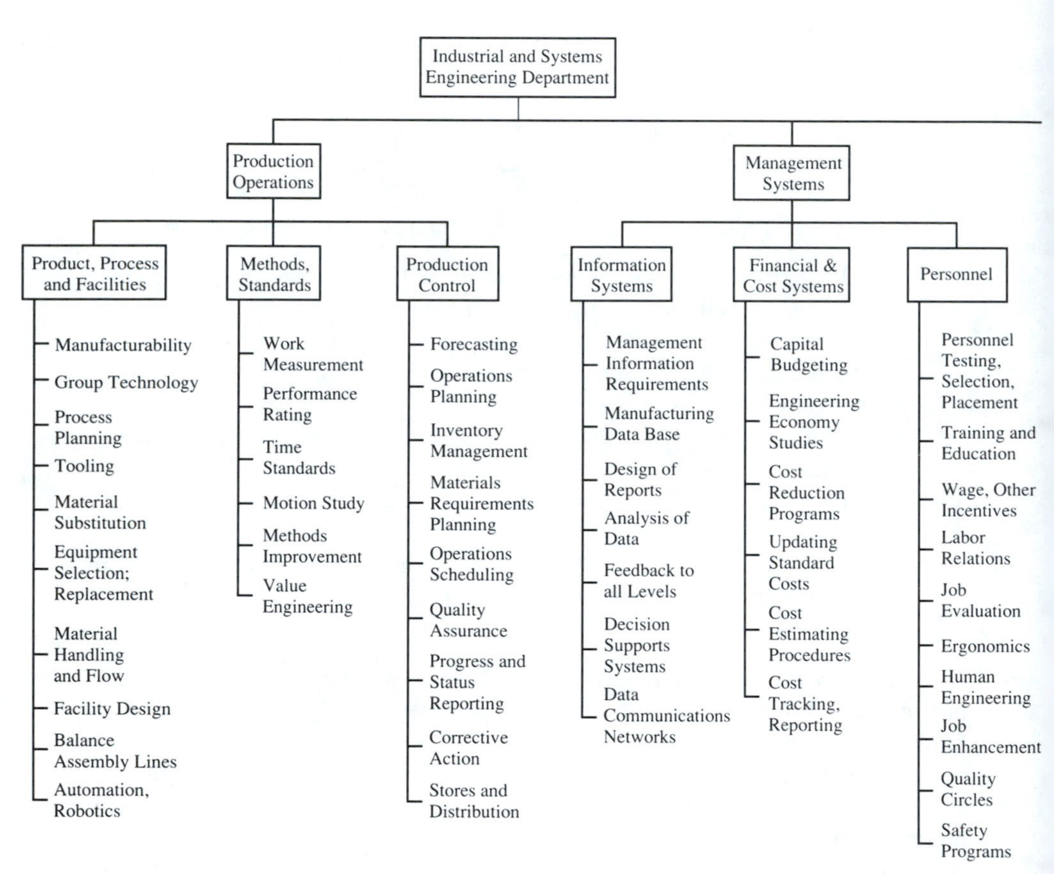

Fig. 2.2. Industrial and systems engineering department organized by major functional groupings.

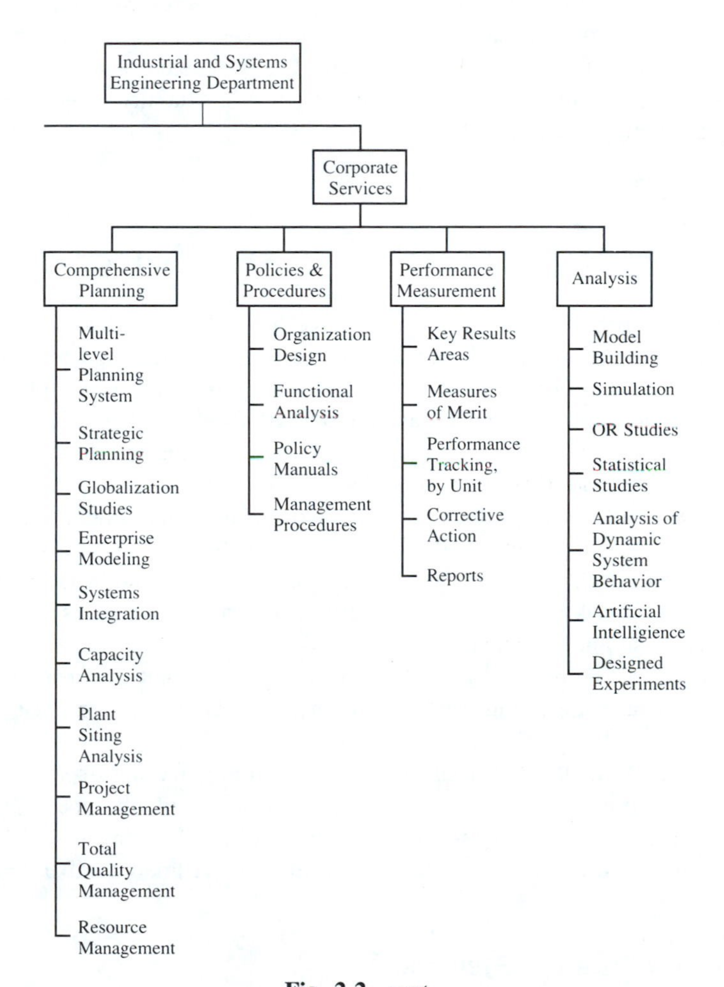
Industrial and Systems Engineering Department
Corporate Services
Comprehensive Planning
Policies & Procedures
Performance Measurement
Analysis
Multi-level Planning System
Strategic Planning
Globalization Studies
Enterprise Modeling
Systems Integration
Capacity Analysis
Plant Siting Analysis
Project Management
Total Quality Management
Resource Management
Organization Design
Functional Analysis
Policy Manuals
Management Procedures
Key Results Areas
Measures of Merit
Performance Tracking, by Unit
Corrective Action
Reports
Model Building
Simulation
OR Studies
Statistical Studies
Analysis of Dynamic System Behavior
Artificial Intelligience
Designed Experiments

Fig. 2.2. *cont.*

2.5. Internal Organization of the I.&S.E. Function

For industrial engineering departments that are very small, internal organization is not a problem. Its members simply take on problems as they come and assist each other in their work. Larger departments should be organized into a formal structure.

The internal organizational structure of an industrial engineering department should be *designed* by using basic principles of organizational design. The total scope of activities required to accomplish the industrial engineering mission should be analyzed and formed into logical groupings. These then become the sections within the industrial engineering department. One possible organizational arrangement is shown in Figure 2.2. This arrangement is essentially consistent with the activity groupings in Section 2.3.

2.6. Effectiveness Measures for the I.&S.E. Function

How does an organization determine how effectively its industrial engineering department is functioning? All meaningful effectiveness measures are stated so that they can be used as performance yardsticks against explicitly stated objectives.

The management of an organization must establish the objectives of the industrial engineering function in accordance with the mission the function is expected to accomplish. Objectives should always be stated in measurable terms. The industrial engineering department will be involved in certain ongoing activities of a repetitive nature and certain other one-time activities of a project nature. Clear-cut objectives for all activities should be established before the fact. Periodic measurements of actual accomplishments should be compared to the goals and significant discrepancies should be noted.

Shown in Figure 2.3 are a few typical objectives that might be appropriate for an industrial engineering department. These are obviously for illustrative purposes only.

At least annually, the head of the industrial engineering department should prepare an activity report describing accomplishments of all types. The actual accomplishments should be compared to the goals of the department. This comparison provides an indication of the effectiveness of the department for that time period.

2.7. The Nature of "Systems" [3]

In previous sections of this chapter, we have used the word "system" many times. We discussed topics such as "human activity *systems*," "management control

[3] B. S. Blanchard and W. J. Fabrycky, *Systems Engineering and Analysis*, 2nd ed. (Englewood Cliffs, N.J.: Prentice Hall, 1990).

ON-GOING ACTIVITIES

Achieve cost savings in direct labor equal to three times the departmental budget.

Reduce direct man-hour labor content of production by 10%.

Reduce manufacturing time through improved methods by 7%.

Obtain participation of 50% of workers in cost reduction program.

Reduce the cost of setting time standards by 10%.

Reduce in-process inventory by 8%.

Attain 96% accuracy in management reporting.

PROJECTS

Overhaul unit #3 at a cost not exceeding $300,000.

Rearrange production equipment to achieve a 15% decrease in flow time; complete project no later than November 15 at a cost not greater than $18,000.

Fig. 2.3. Typical objectives of an I.&S.E. department.

systems," and "*systems* design." It is useful to define and illustrate several basic concepts associated with "systems."

2.7.1 Definitions

A *system* may be defined as a set of components which are related by some form of interaction, and which act together to achieve some objective or purpose. Several words used in this definition need to be explained. *Components* are simply the individual parts, or elements, that collectively make up a system. *Relationships* are the cause–effect dependencies between components. The *objective* or *purpose* of a system is the desired state or outcome which the system is attempting to achieve.

Consider the air-conditioning system in a home. Assume that the system uses a heat pump for both heating and cooling, depending on the need. The *components* of this system include the house (walls, ceiling, floors, furniture, etc.), the heat pump, the thermostat, the air within the system, and the electricity that drives the system. The *relationships* between the system components are as follows:

(1) The *air temperature* depends on:
  (a) Heat transfer through the walls, ceiling, floor, and windows of the house.
  (b) Heat input or output due to heat pump action.

(2) The *thermostat* action depends on:
  (a) Air temperature.
  (b) Thermostat setting.

(3) The *heat pump* status depends on:
  (a) Thermostat action.
  (b) Availability of electricity.

Although there are other relationships in this system, the ones listed above serve to give physical meaning to the terms we are using. We will look in greater detail at this system in a later section.

2.7.2 System Classifications

Systems may be classified in a number of different ways. On the following pages we discuss a few classifications that illustrate the similarities and dissimilarities of systems.

- *Natural vs. Man-Made Systems*—*Natural systems* are those that exist as a result of processes occurring in the natural world. A river is an example of a natural system. *Man-made systems* are those that owe their origin to human activity. A bridge built to cross a river is an example of a man-made system.
- *Static vs. Dynamic Systems*—A *static system* is one that has structure but no associated activity. The bridge crossing a river is a static system. A *dynamic system* is one that involves time-varying behavior. The U.S. economy is an example of a dynamic system.
- *Physical vs. Abstract Systems*—A *physical system* is one that involves physically existing components. A factory is an example of a physical system, since it involves machines, buildings, people, and so on. *Abstract systems* are those in which symbols represent the system components. An architect's drawing of a factory is an abstract system, consisting of lines, shading, and dimensioning.
- *Open vs. Closed Systems*—An *open system* is one that interacts with its environment, allowing materials (matter), information, and energy to cross its boundaries. A *closed system* operates with very little interchange with its environment.

2.8. Feedback Control in Systems

Essentially all systems exist to perform some function. A watch is designed to keep track of time. We set the watch and then the internal mechanism causes the hands to move. However, as time goes on, the time portrayed on the face of the watch may be different from the true time, due to imperfections in the mechanism. The watch is called an *open-loop system* because it is not able to provide for its own control or modification. Figure 2.4 shows a diagram of an open-loop system and some of its characteristics.

Many systems that we encounter are able to control or adjust their own

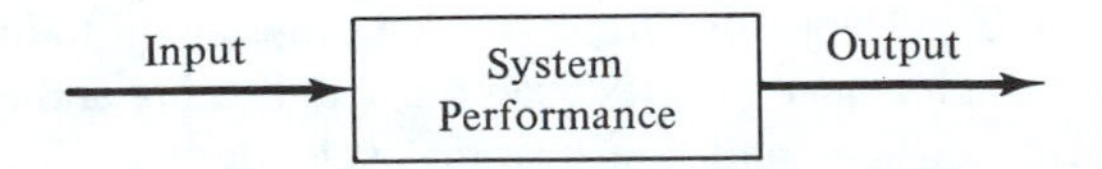

- Not aware of its own performance
- Past action has no influence on future action
- Possesses no means to provide for its own control or modification
- Output = f(input), but input $\neq f$(output)

Examples:
- Toaster – doesn't know whether it burned the toast
- Watch – not self-correcting
- Auto without a driver

Fig. 2.4. Open-loop system.

performance. Such systems are called *closed-loop systems*. Specifically, a closed-loop system is one that controls or adjusts its own performance in response to output data generated by the system itself. The home air-conditioning system described in Section 2.6 is a closed-loop system. Figure 2.5 shows a diagram of a closed-loop system and some of its characteristics.

The only difference between the diagrams of Figures 2.4 and 2.5 is the presence of a "feedback" loop in Figure 2.5. *Feedback* can be defined as the system function that obtains data on system performance (output), compares the actual performance to the desired performance (a *standard* or a *criterion*) and determines the modifications (corrective action) necessary prior to the next execution of system performance. This definition of feedback is long and complex. Indeed, it is common to treat this subject in a semester-long university course, usually in electrical or mechanical engineering curricula.

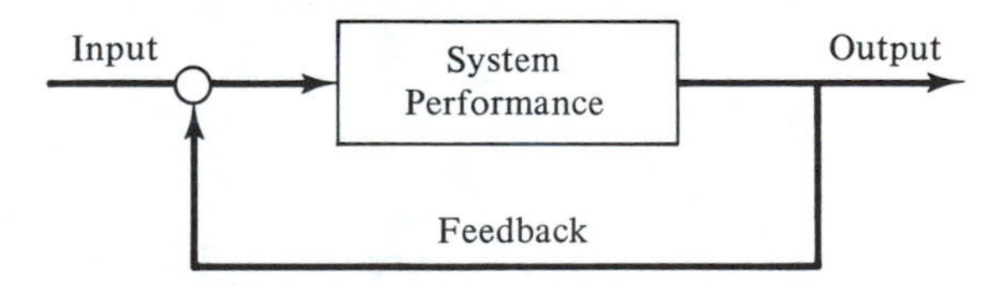

- Aware of and influenced by its own past performance
- Results of past action influences future action
- Senses its performances and automatically makes adjustments
- Output = f(input) and input = g(output)

Examples:
- Toaster and operator – adjust to desired darkness
- Watch and owner – adjust time to "standard"
- Auto with a driver

Fig. 2.5. Closed-loop system.

To illustrate the concepts just discussed, consider again the home air-conditioning system, in which a heat pump is used for both heating and cooling. For this illustration, we will consider only the heating cycle. It is desired to maintain a temperature of 70°F. To avoid having the unit cycle on and off too frequently, the thermostat is set to allow fluctuations between 67 and 73°F. Such a closed-loop feedback system can be illustrated as shown in Figure 2.6. The system illustrated is somewhat simplified. For example, the decision process would probably protect against turning the unit on if there was no power. It might also have some time delays built in relative to the frequency of switch activation.

Our interest in the concept of feedback is a very specialized one. At the beginning of this chapter, we established that industrial and systems engineers design systems at two levels. The first level, *human activity systems*, is concerned with the physical environment in which human activity occurs. The second level, *management control systems*, is concerned with procedures for planning, measuring, and controlling all activities within an organization.

When we design management control systems, we must explicitly provide for information feedback to the decision processes that control and adjust ongoing activities. For example, in a production process, we establish a production plan that specifies the desired outputs to be achieved through time. We then implement the plan. After a short period of operation (say, one week), we measure actual outputs and compare them to the desired outputs (the *plan* or *standard*). If the actual outputs are within an acceptable range from the desired outputs, we make no modifications

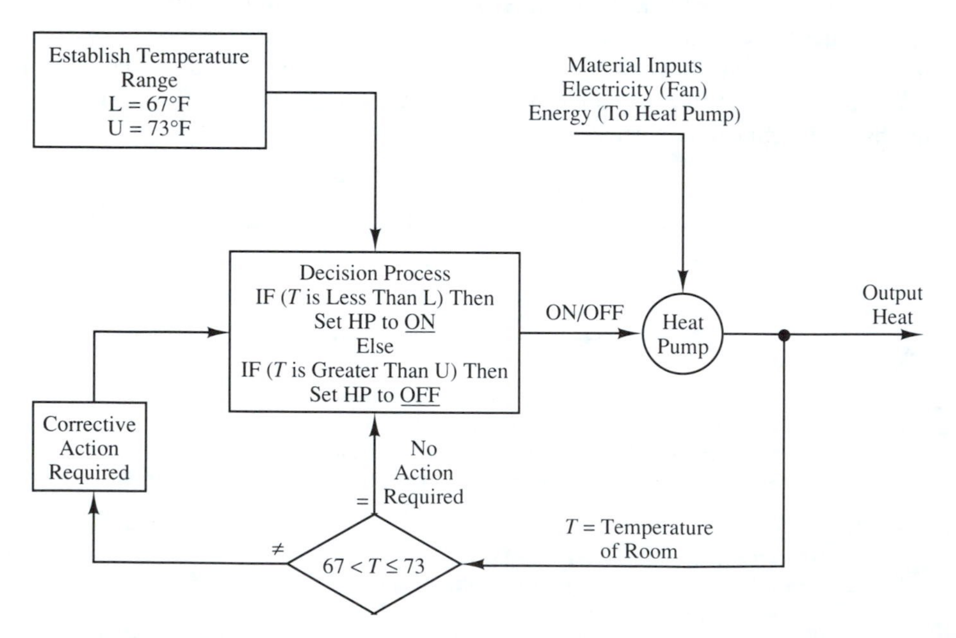

Fig. 2.6. Feedback-control system for heat pump system; heating cycle only.

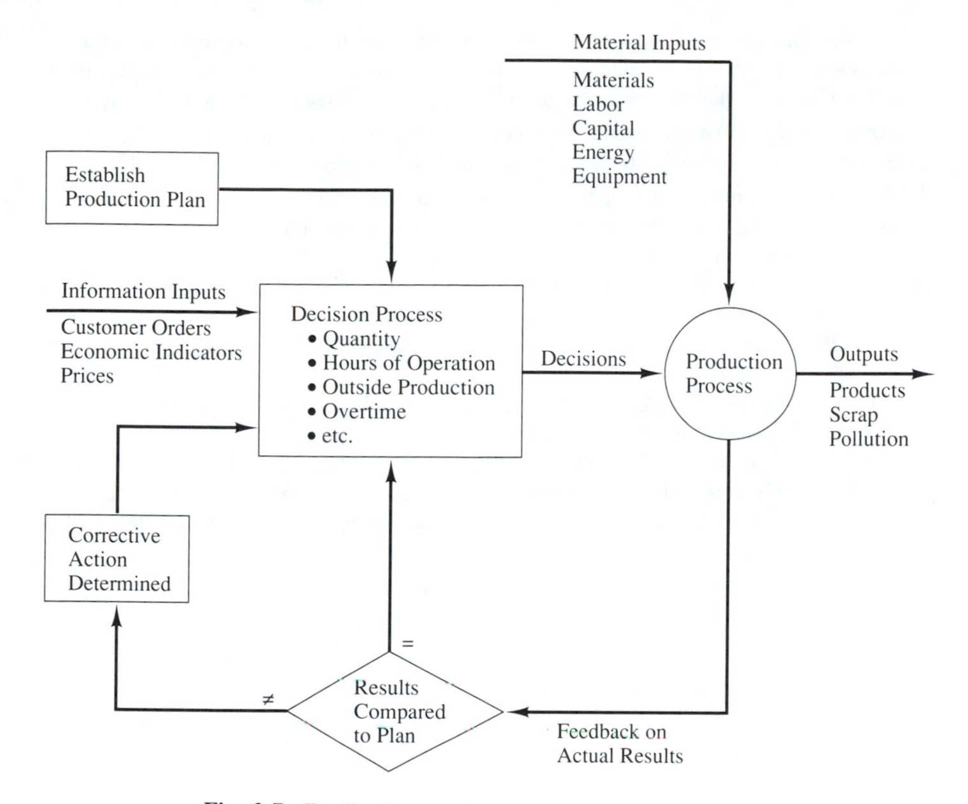

Fig. 2.7. Feedback-control system for production process.

to the system operation. If, however, the actual outputs are significantly different from the desired outputs, we must decide on appropriate corrective action before proceeding to the next execution of the system.

The key concept to the successful operation of a management control system is that of feedback. This process is illustrated in Figure 2.7 for a typical production process. Notice that this feedback system is more complex than the one in Figure 2.6, due to the greater number of decisions required in a production process.

A more complete treatment of systems concepts is presented in Chapter 18. Then, in Chapter 19, we explore the topic of management systems design in much greater depth.

DISCUSSION QUESTIONS

1. Analyze your university according to the two-level conceptual framework described in Section 2.2. Specifically, describe all the elements of the "human activity system" and of the "management control system" of your university.

2. Suppose that your university hires an industrial engineer to help it become more efficient and more effective. For such an environment, which of the activities described in Section 2.3 would be of the greatest importance? Defend your answer with a logical argument.
3. For the I.&S.E. function described in Question 2, specify a set of appropriate effectiveness measures, using Section 2.6 as a general guide. Be as specific as possible.
4. Represent each of the following systems in the context of Figure 2.6:
   (a) A thermostatic temperature control system for a house; the objective is to maintain the temperature within a specified range (L,H).
   (b) A guided missile (the guidance and control system is on-board the missile) being launched at a target; the objective is to hit within 20 feet of the center of the target.
   (c) A bullet from a rifle being shot at a bulls-eye target; the objective is to hit as near the target center as possible.
   (d) Filling a gas tank at a service station; the objective is to fill the tank as full as possible but to stop on an even $0.05 amount.
   (e) A carpenter driving a nail; the objective is to drive the nail as straight as possible.
   (f) A person taking a shower; the objective is to maintain a constant temperature of the water striking the person's body. The cold water temperature is constant, but the hot water temperature fluctuates.

PART 2 INDUSTRIAL AND S
ENGINEERING MET

CHAPTER 3

Manufacturing Engi

3.1. Introduction

Manufacturing engineering may be defined as *designing the production process for a product*. Although there is a large difference of opinion on exactly what is included in the design of a production process, almost all people would generally agree with the above definition. Production or manufacturing engineering includes all considerations pertaining to the *process* of production. This includes such functions as the following:

(1) Evaluating the manufacturability of the product.
(2) Selecting processes and setting process parameters such as cutting tool material, size, and shape; cutting speed; depth of cut; and so on.
(3) Designing work-holding devices (jigs and fixtures) to secure and control the position of the workpiece during manufacture.
(4) Estimating the cost of manufacturing the part.
(5) Assuring the quality of the part produced.

The rate of technological change in manufacturing engineering is phenomenal and likely to increase. ''Keeping up'' is already a very difficult task. The future is likely to see a progressive manufacturing engineer spending 15 to 25% of his or her time simply studying new technology. Computerization, integration of controls,

ation of manufacturing activities are areas where change pid. It is important to note, however, that in spite of the new ics still apply. For example, the most sophisticated computer- e center still performs its functions according to basic metal cutting al for the manufacturing engineer to know and understand the basics ring engineering.

chapter discusses some of the primary areas of manufacturing engi- . Since much industrial engineering has been applied to the metal working try, we concentrate on that particular industry. The reader should be aware t industrial engineering techniqués can be applied in any operating system, such as manufacturing, service, and governmental activities. Fully 60% of the American work force functions in service and governmental areas; but the limited length of this chapter prohibits any detailed discussion of those areas.

3.2. Product–Production Design Interaction

Product design requires that a person develop and evaluate the ability of the part to perform its intended function. Part characteristics such as size, shape, strength, reliability, safe operating range, and so on, are evaluated using knowledge of physics, strength of materials, tribology, and so on, often using computerized analysis. Manufacturing engineering develops and evaluates the cost of producing the part and uses knowledge of the relative cost, capabilities, and limitations of the various alternative processing methods available to produce the particular part shape, as well as detailed knowledge of cutting tools, machine tools, skill levels of workers, similarity to other parts being produced, and so on. Unfortunately, few industrial engineers are trained in product design and few product designers are trained in manufacturing engineering. This makes their interaction exceptionally important.

In every manufacturing operation, some variation in the size of the individual parts produced by the process will inevitably occur due to a variety of causes, such as tool wear, operator error, and material variations. The range of part sizes that can be used without compromising part function or reliability, that is, the variation in each dimension of the product that can be tolerated, is referred to as the tolerance for the product. The part designer, who is most concerned with product function, would want the tolerances set as small as possible to assure that the part will function without any problems. The manufacturing engineer, who is most concerned with product cost, wants the largest possible tolerances to be specified, as this often gives him or her a wider choice of processes to be used in manufacturing the part. This wider choice will often result in a reduction in product cost. Sometimes, a product designer may specify very tight tolerances on the product because he or she does not realize the cost of machining to exceptionally tight tolerances or the inability of machines to produce unusual configurations. Often, the manufacturing engineer will assume that these tight tolerances or unusual configurations are necessary (when in fact they are not) and design the process to produce them. This adds unnecessary cost to the product.

Ideally, a manufacturing engineer should work with the product designer from the very beginning to ensure producibility. If it is not possible to have this early interaction, the manufacturing engineer should inform the design engineer of unusually costly operations. By having this information, the design engineer can frequently avoid certain costly operations. This interaction must happen and should occur as early as possible in the design of products.

Employee participation is also vital in this interaction phase. No one knows the details of a job better than the person doing it, so this employee feedback input must be encouraged for industry to remain competitive. In fact (as will be seen later), the use of employee skills in some type of participative management may be the most exciting trend today.

> It is difficult to overemphasize the value of input from the hourly shop floor workers in planning and implementing automation systems. No one knows the nitty-gritty of a job better than the person who does it.[1]

3.3. Process Engineering

Process engineering is concerned with the design of the actual process to be used in the manufacture of the product. In designing the processes to be used, a six-step sequence should be undertaken: defining the product structure and specifications, assessing each component's manufacturability, listing the different processes capable to manufacturing the component, evaluating the cost of each of the alternative processes, determining the sequence in which the operations are to be performed, and documenting the process.

3.3.1 Defining Product Structure and Specifications

Product structures are often shown in a hierarchical chart that shows all of the subassemblies, sub-subassemblies, components, and raw materials that comprise the product. Figure 3.1 shows the product structure for a product assembled from two subassemblies (S_i), each of which is composed of sub-subassemblies (SS_j), components (C_k), and raw materials (R_l).

This type of chart clearly defines "what goes into what" and on which of the five levels the item belongs. More complex products would have more levels in their product structures. In Figure 3.1 each horizontal line indicates that an assembly operation must be designed to join its constituent parts. For example, the horizontal line linking subassemblies 1 and 2 indicates that a process to connect these two subassemblies must be designed. Note that sub-subassemblies, components, and/or raw materials can "go into" the subassembly; the hierarchy of the product structure dictates only that the horizontal line be above the highest-level item (subassembly,

[1]Frank Curtin, "Automating Existing Facilities: GE Modernizes Dishwasher, Transportation Equipment Plants," *Industrial Engineering*, Sept. 1983, pp. 32–38.

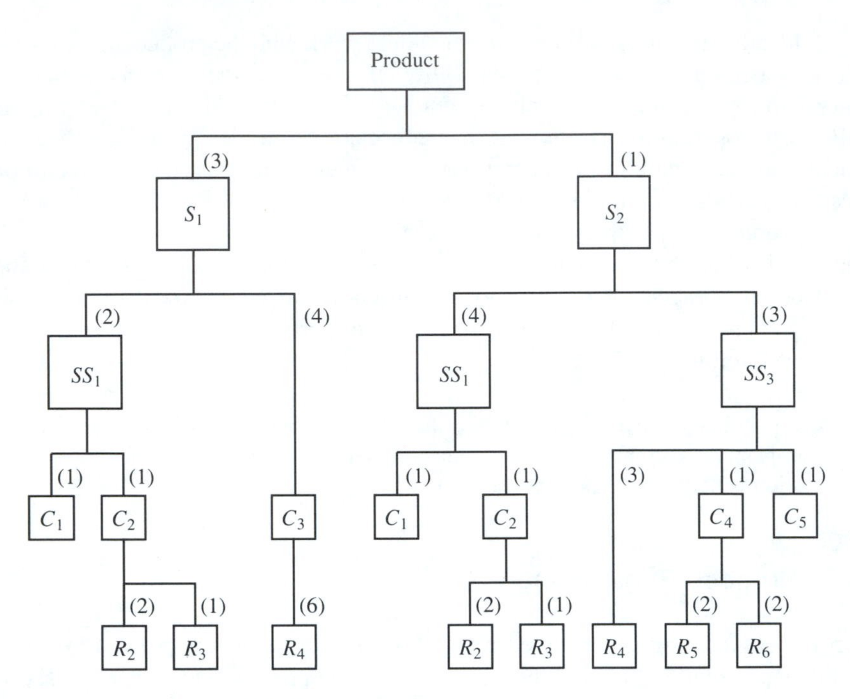

Fig. 3.1. Example product structure.

component, raw material) in the assembly. Also shown in Figure 3.1 are components that are purchased already fabricated and/or assembled and thus have no raw materials that must be processed by this facility. Vertical lines in the product structure indicate that the lower-level item is modified by some operation to transform or process it into the higher-level item. In industrial applications, each item in the product structure is identified by a unique number and incorporated in this number is the level of the item. This numbering is helpful when scheduling the production of the item, as we shall see in Chapter 7.

The number of each item required to make one of the next-higher-level item is also shown in the product structure. Figure 3.1 shows that three of subassembly 1 are required to make one unit of the product and that two sub-subassembly 1s are needed for each subassembly 1. This means that six units of sub-subassembly 1 are needed to make our product. These data are used when determining the number of each item that will be required to meet the sales forecast for our product.

This information can also be shown in bills of material. A *bill of material* contains the number and name of all parts, the source of each part (purchase or manufacture), and the number and name of the completed product. Normally, only those parts that go directly into that product are shown. For example, if a subassembly is used, the number and name of that subassembly are shown and a separate bill of material is developed for that subassembly. Figure 3.2 is a sample bill of material. In this figure, if the shelf (stock number 3) were a subassembly consisting

BILL OF MATERIAL			
Product Description: Bookcase, metal, 3 shelves. *Stock Number*: 1			
Component		Quantity (Amount) Required	Source
Stock No.	Description		
3	Shelf	3	Manufacturing
4	Leg	4	Manufacturing
5	Inserts	8	Purchasing
6	Screws	12	Purchasing

BILL OF MATERIAL			
Product Description: Bookcase, metal, 6 shelves. *Stock Number*: 2			
Component		Quantity (Amount) Required	Source
Stock No.	Description		
3	Shelf	6	Manufacturing
4	Leg	8	Manufacturing
5	Inserts	8	Purchasing
6	Screws	24	Purchasing
7	Connectors	4	Purchasing

BILL OF MATERIAL			
Product Description: Shelf *Stock Number*: 3			
Component		Quantity (Amount) Required	Source
Stock No.	Description		
8	Sheet metal	3 square feet	Purchasing

BILL OF MATERIAL			
Product Description: Leg *Stock Number*: 4			
Component		Quantity (Amount) Required	Source
Stock No.	Description		
8	Sheet metal	2 square feet	Purchasing

Fig. 3.2. Bill of material. [Taken with permission, from Lawrence E. Doyle et al., *Manufacturing Processes and Materials for Engineers*, 2nd ed. (Englewood Cliffs, N.J.: Prentice Hall, 1969).]

of perhaps the basic shelf, a laminated walnut top, and molding, then the bill of material for the shelf would show these three items.

Nowhere have the powers of computers been more evident than in the generation of these forms (process sheets, route sheets, bills of material) and their subsequent use in production control. Many on-line computerized control systems generate the route sheets and trace the product as it flows through the shop. Recent research has even demonstrated the possibility of the computer going from product design through process engineering, through recognition of shapes and forms and "artificial intelligence." In artificial intelligence, the computer learns from past experiences until it actually is capable of designing the production process. The manufacturing engineer should watch this development carefully.

Once all of the items that "go into" the product have been enumerated using the product structure or bill of material, the specifications associated with each item must be developed. These specifications would detail the desired material, length, width, height, diameter, material, surface finish, and tolerances for the item, which will assure that it will perform its intended function. This will enable the evaluation of the item's manufacturability and development of a list of the processes that are capable of producing this item.

3.3.2 Assessing Manufacturability

The manufacturability of an item is a measure of the relative ease or difficulty of producing the item within tolerance. The tighter the tolerances, in general, the more expensive the item will be to produce (see Figure 3.3). Also contributing to the manufacturability of the part is the incorporation of standard-sized features in the part. For example, if a hole in a part is specified to be a size that can be produced by a commonly available drill size, its cost to drill would be relatively low compared to the cost to drill a nonstandard hole size. Similarly, if at all possible, the part material should be specified to be a commonly available material. Only when particular part characteristics are not available from a commonly used material should an exotic material be specified. Each feature of each item in the product must be examined to determine if its function can be achieved in a manner that will allow consideration of a greater number of processing alternatives, some of which may be less expensive.

3.3.3 Determining Processes Capable of Producing the Part

By examining the specifications for each item and the capabilities of the various processes, a list of candidate processes can be developed. In developing this list, the following must be considered:

(1) The compability of the material and the process.
(2) The ability of the process to produce the desired tolerances.
(3) Part (re)design required to make it easier to produce by this process.

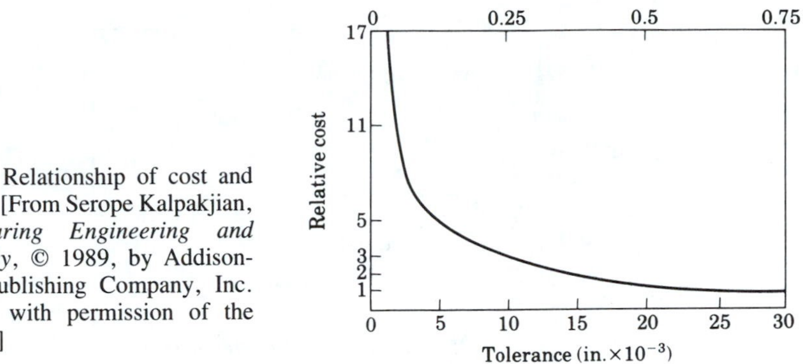

Fig. 3.3. Relationship of cost and tolerance. [From Serope Kalpakjian, *Manufacturing Engineering and Technology*, © 1989, by Addison-Wesley Publishing Company, Inc. Reprinted with permission of the publisher.]

(4) The availability of the process.
(5) The tooling, and so on, necessary to make the part using the process.

Use of tables such as those shown in Figures 3.4 through 3.6 enables the manufacturing engineer to quickly generate a complete list of available processes.

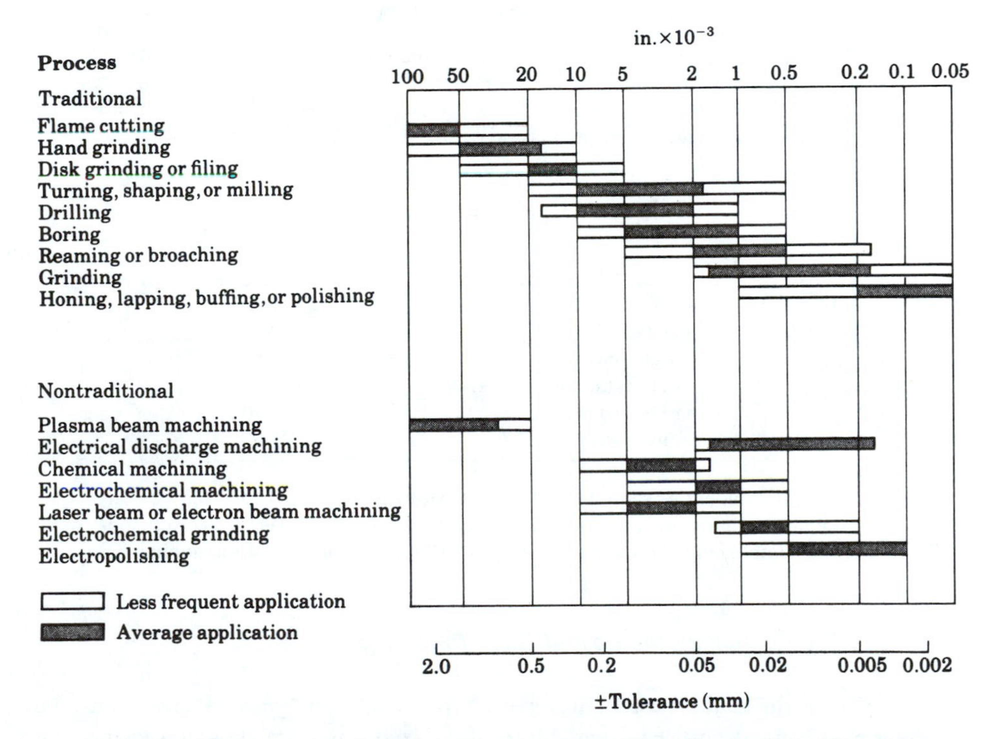

Fig. 3.4. Tolerances produced by various processes. [From Serope Kalpakjian, *Manufacturing Engineering and Technology*, © 1989, by Addison-Wesley Publishing Company, Inc. Reprinted with permission of the publisher.]

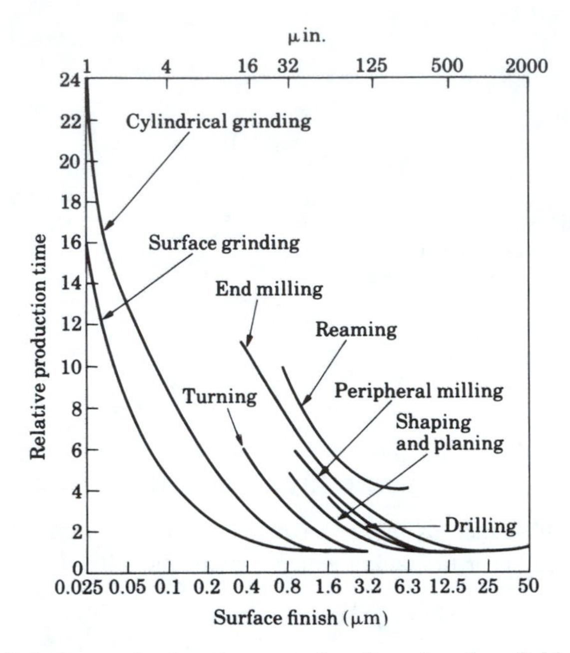

Fig. 3.5. Relative production time as a function of surface finish produced by various manufacturing methods. [From Serope Kalpakjian, *Manufacturing Engineering and Technology*, © 1989, by Addison-Wesley Publishing Company, Inc. Reprinted with permission of the publisher.]

Process	*Scrap (%)*
Machining	10–60
Hot closed-die forging	20–25
Sheet-metal forming	10–25
Cold or hot extrusion, forging	15
Permanent-mold casting	10
Powder metallurgy	5

Fig. 3.6. Amount of scrap produced by various processes. [From Serope Kalpakjian, *Manufacturing Engineering and Technology*, © 1989, by Addison-Wesley Publishing Company, Inc. Reprinted with permission of the publisher.]

3.3.4 Evaluating the Cost of Each Process

To evaluate the cost of using each process, two types of costs must be determined: fixed costs and variable costs. Fixed costs are those costs that are constant regardless of the number of parts made using the process. Examples of fixed costs are the purchase cost of the machine(s) used in the process, the cost of installing the machine(s), the cost of designing and fabricating the work-holding devices, and the cost of the space that the machine(s) occupy. Variable costs are those costs that vary directly (i.e., proportionally) with the number of parts produced. Examples of variable costs include the cost of the machine operator's time while making the part, the cost of running the machine(s) (e.g., electricity, natural gas, other fuels), the cost of all cutting tools that are dulled or otherwise consumed during the manufacture of the part, and the cost of the material used in manufacturing the part.

Fixed costs can be estimated by contacting vendors and/or users of the various equipment that is being considered as well as cost data developed by the firm's accounting department. Estimates of the variable cost rate (i.e., dollars/hour) can be obtained from the same sources. Expressing this in an equation, we obtain

$$\text{total cost} = \text{fixed cost} + \text{variable cost/unit} \times \text{number of units}$$

This relationship is also shown in Fig. 3.7.

If a number of alternative processes are being considered, their costs can be plotted on the same graph and the range of production quantities for which each process is less expensive can be determined. For example, suppose that three processes are under consideration and the following costs apply:

	Process A	*Process B*	*Process C*
Fixed cost	$ 0	$2,000	$8,000
Cost/unit	10	8	5

If we graph the costs, as shown in Figure 3.8, we can see that if we need to produce fewer than 1,000 units, process A should be used. If more than 2,000 units are needed, process C should be used. For production requirements between 1,000 and 2,000, process B should be used. The production requirement of 1,000 is said to be the break-even point between process A and process B and is the production

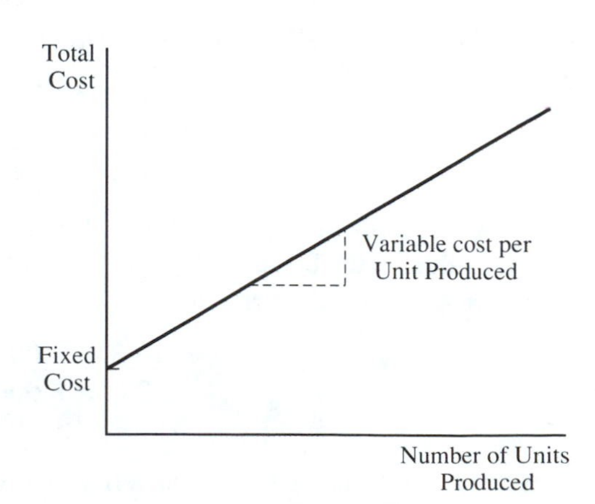

Fig. 3.7. Total cost as a function of the number produced.

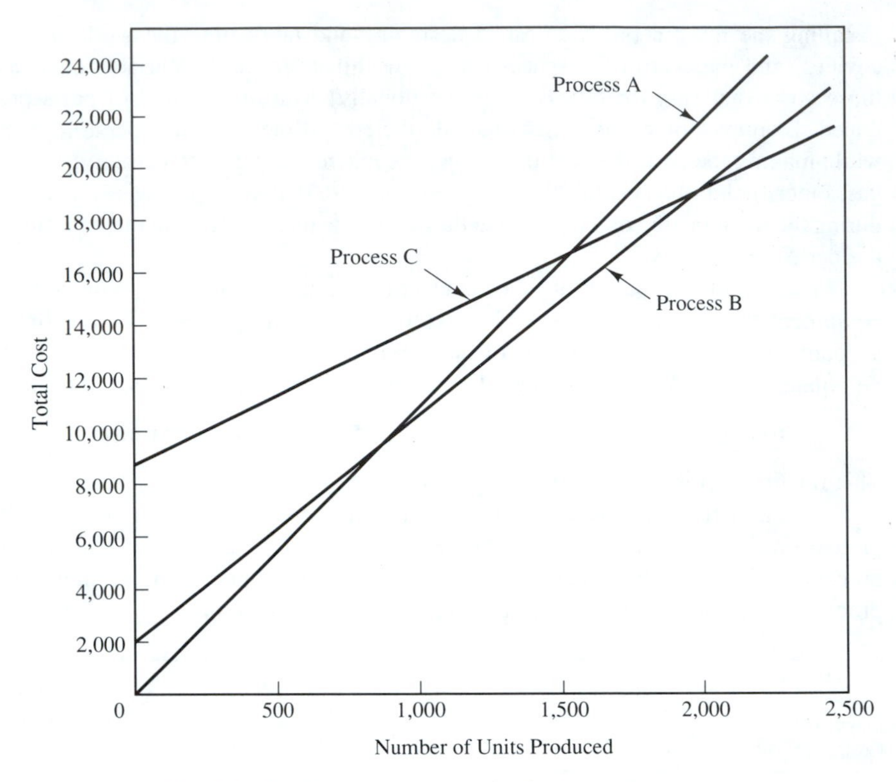

Fig. 3.8. Total cost versus quantity for three processes.

volume at which the total manufacturing costs are equal for the two processes. Production quantities below this break-even point can be produced at a lower cost using process A, and quantities above this break-even point can be produced at a lower total cost using process B. Similarly, 2,000 is the break-even point for processes B and C.

We could also have used the two total cost equations to determine the break-even points:

$$\text{total cost}_A = 0 + 10X$$
$$\text{total cost}_B = 2{,}000 + 8X$$

where X is the quantity to be produced. Realizing that total costs for both processes must be equal at the break-even point, we set the equations equal to each other and solve for X, the break-even point.

$$0 + 10X = 2{,}000 + 8X$$
$$2X = 2{,}000$$
$$X = 1{,}000$$

Similar calculations can be used to verify that the break-even point between process B and process C is 2,000.

The variable cost per part reflects the time that it takes to manufacture one part using the process and the costs that are incurred during this period of time.[2] This variable cost for a metal cutting operation has four components: the cost of labor (operator) and the machine while the machine is cutting, the cost of the tool, the cost of labor and the machine while sharp tools are being mounted, and the cost of labor and the machine while the workpiece is being loaded and unloaded. These costs can be expressed as

$$\text{variable cost} = T_c \times r_{l,m} + r_t \times \frac{T_c}{T} + T_{dt} \times r_{l,m} \times \frac{T_c}{T} + r_{l,m} \times T_{dw} + r_{l,m} \times T_a$$

where T_c = cutting time/part (min)
T = tool life (i.e., amount of time a tool can cut before dulling) (min)
$r_{l,m}$ = cost (dollars/min) of running the machine, including operator cost
T_{dt} = time (min) it takes to put a sharp tool on the machine (downtime for tool change)
r_t = cost of one sharp tool
T_{dw} = time (min) to load/unload one workpiece (downtime for workpiece change)
T_a = time (min) to adjust or position tool on each workpiece

This cost is dependent on the process parameters that are selected for the operation. The need to define the relationship of the cutting speed and tool life was addressed by Federick W. Taylor, the father of industrial engineering, early in the twentieth century.[3]

He knew that the time it takes to machine a part can be determined by calculating the length of the cut(s) that are required to machine the part to the desired size. For example, if a shallow groove is to be cut into a 36-inch-long workpiece using a single-point cutting tool operating at 20 ft/min, it would take 9 seconds to complete the cut. If 20 grooves were required on the part, 3 minutes of cutting would be required to manufacture the part. If one wanted to reduce the processing time, one could increase the cutting speed. The problem with this approach is that when the cutting speed is increased, the tool wears out or dulls much more quickly. To determine the optimum cutting speed, the relationship between the speed at which a cutting tool was being operated and how long the tool would cut until it was too dull to use was needed. This relationship is described by the following formula, known as *Taylor's tool life equation*:

[2]Vidosic, Joseph P., *Metal Machining and Forming Technology* (New York: Ronald Press, 1964), p. 329.

[3]Taylor, F. W., "On the Art of Cutting Metals," *Transactions of ASME*, 28 (1906), p. 31.

$$C = VT^n$$

where C = cutting speed (ft/min) at which the tool would dull after one minute of use
V = cutting speed (ft/min) at which the process will be operated
T = tool life (min) that can be expected when the tool is operated at cutting speed V
n = a constant that is dependent on the material being cut and the cutting tool material

If we define the length of the cuts on a particular workpiece to be L, $T_C = L/V$, and if we rearrange the terms in the Taylor tool life equation,

$$T = \left(\frac{C}{V}\right)^{1/n}$$

we can substitute these two variables into our variable cost equation:

$$\text{cost/part} = \frac{L}{V} \times r_{l,m} + r_t \times \frac{L}{V} \times \left(\frac{C}{V}\right)^{-1/n} + T_{dt} \times r_{l,m} \times \frac{L}{V} \times \left(\frac{C}{V}\right)^{-1/n} + r_{l,m} \times T_{dt} + r_{l,m} \times T_a$$

Since this is now an equation involving only the single variable V, we can differentiate it with respect to V, set the first derivative equal to 0, and solve for V. When we do,

$$V = C\left(\frac{n}{1 - n}\right)^n \left(\frac{r_{l,m}}{r_{l,m}T_{dt} + r_t}\right)^n$$

which is the optimum cutting speed, that is, the cutting speed that minimizes the cost of cutting each workpiece and hence minimizes the cost of producing the product on the machine in question.

Suppose, for example, that a 2-foot section of a 3-in.-diameter 1020 steel part is to be machined using a high-speed steel tool. Assume that:

$$C = 225 \text{ ft/min} \qquad r_{l,m} = 1 \text{ \$/min} \qquad r_t = \$4$$
$$n = 0.105 \qquad T_{dt} = 3 \text{ min}$$

$$\begin{aligned} V &= 225\left(\frac{0.105}{1 - 0.105}\right)^{0.105} \left[\frac{1}{(1)(3) + 4}\right]^{0.105} \\ &= (225)(0.7976)(0.8152) \\ &= 146.2 \text{ ft/min} \end{aligned}$$

Further, the time to cut the part using a feed rate (the translation of the tool per revolution of the workpiece) of 0.05 in./rev:

$$\text{No. revolutions to cut 24 in.} = \frac{24}{0.05} = 480$$

$$\text{rev/min at cutting speed of 146 ft/min} = \frac{146 \text{ ft/min}}{\pi\left(\frac{3}{12}\right) \text{ ft/rev}} = 186 \text{ rev/min}$$

$$\text{time to cut } (T_c) = \frac{480}{186} = 2.58 \text{ min}$$

$$\text{tool life } (T) = \sqrt[n]{\frac{C}{V}} = \sqrt[0.105]{\frac{225}{146}} = 61.5 \ min$$

$$\text{length of the cut} = \pi(3)\left(\frac{24}{0.05}\right) = 4{,}523 \text{ in.} = 377 \text{ ft.}$$

and if $T_{dt} = 3$ min, $T_{dw} = 1$ min, $r_{l,m} = \$1/\text{min}$, $T_a = 2$ min, and $r_t = \$1/\text{edge}$:

$$\begin{aligned} \text{total cost} &= \frac{L}{V} r_{l,m} + r_t \frac{L}{V}\left(\frac{C}{V}\right)^{-1/n} \\ &+ T_{dt}\, r_{l,m} \frac{L}{V}\left(\frac{C}{V}\right)^{-1/n} \\ &+ r_{l,m} T_{dt} + r_{l,m} T_a \\ &= \$7.87 \end{aligned}$$

3.3.5 Determining the Sequence of Operations

The determination of the sequence in which the operations that will be used to transform the material into its desired final shape depends on a number of considerations, and a full discussion of them is beyond the scope of this book. The student should be aware, however, that the sequence is developed based on (1) minimization of part handling (the part is routed, if possible, along the shortest path from machine to machine without backtracking); (2) assuring that no succeeding operation adversely affects previous operations by leaving burrs, debris, scratches, and so on; and (3) performing as many operations on each machine as possible so that close tolerances can be maintained and quality assured.

3.3.6 Documenting the Process

After the manufacturing engineer has reviewed the various processes and has decided on the most economical and effective procedure or set of processes, he or she must then find a way of recording this information for later use in the shop. One way is to construct an *operations process chart*.

An operations process chart depicts the flow of the material through the various processes. It shows only operations and inspections; it does not show transportations, storages, or delays. These are shown in flow process charts that will be discussed later. Using the following symbols, a manufacturing engineer can construct an operations process chart (see Fig 3.29 for an example.)

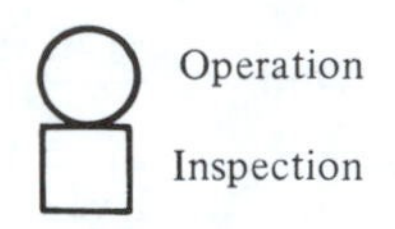

In the operations process chart, a complete unit is shown on a page (whenever possible). The chief component starts at the top and subsidiary components are added as the flow progresses to the right and down. (This will be discussed again in Chapter 5 and an example will be presented later in this chapter.)

The next step is to develop a route sheet that shows the sequence of operations, a description of each, and the various machining information necessary. A *route sheet* is a set of instructions to the plant showing how to manufacture the product. Figure 3.9 is an example of a route sheet. One such sheet would be required for each part in the operations process chart. The exact information shown varies between companies, but a good rule of thumb is to provide enough information to be able to produce the part including machine numbers, tool numbers, directions, and perhaps times for setup of the equipment and actual run time required. Figure 3.9 is not only a routing sheet, but also a production order for use in dispatching an order to the shop.

3.4. Industrial Processes

There is a wide variety of processes available to the manufacturing engineer to transform material into the desired size and shape. While a large number of materials are used in products, we will limit our discussion to those used for production of metal parts. These processes include refining and alloying of the metal, casting, forming, metal cutting, welding, assembly, and finishing.

3.4.1 Refining and Alloying

The general procedure for almost all metals is to begin with ore in its natural state and send it through some process to improve its usefulness. Steel, for example, is extracted as iron ore, is sent through a blast furnace to produce pig iron, and then it is sent to the steel mill where it is made into steel by the open hearth, basic oxygen, or electric furnace process. The type of steel produced is determined by the temperature of the furnace and its chemical composition.

Since all metals do not have the same properties, the proper choice of the metal to be used in a product depends on the functions the part must perform. This choice occurs in the production–product design interaction. To further compound the problem, metals emerging from the primary operations usually do not have the properties necessary. This means that some sort of metallurgical transformation is

TR

W-WORK ORDER
R-ROUGH STORES
F-FINISHED STORES

PRODUCTION WORK ORDER

TOTAL COPIES | SHEET NO. | MIN. LOT 72 | PRACT. LOT 4400 | PART NO. 8B-3381

EFFECTIVE DATE IF NOT STARTED

OPER. NO.	MACHINE: LOC. CLASS	MACHINE: M. NO.	BASE TIME: HRS. PER SET-UP	BASE TIME: HRS. PER PIECE	BURDEN HRS.	PERFORM. HRS.	OPERATION
1	X18B A2C	1939	1.00	.0009 1111			CHAMFER & CUT OFF
2	X23D G2R	1383	.10	.0025 400			REMOVE BURRS
3	X19C G20C	1997	.50	.0061 164			GRIND O.D. (2) PASSES
4	X19C M1B	7080	.70	.0161 62			STRADDLE MILL TONGUE
5	X19C M1B	7080	.80	.0141 71	.80 .0141	.80 .0141	MILL KEYWAY
6	X23D G2R	1383	.10	.013 77			REMOVE BURRS & CHAMFER TONGUE
7	X19C D6S	6076	.30	.0167 60			DRILL HOLE
8	X22D E1B	B131	.20	.0050 200			REMOVE BURRS

PARTS GROUP NO. 49

FILE	HT	I	C	W	R	F

COMP. PARTS: 4B-588, 4B-608, 4B-3429, 7B-4920, 3B-6794, 4B-8735

MATERIAL: C.T. #1E-40 C.F. STEEL 5/8" RD. M.L. (N.C.)

PART NAME: OIL PUMP GEAR SHAFT

QTY. PER PC. .505 FT.

5

DELIVER TO

ORDER DATE	SCHED. LOT SIZE	SCHEDULE	CSTGS. REQ'D

8B-3381
PART NO.

GG:MKG:CK

Fig. 3.9. Sample route sheet. [Taken with permission, from Lawrence E. Doyle et al., *Manufacturing Processes and Materials for Engineers*, 2nd ed. (Englewood Cliffs, N.J.: Prentice Hall, 1969).]

required to improve the hardness, strength, workability, or other properties of the metal. Much of this is done by heat treating the metal to alter its properties. In some cases, it may be necessary to combine two or more metals to obtain better properties. These combinations are called *alloys*. The production engineer must be aware of the different types of metals available and the particular alloys that can be made and their properties.

Because the properties of the various metals are very important, testing for these properties is of primary importance. Some of the properties that are often tested are:

(1) *Tensile Strength*—The ability of the metal to withstand elongation forces.
(2) *Hardness*—The ability of the metal to withstand penetration forces.
(3) *Impact Resistance*—The ability of the metal to absorb energy.
(4) *Malleability*—The ability of the metal to be shaped or extended.
(5) *Fatigue Resistance*—The ability of the metal to withstand repeated applications of a force.
(6) *Corrosion Resistance*—The ability of the metal to withstand corrosive forces.

The production engineer must understand these and other tests in order to select the correct material.

3.4.2 Casting

The first step in production is to obtain an approximate shape that, if necessary, can later be refined or finished more smoothly. One way of doing this is through casting. Casting is the process of forming objects by pouring liquid material into a mold and allowing the material to solidify. Casting is an old process, but it is still the most economical and effective method for obtaining many desired shapes.

Mold design is one of the most important parts of casting and the correct design of the mold is both a science and an art. Although many types of molds exist, by far the most predominant is a sand mold, especially for iron and steel.

Figure 3.10 illustrates a typical sand mold with its principal parts. Each part is described below.

Flask—Frame in which the mold is made.
Cope—Top section of the flask.
Drag—Bottom section of the flask.
Cheek—Intermediate sections of the flask (when required).
Riser—Reservoir of liquid metal to supply additional liquid metal as the metal solidifies and contracts.
Pouring Basin—Area into which molten metal is poured.
Sprue—Channel for vertical delivery of molten metal.
Runner—Channel for horizontal delivery of molten metal.
Gate—Opening through which metal enters the mold.

The correct relative and absolute shapes and dimensions of each of the above parts are highly critical in the design of a good mold. If the manufacturing engineer

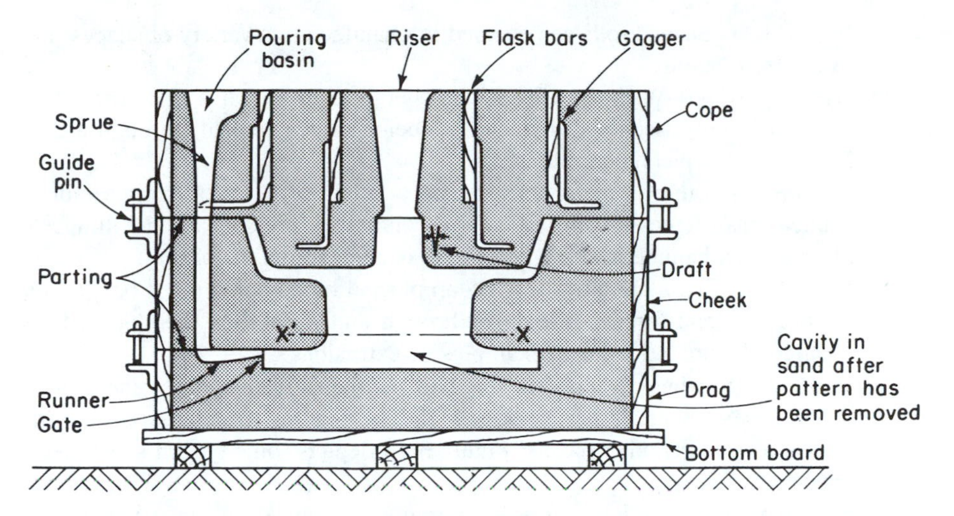

Fig. 3.10. Cross-sectional view of a three-part mold. [Taken with permission, from Lawrence E. Doyle et al., *Manufacturing Processes and Materials for Engineers*, 2nd ed. (Englewood Cliffs, N.J.: Prentice Hall, 1969).]

is to be working with castings, he or she should be cognizant of these and should be able to design good molds.

Obviously, this type of sand casting is expensive when large quantities of products are to be made, as the mold is destroyed each time a casting is made. Consequently, many specialized types of casting with permanent molds have been developed. The subject is too long and complex for this book, but some of the specialized types are centrifugal casting, die casting, gravity or permanent mold casting, pressed casting, ceramic shell casting, plaster mold casting, and shell molding casting.

3.4.3 Metal Forming

Another type of primary metal working process is metal forming. Here, the metal is worked under pressure to form a desired shape, to improve the physical properties, or both. This working can be either *hot* or *cold*. In hot working, the metal is worked at a temperature above its recrystallization temperature so that certain properties are enhanced. For example, in hot working, the metal working requires less pressure, and unusual shapes can be made more easily. Cold working takes place at a temperature below the metal's recrystallization temperature. This is done to hold close tolerances, to produce good surface finishes, and to enhance other physical properties.

Some of the processes that are classified as metal forming operations are listed below with short descriptions. Many of these processes can be performed in either hot or cold states, depending on the needs.

Rolling—Rolling is a squeezing operation in which the metal is elongated and/or widened as it passes through two or more rollers, as depicted in

Figure 3.11. Shaped rolls can be used to manufacture a variety of shapes, for example, I-beams.

Wire Drawing—Wire drawing is an operation in which a wire or rod is reduced in cross-sectional diameter by being drawn (pulled) through a die. Figure 3.12 depicts the process.

Forging—Forging is metal forming by a single or a series of intermittent applications of pressure instead of by continuous pressure as in rolling. A blacksmith's hammering of a horseshoe is an example of forging.

Extrusion—Extruding a metal is accomplished by compressing it beyond its elastic limit and forcing it to flow through and to take on the shape of an opening. Figure 3.13 shows examples of extrusions.

Bending—Bending is applying force to permanently distort the metal to a preconceived shape. It may be done by a tube bending machine, a press with V dies, or several other means. Figure 3.14 depicts some typical sheet metal bends.

Drawing and Stretching—These operations produce seamless vessels by applying pressure and forcing the sheet metal to take on the shape of a preconceived form. As in wire drawing, some stretching of the material occurs. Figure 3.15 is an example of a simple drawing operation.

3.4.4 Metal Cutting

After the primary shape forming, the next step is often metal cutting. Metal cutting is the removal of metal by using a cutting tool. The objective is to obtain desired shapes, close tolerances, and specified surface finishes. There are many types of metal cutting operations. Some of the better known are listed below with short descriptions:

Shearing—Shearing is the cutting of sheet metal by applying pressure and forcing the metal between two sharp edges. Blanking, parting, punching,

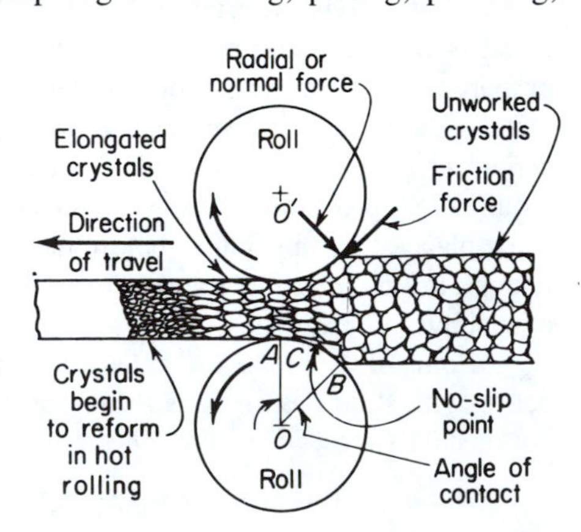

Fig. 3.11. Rolling operations. [Taken with permission, from Lawrence E. Doyle et al., *Manufacturing Processes and Materials for Engineers*, 2nd ed. (Englewood Cliffs, N.J.: Prentice Hall, 1969).]

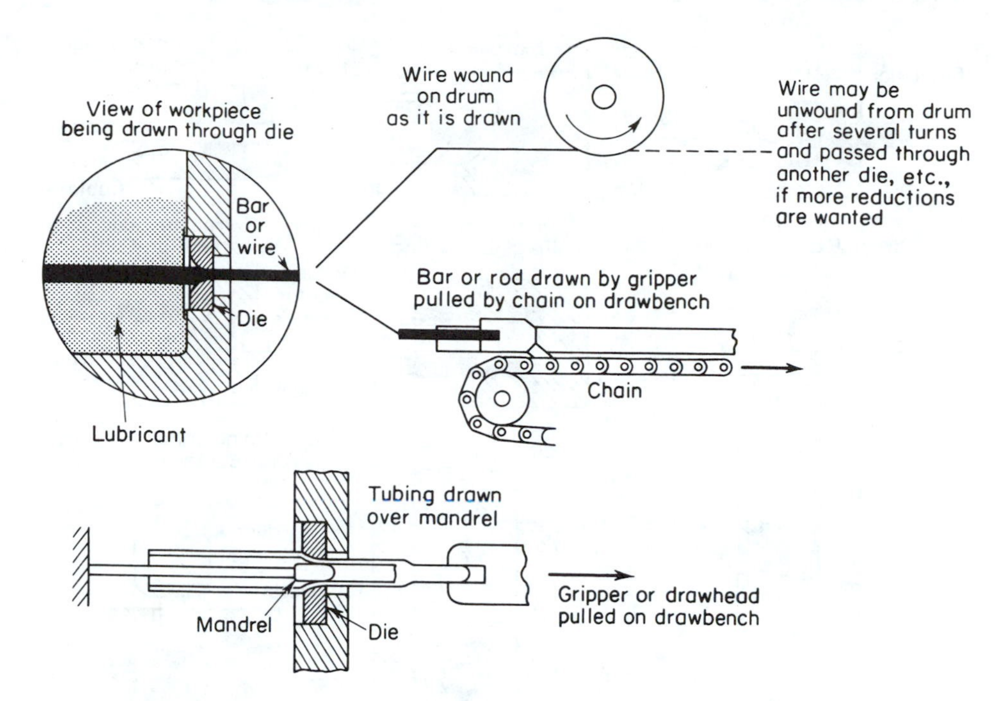

Fig. 3.12. Drawing operations. [Taken with permission, from Lawrence E. Doyle et al., *Manufacturing Processes and Materials for Engineers*, 2nd ed. (Englewood Cliffs, N.J.: Prentice Hall, 1969).]

nibbling, notching, and shearing along a straight line are all shearing operations. Examples are shown in Figure 3.16.

Turning—Turning is the process of rotating the workpiece and cutting material by feeding a cutting tool into the workpiece and/or the workpiece into the tool. These operations are generally performed on a lathe and may include facing, contour turning, threading, parting, drilling, knurling, and others.

Drilling—Drilling is the opening, enlarging, or finishing of a hole. It is generally associated with the process in which the tool turns and the workpiece is stationary. When the workpiece turns, it is known as a *turning operation* (see preceding page). The operation is generally performed on a drill press or boring machine.

Shaping and Planing—These are operations in which surfaces are cut by a reciprocating action. Generally, the surfaces are flat, but curved surfaces can also be machined by the processes. In shaping, the tool reciprocates while the workpiece is stationary. In planing, the workpiece reciprocates while the tool is stationary. Figure 3.17 is an illustration of a shaper in operation.

Milling—Milling is the process by which a revolving cutting tool with a number of teeth takes intermittent, successive cuts. Usually the workpiece is fed into the tool, but this is sometimes reversed. Figure 3.18 depicts some common milling operations.

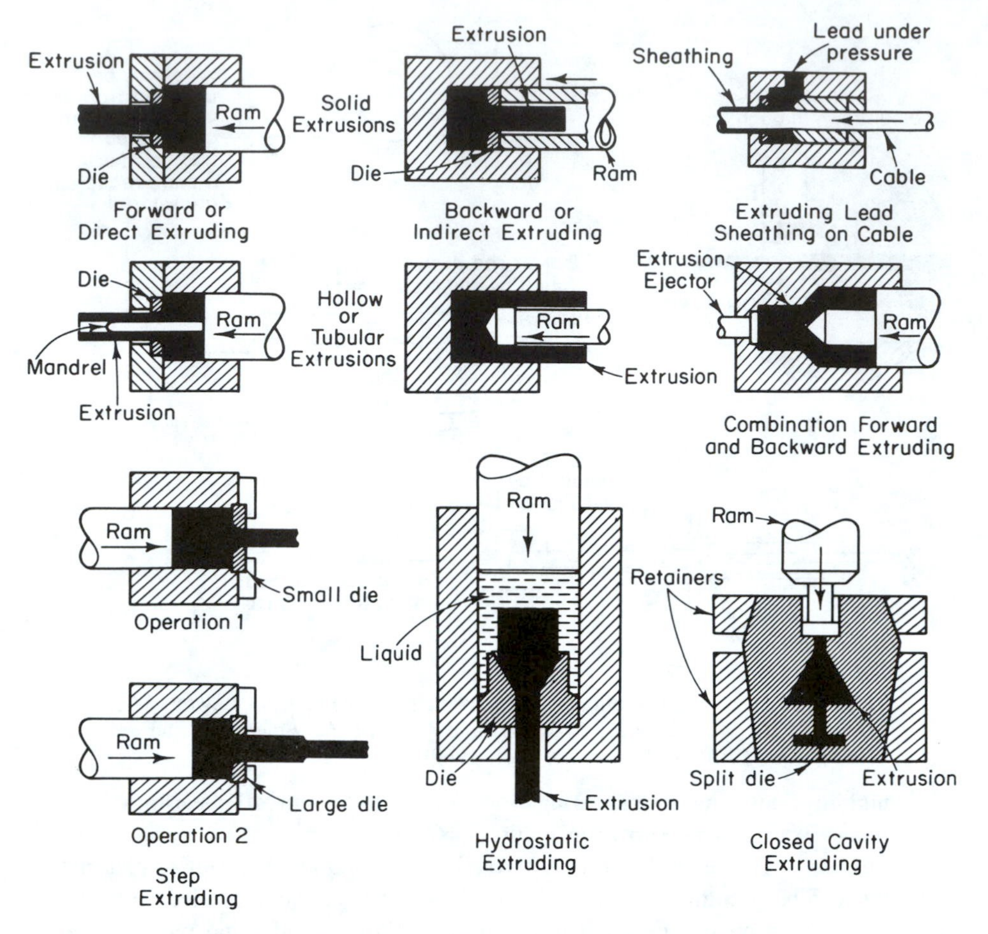

Fig. 3.13. Extruding. [Taken with permission, from Lawrence E. Doyle et al., *Manufacturing Processes and Materials for Engineers*, 2nd ed. (Englewood Cliffs, N.J.: Prentice Hall, 1969).]

Broaching—Broaching is similar to milling in that a series of teeth are used to take successive cuts on the workpiece. The broaching tool does not revolve, however; it is pushed or pulled. Figure 3.19 shows the basics of a broaching operation.

Sawing–Filing—Sawing and filing also remove metal by taking cuts with successive teeth on a tool that is pushed or pulled. In sawing, however, large amounts of metal can be removed without reducing it to chips. Figure 3.20 is a picture of a power hacksaw in operation.

Grinding—Grinding is the removal of metal in small pieces by the continuous action of an abrasive material on the workpiece. Grinding may be necessary on materials too hard for other methods, to improve surface finishes, or to hold very close tolerances. The small grinder in a garage used to sharpen hand tools is a familiar example.

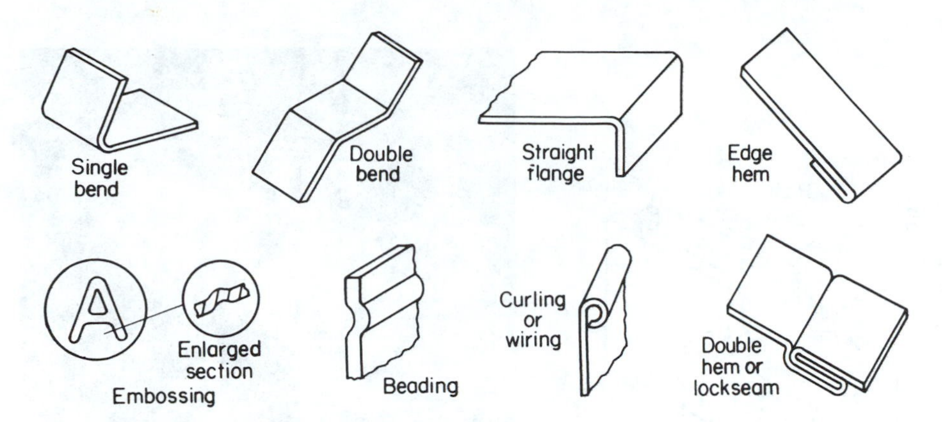

Fig. 3.14. Examples of sheet metal bends. [Taken with permission, from Lawrence E. Doyle et al., *Manufacturing Processes and Materials for Engineers*, 2nd ed. (Englewood Cliffs, N.J.: Prentice Hall, 1969).]

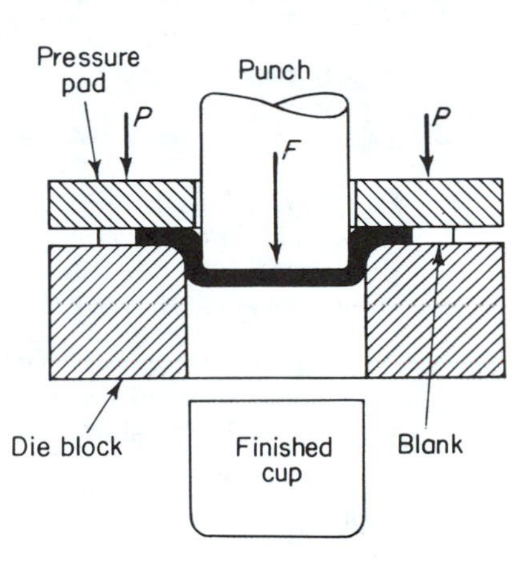

Fig. 3.15. Drawing of a cup. [Taken with permission, from Lawrence E. Doyle et al., *Manufacturing Processes and Materials for Engineers*, 2nd ed. (Englewood Cliffs, N.J.: Prentice Hall, 1969).]

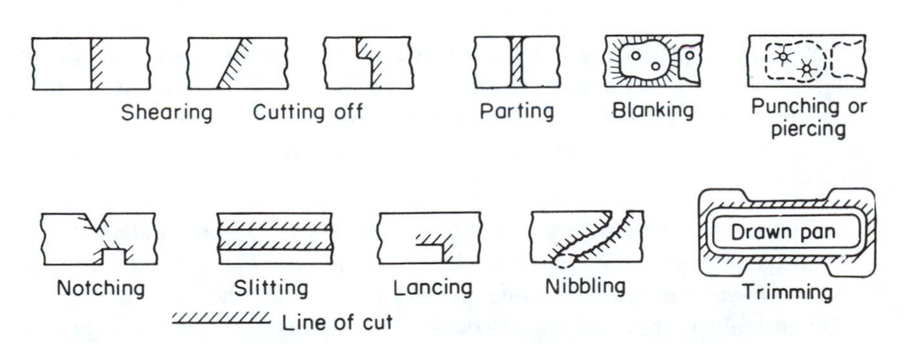

Fig. 3.16. Typical shearing operations. [Taken with permission, from Lawrence E. Doyle et al., *Manufacturing Processes and Materials for Engineers*, 2nd ed. (Englewood Cliffs, N.J.: Prentice Hall, 1969).]

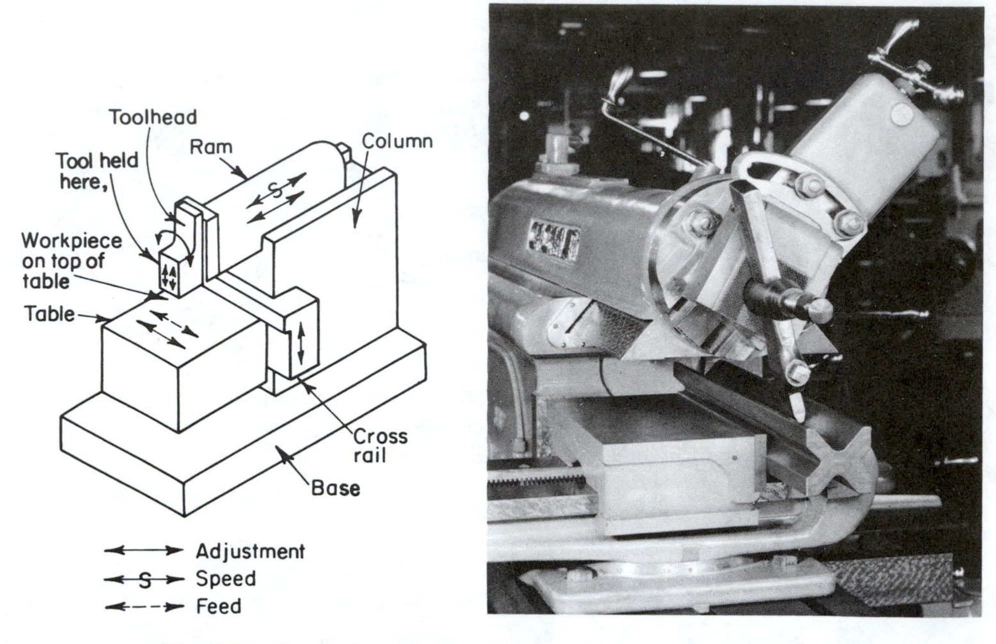

Fig. 3.17. Shaper in operation. [Taken with permission, from Lawrence E. Doyle et al., *Manufacturing Processes and Materials for Engineers*, 2nd ed. (Englewood Cliffs, N.J.: Prentice Hall, 1969).]

Other Operations—There has been a great deal of work directed toward the development of more sophisticated methods of metal removal. Examples are electrodischarge machining, electrochemical machining, electrolytic grinding, photoetching, and chemical milling. These will not be discussed in this chapter.

3.4.5 Welding

Welding is a process by which two pieces of the same metal are made to bond through the application of heat or pressure or both. The process actually fuses metals together by forcing the two parent parts to melt and coalesce. Some of the broad classifications and more popular types of welding processes are listed on the following pages.

Electric Arc—Electric arc welding is a process by which an electric arc is maintained between an electrode and the workpiece or between two electrodes. The arc generates immense quantities of heat causing the metal to melt and, upon cooling, cause the weld to occur. In some cases, the electrode provides the filler metal and sometimes a slag or protective covering over the weld.

Resistance Welding—In resistance welding, an electric current is passed

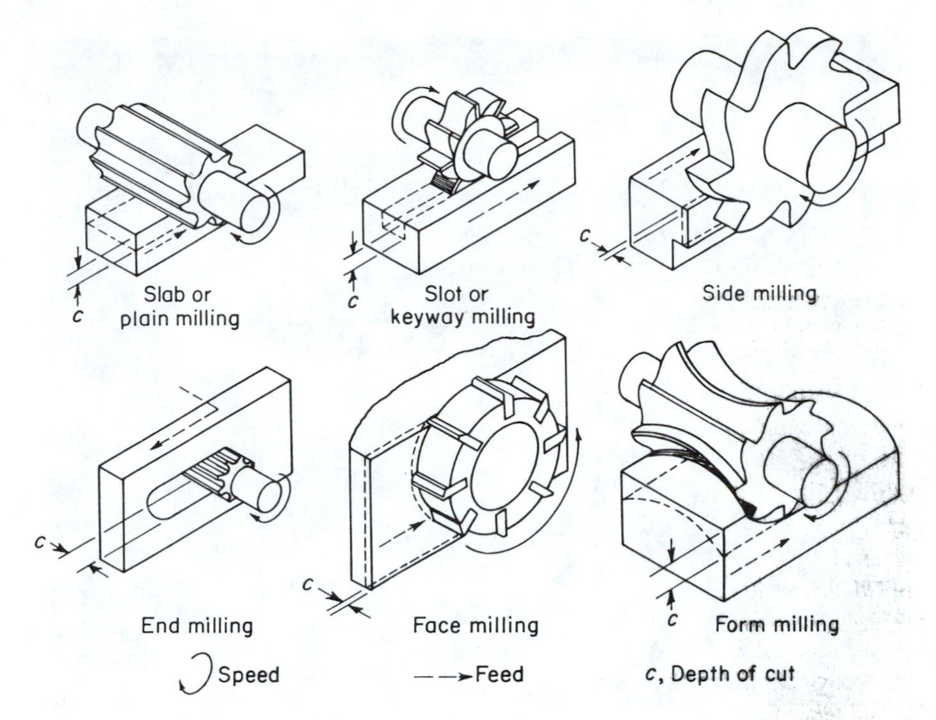

Fig. 3.18. Common milling operations. [Taken with permission, from Lawrence E. Doyle et al., *Manufacturing Processes and Materials for Engineers*, 2nd ed. (Englewood Cliffs, N.J.: Prentice Hall, 1969).]

through two pieces of metal that are held together under pressure. The electric resistance across the surface discontinuity generates heat and the pressure causes the two metals to bond. Spot and seam welding are two popular examples of resistance welding. Figure 3.21 is a diagram of a spot welder.

Beam Welding—This process supplies the heat by bombarding the workpiece with a concentrated beam of electrons (electron beam welding) or by concentrating a high-energy light beam on the workpiece (laser welding). Both are available commercially but have somewhat limited applications at this point in time.

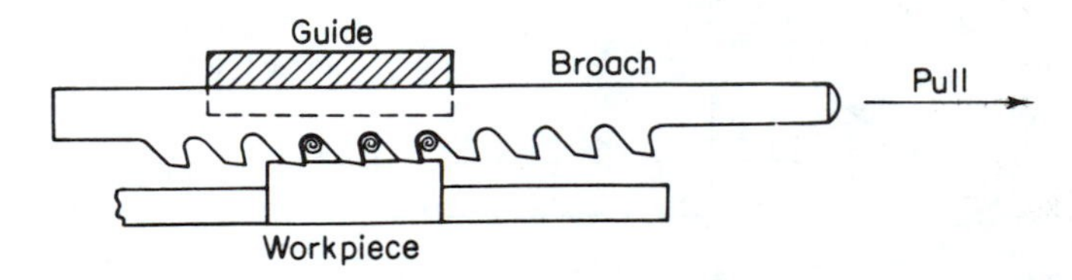

Fig. 3.19. Broaching operation. [Taken with permission, from Lawrence E. Doyle et al., *Manufacturing Processes and Materials for Engineers*, 2nd ed. (Englewood Cliffs, N.J.: Prentice Hall, 1969).]

Fig. 3.20. Power hacksaw. [Taken with permission, from Lawrence E. Doyle et al., *Manufacturing Processes and Materials for Engineers*, 2nd ed. (Englewood Cliffs, N.J.: Prentice Hall, 1969).]

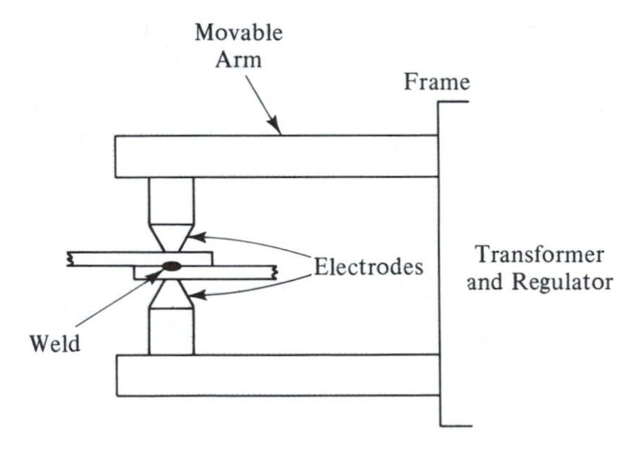

Fig. 3.21. Diagram of a spot welder. [Taken with permission, from Lawrence E. Doyle et al., *Manufacturing Processes and Materials for Engineers*, 2nd ed. (Englewood Cliffs, N.J.: Prentice Hall, 1969).]

Thermit Welding—Thermit welding is a process by which a cavity is formed by the pieces to be welded and molten metal poured into the cavity to make the weld. The molten metal is usually iron, steel, or copper and the welding process is particularly useful for large parts.

Pressure Welding—In pressure welding, the two parts are made to join by applying sufficient pressure with or without an external heat source. The pressure must be high enough to *squeeze* out all imperfections. Friction welding is a good example of pressure welding.

Gas Welding—In gas welding, a combustible gas is burned with air or oxygen in a concentrated flame. This flame is used to heat the parent pieces and filler for the actual weld.

Brazing and Soldering—In brazing and soldering, the filler metal is melted and added to the weld area. The filler is chosen so that its melting point is below that of the parent metal. Consequently, the parent pieces never melt, and when the filler material cools and solidifies, it acts as a (metallic) glue to hold the parts together.

3.4.6 Assembly

Assembly is the process by which the various parts and subassemblies are brought together to form a completed assembly or product. Since there are numerous methods of assembly operations, they will not be covered in detail.

For example, welding is sometimes called an assembly process although assembly usually implies less permanent bonds. Gluing, riveting, screwing, and force fitting are familiar examples of assembly processes. The manufacturing engineer must be familiar with these various processes and understand the relative advantages and disadvantages of each process.

Organizationally, the assembly may be done in a *batch* process whereby assemblies are made intermittently or in a *continuous* process such as assembly lines. The manufacturing engineer must be familiar with these concepts; he or she must know how to set up and balance assembly lines, and so on.

3.4.7 Finishing

Finishing is a set of processes in which the material, either as a completed product, subassembly, or component part, is made more effective or presentable through the external application of energy and/or other materials.

Honing and lapping are two processes in which an abrasive material is rubbed over the metal to produce a fine surface. Honing and lapping are similar to grinding except that the speed is slower and the part is never overheated. The purpose is to improve the surface finish. Other more familiar finishing operations are polishing, buffing, brushing, and tumbling.

Another form of finishing is surface cleaning and coating. The purpose of cleaning is to remove dirt and scale that may be contaminating the workpiece, usually prior to coating. Familiar cleaning operations are spraying, vapor degreasing, and dip baths (such as pickling with an acid solution).

The purpose of coating is to provide decoration, protection, texture, and other desirable properties. Painting and plating are two of the most common coating techniques. Metallizing (spraying of the surface with an atomized molten metal) and phosphatizing and chromating (formulation of a noncorrosive surface over the metal that is also a good paint bond) are examples of less familiar coating operations.

The manufacturing engineer must be familiar with these cleaning and coating techniques and the advantages and disadvantages of each. He or she must know how to set up the processes and estimate the cost of each.

3.5. Ancillary Functions

The production or manufacturing engineer performs many other functions other than the process selection just discussed. Some of these are tool, fixture, and jig design, cost estimating, maintenance systems design, and packaging systems design. Each of these is discussed below.

3.5.1 Tool, Jig, and Fixture Design

Tool design is the designing of tools for the most effective and economical operation. Although it is one of the most important areas of production–manufacturing engineering, tool design is overlooked (except in passing) by almost all industrial engineering curricula because it is so large and complex (and often an art as opposed to a science) that it would be very difficult to cover this subject adequately. Correct die design for a stamping operation could be a course in itself. Obviously, it would be practically impossible to thoroughly cover tool design for all metalworking operations, not to mention the basic operations in other industries.

Figure 3.22 gives an indication of the many parameters involved in designing a very simple cutting tool. Each of the angles shown must be selected based on the material being cut, the material of the cutting tool, and the cutting conditions. Changes of one degree in these angles can significantly affect the cutting operation. Imagine, then, the complexities and problems in designing a progressive (several stages or operations sequentially performed by one die) blanking, piercing, and forming die such as shown in Figure 3.23. Many industrial engineers become tool designers who learn almost all their skills on the job.

Similarly, jig and fixture design is an important phase of production–manufacturing engineering passed by in many industrial engineering curricula for the same reasons discussed above. A fixture is a device used to hold and locate a workpiece. The vise found in home workshops is a simple example of a fixture. In highly repetitive and exacting work, fixture design can be very important. A jig is similar to a fixture except that it also guides the tool into the workpiece. Figure 3.24 is an

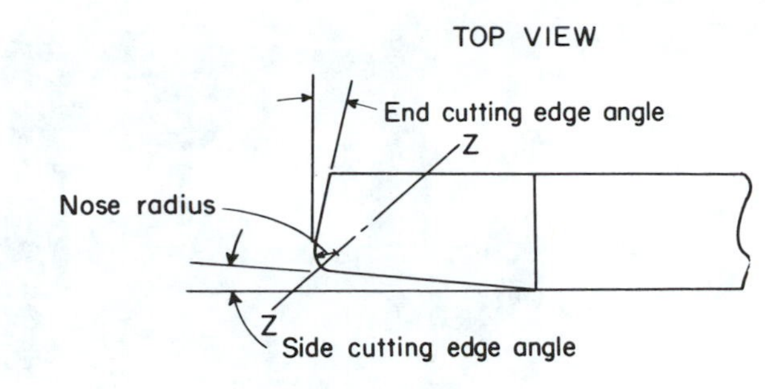

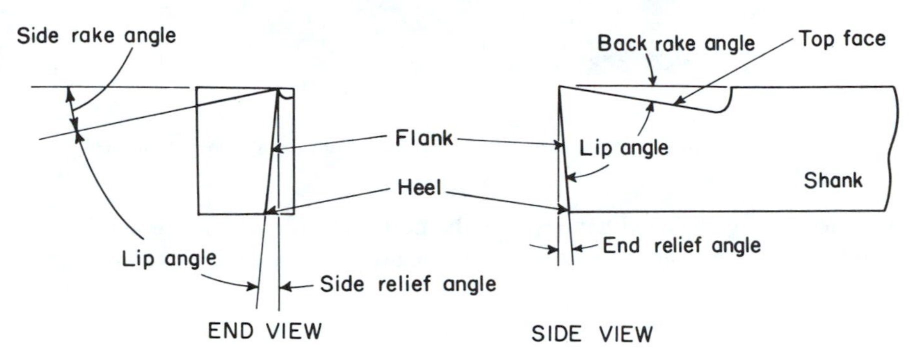

Fig. 3.22. Cutting tool design considerations.

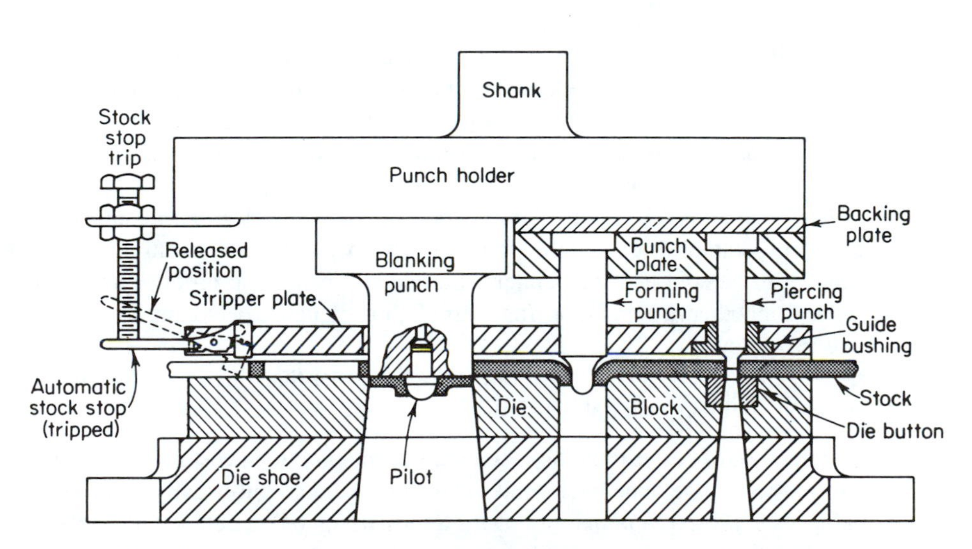

Fig. 3.23. Progressive die example. [Taken with permission, from Lawrence E. Doyle et al., *Manufacturing Processes and Materials for Engineers*, 2nd ed. (Englewood Cliffs, N.J.: Prentice Hall, 1969).]

Fig. 3.24. Jig to line-bore six holes in a master rod for an aircraft engine.

example of a jig. Jigs and fixtures must be designed for each unique application. Consequently, the designer must be highly creative and knowledgeable in the area.

3.5.2 Cost Estimating

Of primary importance to the success of any manufacturing endeavor is cost estimating. *Cost estimating* is the determination of the total manufacturing cost and consists of the following three items:

- *Material—Direct material* is any material whose cost is directly chargeable to a product. Examples are the wood that goes into a wooden pencil and into the paper that makes up the pages of this book. *Indirect material* is all other material cost, for example, cleaning compounds for sweeping floors and paint thinner for cleaning brushes.
- *Labor—Direct labor* is any labor whose cost is directly chargeable to a product. Examples are a lathe operator who performs an operation on the product to be sold and the painter who actually paints the product. *Indirect labor* is all other factory labor, for example, maintenance personnel, foremen, and inspectors.
- *Overhead*—Overhead is all cost of production other than direct labor or direct material. Indirect material and indirect labor are overhead items.

The purpose of cost estimating is to estimate the product's cost prior to actual production. This can be done by several means or methods, but usually some type of planning sheet is used. A *planning sheet* is a form designed for detailed processing of the product through the various stages of manufacture, including all equipment and tool requirements. Costs are added to the planning sheet.

The actual form used varies widely between companies. Usually, enough detail is provided to be able to generate the routing sheets. Figure 3.25 is one

Date ______________ Part Name ______________ Part Number ______________

For use in final product ______________ Number per unit of final product ______________

Operation Number	Operation Description	Equipment	Tooling	Direct Material	Direct Labor	Equipment Availability Cost
			Totals			

Fig. 3.25. Planning sheet form.

example of such a form. Notice that no provisions are made for material handling or other indirect costs because these can be added on at the end by use of an overhead factor (e.g., 150% of direct labor cost) or space can be provided in the form and actual overhead cost calculated. The latter is more accurate, but it is also more difficult.

3.5.3 Maintenance Systems Design

Almost all the machines necessary for the processes discussed in Section 3.3 are very expensive and many are sensitive. Periodic care and maintenance of these machines and other equipment are exceptionally important. Periodic maintenance is called *preventive maintenance*. Some of the equipment is certain to break down occasionally and repair is necessary. This repair is called *emergency maintenance*.

It should be obvious that emergency maintenance could be the more costly of the two because of production interruptions, rescheduling of jobs and labor, and so on. The primary purposes of preventive maintenance, then, are to prolong the life of the equipment and to minimize the amount of emergency maintenance. Proper preventive maintenance such as lubricating and bearing inspection can minimize the number of emergency shutdowns.

Figure 3.26 is an example checklist used for periodic preventive maintenance of automotive vehicles. The design of these checklists and the determination of frequency of their use are often functions of the production–manufacturing engineer. In both types of maintenance, the determination of the maintenance staff and the equipment required is exceptionally important. Here, the production–manufacturing engineer may use forecasting methods, operations research (to be discussed), and other tools of industrial engineering to design the system.

Automotive Vehicles
3,000 Miles or 3-Months
Preventive Maintenance Service

Vehicle No. ____________

Date	Make and Model			Miles	
Service: Fill Battery	Radiator Liquid	Fill Windshield Washer.	Check Engine Oil Level	Tire Pressure	Anti Freeze

Item No.	Description	Status	Corrective Action Including Material	Initials	Hours
A.	Interior				
1.	Foot and Hand Brake				
2.	Ignition Switch				
3.	Starter				
4.	Clutch Pedal				
5.	All Dash Instruments				
6.	Horn				
7.	Windshield Wipers and Washers				
8.	Gear Shift Loose or Broken				
9.	Steering Loose Motion				
10.	Fire Extinguisher Check Seal and Bracket				
B.	Exterior				
1.	Mirror and Brackets				
2.	Doors and Latches				
3.	Chock Blocks				
4.	Tires–RR LR RF LF				
	Underneath Check Exhaust System				
	Under Hood Check Leaks, Hoses, Water, Fuel, Oil and Brake Fluid				
	Lights Check all Lights and Reflectors				

Foreman Signed ____________ Date ________

Fig. 3.26. Automotive vehicles 3,000 miles or 3-month preventive maintenance service.

3.5.4 Packaging Systems

Packaging is the last phase of what we have been calling production–manufacturing engineering. Here, the packaging of the finished product must be designed to protect it during transit and to ensure that the product will arrive undamaged. The package designer must be familiar with the various materials for packaging, the techniques of package design, and the cost of various types of packages. He or she must also know how to set up workstations in order to obtain maximum effectiveness and efficiency.

Packaging is becoming a large area in itself. This may be seen by observing the interest in packaging programs throughout the country and the high demand for good package engineers. "The best package is no package" is an often heard quote. This implies that when economically feasible the product should be designed to withstand the perils of handling during shipment. When this becomes infeasible, a protective package is required. Other purposes of packaging include ease and aesthetics of product display.

3.6. Example

The Wheatley Company, a progressive pump and valve manufacturing company located in Tulsa, Oklahoma, has decided to start production on a new check valve, pictured in Figure 3.27.

The steps in the process engineering of this product are depicted below.[4] Only the bill of material, operation process chart, and a sample routing sheet are shown. The bill of material (highly simplified) is shown in Figure 3.28. The operations process chart is developed next. It is shown in Figure 3.29 using fictitious machine numbers.

After the operations process chart has been prepared, the planning sheet is usually completed. Here, all details on material handling equipment, cost and time estimates, and so on are given. Next, a routing sheet is prepared for each part requiring processing. The routing sheet for the body only is shown in Figure 3.30. The form is such that it can be reproduced later and used as a production order, which instructs the shop to initiate production. Separate routing sheets would also be prepared for the cover, clapper, and arm, but they are not shown here.

The next step is to be sure that all necessary tools, jigs, and fixtures are available, design them when necessary, and issue production orders. Of course, packaging and maintenance systems must also be designed.

3.7. Computer Applications

Computers have changed the nature of production processes in many ways. They are often used to control the production process. For example, a minicomputer can be programmed to automatically adjust the process through a feedback mechanism. This is common in some process industries, such as paper mills. Although numerical control (the use of punched tape to control the operation of a machine) is not a

---

[4]Routing information was supplied by the Wheatley Company. The forms used are those of the authors, not the company. The authors are deeply indebted to the Wheatley Company for their aid and generosity in providing this example.

Fig. 3.27. Check valve.

Bill of Material

Model: Check Valve Type X Design Date ____________ Sheet 1 of 1

Item	Quantity	Description	Source	Part No.
1	1	Body	Manufacturing	227
2	1	Cover	Manufacturing	275
3	8	Stud	Purchasing	271
4	8	Nut	Purchasing	243
5	1	Gasket, RTJ	Purchasing	353
6	1	Seat	Purchasing	236
7	1	O-Ring	Purchasing	201
8	1	O-Ring (Seat Backing)	Purchasing	301
9	1	Clapper	Manufacturing	237
10	1	Washer	Purchasing	220
11	1	Pin, "Roll Pin"	Purchasing	406
12	1	Arm	Manufacturing	221
13	1	Arm Pin	Purchasing	406
14	1	Plug, Arm Pin	Purchasing	420
15	1	Name Plate	Purchasing	282
16	2	Screw, Drive	Purchasing	320

Fig. 3.28. Bill of material for check valve.

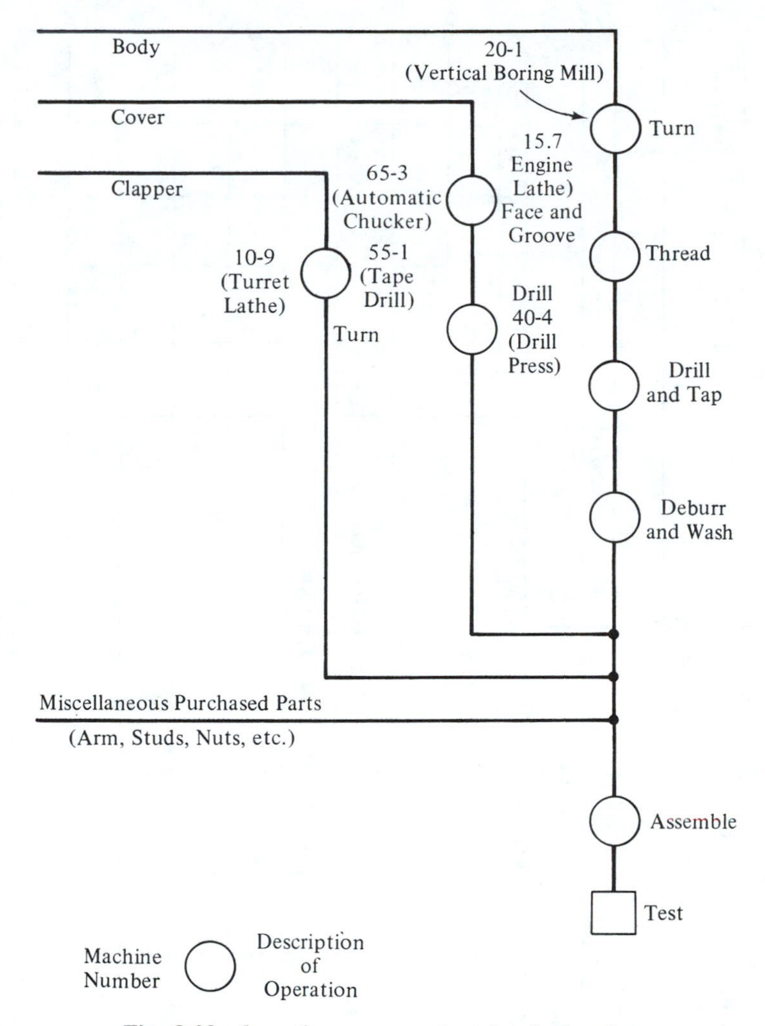

Fig. 3.29. Operations process chart for check valve.

computerized process, computer programs do exist that will translate the part's requirements into a numerical control language and automatically punch the tape (a very tedious process when done by hand).

It is often helpful to use a computer to set the parameters of the production process. At the extreme, the computer can be told what the part is, its dimensions, requirements, and so on, and it can completely design the process, including the order of operations, all feeds and speeds, and so forth. This last application is currently being researched and has, in fact, been done for some parts by using a mathematical programming formulation (mathematical programming will be discussed in Chapter 14).

PRODUCTION ORDER NO.	PART OR ASS'Y. NUMBER	DWG OR MAT'L. LIST NO
2577	004-019577-227	4-19577

SALES/ASS'Y. ORD. NO.	DESCRIPTION-PART OR ASS'Y. NAME
SALES ORDERS	BODY -3" 5000# DPS CHK. VLV.

QUANT ORDRD	FIN. PCS: 50	RAW MATL: 50	MATL. UNITS/PC.: 1 PC.	PATTERN OR RAW MAT'L NO: 19557
SCHEDULE DATES	WRITTEN: 3-12-74	RAW MATL:	ALLOW FOR CUTOFF: NONE	RAW MAT'L. DESCRIPTION: CAST STEEL
	START: 15	FINISH: 8	MAT'L. UNITS: 1 PC.	SPECIFICATION OR FORM: API TYPE 2

SCHEDULE MASTER

PARTIAL COMPLETIONS

DATE	QUANTITY	DATE	QUANTITY	DATE	QUANTITY

CLOSED		COMPLETE		INCOMPLETE	
DATE	TOTAL QUANTITY		CLOSED BY-NAME		

ORDER PRODUCTION

SEQ	START DATE	SCHEDULED HOURS TOTAL	SCHEDULED HOURS SET UP	SCHEDULED HOURS RUN/PC	TOOL NUMBER	MACHINE NUMBER	OPN NO	OPERATION	PIECES SCRAPD	PIECES REJECTD	PIECES ACCEPTD	PIECES TOTAL	INSPCTED /DATE
10			.5	1.5		20-1	5	TURN					
20			1.2	4.0		15-7	10	TURN & THREAD					
30			1.3	2.0		40-4	15	DRILL & TAP					

Fig. 3.30. Routing sheet—body.

The trend today is to use computers and production control techniques to integrate islands of automation in search of a totally automated factory. This means that computers will schedule, sequence, and control machine centers, robots, and automatic machining centers. The future is bright and unlimited for computer usage in manufacturing engineering.

DISCUSSION QUESTIONS

1. In the discussion of metalworking processes we mentioned that there were many other processes in industry and government. List at least 10 other processing areas (e.g., plastics, refuse collection, landfill operations).
2. Discuss what product–production design interaction is. Why is it necessary and how should it occur?
3. List the basic types of metalworking discussed. Give an example of each.
4. We implied that production–manufacturing engineering is a part of industrial engineering. Defend or disagree with this viewpoint.
5. Distinguish between wire drawing and drawing of a cup-shaped object.
6. Give an example of all types of metal cutting operations (shearing, turning, drilling, shaping and planing, milling, broaching, sawing–filing, and grinding).
7. Why are brazing and soldering popular in the electronics field?
8. Find an example of a jig or a fixture and describe it.
9. Discuss the trade-offs between emergency maintenance and preventive maintenance.
10. What is artificial intelligence? How is it being utilized (researched) in production engineering?
11. Discuss the concept of integrated islands of automation.

PROBLEMS

1. Draw an operations process chart for the preparation of a strawberry refrigerator pie. The recipe is given below (courtesy of Mrs. David F. Byrd).

   Ingredients:
   1 cup sugar
   1 package frozen strawberries (1½ cups)
   2 well-beaten eggs
   1 package strawberry Jello
   1 can evaporated milk

   Thaw and drain strawberries to get about ¾ cup liquid. Blend liquid of strawberries, with sugar and eggs. Boil at least 1 minute. Add Jello, mix, and let cool in refrigerator. Add strawberries to mixture. Whip cold milk and fold into mixture. Pour into two 9-inch pie shells, place in refrigerator to cool, and serve.

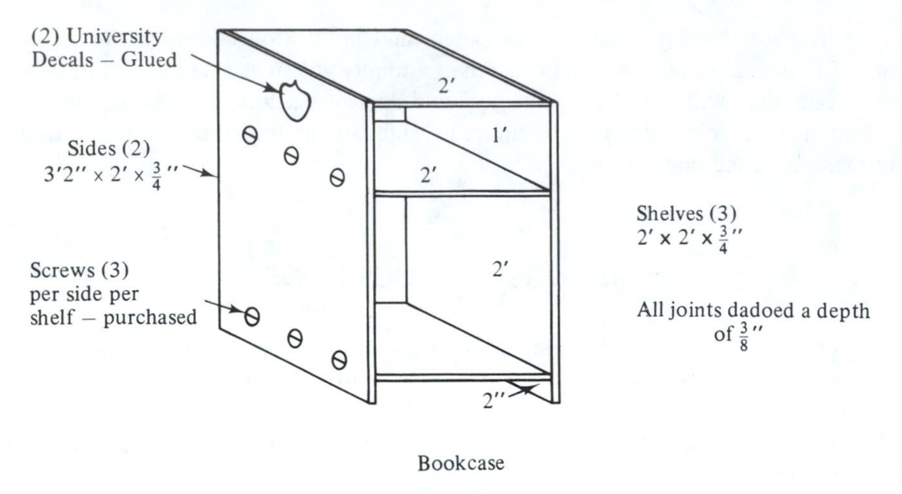

Bookcase

Fig. P3.2.

2. Prepare a bill of material for the product shown in Figure P3.2.

3. Prepare planning sheets for the bookcase in Problem 2. Assume that a skillsaw is used to rip the sides and shelves out of a 4-ft × 8-ft plywood sheet. A good-quality glue and 2-in. roundhead woodscrews are used in the assembly. A radial arm saw is available for dadoing, and all screws and decals are purchased. Do not attempt to derive time or cost figures.

4. Prepare routing sheets for each of the parts in Problem 2.

5. Write a simple FORTRAN program that calculates the total cost of a product given the direct material and direct labor in each of its component parts or assemblies (allow for multiple parts per product unit). Assume that the overhead cost is a fixed percentage of the total direct labor cost.

6. Either use the program in Problem 5 or manually determine the total cost of the completed product, the parts of which are listed below. Assume that the final assembly cost is $5.00 per unit (all direct labor) and that the overhead is 135% of direct labor cost.

Part	*Parts/Unit*	*Direct Material (Cost per Part)*	*Direct Labor per Part*
A	1	$1.00	$ 5.00
B	2	2.00	1.00
C	1	1.50	0.50
D	3	5.00	0.75
E	1	3.00	0.60
F	1	1.50	10.00
G	2	0.75	7.00
H	1	0.50	6.00

7. Find a simple metal part or product, and using the information presented in this chapter, develop the following:
   (a) Bill of material
   (b) Operations process charts.
   (c) Route sheets (use fictitious machine numbers).

8. Prepare an operations process chart for the task of washing your car. Start with obtaining a sponge, a bucket, a hose, and soap. Use your present procedure. By examining the procedure, can you find a better way?

9. By asking graduate students or faculty, find one way that your department is applying a computer to production engineering.

CHAPTER 4

Facilities Location and Layout

4.1. Introduction to Facilities Location

A *facility* is something built or established to serve a purpose. *Facilities management* is a location decision for that facility and the composition or internal layout of the facility once located. This chapter discusses the location and layout of a facility. A sample location problem which is used in subsequent discussions is given in Example 4.1.

Example 4.1 Location of a Manufacturing Facility

Management of the Waterstill Manufacturing Company has decided to embark upon a manufacturing facility expansion. The expansion is necessary because of an increase in sales of its product—water softeners. There are currently two facilities, one in Tulsa, Oklahoma, and the other in Tempe, Arizona, but there is no additional space available at either site. Furthermore, the company has warehouses in Washington, D.C., Cleveland, Ohio; Lincoln, Nebraska; San Francisco, California; and Phoenix, Arizona. They feel that transportation costs might be reduced if a new plant at a different location were constructed. Figure 4.1 shows the existing sites.

Current sales suggest the need for a 10% plant expansion, but forecasts over the next 10 years indicate a steady increase up to double the current sales volume. Then the volume is expected to level off. Because of the relatively unskilled labor requirement and desired

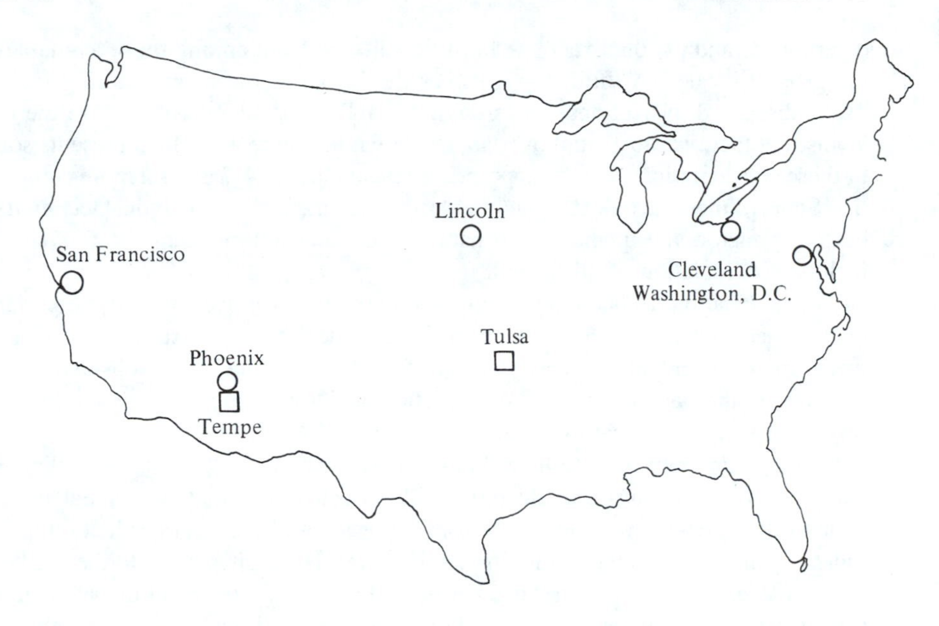

□ Existing Manufacturing Facilities

◯ Warehouses

Fig. 4.1. Waterstill Company—manufacturing and wholesale facilities.

stability of the labor force, a rural location is preferred for the new manufacturing facility. As implied by the warehouses, the market is distributed evenly throughout the country and raw materials availability is no problem. Transportation costs are proportional to the distance traveled, regardless of warehouse location, and it is a company policy to be able to supply any warehouse within 1 week of receipt of an order.

The Waterstill Manufacturing Company has hired your company, the Whitehurst Cowboy Consulting Firm, to aid them in their search for a suitable location. Much information is supplied above, but the company is completely willing to provide any additional information you feel is necessary. Your job is to determine the best location for the new plant.

4.2. Considerations

Although the information provided in the example is certainly not enough to locate the new facility, it does give an idea of the immense complexity of such an undertaking. The problem is so immense and so complex that there is no one "best way" of carrying on such a study. There are analytical procedures to aid in the process (these will be discussed in Section 4.3), but the decision must also include factors that are difficult if not impossible to quantify.

The decision usually is made in two stages: (1) the general location must be determined, and (2) the exact site must be selected from among those available in the general locale. (A third step involves the layout of the manufacturing facility itself, discussed in Sections 4.4 through 4.7.) The considerations vary in the two levels, but there is substantial overlap, as shown in Figure 4.2 which presents some of these considerations and the appropriate stages. Figure 4.7 gives a more complete list. Some of these factors are quantifiable and are amenable to analytical techniques, but other factors are qualitative in nature. Any decision on plant location should include consideration of all factors.

The location decision is not only complex; it is also exceptionally important. Once a decision has been made on a site and the facility constructed or located there, management most likely will "live with" that site for a substantial time because of the capital costs of construction and relocation. Consequently, it is important that the decision be a good one.

It is important to note here that the best location is very case specific. For example, a potato chip manufacturer will probably find that movement of raw material (potatoes) is inexpensive, but transportation of the final product (chips) is more expensive. Thus the optimum may involve several sites close to the markets.

A steel mill, on the other hand, will probably find movement of raw material (ore) more expensive than that of final product (steel coils or sheets). Thus the best might be one steel mill close to the mines with good rail access and inexpensive energy.

A typical team studying location possibilities might include accountants, lawyers, marketing experts, various consultants, executives, and industrial engineers. With this entourage of talent, the industrial engineer usually is not asked to be knowledgeable in all areas. Instead, the industrial engineer may be called upon to

Location Factor	*Territory Selection*	*Selection of Site and Community*
1. Market	*	
2. Raw Materials	*	
3. Transportation	*	*
4. Power	*	*
5. Climate and Fuel	*	
6. Labor and Wages	*	*
7. Laws and Taxation	*	*
8. Community Services and Attitude		*
9. Water and Waste		*

Fig. 4.2. Consideration in the two stages of plant location. [From *Plant Layout and Design* © James M. Moore, 1962. Reprinted with permission of the publisher. (New York: Macmillan Publishing Company, 1962.)]

provide talent and knowledge in the actual operations related areas and in analytical techniques.

Industrial engineering training provides the I.E. with a solid math background that includes analytical location techniques. This chapter discusses some of these location techniques. The chapter is neither a complete course in location analysis nor a complete course in analytical techniques for location analysis. It is a brief introduction that demonstrates some ways that the industrial engineer applies his or her talents in searching for the best location.

4.3. Analytical Techniques

Analytical techniques are solution procedures that involve careful examination of the problem, usually by breaking the problem into components and utilizing mathematics to search for a good solution. Of course, any analytical technique can only consider quantifiable areas and few, if any, are capable of handling all the quantifiable considerations in one model. Thus, as always, analytical tools are only aids in the decision process; they are not the process itself.

Many different plant location problems can be defined based on varying criteria and parameters. Each formulation usually requires a unique solution procedure or at least a modification of another solution procedure. Since it is impossible to mention every problem, only a few will be covered here. First, possible ways of stratifying plant location problems are mentioned.

Usually, the criterion or objective is to minimize some cost function. Often, the distance traveled is chosen as the cost function. Thus, the objective becomes one of minimizing total distance traveled. Sometimes the objective, or at least a consideration, may be to minimize the maximum distance traveled (as, for example, in locating fire stations or hospitals where the maximum or worst response distance is important).

Another possible stratification is in the distance measure used. Sometimes the straight-line or Euclidean distance may be more appropriate. If two facilities are located at points represented by (X_1, Y_1) and (X_2, Y_2), then the Euclidean distance between the two is

$$[(X_1 - X_2)^2 + (Y_1 - Y_2)^2]^{1/2}$$

while the rectilinear distance is

$$|X_1 - X_2| + |Y_1 - Y_2|$$

The two are demonstrated in Figure 4.3. A rectilinear measure is often used when the streets are laid out in grids, as in an urban setting (Philadelphia, Pennsylvania, in Figure 4.3). An Euclidean measure is used when a straight-line distance is more appropriate, as in interstate or intercity travel (Columbus to Akron in Figure 4.3). Still another classification scheme is on the number of facilities to be located—single vs. multiple. For example, the location of one fire station in Manhattan,

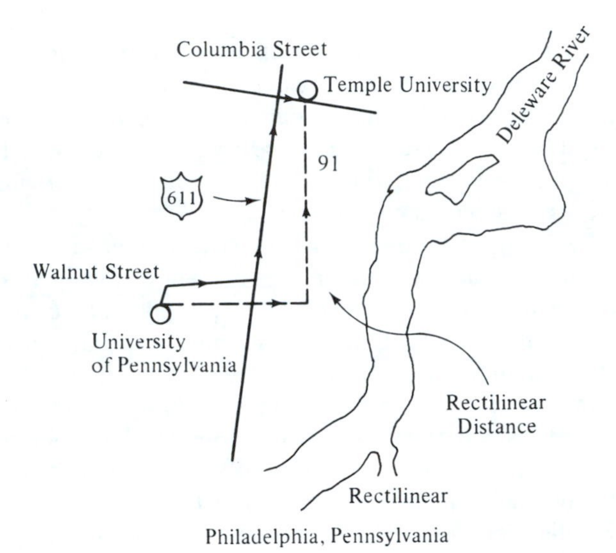

Fig. 4.3. Euclidean and rectilinear distance examples.

Kansas, may require a different technique from the location of ten fire stations in Oklahoma City, Oklahoma.[1]

4.3.1 Transportation Method of Linear Programming

It is relatively simple to formulate a manufacturing plant location problem as a transportation linear programming problem involving minimization of total cost to distribute the product. (Linear programming will be discussed in depth in Section 14.8.) In this case, the transportation linear programming formulation is used to determine the best distribution pattern for the plant in a certain location. Then, total

[1]For amplification, see Richard L. Francis and John A. White, *Facility Layout and Location: An Analytical Approach* (Englewood Cliffs, N.J.: Prentice Hall, 1974), pp. 21–26.

cost can be calculated, various locations tried, and the location with the least total cost chosen.

For example, suppose that the Waterstill Manufacturing Company has narrowed its choice of locations to two—Youngstown, Ohio, and Sparta, South Carolina. Assume further that the costs of manufacturing and distribution to the various warehouses can be calculated as shown in Table 4.1, in which information was obtained through careful examination of forecasts and production and distribution costs.

To formulate this as a transportation linear programming problem, it first must be realized that there are two alternatives: (1) locate the plant in Youngstown or (2) locate the plant in Sparta. There will, therefore, be two transportation problems, one with the plant in Sparta and one with the plant in Youngstown.

Consider the cost of production and distribution. If the Tulsa plant produced and shipped a softener to Washington, the total cost would be

$$\underset{\text{Production}}{\$75.00} + \underset{\text{Distribution}}{\$5.00} = \underset{\text{Total Cost per unit}}{\$80.00}$$

The cost in the transportation matrix would then be \$80.00 for the Tulsa–Washington combination. For Tulsa–Cleveland, it would be

$$\$75.00 + \$3.00 = \$78.00$$

while for Tempe–Lincoln, it would be

$$\$70.00 + \$3.50 = \$73.50$$

If we continue this for all combinations and remember that the plants cannot exceed capacity and that the warehouses do not need more than their demand, the transportation problem, with the plant in Youngstown, can be formulated as in Figure 4.4.

Table 4.1. PRODUCTION AND DISTRIBUTION COSTS PER WATER SOFTENER

To / *From*	*Washington*	*Cleveland*	*Lincoln*	*San Francisco*	*Phoenix*	*Weekly Capacity*	*Production Costs*
Tulsa	5.00	3.00	2.00	3.00	2.00	7,000	\$75.00
Tempe	6.50	5.00	3.50	1.50	.20	5,500	\$70.00
Youngstown	1.50	.50	1.80	6.50	5.00	12,500	\$70.00
Sparta	3.80	5.00	8.00	7.50	8.00	12,500	\$67.00
Forecasted Weekly Demand	5,000	6,000	4,000	7,000	3,000		

From \ To	Washington	Cleveland	Lincoln	San Francisco	Phoenix	Plant Capacity
Tulsa	80.00	78.00	77.00	78.00	77.00	7,000
Tempe	76.50	75.00	73.50	71.50	70.20	5,500
Youngstown	71.50	70.50	71.80	76.50	75.00	12,500
Demand (Weekly)	5,000	6,000	4,000	7,000	3,000	Total 25,000

Fig. 4.4. Transportation formulation—plant in Youngstown.

Here the two existing plants are shown as sources along with the potential plant in Youngstown.

The number in the upper right-hand corner of each square indicates the total cost of production and distribution per unit for that combination. The numbers under plant capacity are the production capability for each plant and the numbers in the row for demand are the demand at the warehouses. The total capacity equals total demand; this must be true for the procedure to work. If total demand does not equal total capacity, an artificial plant (warehouse) with zero cost and the necessary capacity (demand) can be used to achieve the required balance.[2]

To record a solution, numbers must be added to the various blocks to indicate the number of units involved in that transfer. For example,

From \ To	Washington
Tulsa	80.00 500

implies that 500 units are being produced in Tulsa and shipped to Washington at a total cost of 500 units × $80.00/unit = $40,000.

There are several ways of solving this problem. Some give optimal solutions with a relatively high cost of time and effort and others yield very fast, good but not necessarily optimal solutions with a relatively low cost. The latter procedures are called *heuristic*; the former procedures are called *optimal seeking*. An optimal seeking method is described in Section 14.8.2, but for this chapter a heuristic method that quickly yields a good solution is presented. This method is the *least cost assignment* routine.

[2]This will be discussed in further detail in Chapter 14.

The first step in the least cost assignment procedure is to pick the least costly element and assign as much as possible to that combination remembering that demand or capacity cannot be exceeded. Then the demand and capacity are corrected to show partial assignment and the procedure continues. For example, the Tempe–Phoenix combination is the least expensive at $70.20 per unit. Tempe can produce 5,500 units, but Phoenix only needs 3,000, so 3,000 units should be assigned to that combination. Now, Phoenix has 0 units demand and Tempe has 2,500 units capacity. Because Phoenix does not need any more water softeners, that column can be dropped from consideration. The next cheapest is Youngstown–Cleveland at $70.50 per unit. Six thousand units can be assigned there and the Cleveland column can be dropped from consideration. If we continue in this manner, the final solution would be as shown in Figure 4.5, in which the circled numbers imply the order in which the combinations were chosen.[3]

The total cost of this solution (which may or may not be optimal) is

$$2{,}500\,(77) + 4{,}500\,(78) + 2{,}500\,(71.50) + 3{,}000\,(70.20) + 5{,}000\,(71.50) + 6{,}000\,(70.50) + 1{,}500\,(71.80) = \mathbf{\$1{,}821{,}050}$$

The corresponding formulation and solution for the plant located in Sparta are given in Figure 4.6. The total cost of this solution (which again may or may not be optimal) is

$$4{,}000\,(77) + 3{,}000\,(78) + 2{,}500\,(71.50) + 3{,}000\,(70.2) + 5{,}000\,(70.80) + 6{,}000\,(72) + 1{,}500\,(74.50) = \mathbf{\$1{,}829{,}100}$$

Both solutions (Figures 4.5 and 4.6) were obtained using heuristic rather than optimal seeking methods. Thus the solutions are probably "close to optimum" rather than optimum. In Section 14.8.2 we take the solution in Figure 4.6 and show how to obtain the true optimum.

The total weekly cost of production and shipment of Youngstown is *$1,821,050*, which is $8,050 per week *more economical* than the total cost of production and shipment at Sparta. This implies that if all other factors are equal, the plant should be built in Youngstown.

All other factors are probably not equal, however. For example, the wages have been accounted for in the transportation model, but the labor market has not. Is the needed labor available? Are the specialized trades available? What is the attitude of local labor? All of these are questions that must be answered before a decision can be made. The costs developed in Figures 4.5 and 4.6 must be considered along with all the other factors in the proper perspective.

If the entries in the matrix of Figure 4.4 were profit instead of cost, then maximum profit assignment should be used. That is, each assignment is made to

[3]In step 3 either the Tempe–San Francisco or the Youngstown–Washington combination can be chosen since both are the next cheapest. The Youngstown–Washington combination was chosen since 5,000 units can be assigned there and only 3,000 can be assigned to the Tempe–Phoenix combination. This is one way of breaking ties.

From \ To	Washington	Cleveland	Lincoln	San Francisco	Phoenix	Plant Capacity
Tulsa	80.00	78.00	(6) 77.00 2,500	(7) 78.00 4,500	77.00	7,000
Tempe	76.50	75.00	73.50	(4) 71.50 2,500	(1) 70.20 3,000	5,500
Youngstown	(3) 71.50 5,000	(2) 70.50 6,000	(5) 71.80 1,500	76.50	75.00	12,500
Demand (Weekly)	5,000	6,000	4,000	7,000	3,000	25,000

Fig. 4.5. Solution to transportation problem using least cost assignment—Youngstown.

the remaining feasible combinations with maximum profit. Capacities and demands are adjusted as before and the procedure is continued.

4.3.2 Multiple Objectives

The preceding section demonstrated that there are many factors to be considered in the final location decision. One technique for analytically considering many factors is the use of a method for decision making with multiple objectives. The method, which is very similar to the point rating plan of job evaluation (to be discussed in Chapter 6), is needed because so many factors, such as local climate, local attitude, fire and police protection, and so on, cannot be readily quantified. Yet they are important in the location decision.

As an example, suppose that the Waterstill Manufacturing Company has narrowed its choice down to two locations, city A and city B. All cost calculations

From \ To	Washington	Cleveland	Lincoln	San Francisco	Phoenix	Plant Capacity
Tulsa	80.00	78.00	(6) 77.00 4,000	(7) 78.00 3,000	77.00	7,000
Tempe	76.50	75.00	73.50	(3) 71.50 2,500	(1) 70.20 3,000	5,500
Sparta	(2) 70.80 5,000	(4) 72.00 6,000	75.00	(5) 74.50 1,500	75.00	12,500
Demand (Annual)	5,000	6,000	4,000	7,000	3,000	25,000

Fig. 4.6. Solution to transportation problem using least cost assignment—Sparta.

have been made and there is no clear-cut distinction. In fact, for simplicity, assume that all costs are equal at the two locations. How can the decision be made?

The first step is to make a list of all important factors. There are several lists available—Reed gives a list of 21 noncost factors and Moore[4] gives a list of 36. The list from Reed is reproduced in Figure 4.7.

It is important that the list contain all noncost factors exerting an influence on the decision to insure that no significant items are omitted.

The next step is to assign relative point values for each of the factors for the specific company and plant to be located. The values may vary somewhat between companies and within one company between different plants. As an example, the values shown in Figure 4.8 are suggested by Reed as maximum for the 21 factors listed in Figure 4.7. This is not to imply that all companies should use these values; but they are presented as an example.

If we assume that the Waterstill Company uses this scheme, the next step is to assign degrees and points within each factor. Usually, from 4 to 6 degrees are used with a linear assignment of points between degrees. As an example, Figure 4.9 shows the degrees for factor 16 (community attitude) (this list is drawn from Reed, whose book develops degrees and points for each factor).

At this point Waterstill has its evaluation scheme completely defined, so it now must assign each of the two locations (A and B) degrees and corresponding points for each factor. The methods used to obtain the degrees for each location vary substantially. Usually, a team of experts does the rating and management has the right to veto or change the results. Figure 4.10 shows hypothetical results. In factor 16 Waterstill has decided that city A has a cooperative labor force (degree 3), but city B has a noncooperative labor force (degree 2).

Waterstill and the Whitehurst Cowboy Consulting Firm now compare these results with the cost calculations and make a decision. City A has a total point value of 719 compared to 643 for city B. City A would probably be preferred since all cost calculations were assumed equal. If they had not been, some trade-off in the decision process would have to occur.

4.3.3 Mathematical Programming (Optional)

Much work has recently been done toward optimally determining the number, location, and sizes of plants (warehouses) through what is known as *mathematical programming*. Mathematical programming consists of representing the cost of a system by a mathematical expression, constraining the realm of possible solutions when necessary, and searching over the realm of the solutions to find the optimum point (minimum or maximum).

In the case of plant location, an expression specifying the cost of the system is generated and constraints are incorporated to insure feasibility. In the model to

---

[4]James M. Moore, *Plant Layout and Design* (New York: Macmillan Publishing Company, 1962), pp. 48–58.

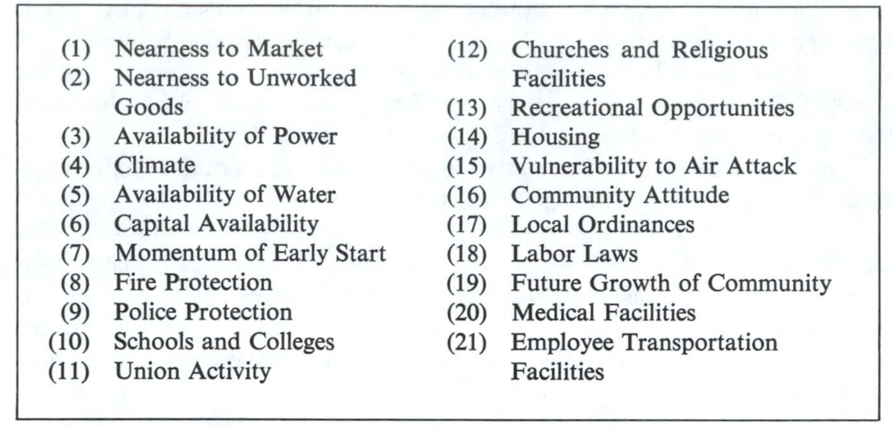

(1) Nearness to Market
(2) Nearness to Unworked Goods
(3) Availability of Power
(4) Climate
(5) Availability of Water
(6) Capital Availability
(7) Momentum of Early Start
(8) Fire Protection
(9) Police Protection
(10) Schools and Colleges
(11) Union Activity
(12) Churches and Religious Facilities
(13) Recreational Opportunities
(14) Housing
(15) Vulnerability to Air Attack
(16) Community Attitude
(17) Local Ordinances
(18) Labor Laws
(19) Future Growth of Community
(20) Medical Facilities
(21) Employee Transportation Facilities

Fig. 4.7. List of 21 noncost factors in plant location. [Extracted with permission from Rudell Reed, Jr., *Plant Layout: Factors, Principles, and Techniques* (Homewood, Ill.: Richard D. Irwin, Inc., 1961), p. 361.]

Factor—Value	*Factor—Value*	*Factor—Value*
1—280	8—10	15—10
2—220	9—20	16—60
3—30	10—20	17—50
4—40	11—60	18—30
5—10	12—10	19—30
6—60	13—20	20—10
7—10	14—10	21—20

Fig. 4.8. Maximum point values for each factor.

Degrees		*Point Assignment*
0	Hostile, Bitter, Noncooperative	0
1	Parasitic in Nature	15
2	Noncooperative	30
3	Cooperative	45
Maximum	Friendly and More Than Cooperative	60

Fig. 4.9. Degress and points for factor 16 (community attitude).

	CITY A		CITY B	
Factor	*Degree*	*Points*	*Degree*	*Points*
1	Maximum	280	3	168
2	4	176	4	176
3	2	12	4	24
4	0	0	4	24
5	4	8	2	4
6	3	36	4	48
7	2	4	1	2
8	Maximum	10	2	4
9	4	16	2	8
10	2	8	3	12
11	3	36	3	36
12	2	6	2	6
13	3	15	Maximum	20
14	4	8	0	0
15	1	2	2	5
16	3	45	2	30
17	2	20	4	40
18	3	23	1	8
19	0	0	3	18
20	1	2	1	2
21	4	12	2	8
Total		719		643

Fig. 4.10. Hypothetical results on plant location.

be described a number of potential plant sites are first determined. Locating a plant at a particular site involves a fixed cost (construction, insurance) and a variable cost of supplying customers. The model, when solved, yields the optimum number, location, and size of the plant. Chapter 14 will discuss mathematical programming solution procedures. This chapter only presents the formulation of the problem.

The following is taken from Efroymson and Ray,[5] but many other authors treat similar problems.

The following notation is necessary:

m = number of customers
n = number of potential plant sites (determined in advance)
k_j = fixed cost of opening plant j
x_{ij} = percent or fraction of customer i's demand that is satisfied by factory j

[5]M. A. Efroymson and T. L. Ray, "A Branch and Bound Algorithm for Plant Location," *Operations Research*, 14 (1966), 361–368.

$$y_j = \begin{cases} 0 \text{ if a plant is not located at } j \\ 1 \text{ if a plant is located at } j \end{cases}$$

c_{ij} = cost of supplying the total demand of customer i by plant j

If all of customer i's demand were supplied by plant j, the variable cost of shipment would be just c_{ij}. Since fractional deliveries are allowed, the total variable cost for the amount customer i receives from plant j becomes $c_{ij}x_{ij}$, and the total variable cost for all customers and plants is given by

$$\sum_{i=1}^{m}\sum_{j=1}^{n} c_{ij}x_{ij} \tag{4.1}$$

($\sum_{i=1}^{m}$ is a symbol signifying "the summation of m elements in the order of 1 through m.")

If a plant is located at site j, a fixed cost of k_j is incurred. If not, no cost is incurred. This can be formulated as k_jy_j since y_j is 1 only if a plant is located at site j and 0 otherwise. The total fixed costs of locating all plants then are given by

$$\sum_{j=1}^{n} k_jy_j \tag{4.2}$$

The total cost is the sum of Equations (4.1) and (4.2), which yields

$$\text{total cost } TC = \sum_{i=1}^{m}\sum_{j=1}^{n} c_{ij}x_{ij} + \sum_{j=1}^{n} k_jy_j \tag{4.3}$$

Some constraints are necessary. First, a customer cannot be supplied from j unless a plant is located at j, and the total amount supplied by a plant located at j cannot exceed the total demand. This is represented by Equation (4.4) (remember that x_{ij} is a fraction and y_j is either 0 or 1). The lowercase letter, n, at the right-hand side of the expression indicates the total number of inequalities (i.e., j = 1 . . . , n implies that there are n expressions where j takes the values 1 through n. $\sum_{i=1}^{m} x_{i1} \geqslant my_1$; $\sum_{i=1}^{m} x_{i2} \geqslant my_2$; and so on).

$$\sum_{i=1}^{m} x_{ij} \leqslant my_j \qquad (j = 1, \ldots, n) \tag{4.4}$$

Next, all customers must be supplied. This can be represented by

$$\sum_{j=1}^{n} x_{ij} = 1, \qquad (i = 1, \ldots, m) \tag{4.5}$$

Finally, all x_{ij} must be 0 or positive and all y_j must be 0 or 1. These can be represented by

$$y_j = (0,1) \quad (j = 1, \ldots, n) \tag{4.6}$$
$$x_{ij} \geqslant 0 \quad (i = 1, \ldots, m \text{ and } j = 1, \ldots, n) \tag{4.7}$$

Putting all these into one model yields

$$\text{minimize } TC = \sum_{i=1}^{m}\sum_{j=1}^{n} c_{ij}x_{ij} + \sum_{j=1}^{n} k_j y_j$$

subject to

$$\sum_{i=1}^{m} x_{ij} \leqslant my_j \; (j = 1, \ldots, n)$$
$$\sum_{j=1}^{n} x_{ij} = 1 \; (i = 1, \ldots, m)$$
$$y_j = (0,1) \quad \forall j$$
$$x_{ij} \geqslant 0 \forall i,j$$

$\forall j$ simply means "for each j" or "for every j" and $\forall ij$ means "for each i and j." The term *subject to* means that the realm of possible solutions is restricted by the constraints that follow.

Since the equation or objective function is for total cost, the goal is to minimize it subject to the constraints. This can be done by using the techniques of linear programming and branch and bound search, but they will not be covered here. Instead, an example is given.

Example 4.2

Suppose that the Whitehurst Cowboy Consulting Firm decided to use such a model in its efforts to aid another company, Indeng Enterprises. This company wants to start a new product line; so it has to construct one or more manufacturing plants capable of supplying its four warehouses located in Norfolk, Virginia; Chicago, Illinois; Seattle, Washington; and Baton Rouge, Louisiana. The company has determined three potential sites. They are in Baton Rouge, Louisiana; Cookeville, Tennessee; and Kansas City, Kansas. The situation is shown in Figure 4.11.

To aid Indeng Enterprises, the Whitehurst Cowboys would first have to determine some cost information at the three potential sites. Specifically, the cost of supplying the demand at any location by a plant must be determined. That information is shown in Table 4.2.

It is relatively simple to develop the mathematical formulation when given the information shown in Table 4.2. For example, x_{11} is the percentage of total goods supplied to warehouse 1 from potential plant 1. Therefore, the cost coefficient in the objective function is the cost of supplying warehouse 1 from plant 1 or \$85,000. Similarly, the cost of supplying warehouse 1 from potential plant 2 is \$75,000. The first two terms in the cost expression are then

$$85{,}000\ x_{11} + 75{,}000\ x_{12}$$

and the total variable cost is

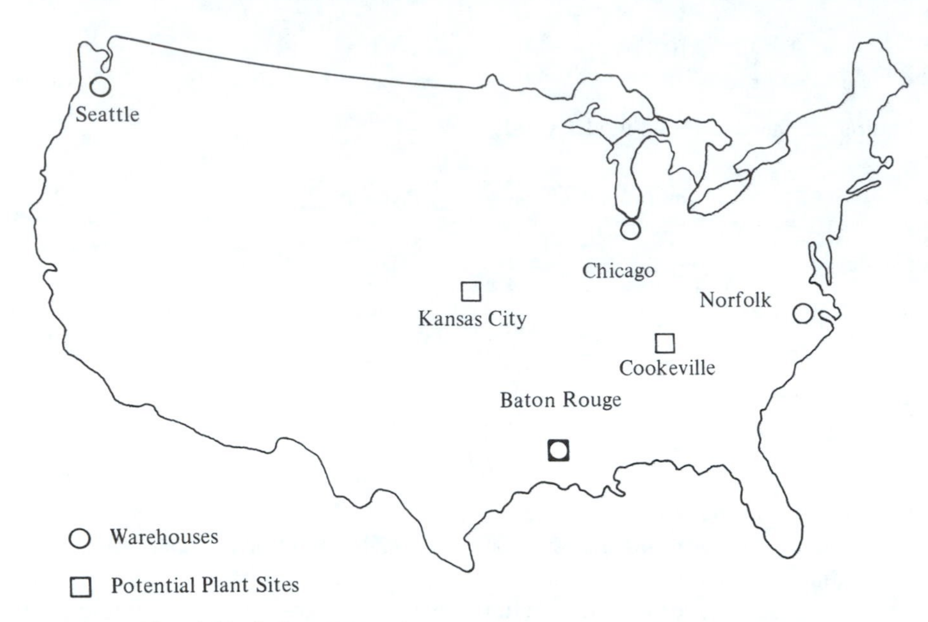

Fig. 4.11. Indeng Enterprises warehouses and potential plant sites.

$$\begin{aligned}85{,}000\,x_{11} &+ 75{,}000\,x_{12} + 100{,}000\,x_{13} + 95{,}000\,x_{21}\\ &+ 85{,}000\,x_{22} + 40{,}000\,x_{23} + 200{,}000\,x_{31}\\ &+ 185{,}000\,x_{32} + 185{,}000\,x_{33} + 10{,}000\,x_{41}\\ &+ 70{,}000\,x_{42} + 60{,}000\,x_{43}\end{aligned}$$

Table 4.2. COST INFORMATION FOR INDENG ENTERPRISES

From / *To*	*Potential Plant Site*	*Baton Rouge*	*Cookeville*	*Kansas City*
Warehouse Site		(1)	(2)	(3)
Norfolk (1)		85,000	75,000	100,000
Chicago (2)		95,000	85,000	40,000
Seattle (3)		200,000	185,000	185,000
Baton Rouge (4)		10,000	70,000	60,000
Fixed Cost to Open Plant		60,000	50,000	65,000

Since the fixed cost associated with opening plant 1 is \$60,000, the first term in the expression for fixed cost is \$60,000 y_1 and the total fixed cost is

$$\$60{,}000\ y_1 + \$50{,}000\ y_2 + \$65{,}000\ y_3$$

The total cost is the sum of fixed cost and variable cost or

$$\begin{aligned}85{,}000\,x_{11} &+ 75{,}000\,x_{12} + 100{,}000\,x_{13} + 95{,}000\,x_{21} + 85{,}000\,x_{22}\\ &+ 40{,}000\,x_{23} + 200{,}000\,x_{31} + 185{,}000\,x_{32} + 185{,}000\,x_{33}\\ &+ 10{,}000\,x_{41} + 70{,}000\,x_{42} + 60{,}000\,x_{43} + 60{,}000\,y_1\\ &+ 50{,}000\,y_2 + 65{,}000\,y_3\end{aligned}$$

The first set of m constraints ($m = 3$) states that the total amount shipped from any plant cannot exceed the total demand or that a plant must be open (i.e., $y_j = 1$) before any shipment can occur. Those constraints are

$$\begin{aligned}x_{11} + x_{21} + x_{31} + x_{41} &\leqslant 4y_1\\ x_{12} + x_{22} + x_{32} + x_{42} &\leqslant 4y_2\\ x_{13} + x_{23} + x_{33} + x_{43} &\leqslant 4y_3\end{aligned}$$

The next set of constraints requires that all demands must be satisfied.

$$\begin{aligned}x_{11} + x_{12} + x_{13} &= 1\\ x_{21} + x_{22} + x_{23} &= 1\\ x_{31} + x_{32} + x_{33} &= 1\\ x_{41} + x_{42} + x_{43} &= 1\end{aligned}$$

Finally, the 0, 1 and non-negativity contraints are

$$\begin{array}{lllll} y_1 = 0, 1 & x_{11} \geqslant 0 & x_{21} \geqslant 0 & x_{31} \geqslant 0 & x_{41} \geqslant 0\\ y_2 = 0, 1 & x_{12} \geqslant 0 & x_{22} \geqslant 0 & x_{32} \geqslant 0 & x_{42} \geqslant 0\\ y_3 = 0, 1 & x_{13} \geqslant 0 & x_{23} \geqslant 0 & x_{33} \geqslant 0 & x_{43} \geqslant 0\end{array}$$

Putting all this together yields

$$\begin{aligned}\text{Minimize } TC = {}& 85{,}000\,x_{11} + 75{,}000\,x_{12} + 100{,}000x_{13}\\ &+ 95{,}000\,x_{21} + 85{,}000\,x_{22} + 40{,}000\,x_{23}\\ &+ 200{,}000\,x_{31} + 185{,}000\,x_{32} + 185{,}000\,x_{33}\\ &+ 10{,}000\,x_{41} + 70{,}000\,x_{42} + 60{,}000\,x_{43}\\ &+ 60{,}000\,y_1 + 50{,}000\,y_2 + 65{,}000\,y_3\end{aligned}$$

subject to

$$\begin{aligned}x_{11} + x_{21} + x_{31} + x_{41} &\leqslant 4y_1\\ x_{12} + x_{22} + x_{32} + x_{42} &\leqslant 4y_2\\ x_{13} + x_{23} + x_{33} + x_{43} &\leqslant 4y_3\\ x_{11} + x_{12} + x_{13} &= 1\\ x_{21} + x_{22} + x_{23} &= 1\\ x_{31} + x_{32} + x_{33} &= 1\\ x_{41} + x_{42} + x_{43} &= 1\end{aligned}$$

$$y_1 = 0, 1, \quad y_2 = 0, 1, \quad y_3 = 0, 1$$

$$x_{11} \geqslant 0, \quad x_{12} \geqslant 0, \quad x_{13} \geqslant 0, \quad x_{21} \geqslant 0, \quad x_{22} \geqslant 0, \quad x_{23} \geqslant 0$$
$$x_{31} \geqslant 0, \quad x_{32} \geqslant 0, \quad x_{33} \geqslant 0, \quad x_{41} \geqslant 0, \quad x_{42} \geqslant 0, \quad x_{43} \geqslant 0$$

The solution procedure for this formulation is simple, but development of it is not really necessary for understanding the model. Those interested should consult Efroymson and Ray (footnote 5).

The optimal solution to this problem obtained through the use of the solution procedure or exhaustive enumeration in this simple problem is

$$y_1^* = 1, \quad y_2^* = 0, \quad y_3^* = 1$$
$$x_{11}^* = 1, \quad x_{23}^* = 1, \quad x_{33}^* = 1, \quad x_{41}^* = 1$$
$$x_{12}^* = x_{13}^* = x_{21}^* = x_{22}^* = x_{31}^* = x_{32}^* = x_{42}^* = x_{43}^* = 0$$

This says that plants are to be constructed in locations 1 and 3. Warehouses 1 and 4 are to be serviced by plant 1, and warehouses 2 and 3 are to be serviced by plant 3. According to this model, Indeng Enterprises should then locate plants in Baton Rouge, Louisiana, and in Kansas City, Kansas.

4.3.4 Public-Sector Location Problems

There has been much interest recently in the use of industrial engineering in the location of public-sector service agencies. Fire stations, hospitals, emergency centers, equipment yards, and landfills are just a few that have been discussed. Often, the criterion is different from that for private-sector problems and the constraints may differ. The objective in the private sector is usually (and has been in all examples to this point) to minimize some total cost function often represented by time or distance. The distance measure is frequently Euclidean.

In the public sector the objective is frequently to minimize some maximum response time, for example, in time to get to a fire or emergency. Also, since almost all problems are in an urban setting, the distance measure is often rectilinear. Of course, there may be problems in the private sector whose objectives are to minimize some maximum response time with a rectilinear measure and vice versa, but, in general, the foregoing observations hold.

A typical public-sector location problem is given in Example 4.3. It will be analyzed and solved later.

Example 4.3

The Whitehurst Cowboy Consulting Firm has been contacted by the American Red Cross of Oklahoma City, Oklahoma. A new blood bank is to be constructed somewhere in the city that will serve 15 existing hospitals and research centers. A given number of trips weekly are to be made by a truck from the blood bank to a given hospital or center and then back to the blood bank. The number of trips between the blood bank and any given facility is relatively constant and known, but the number varies between facilities. The Red Cross wants to locate the blood bank such that the *total distance* traveled by the truck is minimized. The details, including map coordinates and the number of trips weekly, are shown in Table 4.3. Figure 4.12 is a map showing approximate locations of all hospitals and research centers.

The problem described in Example 4.3 is a single-facility location problem with 15 existing facilities. The objective given by the Red Cross is to minimize the total distance traveled by the truck. Since the problem occurs in an urban setting, the distance measure is most likely rectilinear (which it is in this case).

This location problem is one of the easiest to solve. An algorithm for the solution is given below. The proof is not shown, but Francis and White (footnote 1) give a complete proof and discussion of the more complicated problems.

Solution Procedure. Location of a new facility with respect to several existing facilities (rectilinear distances assumed):

(1) Arrange all facilities in increasing order of the X coordinates. Add up the number of trips for that same order. The optimum X location is such that no more than half the trips are to the left and no more than half to the right.
(2) Repeat for the Y coordinate such that no more than half the trips are above and no more than half below.
(3) The X and Y coordinates just chosen are the optimum locations. The price of land, neighborhood, and other parameters must now be considered.

Table 4.3. INFORMATION FOR THE BLOOD BANK LOCATION STUDY

Hospital or Research Center	*Map Location*	*Number of Trips Weekly*
1. Hillcrest	(22.5, 5.5)	10
2. South Community	(26.5, 9.5)	10
3. Capitol Hill	(29.5, 14.0)	5
4. St. Anthony	(28.5, 24.0)	3
5. Bone and Joint	(28.5, 24.5)	1
6. Mercy	(29.0, 25.0)	3
7. Polyclinic	(30.0, 25.0)	1
8. Wesley	(29.5, 24.5)	1
9. University	(32.0, 24.5)	3
10. University School of Medicine	(32.0, 25.0)	1
11. Oklahoma Medical Research Institute	(32.5, 25.0)	1
12. Veteran's Administration Hospital	(33.0, 25.0)	1
13. Crippled Children's	(32.5, 24.5)	1
14. Deaconess	(16.0, 36.5)	5
15. Baptist Memorial	(18.0, 37.5)	4

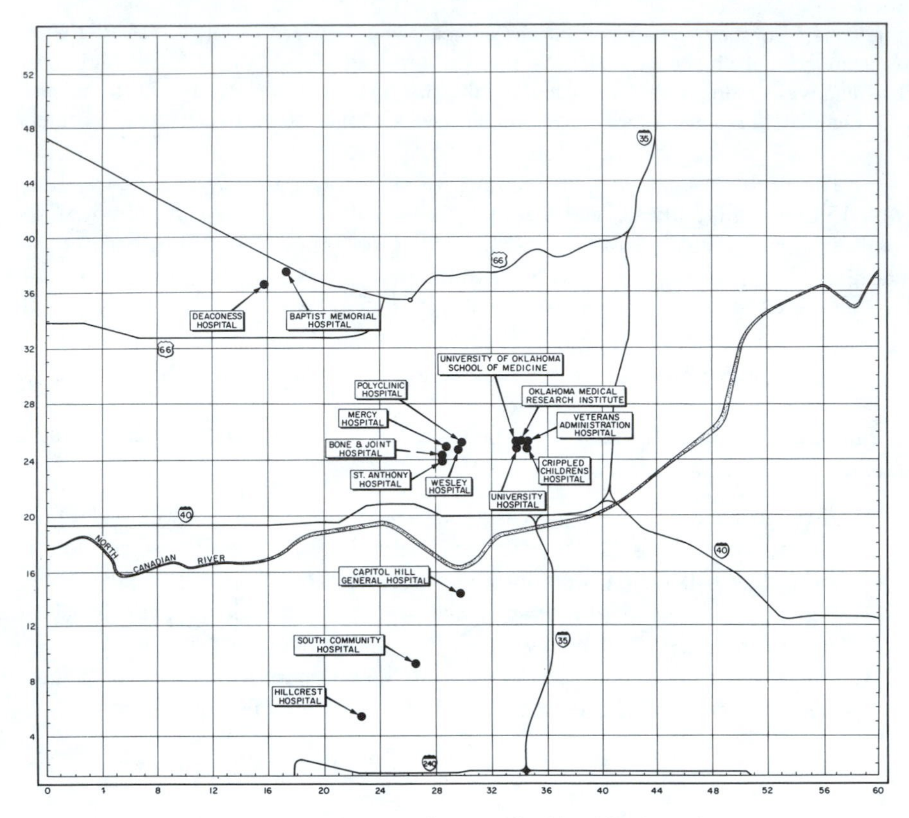

Fig. 4.12. Oklahoma city map for blood bank study.

In the example problem the calculations for the X coordinate are as follows:

	X Coordinate										
	16.0	*18.0*	*22.5*	*26.5*	*28.5*	*29.0*	*29.5*	*30.0*	*32.0*	*32.5*	*33.0*
Number of trips	5	4	10	10	4	3	6	1	4	2	1
Cumulative	5	9	19	29	33	36	42	43	47	49	50

The total number of trips is 50. The optimum x location then is that location where no more than 25 trips are to the left and no more than 25 are to the right. This occurs exactly at

$$X^* = 26.5$$

There are 19 trips to the left of 26.5 and $4 + 3 + 6 + 1 + 4 + 2 + 1 = 21$ trips to the right of 26.5.

For the Y coordinate, the same procedure is required:

	X Coordinate							
	5.5	*9.5*	*14.0*	*24.0*	*24.5*	*25.0*	*36.5*	*37.5*
Number of trips	10	10	5	3	6	7	5	4
Cumulative	10	20	25	28	34	41	46	50

There are 50 total trips. At

$$Y^* = 14.0$$

20 trips are to the left and exactly 25 trips are to the right of the blood bank. This implies that the optimum location is at 14, 0. At

$$Y^* = 24.0$$

however, exactly 25 trips are to the left and 22 trips are to the right of the facility. This implies that the optimum location is at 24.0.

In this example the optimal solution occurs at any point in the following range of Y values:

$$14.0 \leqslant Y^* \leqslant 24.0$$

The facility may then be located at (26.5, 14.0), (26.5, 24.0), or any point between. Of course, it is possible for the optimum location of such a facility to be a point, a line, a square, or rectangular area, depending on the details of the specific problem. In this case, the answer is a straight line since the optimal location on the X axis was exactly 26.5 but that on the Y axis could vary between 14.0 and 24.0.

As the algorithm says, the next step is to consider other factors, for example, the price of the land, the neighborhood, and cost of construction. The American Red Cross should consider all these factors before making a final decision.

4.4. Introduction to Facilities Layout

Sections 4.1 through 4.3 discussed the problem of locating facilities. Sections 4.4 through 4.7 discuss the internal layout of the facility. A layout problem is one of *locating areas*; therefore, facility location and layout are often covered in the same course and/or textbook.

The planning phase of plant layout is exceptionally important. Since an organization normally must live with the layout for a long time, any mistakes in the actual layout can be very costly. These mistakes should be made on paper long before the physical movement of equipment begins. Therefore, careful layout planning is of primary importance.

Almost always the objective in a facility layout study is to minimize total cost, but the term *total cost* is very difficult to define. Many elements comprising total cost are highly complex and obscure. For example, the following list shows just a few of these cost elements:

(1) Construction cost.
(2) Installation cost.
(3) Material handling cost.
(4) Ease of future expansion.
(5) Production cost.
(6) Machine downtime cost.
(7) In-process storage cost.
(8) Safety cost.
(9) Ease of supervision.

Even if this list were exhaustive (which it is not), to develop a total cost function composed of these nine items would be extremely difficult.

Industrial engineers working on facility layout problems often choose as their objective the minimization of material handling cost associated with the layout. The cost of handling materials in production operations often accounts for from 30 to 95% of total production costs.

The choice of minimizing material handling cost is made for several reasons. First, material handling costs are often large and exist year after year. This is contrary to many other costs, such as construction, which occur only once or infrequently.

Second, material handling cost is easily quantifiable. Usually, the cost is proportional to distance moved, and measurement of the distances is a simple matter when given a layout of the facility.

Third, material handling cost is often that cost most affected by the layout itself. Since material handling cost is proportional to distances traveled, the layout has a pronounced effect on the amount of cost.

Because of this pronounced impact on material handling cost, it is important that the layout and material handling systems design be done simultaneously or at least with significant feedback loops. To do either without feedback to the other almost always leads to suboptimization.

Similarly, layout and resource management (especially energy and water) have a dramatic impact on each other, as will be seen in Chapter 12. A good layout will consider resource management using the principles discussed in Chapter 12.

There are two essential types of layout flow: product and process. A *product layout flow* is used when there is a large volume of one or similar products. In this the machinery or equipment is arranged so that there is a continuous flow of material in an orderly fashion throughout the plant. Paper mills, dairies, cement factories, and automotive assembly plants are good examples of product layout. The primary objective in such cases is to minimize material handling cost by properly arranging the equipment in the processing sequence. Many types of flow are possible. Figure 4.13 suggests only a few.

A *process layout flow* is needed for operations involving a small volume of many different products. This type of operation is called *job-lot manufacturing*. Since there is no mass production, there can be no smooth, continuous flow lines.

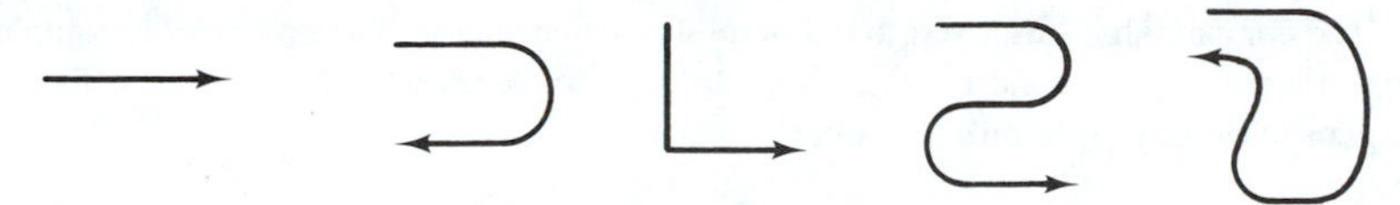

Fig. 4.13. Product flow lines.

Instead, the flow lines vary according to the particular items being produced at a particular time. In this situation it is logical to place similar machines together. For example, all lathes would be in one area, all drill presses in another, and so on.

The following example is intended to place the facility layout problem in perspective.

Example 4.4

The Whittemore Dairy of Floyd, Virginia, has decided to move its operations to a more modern plant. Remodeling and relayout of the old facility were considered but ruled out because of the age of the old building. In the new facility, four products will be produced. The four, with forecasted quantity and other general data, are shown in Table 4.4. To produce these products, the following equipment is needed:

(1) Pasteurizer (one).
(2) Homogenizer (one).
(3) Pasteurization storage tank (one).
(4) Filling machine (three) for placing products in cartons.
(5) Cooler (one).
(6) Laboratory equipment (one complete lab) for testing products.
(7) By-product mixing vat and other equipment for producing cottage cheese and buttermilk.

Table 4.4 PRODUCT QUANTITY DATA FOR WHITTEMORE DAIRY

Product	*Present Quantity*	*Growth*	*General*
Milk	4,200 gal per day	Annual rate of 3%	Packed in $\frac{1}{2}$-pint, 1-quart, and 5-gal boxes. Received in 4,000–6,000-gal trucks early in day—stored until used. Approximately 1 truck incoming and 20 trucks outgoing daily (all products). Normal operation is 14 hours (all products).
Juice	1,000 gal per week	Annual rate of 5% for 3 years; 3% thereafter	See *milk* for packing descriptions.
Cottage cheese	1,000 lb per week	Annual rate of 1%	Slow growth with possible phasing out. See *milk* for packing descriptions.
Buttermilk	300 gal	Annual rate of 3%	See *milk* for packing descriptions.

The company has hired you to aid in the layout endeavor. The company is completely willing to work with you and provide any data you feel necessary. Your job is to locate the equipment in order to minimize total costs.

4.5. General Considerations

Ideally, the objective in this problem is to minimize total cost, but you find that defining total cost is a very difficult task, as discussed in Section 4.4. Therefore, you assume that the objective is to minimize material handling cost by placing interacting departments as close together as possible. In an industry such as a dairy, the distances between departments may not be as important as in other industries because almost all material handling up to the packing stage is done through pipes. The longer the pipes, however, the more equipment and energy are required. Therefore, minimizing distances will minimize material handling cost.

Although not particularly true for the Whittemore Dairy, often the process used and the layout are very interdependent. That is, the choice of a particular process could dictate the total layout of the plant. In these cases, the layout is relatively simple once the process has been chosen. For the Whittemore Dairy, this is assumed not to be true. Therefore, a total layout analysis is necessary.

There are three essential types of layout: product, process, and fixed-position. *Product layout* is useful in a mass-production situation. *Process layout* is necessary in the other extreme—job-lot production. *Fixed-position layout* is useful when the product is so extremely large that it is easier to bring the workstation to the product rather than the product to the workstation. Examples of fixed-position layout include larger shipbuilding and airplane manufacturing. Here, since the ship or airplane is too large to be moved around the shop, the various stages of manufacture (particularly assembly) are performed in one place by bringing all tools to the plane or ship.

Product layout and process layout have their advantages and limitations, as demonstrated in Table 4.5. (Fixed-position layouts are seldom used, but when they are used, it is because other alternatives are infeasible.)

In the Whittemore Dairy example there are only a few products produced and almost all follow the same essential process lines (as will be seen later). For that reason, a product layout is appropriate.

Now, since the construction of the actual layout is a fairly complex problem, how do you go about doing the actual layout?

In your search for procedures you often hear systematic layout planning mentioned. Systematic layout planning is a popular approach that has proven to work well in many situations.

4.6 Systematic Layout Planning

Systematic layout planning (SLP) is an organized approach to layout planning that was developed by Richard Muther and associates. Muther's book describes the

Table 4.5 ADVANTAGES AND LIMITATIONS OF PRODUCT AND PROCESS LAYOUTS*

Product Layout

Advantages
1. Since the layout corresponds to the sequence of operations, smooth and logical flow lines result.
2. Since the work from one process is fed directly into the next, small in-process inventories result.
3. Total production time per unit is short.
4. Since the machines are located in order to minimize distances between consecutive operations, material handling is reduced.
5. Little skill is usually required by operators at the production line; hence, training is simple, short, and inexpensive.
6. Simple production planning and control systems are possible.
7. Less space is occupied by work in transit and for temporary storage.

Limitations
1. A breakdown of one machine may lead to a complete stoppage of the line that follows a machine.
2. Since the layout is determined by the product, a change in product design may require major alterations in the layout.
3. The pace of production is determined by the slowest machine.
4. Supervision is general instead of specialized.
5. Comparatively high investment is required because identical machines (a few not fully utilized) are sometimes distributed along the line.

Process Layout

Advantages
1. Better utilization of machines can result; consequently, fewer machines are required.
2. A high degree of flexibility exists relative to equipment or manpower allocation for specific tasks.
3. Comparatively low investment in machines is required.
4. The diversity of tasks offers a more interesting and satisfying occupation for the operator.
5. Specialized supervision is possible.

Limitations
1. Since longer flow lines usually result, material handling is more expensive.
2. Production, planning, and control systems are more involved.
3. Total production time is usually longer.
4. Comparatively large amounts of in-process inventory result.
5. Space and capital are tied up by work in process.
6. Because of the diversity of the jobs in specialized departments, high grades of skill are required.

* This table is extracted with permission from Richard L. Francis and John A. White, *Facility Layout and Location: An Analytical Approach* (Englewood Cliffs, N.J.: Prentice-Hall, Inc., 1974).

procedure in detail,[6] Francis and White discuss it in their text (footnote 1), and Figure 4.14 depicts the stages in the procedure.

As shown in Figure 4.14, all data must be gathered on current and forecasted production. In the dairy example this has already been accomplished and is presented in Table 4.4. In many studies, a list of from three to five product types will account for from 70 to 80% of the total sales volume. The remaining 20 to 30% normally

[6]Richard Muther, *Systematic Layout Planning* (Boston: Cahners Books, 1973).

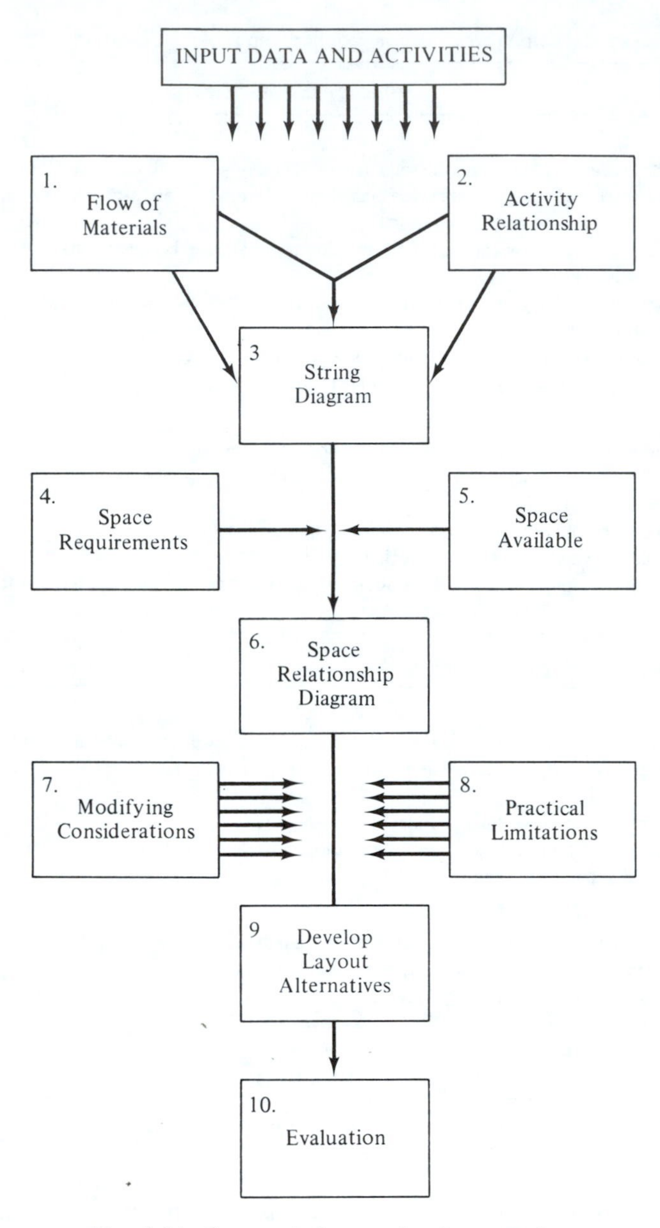

Fig. 4.14. Systematic layout planning procedure.

can be grouped so that only a few product groups need to be considered. Each product group and its respective volume for a projected horizon should be listed. The projected horizon will depend on how often the product or market changes, but usually a detailed projection for each of the next five years is adequate. Any other information that may be important to the layout should be included in the general comments. Information such as the possible phasing out of the manufacture of cottage cheese as shown in Table 4.4 could be very critical to the long-run success of the layout.

After all product quantity data have been gathered, step 1 of the SLP procedure is the preparation of process charts graphically depicting the flow of the material through the plant. If there are just a few products, a separate operations process chart should be made for each. If there are many products, a multiproduct process chart may be used. Both are demonstrated later with the symbols shown in Figure 4.15 used to construct the charts. Actually, an operations process chart depicts only the operations and inspections. A flow process chart, which will be discussed, shows operations and inspections as well as transportations, delays, and storages.

In the operations process chart (discussed briefly in Chapter 3) a line is drawn at the upper right-hand corner of the paper reflecting the start of the largest component. Other parts of materials are fed in from the left. The sequence is from top to bottom and only operations and inspections are shown. Figure 4.16 illustrates an operations process chart for processing milk for Whittemore Dairy. A multiproduct process chart form, shown in Figure 4.17, demonstrates that the only difference is that the multiple products are listed across the top and the flow is from top to bottom only.

Sometimes additional information is needed about a particular phase of the operations. A flow process chart is used for this purpose. A flow process chart is essentially the same as an operations process chart, but a flow process chart shows more information about a specific part or item. Figure 4.18 shows a flow process chart for processing juice in the old dairy. This chart, showing storages and transportations, gives more detail than an operations process chart does. This is a highly simplified flow process chart form. Other more general forms appear in Muther's book and in almost all work simplification books.

In some situations, such as a job shop, it is difficult, if not impossible, to represent all the flow with a few charts like those discussed above. Many products

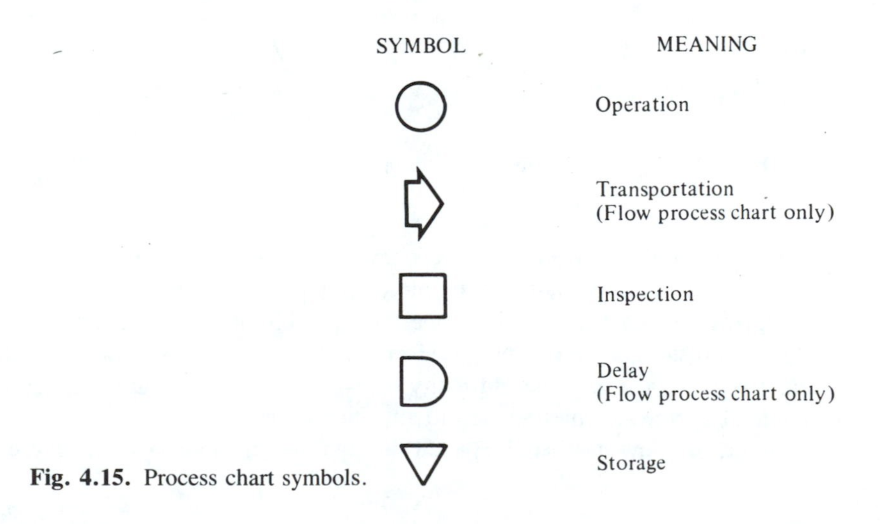

Fig. 4.15. Process chart symbols.

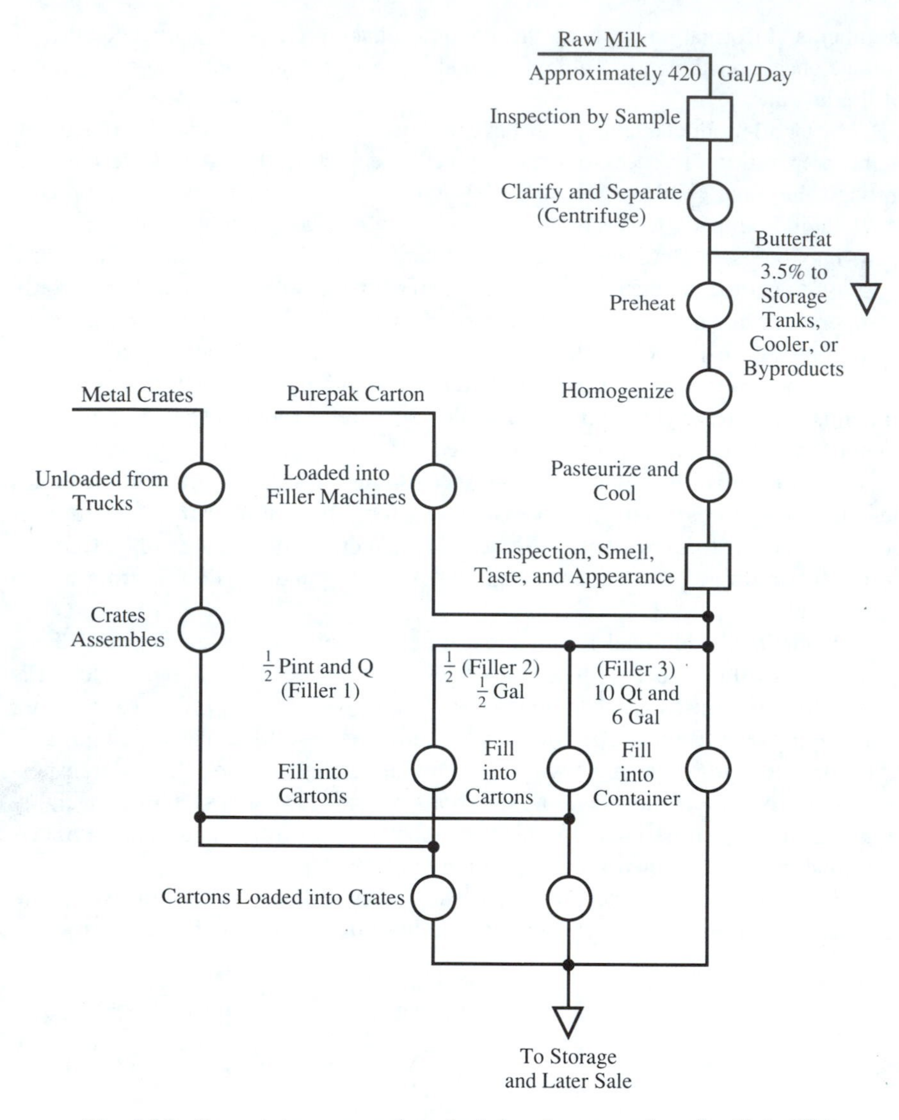

Fig. 4.16. Operations process chart depicting the processing of milk in Whittemore Dairy.

may be produced that require a large number of charts. Instead of preparing all these charts, a from–to chart may be more appropriate.

A *from–to chart* shows the number of trips from one area to another area and is based on historical data or proposed production. The trips can be weighted by product volume or some other difficulty measure if desired. Figure 4.19 shows an example of a from–to chart applied to an office situation.

Now, the flow has been depicted through one or more of the above charts;

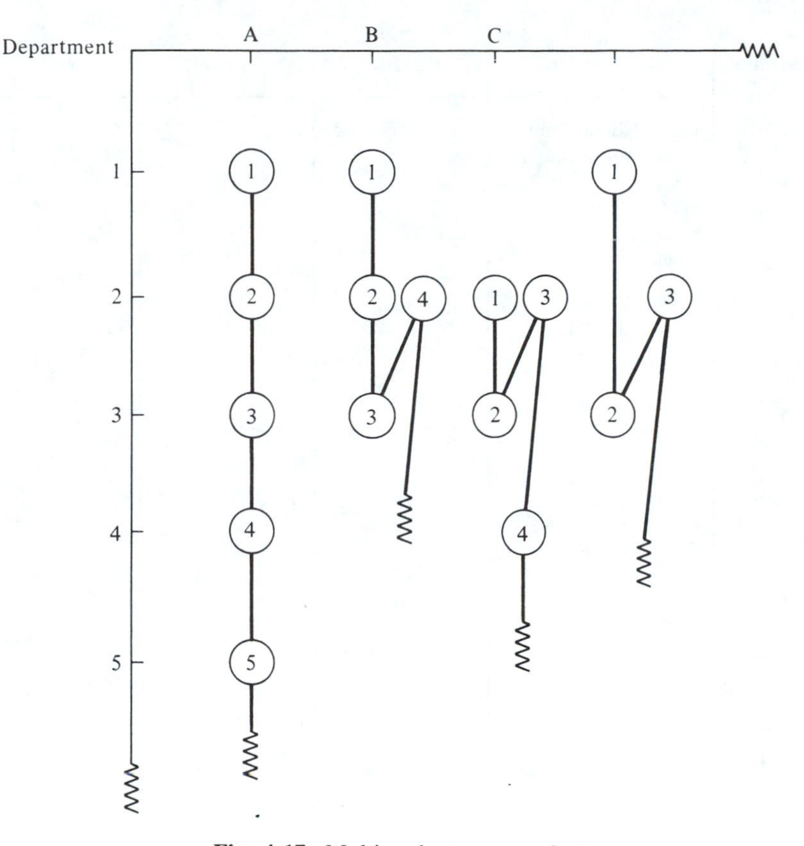

Fig. 4.17. Multiproduct process chart.

often layouts can be constructed on the basis of this flow alone. In the dairy example, almost all areas can now be laid out according to the operations process charts (Figure 4.16) or flow process charts (Figure 4.18). (Actually, several charts are necessary, one for each major product.)

In many situations, however, the flow tells only part of the story, for there are some areas that do not have any product flow and some in which the flow sequence differs for each of several products. In the dairy example the laboratory equipment, rest room, and metal crate areas do not have product flows at all. Also, the flow of cottage cheese and buttermilk is different from that of milk and juice. Therefore, you decide that more study is warranted and you find that an *activity relationship diagram* could be constructed.

Step 2 in SLP is the preparation of an *activity relationship diagram* that shows the desired closeness of departments and areas within the plant. The activity relationship diagram reflects the fact that not all important relationships can be shown by product flows. For example, it is desirable for the quality control lab in a dairy to be situated as close to the processing area as possible and for the rest room facilities to be far away from the mixing vats in the processing area.

Symbol	Description	Distance	Other Factors	Notes
○	Unload Juice from Tank Trucks			
⇨	To Cooler	25 Ft		By Hand Truck. Why Not Pump?
▽	Waiting for Use			Less Than 1 Week
⇨	To Pasteurization Storage Tanks	30 Ft		By Hand Truck. Why Not Pump?
○	Mixed			
□	Check for Quality Control			
⇨	Pumped to Filling Machines	20 Ft		
○	Fill Cartons			
⇨	To Cooler	20 Ft		Powered Conveyor
▽	Stored			Less Than 24 hr Average
⇨	To Truck	15 Ft		Powered Conveyor
○	Loaded and Shipped			

Fig. 4.18. Flow process chart for the processing of juice.

In constructing the chart all departments and areas are listed and a closeness desirability rating is given to each paired combination. Figure 4.20 is a set of ratings as proposed by Muther. Others are possible, but these seem to work very well. For any paired combination then, an A rating means that it is absolutely necessary that the two areas be located adjacent to each other. At the other extreme, an X rating

From \ *To*	*Office Manager*	*Secretary*	*Typist*	*Clerk*	···	*Totals*
Office Manager	—	10	5	1	···	
Secretary	20	—	10	10	···	
Typist	10	30	—	20	···	
Clerk	5	25	10	0	···	
·	·	·	·	·		
·	·	·	·	·		
·	·	·	·	·		
Totals						

Fig. 4.19. From–to chart example (number of trips per day).

means that it is not desirable for the two areas to be adjacent. The rest room and processing area combination is one that would have an X rating in order to eliminate any possibility of their being placed together.

For relationships between areas where there is product flow, the closeness rating can be determined fairly easily by considering the operations process charts, flow process charts, and from–to charts. For relationships involving one or two nonprocessing areas, it is much more difficult to determine these ratings. Often, all involved personnel are asked to give their opinions on what the ratings should be and then an average is taken.

You decide to develop a chart for all areas of this dairy and it is shown in Figure 4.21. It seems best in almost all situations to list all areas[7] and determine the letters by considering the figures developed in the process charts and by holding consultations with the foremen, workers, and others. In this dairy example all ratings were determined either by referring to the charts whenever possible or by working with the personnel involved.

Step 3 involves using the information generated in steps 1 and 2 to prepare a string diagram showing near optimal placement of the facilities without consideration of space requirements. Figure 4.22 illustrates one possible string diagram for the dairy. The placement is done through trial and error. Normally, those areas having an A closeness are shown first and are connected with four straight lines, then E with three straight lines, and so on. When an activity has to be close to several other areas, it can be stretched out or distorted, as shown in Figure 4.22 for storage area 1. The areas may be moved around and interchanged until a final acceptable arrangement is obtained. It is helpful to visualize the straight lines as stretched rubber bands and the jagged lines as coiled springs representing varying attraction and repulsion forces. Therefore, an A rating would imply four rubber bands pulling the areas together while an I rating would imply only two rubber

[7]A typical organization can list all areas and seldom have more than 50. This seems to be an upper limit for effective representation by activity relationship diagrams.

Letter	*Closeness*
A	Absolutely Necessary
E	Especially Important
I	Important
O	Ordinary Closeness O.K.
U	Unimportant
X	*Not* Desirable

Fig. 4.20. Activity relationship diagram symbols.

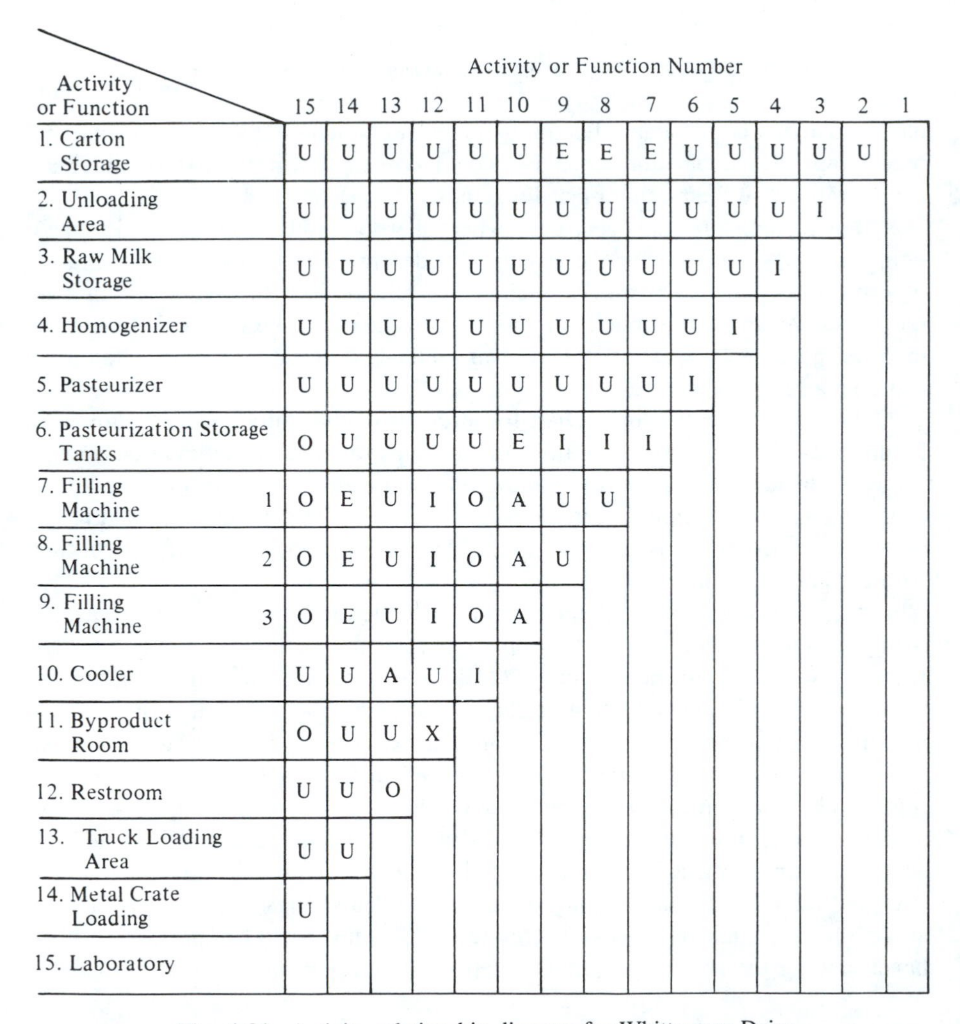

Activity or Function	15	14	13	12	11	10	9	8	7	6	5	4	3	2	1
1. Carton Storage	U	U	U	U	U	U	E	E	E	U	U	U	U	U	
2. Unloading Area	U	U	U	U	U	U	U	U	U	U	U	U	I		
3. Raw Milk Storage	U	U	U	U	U	U	U	U	U	U	U	I			
4. Homogenizer	U	U	U	U	U	U	U	U	U	U	I				
5. Pasteurizer	U	U	U	U	U	U	U	U	U	I					
6. Pasteurization Storage Tanks	O	U	U	U	U	E	I	I	I						
7. Filling Machine 1	O	E	U	I	O	A	U	U							
8. Filling Machine 2	O	E	U	I	O	A	U								
9. Filling Machine 3	O	E	U	I	O	A									
10. Cooler	U	U	A	U	I										
11. Byproduct Room	O	U	U	X											
12. Restroom	U	U	O												
13. Truck Loading Area	U	U													
14. Metal Crate Loading	U														
15. Laboratory															

(Column headings: Activity or Function Number)

Fig. 4.21. Activity relationship diagram for Whittemore Dairy.

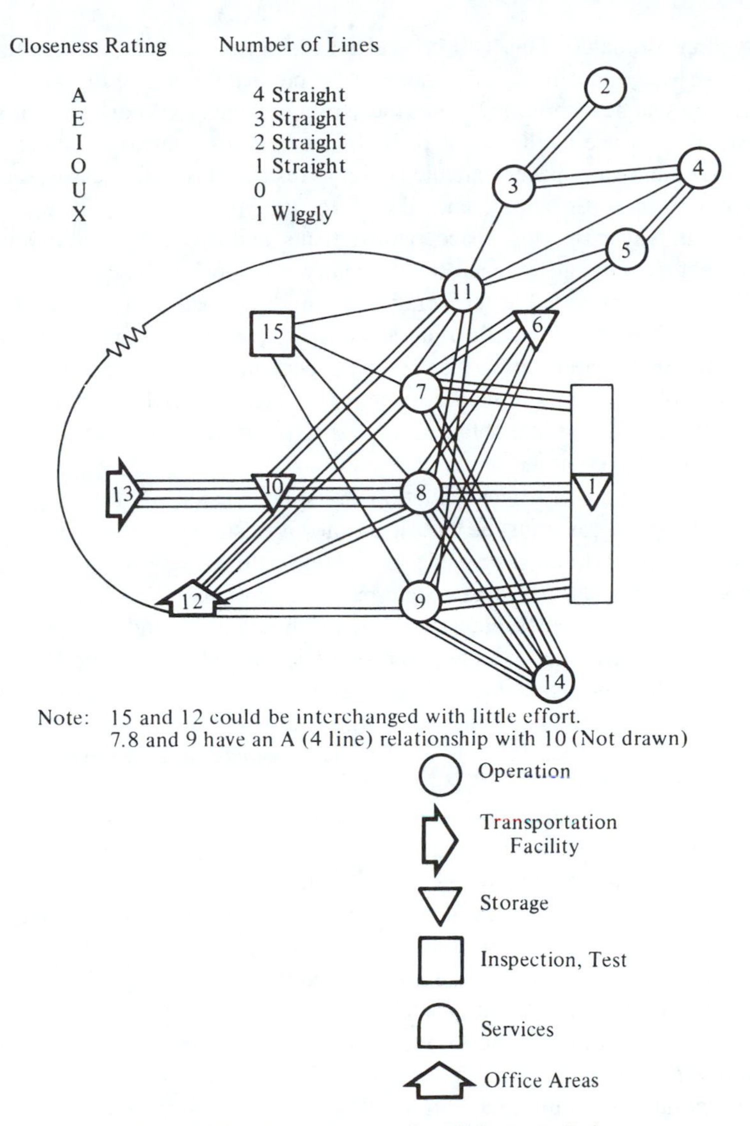

Fig. 4.22. String diagram for Whittemore Dairy.

bands. Many diagrams and arrangements will probably have to be drawn before a good layout is obtained. Normally, two or more alternatives are developed. Essentially, these alternatives each constitute a final layout. Space will have to be added and some modifications made, but the overall picture should not change much. This third step is, therefore, probably the most creative and the most important.

Step 4 may be called the *adjustment step*. Here, adjustments must be made for space needs as related to space availability; therefore, space requirements have to be determined. This can be done through calculations, adjustment of past areas, or just good estimates. This is a critical stage, but for almost all organizations, in-process storage and machine areas can be predicted accurately, so that space requirements can be determined. Once these requirements are known, it is necessary to consider the space available (step 5). In some cases, since the layout must fit existing buildings, the space available is highly restricted. In other cases, the capital budget is the main restriction, and, therefore, the space availability may be less restricted. In any case, the space requirements and space availability must be balanced before moving into step 6—the space relationship diagram.

Here, space is added to the string diagram developed in step 3. By using the space from steps 4 and 5 and the diagram from step 3, a layout is constructed; but many trials usually are necessary before an acceptable layout is developed. Often, a number of blocks representing the space requirements of a given area are shuffled around until good layouts are obtained. (We shall examine alternative space relationship diagrams in section 4.7.)

In steps 7 and 8 practical limitations must be considered. This is particularly true if the string diagram must be massaged into an existing building or preconceived shape, or if allowances must be made for columns, utility connections, and unusual shapes or requirements. At least two alternative layouts should be developed. These alternatives should be block diagrams showing broad areas and general shapes.

Step 9 is to consider the alternatives developed and possibly go back through steps 1 through 5 for detailed layouts of individual areas. At this stage, the two or more alternatives should be examined using some criterion so that the best one can be chosen. Usually, the criterion is a subjective estimate of the combined effects of several criteria such as those discussed in Section 4.4. A decision model such as that developed earlier for decision making with multiple criteria could be useful.

The final step, step 10, is the crucial one: selling the layout not only to management but also to employees. Almost all good products need some selling. Therefore, it is necessary to develop and present a strong case for the layout to top management. The alternatives considered, and the reasons for elimination and subsequent final choice, often make a very effective sales package. The criteria and the reasons why the final layout was chosen should be discussed.

Now that you have gone through this SLP procedure, you have developed several alternative layouts for Whittemore Dairy. You approach management and make a presentation as discussed above. Management is impressed with your outstanding abilities in facility layout and accepts your final proposal.

Given that the overall layout is complete and that management agrees with the layout developed, you must now go back and lay out each department or area in detail. For example, in the Whittemore Dairy the individual pieces of test equipment must be laid out in the laboratory, various pieces of equipment must be located in the by-product processing area, and storage racks and other equipment must be located in the cooler.

To do this, you must repeat the 10-step SLP procedure for each of the departments that must be laid out in detail. Of course, some steps will be simpler than before since almost all of the data have already been gathered. For example, in the laboratory each piece of test equipment and other items needed becomes an activity in a new activity relationship diagram for the laboratory only. Since the product quantity analysis has already been done, few new flow process charts or operations process charts will be necessary.

Suppose, however, Whittemore management tells you that they very much like your ideas but would really like to see more alternatives. One approach for you is to turn to computerized procedures in order to develop additional layouts.

4.7. Computerized Layout Planning (Optional)

One of the advantages of SLP is that alternative layouts can be generated based on different space relationship diagrams developed in step 6. It is logical that the more space relationship diagrams developed, the more likely a good decision can be made. Therefore, computerized layout packages can be a great aid here. Computerized packages take the activity relationship charts developed in step 2 or from–to charts developed in step 1 (depending on which package is used) and generate any specified number of layouts. These can be scored and compared by the computer and/or the layout analyst. The final decision must be made by the manager. *The chief advantage in computerized packages lies in the large number of alternatives generated and in the lack of any preconceived ideas by the computer.*

There are many computerized plant layout packages available and the number is growing steadily. Francis and White discuss three in their text. They are ALDEP, CORELAP, and CRAFT. Basically, ALDEP and CORELAP are "construction" types in that layouts are generated from scratch. CRAFT is an "improvement" type in that an initial layout is required and improvements are made upon it. Some other packages are RUGR, LAOPT, PLANET, LSP, and RMS COMP I. Only one package, ALDEP, will be considered here. The ALDEP program was developed by IBM and was originally presented by Seehof and Evans.[8] ALDEP is chosen for this presentation because it was one of the first packages developed and it is relatively simple and easy to understand.

There are several versions of ALDEP available. The one you choose for Whittemore and, coincidentally, the one used in this chapter, is the one in which the first department is chosen for placement on a random selection basis. Then, the activity relationship chart (Figure 4.23, subsequently called the *Rel chart*) is scanned to find a department with a high (A or E) closeness rating with the department just placed. If one is found, that department is placed and the Rel chart is scanned for another department with a high closeness rating. If at any time, no departments can be found with a high rating, either the layout is complete or another random selection is made.

---

[8] J. M. Seehof and W. O. Evans, "Automated Layout Design Program," *The Journal of Industrial Engineering,* 18 (1967), 690–695.

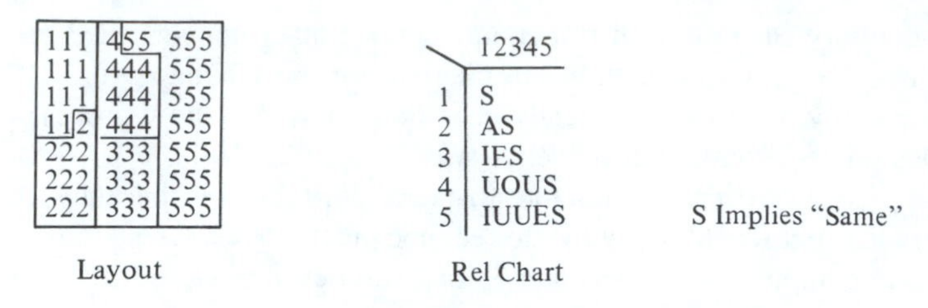

Fig. 4.23. Sample rel chart and layout for ALDEP program.

An important part of the ALDEP program is the scoring of alternative layouts. For this scoring, the program considers immediate adjacencies only and assigns the following values to those adjacencies:

$$A = 64 \qquad O = 1$$
$$E = 16 \qquad U = 0$$
$$I = 4 \qquad X = -1{,}024$$

For example, the simple layout and Rel chart shown in Figure 4.23 has the following immediate adjacencies and scores:

Departments	*Score*	*Departments*	*Score*
1–4	0	3–5	0
1–2	64	4–5	16
2–3	16	3–4	0
2–4	1		

The total points are

$$(0 + 64 + 16 + 1 + 0 + 16 + 0)2 = 194$$

The number of points must be multiplied by 2 since the program considers two-way adjacency (i.e., a 1–4 relationship is also a 4–1 relationship).

ALDEP uses a vertical scanning method of locating departments by laying a vertical strip of a width specified by the user across the depth of the layout. Then the strip is turned around and the process is repeated. As the strip is being laid, the departments are being placed. In the example shown in Figure 4.23 the scanning width is 3 and the first department located is department 1. The others are located in the order 2, 3, 4, 5.

As an example of the use of ALDEP, the Rel chart in Figure 4.21 is to be used as input to the program to lay out the Whittemore Dairy.

The basic inputs to the ALDEP program are the following:

(1) The vertical scanning width. Here a width of 3 is used, but by using other values, different layouts can be obtained.

(2) The Rel chart. The chart shown in Figure 4.21 is used.
(3) The departmental area requirements. These values are shown in Figure 4.24. You developed these by working with Whittemore management and other dairies.

It is also necessary to tell the program what minimum score is acceptable. In this case, a minimum score of 0 is used for the first run in which 20 layouts are done. For the second run, the best score of the first 20 is used as minimum and 20 more are run. The normal procedure is to continue this until either no additional layouts are developed or else a decision is made.

You feed all this into the program and develop the first set of 20 runs. The first layout developed (which is probably not the best) is reproduced in Figure 4.25. The score for this layout is 188, developed by looking at the following relationships:

Relationships	*Points*	*Relationships*	*Points*	
5–10	0	4–9	0	
10–7	64	9–8	0	
7–1	16	8–6	4	
1–11	0	8–15	1	
1–14	0	8–12	4	
1–3	0	6–2	0	
14–11	0	6–13	0	
14–3	0	6–12	0	
3–11	1	12–13	1	94 × 2 = 188
11–4	1	2–13	0	
11–9	1	6–15	1	
4–6	0	15–12	0	
4–8	0	Total	94	

Department	*Area (Ft²)*	*Department*	*Area (Ft²)*
1 Carton Storage	100	9 Filling Machine 3	50
2 Unloading Area	50	10 Cooler	300
3 Raw Milk Storage	100	11 Byproduct Room	300
4 Homogenizer	50	12 Restroom	30
5 Pasteurizer	50	13 Truck Loading Area	40
6 Pasteurization Storage Tank	90	14 Metal Crate Loading	40
7 Filling Machine 1	100	15 Lab	50
8 Filling Machine 2	100		

Fig. 4.24. Departmental area requirements for Whittemore Dairy.

```
TRIAL LAYOUT     18   SCORE =  188


0 0 0 0 0 0 0 0 0 0 0 0 0 0 0 0 0 0 0 0 0 0 0 0 0 0 0 0 0 0 0 0 0 0 0 0 0 0 0 0 0 0 0 0 0 0 0 0 0 0 0 0 0 0
0 5 5 5101010101010 7 7 7 7 7 7 1 1 1 1 1 1111111111111111111111111 4 4 4 4 4 4 6 6 6 6 6 6 2 2 2 2 2 0
0 5 5 5101010101010 7 7 7 7 7 7 1 1 1 1 1 1111111111111111111111111 4 4 4 4 4 4 6 6 6 6 6 6 2 2 2 2 2 0
0 5 5 5101010101010 7 7 7 7 7 7 1 1 1 1 1 1111111111111111111111111 4 4 4 4 4 4 6 6 6 6 5 6 2 2 2 2 2 0
0 5 5 51010101010101010 7 7 7 7 1 1 11414 1111111111111111111111111 4 4 4 8 8 4 6 6 6 6 6 6 2 2 2 2 2 0
0 5 5 51010101010101010 7 7 7 1 1 1141414111111111111111111111111 4 4 4 8 8 8 6 6 6 6 6 6 2 2 2 2 2 0
0 5 5 51010101010101010 7 7 7 1 1 1141414111111111111111111111111 4 4 4 8 8 8 6 6 6 6 6 6 2 2 2 0 0 0
0 5 5 51010101010101010 7 7 7 1 1 1141414111111111111111111111111 4 4 4 8 8 8 6 615 6 6 6 2 2 2 0 0 0
0 5 5 51010101010101010 7 7 7 1 1 1141414111111111111111111111111 4 4 4 8 8 8151515 6 6 6 2 2 2 0 0 0
0 5 5 51010101010101010 7 7 7 1 1 1141414111111111111111111111111 4 4 4 8 8 8151515 6 6 6 2 2 2 0 0 0
0 5 5 51010101010101010 7 7 7 1 1 1141414111111111111111111111111 4 4 4 8 8 8151515 6 6 6 2 2 2 0 0 0
0 5 5 51010101010101010 7 7 7 1 1 1141414 3 3 3111111111111111111 4 4 4 8 8 8151515 6 6 6 2 2 2 0 0 0
0 5 5 51010101010101010 7 7 7 1 1 1141414 3 3 3111111111111111111 4 4 4 8 8 8151515 6 6 6 2 2 2 0 0 0
0 5 5 51010101010101010 7 7 7 1 1 1141414 3 3 3111111111111111111 4 4 4 8 8 8151515 6 6 6 2 2 2 0 0 0
0 5 5 51010101010101010 7 7 7 1 1 1141414 3 3 3111111111111111111 9 9 4 8 8 8151515 6 6 61313 2 0 0 0
0 5 5 51010101010101010 7 7 7 1 1 1141414 3 3 3111111111111111111 9 9 9 8 8 8151515 6 6 6131313 0 0 0
0 5 5 51010101010101010 7 7 7 1 1 1141414 3 3 3111111111111111111 9 9 9 8 8 8151515 6 6 6131313 0 0 0
0 5 510101010101010101010 7 7 7 1 1 11414 3 3 3 3111111111111111111 9 9 9 8 8 8151515 6 6 6131313 0 0 0
0101010101010101010101010 7 7 7 1 1 1 3 3 3 3 3 3111111111111111111 9 9 9 8 8 8151515 6 6 6131313 0 0 0
0101010101010101010101010 7 7 7 1 1 1 3 3 3 3 3 3111111111111111111 9 9 9 8 8 8151515 6 6 6131313 0 0 0
0101010101010101010101010 7 7 7 1 1 1 3 3 3 3 3 3111111111111111111 9 9 9 8 8 8151515 6 6 6131313 0 0 0
0101010101010101010101010 7 7 7 1 1 1 3 3 3 3 3 3111111111111111111 9 9 9 8 8 8151515 6 6 6131313 0 0 0
0101010101010101010101010 7 7 7 1 1 1 3 3 3 3 3 3111111111111111111 9 9 9 8 8 8151515 6 6 6131313 0 0 0
0101010101010101010101010 7 7 7 1 1 1 3 3 3 3 3 3111111111111111111 9 9 9 8 8 8151515 6 6 6131313 0 0 0
0101010101010101010101010 7 7 7 1 1 1 3 3 3 3 3 3111111111111111111 9 9 9 8 8 8 8 8151212 6131313 0 0 0
0101010101010101010101010 7 7 7 1 1 1 3 3 3 3 3 3111111111111111111 9 9 9 8 8 8 8 8121212131313 0 0 0
0101010101010101010101010 7 7 7 1 1 1 3 3 3 3 3 3111111111111111111 9 9 9 8 8 8 8 8121212131313 0 0 0
0101010101010101010101010 7 7 7 1 1 1 3 3 3 3 3 3111111111111111111 9 9 9 8 8 8 8 8121212131312 0 0 0
0101010101010101010101010 7 7 7 1 1 1 3 3 3 3 3 3111111111111111111 9 9 9 8 8 8 8 8121212121212 0 0 0
0101010101010101010101010 7 7 7 1 1 1 3 3 3 3 3 3111111111111111111 9 9 9 8 8 8 8 8121212121212 0 0 0
0101010101010101010101010 7 7 7 1 1 1 3 3 3 3 3 3111111111111111111 9 9 9 8 8 8 8 8121212121212 0 0 0
0 0 0 0 0 0 0 0 0 0 0 0 0 0 0 0 0 0 0 0 0 0 0 0 0 0 0 0 0 0 0 0 0 0 0 0 0 0 0 0 0 0 0 0 0 0 0 0 0 0 0 0 0 0


UNDER THE RULES FOR EVALUATION, THIS LAYOUT DOES NOT SATISFY THE FOLLOWING NECESSARY CLOSE RELATIONSHIPS


 108   110
 109   110
 110   108
 110   109
 110   113
 113   110
```

Fig. 4.25. First ALDEP layout of Whittemore Dairy.

The seventeenth layout (reproduced in Figure 4.26) produces the maximum score, which is 440. This layout has some interesting properties. First, at the extreme left end of the building is the cooler (number 10) with a truck loading area (number 13) for moving material from the cooler to the trucks. This looks good and would work very well for Whittemore Dairy. Also, the product *essentially* starts from the right and flows to the left, which is good.

Finally, the layout brings some interesting points to light. The management always thought that the filling machines should be located together. Yet, neither the layout nor the Rel chart showed this; consequently, management agreed that this was not necessary (the computer is not constrained by any mental blocks or narrow points of view). The laboratory is located in the middle of the building, which is good, but so is the metal crate unloading area, which is bad. You and Whittemore management decide to generate some more layouts utilizing the best parts of this one.

One of the advantages of ALDEP is that areas can be preassigned or fixed.

```
TRIAL LAYOUT    178  SCORE =  440





0 0 0 0 0 0 0 0 0 0 0 0 0 0 0 0 0 0 0 0 0 0 0 0 0 0 0 0 0 0 0 0 0 0 0 0 0 0 0 0 0 0 0 0 0 0 0 0 0 0 0 0
0131313101010101010 7 7 7 7 7 7 4 4 4111111111111111111 5 5 5 5 5 5 3 3 3 2 2 2 8 8 8 1 1 1 6 6 6 6 6 0
0131313101010101010 7 7 7 7 7 7 4 4 4111111111111111111 5 5 5 5 5 5 3 3 3 2 2 2 8 8 8 1 1 1 6 6 6 6 6 0
0131313101010101010 7 7 7 7 7 7 4 4 4111111111111111111 5 5 5 5 5 5 3 3 3 2 2 2 8 8 8 1 1 1 6 6 6 6 6 0
0131313101010101010 7 7 7 7 7 7 4 4 4111111111111111111 5 5 5 5 5 5 3 3 3 2 2 2 8 8 8 1 1 1 6 6 6 6 6 0
0131313101010101010 7 7 7 7 7 7 4 4 4111111111111111111 5 5 5 5 5 5 3 3 3 2 2 2 8 8 8 1 1 1 6 6 6 6 6 0
0131313101010101010 7 7 7 7 7 7 4 4 4111111111111111111 5 5 5 5 5 5 3 3 3 2 2 2 8 8 8 1 1 1 6 6 6 0 0 0
0131313101010101010 7 710 7 7 7 4 4 4111111111111111111 5 514 5 5 5 3 3 3 2 2 2 8 8 8 1 1 1 6 6 6 0 0 0
0131313101010101010101010 7 7 7 4 4 4111111111111111111111414 5 5 5 3 3 3 2 2 2 8 8 8 1 1 1 6 6 6 0 0 0
0131313101010101010101010 7 7 7 4 4 4111111111111111111111414 5 5 5 3 3 3 2 2 2 8 8 8 1 1 1 6 6 6 0 0 0
0131313101010101010101010 7 7 7 4 4 4111111111111111111111414 5 5 5 3 3 3 2 2 2 8 8 8 1 1 1 6 6 6 0 0 0
0131313101010101010101010 7 7 7 4 4 4111111111111111111111414151515 3 3 3 2 2 2 8 8 8 1 1 1 6 6 6 0 0 0
0131313101010101010101010 7 7 7 4 4 4111111111111111111111414151515 3 3 3 2 2 2 8 8 8 1 1 1 6 6 6 0 0 0
0131313101010101010101010 7 7 7 4 4 4111111111111111111111414151515 3 3 3 2 2 2 8 8 8 1 1 1 6 6 6 0 0 0
0101013101010101010101010 7 7 7 4 4 4111111111111111111111414151515 3 3 3 2 2 2 8 8 8 1 1 1 6 6 6 0 0 0
0101010101010101010101010 7 7 7 4 4 4111111111111111111111414151515 3 3 3 2 2 2 8 8 8 1 1 1 6 6 6 0 0 0
010101C101010101010101010 7 7 7 4 4 4111111111111111111111414151515 3 3 3 2 2 2 8 8 8 1 1 1 6 6 6 0 0 0
0101010101010101010101010 7 7 7 4 4 9111111111111111111141414151515 3 3 3 2 212 8 8 8 1 1 1 6 6 6 0 0 0
0101010101010101010101010 7 7 7 9 9 9111111111111111111141414151515 3 3 3121212 8 8 8 1 1 1 6 6 6 0 0 0
0101010101010101010101010 7 7 7 9 9 9111111111111111111141414151515 3 3 3121212 8 8 8 1 1 1 6 6 6 0 0 0
0101010101010101010101010 7 7 7 9 9 9111111111111111111141414151515 3 3 3121212 8 8 8 1 1 1 6 6 6 0 0 0
0101010101010101010101010 7 7 7 9 9 9111111111111111111111111151515 3 3 3121212 8 8 8 1 1 1 6 6 6 0 0 0
0101010101010101010101010 7 7 7 9 9 9111111111111111111111111151515 3 3 3121212 8 8 8 1 1 1 6 6 6 0 0 0
0101010101010101010101010 7 7 7 9 9 9111111111111111111111111151515 3 3 3121212 8 8 8 1 1 1 6 6 6 0 0 0
0101010101010101010101010 7 7 7 9 9 9111111111111111111111111151515 3 3 3121212 8 8 8 1 1 1 6 6 6 0 0 0
0101010101010101010101010 7 7 7 9 9 9111111111111111111111111151515 3 3 3121212 8 8 8 1 1 1 6 6 6 0 0 0
0101010101010101010101010 7 7 7 9 9 9111111111111111111111111151515 3 3 3121212 8 8 8 1 1 1 6 6 6 0 0 0
0101010101010101010101010 7 7 9 9 9 91111111111111111111111111515 3 3 3 31212 8 8 8 8 1 1 1 6 6 1 0 0 0
0101010101010101010101010 9 9 9 9 9 9111111111111111111111111 3 3 3 3 3 3 8 8 8 8 8 8 1 1 1 1 1 1 0 0 0
0101010101010101010101010 9 9 9 9 9 9111111111111111111111111 3 3 3 3 3 3 8 8 8 8 8 8 1 1 1 1 1 1 0 0 0
0101010101010101010101010 9 9 9 9 9 9111111111111111111111111 3 3 3 3 3 3 8 8 8 8 8 8 1 1 1 1 1 1 0 0 0
0 0 0 0 0 0 0 0 0 0 0 0 0 0 0 0 0 0 0 0 0 0 0 0 0 0 0 0 0 0 0 0 0 0 0 0 0 0 0 0 0 0 0 0 0 0 0 0 0 0 0 0


UNDER THE RULES FOR EVALUATION, THIS LAYOUT DOES NOT SATISFY THE FOLLOWING NECESSARY CLOSE RELATIONSHIPS


 108   110
 110   108
```

Fig. 4.26. Computerized layout alternative Whittemore Dairy.

Therefore, you decide to preassign (number 13) exactly as it is and to preassign the unloading area (number 2) to the upper right-hand corner of the building. Since both would be at the top side of the building, road construction would be facilitated.

You generate 20 more runs. One of those runs scores 512 points and seems to offer potential. You subsequently modify it (to round off corners, and so on) and present it to Whittemore management. The semifinished product is shown in Figure 4.27.

Management likes several things particularly about the layout represented in Figure 4.27. The flow moves from right to left without much back-tracking. The unloading, metal crate, truck loading, and carton storage areas are all located on one side of the building, which facilitates road construction. The by-product room is located directly in the middle of the plant, which facilitates whatever manual material handling there is. Management accepts your proposal, rewards you with a fat bonus, and asks you to do the detailed layout.

The decision on the overall layout has been made. You must now go back and repeat the SLP procedure for each of the departments and lay them out in detail as discussed in Section 4.6.

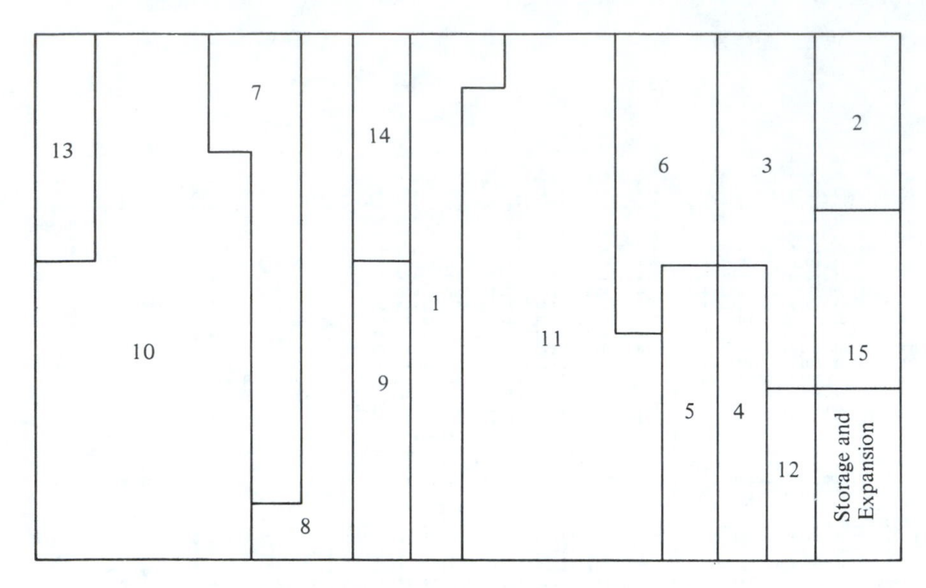

Fig. 4.27. Final Whittemore Dairy layout.

4.8. Impact of Computers

Computers have been used in various areas of location-layout studies with mixed successes. Undeniably, the computer has been immensely useful in storing and manipulating large amounts of data as might be required in location decisions and in manipulating large mathematical models.

Also, computer graphics can be a big aid in location and layout studies. Alternative locations or layouts can be generated and displayed on cathode-ray tubes (CRTs) or drawn in hard copy. Some graphics routines allow manual intervention with a light pen or mouse and rescoring. This can be very helpful and it can mean the difference between a worthless computer exercise and a realistically feasible location or layout. Recently, computer-aided design (CAD) packages have made design of alternatives so easy and the product so professional looking that layout planners often "sketch" many alternatives and seek to improve them. Thus the scope of computerized layout planning is changing.

Finally, research is under way to impart more realism and architectural considerations into layout computer programs with some interesting successes. Future computer utilization in this area seems to be toward interactive routines capable of recognizing realistic opportunities.

DISCUSSION QUESTIONS

1. Discuss the cost items affected by plant layout and how the effect occurs.
2. What differences might exist in applying the systematic layout procedure (SLP) to expansion layouts and relayouts as opposed to original layouts?

3. Almost all of this chapter is oriented toward the use of analytical techniques in plant location. Discuss the place of these techniques in the actual decision process and point out any advantages or disadvantages that are pertinent.

4. Give an example of when a rectilinear distance measure is appropriate and when a Euclidean measure is appropriate.

5. Name the three basic layouts and give an example where each would be applicable.

6. In Figure 4.13 several product flow lines are given that resemble letters of the alphabet (epg., S, O, E, L). List some other possible flow lines that resemble letters.

7. What distinguishes a heuristic from an optimal seeking procedure? Discuss the characteristics of each.

8. Give an example for a facility location problem in which the objective may be to
(a) Minimize total distance traveled.
(b) Minimize maximum distance traveled.

9. Discuss the advantages and disadvantages of product, process, and fixed-position layouts.

10. Distinguish between construction and improvement types of computerized layout procedures. Give an example of each.

11. Can you think of a situation in which the objective might be to maximize the minimum distance between the facilities?

12. What are the two levels of the decision process in locating a facility?

13. Discuss the differences between location problems in the private sector and location problems in the public sector.

14. Discuss the advantages in using the computer in layout design as opposed to using purely manual design.

15. Discuss the advantages of using the computer to score alternative layouts as done in the ALDEP package.

16. Modifying considerations and practical limitations are not considered in SLP until steps 7 and 8. Discuss any advantages and disadvantages to waiting that long before considering these factors.

17. How is facility layout similar to facility location?

18. Usually, facility layout discussion is limited to manufacturing plants. List three nontraditional manufacturing companies where facility layout could have a significant effect. Discuss briefly why you think layout analysis could help.

19. In the discussion on plant location, the point was made that some industries need to be located nearer to raw materials than customers, and vice versa. Give an example of each.

PROBLEMS

1. The American Red Cross of College Town, USA, has hired you to aid in locating a blood bank to serve the five existing hospitals in the city. They wish to locate the bank in order to minimize the total distance traveled by a truck that must visit each hospital a given number of times each week. The truck visits one and only one hospital per trip and a rectilinear distance measure is used. The map coordinates of the hospitals with the number of trips weekly to and from the blood bank are given below. Determine the best location(s) for the blood bank. Draw a graph showing this location(s).

Hospital	*X Coordinate*	*Y Coordinate*	*Number of Trips Weekly*
A	5	80	10
B	40	60	10
C	20	70	15
D	75	20	20
E	60	25	10

2. A wholesaler is considering locating a distribution center in Morgantown, West Virginia, for supplying space heaters. This distribution center is to serve stores in Richmond, Virginia; Philadelphia, Pennsylvania; Cleveland, Ohio; and Knoxville, Tennessee. The wholesaler already has one distribution center in Louisville, Kentucky. The necessary information is given in Figure P4.2. Calculate the total distribution cost for the best distribution pattern. (Set up as a transportation linear programming problem and obtain a solution by using least cost assignments discussed in this chapter. Assume that the new warehouse has been constructed and is available for use.) Total distribution cost per space heater:

From \ To	Richmond	Philadelphia	Cleveland	Knoxville
Louisville	$90	$90	$100	$70
Morgantown	$85	$70	$60	$130

Store:	Richmond	Philadephia	Cleveland	Knoxville
Demand:	10,000	5,000	5,000	10,000

Wholesale center:	Louisville	Morgantown
Capacity:	10,000	20,000

3. In Problem 2 the wholesaler finds that he can enlarge his operations in Louisville to accommodate the necessary 20,000 units at a total yearly cost of $150,000. This includes labor, depreciation, interest, taxes, and any cost other than the distribution cost. If he constructs the new warehouse in Morgantown, the total annual cost would be $200,000. (This is higher than enlarging operations at Louisville since a new building is needed, land purchase is necessary, and all new labor is required.) Should he enlarge the operations in Louisville or should he construct the new warehouse in Morgantown?

4. Design an activity relationship chart for the main office in your department and fill in the desired relationships.

5. The layout shown in Figure 4.27 scored 512 points based on the following values. Show how this score was obtained.

A = 64	O = 1
E = 16	U = 0
I = 4	X = −1,024

6. Given the activity relationship chart in Fig. P4.6, develop a string diagram for a three-bedroom home.

		1	2	3	4	5	6	7	8	9
1	Kitchen	—								
2	Bathroom 1 (Master)	U	—							
3	Bathroom 2	O	U	—						
4	Master Bedroom	U	A	U	—					
5	Bedroom 2	U	U	E	U	—				
6	Bedroom 3	U	U	E	U	U	—			
7	Garage	I	U	U	U	U	U	—		
8	Workshop	I	U	U	U	U	U	A	—	
9	Living Room	E	U	E	U	U	U	U	U	—

7. Extend the string diagram in Problem 6 to a full-fledged layout given neither practical nor modifying restrictions and the following area requirements:

Number	*Area*	*Requirement* (Ft^2)
1	Kitchen	150
2	Bathroom 1 (master)	60
3	Bathroom 2	40
4	Master bedroom	150
5	Bedroom 2	100
6	Bedroom 3	100
7	Garage	350
8	Workshop	100
9	Living room	150

8. Ace Manufacturing Company has decided to start a new product line that is to be marketed through its existing warehouses. One or more new manufacturing plants must be constructed to meet the anticipated demands at the warehouses. Three potential sites have been chosen. The necessary information is given in Figure p. 4.8. Formulate the problem in a nonlinear programming format.

To Warehouse \ From Potential Plant Site	1	2	3
1	$10.00	$8.00	$7.00
2	$7.00	$5.00	$6.00
3	$3.00	$5.00	$4.00
4	$5.00	$2.00	$3.00
Fixed Cost To Open Plant	100,000	70,000	65,000

Transportation Cost Per Unit

Existing warehouses:	1	2	3	4
Demand:	5,000	6,000	3,000	8,000

9. You have decided to construct a plant in city C to supplement the ones you already have in city A and city B. The three plants must serve five existing warehouses as shown in Figure P4.9. Given that the entries in the cost matrix are profits and that you want to determine the allocation to maximize profits, what is the best allocation using maximum profit assignment? (Use a heuristic that you have studied.)

		Warehouse					
	From \ To	V	W	X	Y	Z	Capacity
PLANT	A	10	5	6	3	8	10,000
	B	2	7	4	7	5	15,000
	C	5	5	7	8	9	8,000
	Demand	10,000	7,000	3,000	8,000	5,000	

10. You are considering the possibility of locating a plant in city E. Annually, you sell 100,000 items of a product that weighs 100 pounds. Estimated cost data for distribution are shown in Figure P4.10. The percentage figures are total market share and dollar figures are the shipping cost per ton-mile. Calculate the average distribution cost per ton-mile for the plant located in city E. What is the total distribution cost?

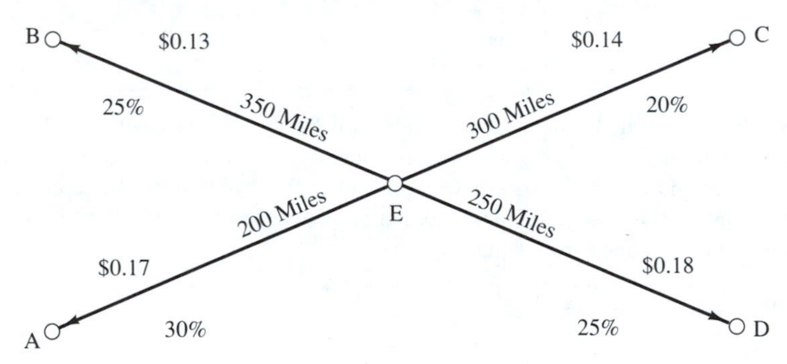

11. Develop a flow process chart for the processing of a book being returned to your library. Start when the book arrives at the librarian's desk and stop when it is back on the shelf. Make suggestions for layout changes. (*Hint:* The chart should follow the book—not the people doing the work.)

12. From the activity relationship chart developed in Problem 4 and the actual space currently occupied, relayout the office.

13. Using the string diagram presented in Figure 4.22, develop another layout of the dairy. Use the area requirements shown in Figure 4.24. Score the layout using the values shown in Problem 1 and the ALDEP scoring procedure. Compare your layout with that shown in Figure 4.27.

14. The Fly By Nite Manufacturing Company produces one product, gudfunutins. A gudfunutin is made up of four parts: A, B, C, D. One gudfunutin requires the assembly of three A's, two B's, one C, and one D. The processing sequence is shown below. If monthly production is 1,000 units, develop a from–to chart showing the total number of units moving between the departments per month.

Part	*Sequence (Department Numbers)*
A	1–2–5
B	2–5–2–3–4
C	1–3–2–3
D	4–3–4–1–2–5

Noncost Factor	*Degree*				*Maximum*
	0	*1*	*2*	*3*	
1. Nearness to hospital	0	70	140	210	280
2. Availability of power	0	15	30	45	60
3. Climate	0	50	100	150	200
4. Police protection from angry fans	0	20	40	60	80
5. Adequate night light	0	60	120	150	200
6. Hostility of fans (lack of)	0	50	100	150	200
7. Nearness to colleges for recuiting	0	5	10	15	20
8. Ability to win games (home crowd enthusiasm, etc.)	0	5	10	15	20

Location	*Noncost Factor*	*Degree*
A	1	3
	2	2
	3	3
	4	2
	5	2
	6	3
	7	2
	8	3
B	1	2
	2	3
	3	1
	4	0
	5	Maximum
	6	Maximum
	7	0
	8	1

15. The Wandering Flyers, a football team, has decided to move its operations to a new locality to encourage attendance at its home games. The team management has narrowed its choice to two sites where it believes attendance will be better. Those sites are locations A and B. Projections for attendance are equal at both sites as well as cost of land acquisition, stadium construction, and so forth. In fact, the Flyers find no difference between the two sites on the basis of cost. There are some noncost factors that the team management feels are important. These factors are shown below with relative point values. Formulate the problem as a decision under multiple objectives and choose a site.

16. Go home and make yourself a hamburger. Carefully construct an operations process chart of your "system." Be sure to include all steps and condiments. Should you change your storage locations if you were making nothing but hamburgers? If so, how?

CHAPTER 5

Material Handling, Distribution, and Routing

5.1. Introduction

This chapter discusses the management of *material movement* in an organization. Material movement occurs anytime a product or part thereof moves or is transported from one place to another. Material movement usually accounts for somewhere between 5 and 90% of total factory cost, with a probable average of 25%. Since little or no work is done on a part or product as it moves, movement in that sense is nonproductive; yet, on the average, a factory spends 25% of its dollars on this movement.

Material movement occurs throughout the entire product manufacturing cycle and both before and after it. Raw materials are usually moved from the natural state to some type of primary operation before finally being moved to the manufacturing plant. After production, the product is moved or distributed to the various users. After the product has completed its useful life, it is either discarded or scrapped. For discarding, one more movement occurs before the material is disposed; but in the case of scrappage, movement occurs back to the primary operation for reclamation, thus completing what might be called a material movement cycle. This can be represented as shown in Figure 5.1. Notice that movement occurs throughout the entire cycle. For example, not only does movement occur between phases, but it also occurs within a phase (e.g., material handling within the plant). This diagram

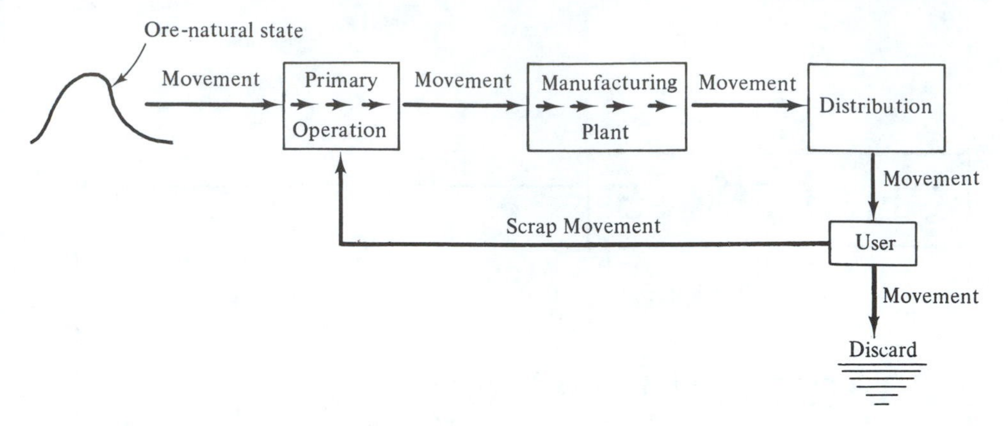

Fig. 5.1. Material movement cycle.

was simplified by showing distribution as one box when actually it could involve several boxes (e.g., wholesaler, retailer, warehouses, regional warehouses, etc.).

Movement is, of course, not restricted to the private sector of the economy. School buses carrying children to and from schools are an exceptionally important material movement; and so are street sweepers, mailmen, and bookmobiles. All these are examples of movement which require energy and therefore should be managed well.

5.2. Material Handling

In this section we present a short discussion on *material handling*. Although material handling is handling any material, any place, and any time, we will limit our discussion here to within-plant movement and leave distribution (outside-plant movement) to be covered in the next section. Before beginning, however, let us go to Corvallis, Oregon, to see how material handling affects Gadgets, Inc.[1]

Example 5.1

Being an astute manager who is very cost conscious, you have always been aware of where your money goes and have tried to cut expenses whenever possible. Not until you made a study of material movement, however, did you realize just how important this area was. You find that movement occurs in getting all your raw materials and purchased parts to the plant and within the plant to the appropriate areas. Within the plant, you might find that material movement accounts for approximately 25% of total factory cost and that little or no productive work is involved in movement. Finally, you might find that distribution costs

[1] In future chapters, we will visit Gadgets, Inc. frequently to see how what we are studying affects the plant. More details on the plant will be given as needed.

are very high and that substantial savings are available through good industrial engineering management of distribution.

After becoming cognizant of all the above, you would probably immediately initiate a study of material handling methods. With the subject matter presented in this chapter, you are able to conduct a detailed thorough study and come up with improved material handling effectiveness at less cost.

5.2.1 Equipment Concepts

The first problem encountered in material handling is the almost impossible task of understanding what types of equipment are available and what are the strengths and weaknesses of each. There are available more than 500 different types and varieties of material handling equipment, with each material handling system usually requiring several different types. This coupled with the rapidly changing technology makes it very difficult to keep up to date in material handling. Some of the basic types of material handling equipment are listed below.[2]

(1) *Conveyors*—Gravity or powered devices used for conveying material from one fixed point to another fixed point (although portable conveyors may be moved from time to time). They are particularly useful for moving homogeneous materials over a route that does not vary and when the material is supplied at a relatively constant rate. Figure 5.2 is an example of an extendible belt conveyor useful for loading and unloading trucks.

(2) *Industrial Trucks*—Vehicles (either manual or powered) used for transporting material over varying paths where usually the load is intermittent. They are especially useful when the nature of the load and the path prohibit the use of conveyors or other mass production oriented devices. Job shops, for example, find industrial trucks particularly useful. Figure 5.3 is a picture of a powered high-lift electric fork truck.

(3) *Cranes and Hoists*—Overhead lifting devices used for moving loads in a fixed area. They are particularly useful for moving intermittent loads when the size or shape varies between loads. They, of course, require some ceiling clearance and adequate support from the roof or building columns. Figure 5.4 is a picture of a coil lifter suspended from an overhead crane.

(4) *Containers and Racks*—Devices used to store and handle bulk materials. Containers are used to unitize quantities of material for storage or use. Racks are storage devices for better use of space (higher utilization of the cube). Figure 5.5 shows a sample storage rack application and Figure 5.6 is a picture of a plywood container being stacked for storage purposes.

(5) *Elevators and Lifts*—Devices used to vertically raise or lower material. Although fork trucks, conveyors, cranes, and hoists also lift, these devices lift

[2]Adapted from James M. Apple, *Lesson Guide Outline on Material Handling Education*, College Industry Committee on Material Handling Education, Pittsburgh, Pa., May 1975, p. 68.

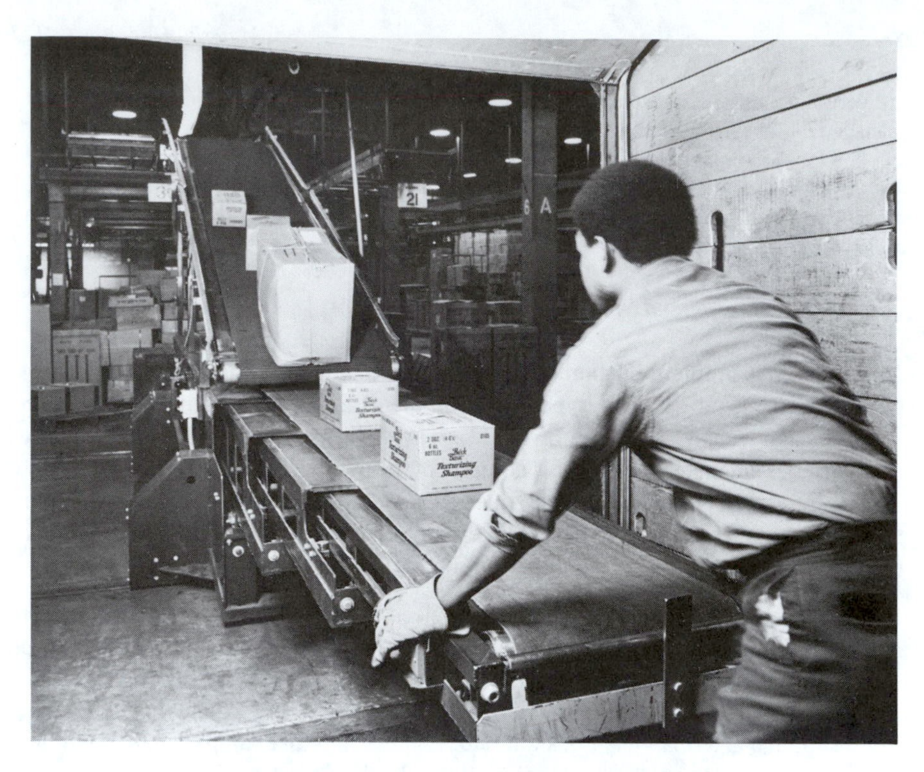

Fig. 5.2. Extendible belt conveyor.

Fig. 5.3. High-lift electric fork truck.

Fig. 5.4. Coil lifter suspended from overhead frame.

Fig. 5.5. Storage rack application.

Fig. 5.6. Stacked plywood pallets.

by applying force from underneath and are fixed in location. Figure 5.7 is a picture of a hydraulic scissor lift.

(6) *Automatic Guided Vehicles (AGVs)*—Driverless vehicles capable of moving along predetermined paths and performing certain prescribed duties. Usually, AGVs are relatively nonintelligent in that they can only go where they are

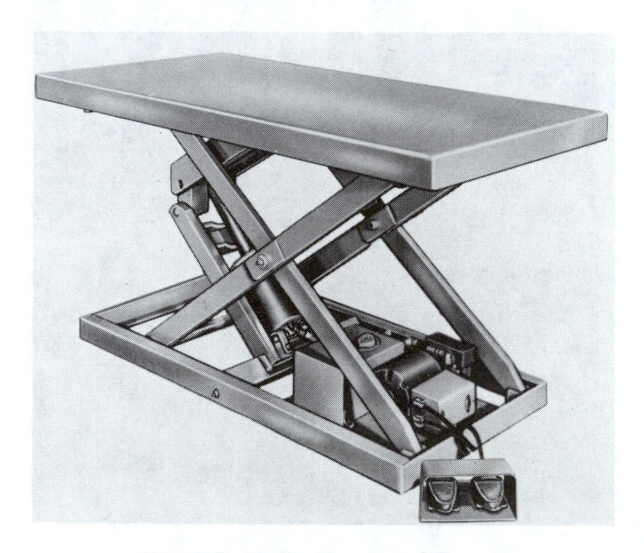

Fig. 5.7. Hydraulic scissor lift.

sent and perform a very limited activity, such as "stop at point A." Figure 5.8 is a picture of an AGV.

(7) *Automatic Storage and Retrieval System (ASRS)*—Storage rack systems where storage locations and inventory records are kept by the control system. When product is needed, the computer automatically finds the storage location and a driverless (usually, crane) system automatically retrieves the product. The reverse happens for products going to storage. These systems are very popular today and are useful in reducing labor costs and in achieving high-density storage. Figure 5.9 is a picture of an ASRS in a flooring manufacturing operation.

The industrial engineer must have a working knowledge of each of these basic types of material handling equipment. With this knowledge, he or she is able to make decisions on the equipment needed and can proceed to do detailed economic analysis.

5.2.2 Principles of Material Handling

Given that you now understand the basic types of material handling equipment, you must still design systems that effectively and efficiently utilize the equipment. Over the years certain principles of material handling that lead to efficient and effective systems have been observed and recorded in list form for use by those involved in material handling systems design.[3]

(1) *Planning Principle*—All material handling systems should be planned, not evolved.

(2) *Systems Principle*—The planning of the system should include as much as possible of the material movement cycle. For example, a vendor could ship material in a container that can be used throughout your production process and perhaps even used as the container for shipment to a customer.

(3) *Material Flow Principle*—Facilities layout and process engineering should assure that materials flow smoothly.

(4) *Simplification Principle*—Apply the concepts of work simplification to material handling systems design. For example, the principles of motion economy should be obeyed.

(5) *Gravity Principle*—Gravity should be used to move material whenever practical. Gravity feed bins and roller conveyors are examples.

(6) *Space Utilization Principle*—Make optimum use of space. High racks for use with narrow aisle stacking trucks make good total "cube" usage.

(7) *Unit Size Principle*—In general, the larger the accumulated load handled, the better. The use of pallets to handle large quantities of properly stacked cartons is an example. Unitization should occur as soon as possible and as long as

---

[3]Extracted, with permission, from James M. Apple, *Material Handling Systems Design* (New York: Ronald Press, 1972).

Fig. 5.8. Automatic guided vehicle.

Fig. 5.9. Automatic storage retrieval system.

possible in the cycle and stacking patterns should be developed for more effective unitization.

(8) *Mechanization and Automation Principle*—Always consider mechanization and automation in the design of the handling system. They may be impractical but they should be considered. Powered conveyors are an example of mechanization and automatic pallet stackers is an example of automation.

(9) *Equipment Selection Principle*—Be careful to consider all aspects of the material to be handled, the move to be made and the methods to be utilized, in selecting equipment. The purchase of fork trucks with a great deal of flexibility for use throughout the plant is an example.

(10) *Standardization Principle*—Standardize material handling methods and equipment. The adoption of a system of pallet stacking patterns and the use of standard-sized pallets are examples.

(11) *Adaptability Principle*—Design methods and equipment so they can perform a variety of tasks. Variable-speed conveyors are examples.

(12) *Dead Weight Principle*—Reduce the ratio of equipment or dead weight to payload. The use of plastic, aluminum, or other lightweight pallets is an example.

(13) *Utilization Principle*—Effectively utilize both manpower and equipment. The use of walkie talkies by fork truck drivers for rapid assignment to new jobs is an example.

(14) *Maintenance Principle*—Plan for preventive as well as emergency maintenance on material handling equipment. The use of a preventive maintenance program and the adequate stocking of replacement parts are examples.

(15) *Obsolescence Principle*—Make economic evaluations periodically to check on replacement of material handling equipment by identical models or newer types. Scheduled replacement policy based on sound engineering economy is an example.

(16) *Control Principle*—Use material handling equipment to aid production control, inventory control, quality control, etc. The use of conveyors for assembly lines to ensure a constant rate is an example.

(17) *Capacity Principle*—Use handling equipment to aid in fully utilizing production capacity. The use of outdoor storage whenever possible and the design of conveyor speeds for optimum feeding of parts are examples.

(18) *Performance Principle*—Measure and monitor the performance of your material handling system. For example, the calculation of a cost per move and the constant monitoring of this should be done.

(19) *Safety Principle*—Design methods and equipment for operator safety. The use of "cages" around and especially over the head of a fork truck operator is an example.

These 19 principles are vitally important to anyone involved in material handling systems design and each principle should be considered in every design endeavor. They are, however, qualitative in nature. Almost all industrial engineers prefer to deal with quantitative techniques. In fact, many quantitative techniques are applicable in the design of material handling systems, as we shall see in the next section.

5.2.3 Quantitative Techniques

In Section 4.3.1 we presented a technique called *transportation programming* that was used in the location of facilities. That same technique can be used in material handling systems design. As an example, a transportation problem can be formulated for the allocation of fork trucks, electric walkie trucks, pallets, and hand trucks to various sites throughout the plant where they are needed. Another technique very similar to transportation programming is called the *assignment technique*. This technique allocates scarce resources just as transportation programming does, but it can only handle one unit at a time.

For example, suppose that you have a fork truck located at each of five different sites throughout the plant, but these sites no longer have need for the trucks. Suppose, further, that there are five additional sites that need fork trucks. Since the distances between the sites vary substantially, you want to assign the fork trucks to the five new sites and minimize the total distance traveled. Let us assume that the data are as given in Figure 5.10.

Since your objective is to assign trucks from existing sites to sites that need trucks, the problem is similar to a transportation problem. Only one truck is available or needed at each site, however; therefore, the assignment problem is simpler than the transportation problem, and it is more easily solved. One alternative solution to the problem (*not optimal*) is given in Figure 5.11.

Notice that this solution has *one and only one* entry in each row and in each

		Locations Needing Trucks				
	From \ To	Shipping (A)	Inspection (B)	Maintenance (C)	Receiving (D)	Dock Storage (E)
Present Locations of Trucks	Tool Room (A)	10	5	7	3	2
	Press Shop (B)	9	4	8	6	5
	Paint Storage (C)	2	1	4	3	2
	Welding (D)	8	7	9	10	6
	Assembly (E)	3	2	5	8	1

Entries in matrix are distances in hundreds (00's) of feet.

Fig. 5.10. Assignment formulation.

	A	B	C	D	E
A				1	
B			1		
C					1
D		1			
E	1				

Assign Fork Truck From A to D
Assign Fork Truck From B to C
etc.

Fig. 5.11. Typical solution to assignment problem.

column. Going back to the original distance matrix, we can subtract a number from every element in a specific row or column without affecting the solution (the value of the solution will be changed by the amount of the number subtracted; but not the solution itself). For example, since the smallest number in the first row is 2, subtract 2 from every entry in that row. This yields

8 3 5 1 0

The smallest number in the second row is 4; subtracting 4 from every entry in that row and repeating for all other rows yields Figure 5.12. Now, if this can be done for rows, it can also be done for columns, as shown in Figure 5.13.

The reduced matrix B is of interest to us for we can now solve the assignment problem and be certain that the solution is a solution to the original one (by our observation that subtracting a number from every entry in any row or column does not affect the solution). Since there are no negative numbers in the reduced matrix B, an assignment that has zero total cost must be optimal. We can see by inspection that assigning the truck from the Tool Room (A) to Receiving (D), the truck from the Press Shop (B) to Inspection (B), the truck from Paint Storage (C) to Shipping (A), and so on, as shown in Figure 5.14, results in a zero cost solution.

This is the optimum solution to our problem, and its cost should be equal to the sum of the numbers subtracted (again by our earlier observation) or

$$\underbrace{2 + 4 + 1 + 6 + 1}_{\text{Rows}} + \underbrace{1 + 0 + 3 + 1 + 0}_{\text{Columns}} = 19$$

	A	B	C	D	E	Smallest number in row
A	10	5	7	3	2	2
B	9	4	8	6	5	4
C	2	1	4	3	2	1
D	8	7	9	10	6	6
E	3	2	5	8	1	1

Original matrix

	A	B	C	D	E
A	8	3	5	1	0
B	5	0	4	2	1
C	1	0	3	2	1
D	2	1	3	4	0
E	2	1	4	7	0

Reduced matrix A

Fig. 5.12. Row reduction of assignment matrix.

Reduced Matrix A

	A	B	C	D	E
A	8	3	5	1	0
B	5	0	4	2	1
C	1	0	3	2	1
D	2	1	3	4	0
E	2	1	4	7	0
	1	0	3	1	0

Smallest number in column

Reduced Matrix B

	A	B	C	D	E
A	7	3	2	0	0
B	4	0	1	1	1
C	0	0	0	1	1
D	1	1	0	3	0
E	1	1	1	6	0

Fig. 5.13. Column reduction of assignment matrix.

Going back to the original matrix, we find that solution is

3	+	4	+	2	+	9	+	1	= 19
A-D		B-B		C-A		D-C		E-E	

which agrees with the solution just obtained.

This procedure is actually the first phase of the *Hungarian method* which will be presented in detail in Chapter 14. More is necessary if a zero cost assignment cannot be found by inspection of the row and column reduced matrix. The first phase is repeated below:

(1) Subtract the smallest number in each row from every entry in that row.
(2) In the matrix resulting from step 1 subtract the smallest number in each column from every entry in that column.
(3) In the matrix resulting from step 2 a zero cost assignment (if it can be made) is optimal. If not, complete the Hungarian method.

These two techniques are, of course, not the only quantitative techniques that can be used in material handling systems design. There are many others, for example, queueing theory, integer programming, linear programming, simulation, and transshipment programming. Many of these will be discussed in the chapters on operations research (Chapters 14 through 17).

Cost matrix

	A	B	C	D	E
A	7	3	2	0	0
B	4	0	1	1	1
C	0	0	0	1	1
D	1	1	0	3	0
E	1	1	1	6	0

Solution

	A	B	C	D	E
A				1	
B		1			
C	1				
D			1		
E					1

Fig. 5.14. Zero cost solution to reduced assignment problem.

5.3. Distribution

Distribution covers all aspects of delivering the final product to the customers. Actually, it is a part of material handling since we defined material handling as handling material any time and any place. For ease of presentation, however, we have chosen to cover distribution separately. A revisit to Corvallis, Oregon, should help set the stage.

Example 5.2

As president of Gadgets, Inc., you are vitally concerned with the marketing activities of your plant. Marketing is a large field in itself covering many aspects of daily operations. One of the biggest areas, however, is in maintaining a good distribution function so that all customers can be assured of good service. This entails rapid delivery of your product and economical operations.

One of the first problems you encounter is that of warehouse location. You have a stated policy of being able to supply every customer within a maximum of 48 hours after receipt of order. Using the material presented in this chapter and Chapter 4, you locate your warehouses accordingly.

Another problem immediately confronts you. How should you deliver the product to the warehouses and from the warehouses to the customers? This necessitates a feasibility study for maintaining your own fleet of vehicles. You conduct this study examining such areas as number and types of vehicles, replacement policies for the vehicles, and maintenance systems for the vehicles. Using the material covered in this chapter and other chapters in this book, you are able to make these studies and corresponding decisions. You are then ready to move into the next area. (To be continued.)

5.3.1 Warehouse Location

In Chapter 4 we discussed several problems of facilities location. For Gadgets, Inc., there is now the problem of locating warehouses such that any customer can be serviced within 48 hours of receipt of order. Since we can correlate time and distance very easily, the objective is similar to minimizing the maximum distance. Since we are also interested in efficient operations, the cost of warehouses is important. Solving a problem like this by using strictly analytical techniques is impossible because the problem is so complex.

Many people turn to *simulation* to solve problems such as this. We shall cover simulation later but we can introduce it here. Simulation is using statistical data on location (of customers in this case) and demand to "try out" various numbers and locations of warehouses. Using the speed of digital computers, we can rapidly simulate many different solutions and management policies and monitor the results. From these results, intelligent decisions can be made.

5.3.2 Operations Management—Routing

Once the warehouses have been located, it is necessary to operate the logistics system. This includes maintaining adequate inventories of finished products and supplying warehouses and customers in an efficient and effective manner. Basic inventory models are developed in Chapter 7, many of which are applicable to this problem, but more advanced models are often required in multilevel systems such as the one we are now discussing. These advanced inventory models are beyond the scope of this text.

We can, however, discuss the delivery question in some depth. Let us assume that we have located the warehouses and have designed our inventory system. We are then concerned with the delivery of goods to warehouses and customers. Questions encountered here include such items as commercial vs. in-house fleet systems, size and composition of in-house fleets, and routing of vehicles to efficiently deliver goods. Except for the routing question, all of these can be handled by using the techniques of engineering economy for decision making among alternatives. We shall discuss routing in some depth.

Since Gadgets, Inc., is small, we can assume that one vehicle can supply all warehouses. This vehicle must start at the main plant, deliver super gadgets to each warehouse, and return to the main plant. There may be several trips involved, for example, visiting warehouses 1, 2, and 5 on the first route and warehouses 3, 4, and 6 on another, and so on, but for simplicity we can assume just one trip.

This is the famous traveling salesman problem which has received considerable research attention in industrial engineering and operations research. In this problem a salesman is located in one city and must visit several other cities. He must start in this one city, travel to all other cities, and return to the starting city while minimizing the total distance traveled.

There are many procedures for solving traveling salesman problems, some of which guarantee optimum solutions and some of which are heuristic. Almost all traveling salesman problems are solved heuristically and yield near optimum answers.

Perhaps the simplest of these heuristics is the ''closest unvisited city'' approach:

(1) Start at any city and visit the closest unvisited city. Continue in this manner until all cities have been visited.
(2) Repeat step 1 for all cities as a starting point. Choose the best of the resulting solutions.

As an example of this formulation and approach, consider the problem of supplying gadgets from Corvallis, Oregon. Let us suppose that there are six warehouses located in Stillwater, Oklahoma; College Park, Pennsylvania; Cleveland, Ohio; Bozeman, Montana; Tampa, Florida; and San Diego, California. Further, assume that one truck can supply all warehouses in one trip. The objective is to leave Corvallis and visit each of the six warehouses before returning to Corvallis while traveling a minimum distance. Since the truck driver has supplied us with the mileages (given in Figure 5.15), we are ready to start the solution procedure.

First, however, let's make some observations on this distance matrix. The truck driver has five friends scattered all around the country. Therefore, the mileages

From \ To	A	B	C	D	E	F	G
(A) Corvallis	∞	14	21	20	6	24	9
(B) Stillwater	14	∞	10	9	9	10	11
(C) College Park	21	10	∞	1	15	9	21
(D) Cleveland	20	9	1	∞	14	9	20
(E) Bozeman	6	9	15	14	∞	19	9
(F) Tampa	24	10	9	9	19	∞	21
(G) San Diego	9	11	21	20	9	21	∞

Entries Are in Hundreds (00's of Miles)

Fig. 5.15. Travel distances for Gadgets, Inc.

given above are only approximate and correspond to the routes he would take. The distance matrix has ∞'s down the main diagonal of the matrix, signifying that the truck cannot go from a city back to the same city. The distance matrix is symmetric; that is, the mileage from city i to city j is the same as the mileage from city j to city i. Finally, using our heuristic procedure, there will be seven routes (one for each starting point) and we will choose the best of these seven.

If we start with Corvallis, the closest city is Bozeman (E)—600 miles; the closest unvisited city to Bozeman is either Stillwater (B) or San Diego (G) (we will have to try both). Assuming that Stillwater is selected, the next closest is Cleveland (D)—900 miles. Continuing, we obtain the following:

City	A – E – B – D – C – F – G – A	
Mileage	6 9 9 1 9 21 9	= 64

Going back and trying San Diego in place of Stillwater, we obtain

City	A – E – G – B – D – C – F – A	
Mileage	6 9 11 9 1 9 24	= 69

Using city B as a starting point, we obtain

City	B – D – C – F – E – A – G – B	
Mileage	9 1 9 19 6 9 11	= 64

and (a tie occurred at the first step)

City	B – E – A – G – D – C – F – B	
Mileage	9 6 9 20 1 9 10	= 64

Notice that these are feasible tours even though it looks like we start at B. Actually, the tour (for the second one) would start at A and go A-G-D-C-F-B-E-A. (It makes no difference where one starts on a circle.)

Using the rest of the cities as starting points, we obtain

City	C – D – B – E – A – G – F – C	
Mileage	1 9 9 6 9 21 9	= 64

City	C – D – F – B – E – A – G – C	
Mileage	1 9 10 9 6 9 21	= 65

City	D – C – F – B – E – A – G – D	
Mileage	1 9 0 9 6 9 20	= 64
City	E – B – D – C – F – G – A – E	
Mileage	9 9 1 9 21 9 6	= 64
City	E – G – A – B – D – C – F – E	
Mileage	9 9 14 9 1 9 19	= 70
City	F – C – D – B – E – A – G – F	
Mileage	9 1 9 9 6 9 21	= 64
City	F – D – C – B – E – A – G – F	
Mileage	9 1 10 9 6 9 21	= 64
City	G – A – E – B – D – C – F – G	
Mileage	9 6 9 9 1 9 21	= 64
City	G – E – A – B – D – C – F – G	
Mileage	9 6 14 9 1 9 21	= 69

The best of these has a distance of 64 (6,400 miles) and is generated by several routes. This would quickly lead us to conclude that this distance must be optimal and is generated by several routes. Actually, the optimal is 60 which is generated by the route following:

> Corvallis (A)–San Diego (G)–Stillwater (B)–Tampa (F)–College Park (C)–Cleveland (D)–Bozeman (E)–Corvallis (A)

This route is shown in Figure 5.16, and of course, the route could be run in reverse order.

Two observations can be made here. First, the heuristic solution procedure did not give us the optimal answer, but it did give us some good starting solutions and helped us examine the problem. As always, the analytical technique is only an aid to the decision process, not a substitute for it.

Second, almost all traveling salesman routing problems can be solved manually and will yield very good solutions if the stops are shown on a map or some physical display. Figure 5.16 dramatically demonstrates that the solution is obvious. Of course, since in almost all cases the optimum solution is not so obvious, analytical procedures could be a great help. Nevertheless, all routing problems should be displayed on a map so that solutions can be seen.

Now, all this works very well for a traveling salesman problem, but what happens when more than one truck is required? That is, what if two or more trucks are available to make the circuit? First, if two or more trucks are available but one could handle the entire load (capacity is no problem), then the formulation is called a *multiple traveling salesmen problem*. There are solution procedures designed to handle multiple traveling salesmen problems, but like the traveling salesman problem, there are no efficient ways of obtaining the exact solution.

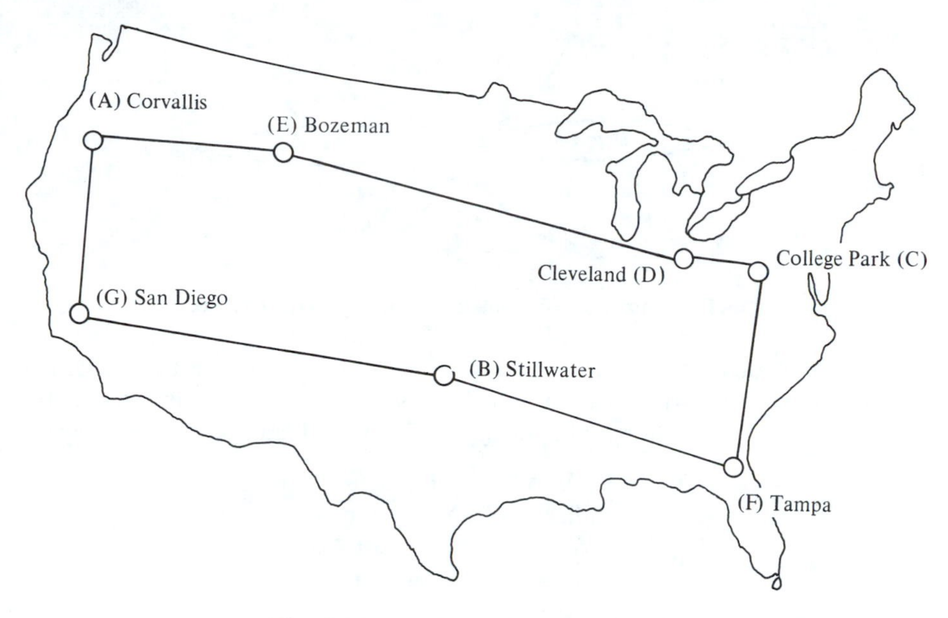

Fig. 5.16. Traveling salesman solution.

Usually, however, if there are two trucks, it means that one truck cannot handle the entire load; therefore, two trucks are required even if the solution with two trucks is more costly than that with one truck. This is also true for problems requiring more than two trucks.

This problem is called a *transportation routing problem* and is very similar to a multiple traveling salesmen problem except that there are capacity constraints on the vehicles. There are many solution procedures available for this problem, but, as with the traveling salesman problems, there are no efficient and effective techniques available today. In fact, very few transportation routing problems can be solved optimally because of the computational time and storage requirements on computers. Therefore, almost all problems are solved by using a heuristic procedure and there are many procedures available. One of the oldest, probably the most used, and perhaps conceptually the simplest is that developed by Clark and Wright.[4]

This procedure starts with the unrealistic assumption that each of the N stops or cities is serviced by a separate vehicle that starts from home base or the depot, goes to the stop and performs the service (epg., delivers the goods), and returns to home base. Figure 5.17 depicts this situation. (It is important to realize that even though this assumption is made at the beginning, few, if any, final solutions will have such a situation.)

The next step in the Clark–Wright procedure is to calculate the savings

[4]G. Clark and J. Wright, ''Scheduling of Vehicles from a Central Depot to a Number of Delivery Points,'' *Operations Research*, 12 (1964), 568.

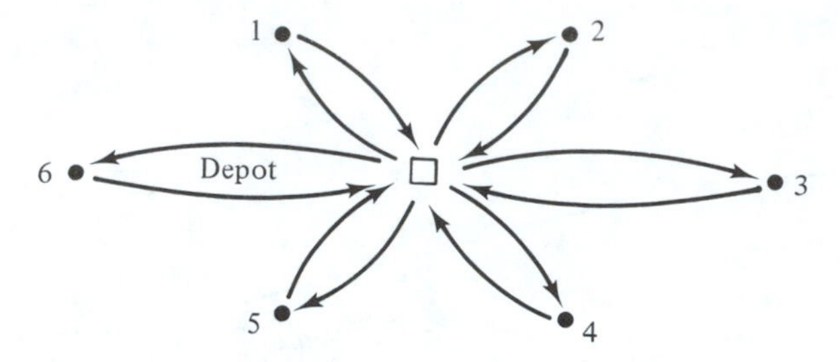

Fig. 5.17. Initial route formulation for Clark–Wright procedure.

available by combining two stops and forming one tour instead of two tours. For symmetric problems (i.e., the distance from stop i to stop j is the same as the distance from stop j to stop i), that savings is S_{ij} for combining stops i and j, where

$$S_{ij} = C_{oi} + C_{oj} - C_{ij}$$
$$C_{oi} = \text{distance from depot to stop } i$$
$$C_{oj} = \text{distance from depot to stop } j$$
$$C_{ij} = \text{distance from stop } i \text{ to stop } j$$

This is demonstrated in Figure 5.18.

The Clark–Wright procedure then orders or ranks the savings in a list of decreasing magnitude such that the first on the list is the combination that yields the greatest saving, the second is the combination that yields the second greatest saving, and so forth on down the list. The procedure starts with the first combination on the list and makes the two stops into one route (if the constraints allow that combination) and continues down the list until the solution is complete.

As an example of this situation, consider your problem of Gadgets, Inc., as shown in Figure 5.16. This time, however, let us assume that there is enough demand on the warehouses so that at least two trucks are required. That demand is given in Table 5.1. Assume that each truck can carry 25,000 gadgets. Since total demand is for 42,000 gadgets, at least two trucks (or two visits by the same truck) are required.

The first step is to calculate the savings and rank them. That is done below. Notice that in this formulation the depot (Corvallis) has been removed so that there are only six stops.

From \ To	B	C	D	E	F	G
B	∞	10	9	9	10	11
C	10	∞	1	15	9	21
D	9	1	∞	14	9	20
E	9	15	14	∞	19	9
F	10	9	9	19	∞	21
G	11	21	20	9	21	∞

From \ To	B	C	D	E	F	G
A	14	21	20	6	24	9

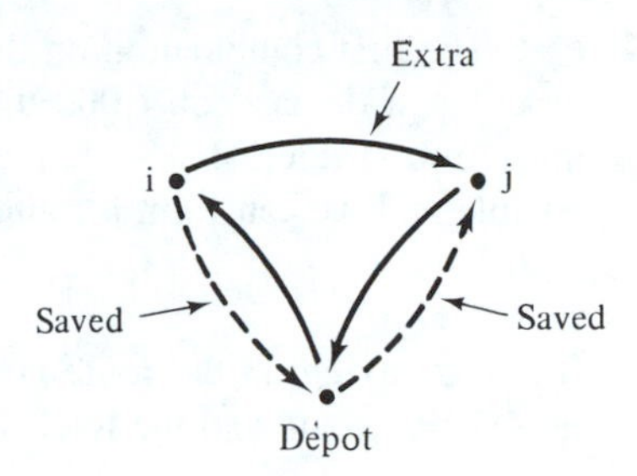

Fig. 5.18. Saving for combining stops *i* and *j*.

The saving for combining B and C is

$$AB + AC - BC = 14 + 21 - 10 = 25$$

For B and D, the saving is

$$AB + AD - BD = 14 + 20 - 9 = 25$$

The results for all possible combinations are as follows:

Stops	*Saving*	*Stops*	*Saving*	*Stops*	*Saving*
BC	25	CD	40	DF	35
BD	25	CE	12	DG	9
BE	11	CF	36	EF	11
BF	28	CG	9	EG	6
BG	12	DE	12	FG	12

Reranking these yields

Stops	*Saving*	*Stops*	*Saving*	*Stops*	*Saving*
1 CD	40	6 BD	25	11 BE	11
2 CF	36	7 BG	12	12 EF	11
3 DF	35	8 CE	12	13 CG	9
4 BF	28	9 DE	12	14 DG	9
5 BC	25	10 FG	12	15 EG	6

Table 5.1. DEMAND FOR GADGETS

	Stop	*Demand (gadgets)*
G	San Diego	10,000
B	Stillwater	5,000
F	Tampa	6,000
C	College Park	7,000
D	Cleveland	10,000
E	Bozeman	4,000

The first combination on the list is CD with a saving of 40. Stop C (College Park) has a demand of 7,000 units; stop D (Cleveland) has a demand of 10,000 units (from Table 5.1). Since the truck can hold 25,000 units, this combination is feasible and we can form a route consisting of

Depot–C–D–depot or depot–D–C–depot

The next saving on the route is for stops C–F. This would add another stop F to the existing route and the total demand would be

$$17{,}000 + 6{,}000 = 23{,}000 \text{ units}$$

Since the truck can handle this load, we now have one route consisting of

Depot–F–C–D–depot or depot–D–C–F–depot

The next saving on the list is for stops D and F. Since both D and F are already on the same route, the link is infeasible.

The next saving is for stops B and F. This would add another stop to the existing route and would create a total demand of

$$23{,}000 + 5{,}000 = 28{,}000 \text{ units}$$

Since the truck can handle only 25,000 units, this combination is infeasible.

The next combination on the list is for stops B and C, but since stop C is in the *middle* of the existing route, this link is infeasible. The next combination is also infeasible because it would combine stops B and D and would create a total demand of 28,000 units.

Next, stops B and G can be combined for a saving of 1,200 miles. Since the total load is only 15,000 units, we can create a new route that yields the following two routes:

Depot–F–C–D–depot or vice versa

and

Depot–B–G–depot or vice versa

The next two combinations (C and E; D and E) are both infeasible. The third is for stops F and G. Since this would combine the above two routes and would yield a total demand exceeding the truck capacity, this combination is also infeasible.

Then, stops B and E can be combined for a saving of 1,100 miles. This would add a third stop to the second route and would create a total demand of

$$15{,}000 + 4{,}000 = 19{,}000 \text{ units}$$

which is feasible.

Since this connects all stops (no combination of two routes is feasible), we can stop with the two routes being

Depot–F–C–D–depot or vice versa

and

Depot–E–B–G–depot or vice versa

The total distance traveled is

$$\underbrace{24 + 9 + 1 + 20}_{\text{Route 1}} + \underbrace{6 + 9 + 11 + 9}_{\text{Route 2}} = 89 \quad \text{or} \quad \mathbf{8{,}900\ miles}$$

The associated routes are shown in Figure 5.19.

Naturally, more distance is traveled than in the case of the traveling salesman problem because now there are two trucks. Since two trucks are required, the solution shown in Figure 5.19 is about the best that can be done, but the Clark–Wright procedure does not guarantee that the solution is optimal.

5.3.3 Routing in the Public Sector

Recently, industrial engineers have begun to solve more problems in public-sector activities. School bus routing is one area that has and still is receiving a great deal of attention. The average school bus costs well in excess of $2.00 per mile to run exclusive of labor cost. Therefore, any savings in distance traveled would be helpful.

The same techniques used for private sector routing of vehicles can also be used for public sector routing. Particularly, the Clark–Wright approach has been used for school bus routing.

Another approach has been developed for use for smaller or more rural school districts that may not have access to computers.[5] The procedure works this way:

(1) Obtain a large map of the school district.
(2) Plot stops and school locations with straight pins.
(3) Starting at any convenient spot, make a sweep about the schools so that the area around an individual group of schools can be divided into wedges, each of which represents one bus route (Figure 5.20). "Close" each wedge as soon as enough students to fill one bus have been identified.
(4) In each wedge, start with the bus stop farthest from the school and determine whether transferring that stop to a neighboring wedge can yield any saving in distance. If, for example, stop 4 is in wedge E, route the buses in wedge D or A so that they pick up at that stop. Calculate the distance saved by placing stop 4 in D or A and include the stop in whichever wedge gives minimum distance. During this "interchanging" keep a close count on the number of students involved. Although the buses in a wedge can exceed capacity during interchanging, make sure that numbers are leveled off before drawing final routes.

---

[5] J. A. Robbins and W. C. Turner, "No Computer? Try Wedging Bus Routes," *Nation's Schools and Colleges*, May 1975, pp. 44–45.

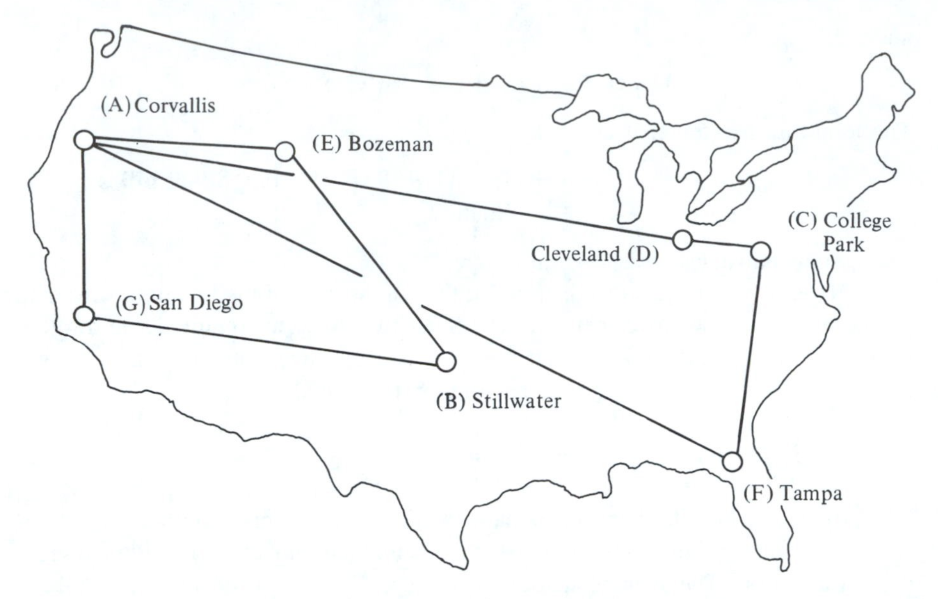

Fig. 5.19. Transportation routing solution.

(5) Once a stop has been interchanged, go through the procedure again, starting with the farthest stop left in a particular wedge. It may now be beneficial to interchange stops that were not feasible to transfer before—because interchanging stops in any two wedges may have a domino effect on the entire system.

(6) When it's no longer advantageous to interchange, begin with the stop farthest from the school in each wedge and route the buses into the schools (Figure 5.21).

School buses in a sample rural district with two elementary schools and one high school were rerouted by this method. Before the study was made 14 buses traveling 445 miles per day served the schools.

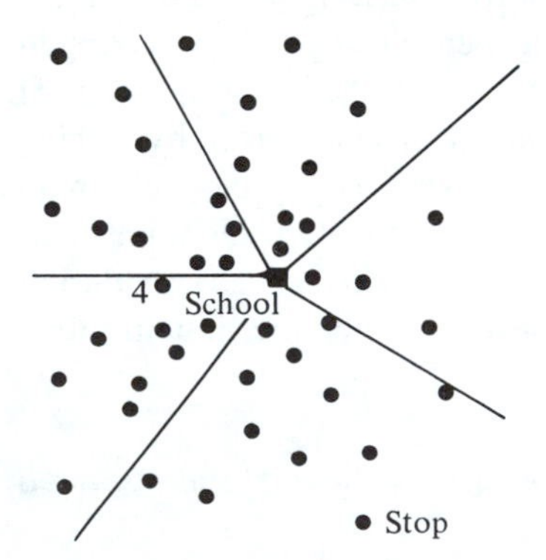

Fig. 5.20. Wedging for school bus routing.

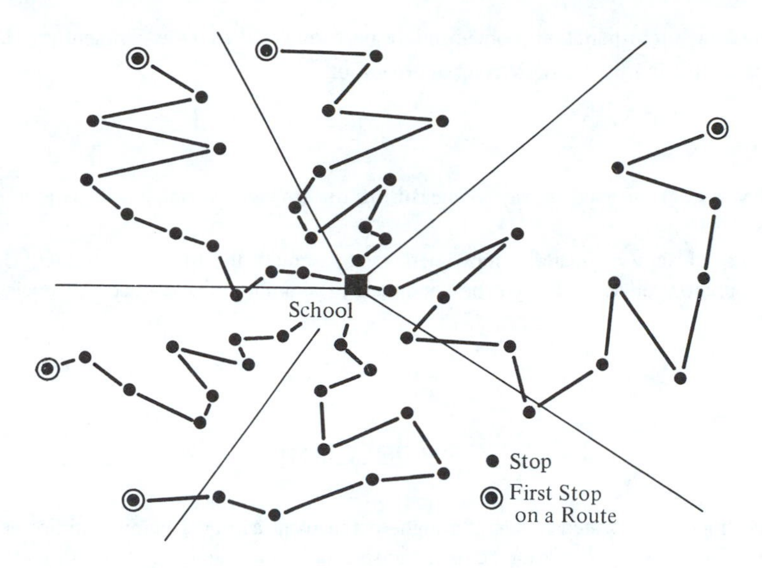

Fig. 5.21. Routing with wedges.

When the simplified procedure described was used, it was determined that 14 buses were indeed necessary, but that savings could be made in distance traveled. The new routing resulted in a total daily distance of 377 miles—a saving of 15.3%, or 68 miles per day. Maximum mileage on any single route did not change substantially, and students did not have to ride buses any longer than they did before. In another rural district the same procedure saved approximately 10% of the total mileage formerly used.

This area of routing in the public sector is a problem of growing concern because of the increased emphasis on productivity and energy conservation. It is attracting a great deal of attention in many industrial engineering programs.

DISCUSSION QUESTIONS

1. What is material handling? Discuss its importance in industry.
2. Name the classifications or types of material handling equipment discussed in this chapter and give an example of each.
3. List the 19 principles of material handling and give an example of each.
4. Discuss why you think "minimizing material handling cost" is often chosen as the criterion for plant layout studies.
5. Describe the traveling salesman formulation and give two examples of when it is applicable.
6. Discuss why you think manual procedures (using maps) often come up with as good if not better solutions than computerized procedures.

7. State how a transportation problem might be formulated as an assignment problem.
8. Instead of using the Clark–Wright approach of

$$S_{ij} = C_{oi} + C_{oj} - C_{ij}$$

can you think of another way of measuring the saving associated with combining two links?

9. Figure 5.1 shows a material movement cycle through the life of a product. Using an appropriate product that might be contained in aluminum cans, trace this cycle for the can.

PROBLEMS

1. Fork trucks are located at sites throughout the plant as noted below and are needed at other sites as shown. Formulate as a transportation programming problem and obtain a solution by using the least cost method (see Figure P5.1).

Site where fork trucks are available:

Site	A	B	C	D
Number Available	3	1	2	1

Sites where fork trucks are needed:

Site	X	Y	Z
Number Needed	4	2	1

Distance Matrix (Entries in $)

From \ To	X	Y	Z
A	5	4	2
B	2	1	2
C	5	4	5
D	3	4	6

Fig. P5.1 Distance matrix (entries in $).

2. Formulate Problem 1 as an assignment problem. Try to obtain a solution and compare that to the solution in Problem 1. If you cannot solve, stop.
3. A fork truck with a particular type grab is currently located at the tool room, and it must visit the machine shop, paint room, raw materials storage, and finished goods storage. Given the distances in Figure P5.3, formulate as a traveling salesman problem and determine a solution. How far does the fork lift travel? It must return to the tool room.

From \ To	A	B	C	D	E
(A) Tool Room	∞	4	10	8	7
(B) Machine Shop	4	∞	12	7	6
(C) Paint Room	10	12	∞	5	4
(D) Raw Materials Storage	8	7	5	∞	3
(E) Finished Goods Storage	7	6	4	3	∞

Fig. P5.3

4. Ace Company has a central warehouse in city A. Each morning, one or more trucks starts at the warehouse and visits each of 10 stores located in the state (see Figures P5.4a and P5.4b). Given the information below, determine the number of trucks required and a good route(s). Use the Clark–Wright method. How far do the truck(s) travel?

From \ To	1	2	3	4	5	6	7	8	9	10	
A	5	10	15	8	6	4	20	13	10	7	Distance in Miles

Fig. P5.4a

From \ To	1	2	3	4	5	6	7	8	9	10
1	∞	10	9	7	6	8	5	9	10	7
2	10	∞	2	3	8	5	4	7	9	7
3	9	2	∞	15	10	9	7	3	8	10
4	7	3	15	∞	2	15	4	11	9	8
5	6	8	10	2	∞	5	6	8	7	2
6	8	5	9	15	5	∞	4	5	7	6
7	5	4	7	4	6	4	∞	1	4	3
8	9	7	3	11	8	5	1	∞	2	5
9	10	9	8	9	7	7	4	2	∞	10
10	7	7	10	8	2	6	3	5	10	∞

Distance in Miles

Fig. P5.4b

Site number	1	2	3	4	5	6	7	8	9	10
Demand (units)	1,000	1,500	1,200	2,000	500	400	1,700	1,100	900	1,800

Each truck can hold 6,000 units.

5. Company A attempts to maintain good inventories on its steel pallets because the pallets are so expensive. One way they do this is by knowing where each pallet is at a certain time. A typical problem they might face is given a certain quantity of pallets at each of five locations as shown below and a certain quantity needed at each of six locations as shown below. What is the best method of moving from the five sources to the six destinations where the pallets are needed? The cost *per pallet* for each move is shown in Figure P5.5.

	Source					
	A	*B*	*C*	*D*	*E*	
Pallets available	100	150	50	75	20	
Destinations	1	2	3	4	5	6
Pallets needed	100	75	50	50	40	80

Cost per Pallet

From \ To	1	2	3	4	5	6
A	5	4	6	7	3	7
B	2	1	3	2	2	4
C	10	15	12	11	9	8
D	2	5	5	4	6	6
E	7	3	1	8	9	7

Fig. P5.5

6. Rework Problem 5 if destination 6 needed 100 pallets.

7. Rework Problem 4 if there are four trucks available as shown below.

	Truck			
	1	*2*	*3*	*4*
Capacity (units)	2,000	10,000	7,000	5,000
Cost per mile	$0.10	$0.20	$0.15	$0.12

8. A company has a warehouse at site X that must serve sites A, B, C, and D: The distances are shown below. Solve using Clark–Wright. There is no capacity problem, so distance traveled is the sole objective.

	A	*B*	*C*	*D*
A	∞	9	8	6
B	9	∞	4	3
C	8	4	∞	5
D	6	3	5	∞

	A	*B*	*C*	*D*
X	5	10	2	3

9. Take your local paper and find the location of five garage sales. Plot those on a map. Formulate and solve as a traveling salesman problem. Use a closest unvisited site heuristic.

CHAPTER 6

Work Design and Organizational Performance—Work Measurement[1]

6.1. Introduction

"Managers at all levels and in all organizations are striving to achieve results through people."[2] To do this all sorts of control systems have been designed to measure and enhance organizational performance. Unfortunately, many of the popular control systems monitor only a small part of total system performance. Recently, however, great strides have been made toward the design of effective and efficient control systems. Industrial engineers are leading the way.

In the first edition, this chapter focused on work measurement and design at the workplace. That was an accurate portrayal of industrial engineering involvement at the time but was microscopic or myopic in that it did not measure organizational performance. It did measure individual efficiency. In the second edition we modernized that a bit by talking more about true "productivity," as explained below. Thus, the area is becoming more system or total system performance oriented.

Now, the third edition recognizes that this trend is continuing and perhaps

---

[1]The authors are deeply indebted to Dr. D. Scott Sink and Ms. Leva K. Swim for their guidance and help with this chapter.

[2]D. Scott Sink, "Organizational System Performance: Is Productivity a Critical Component?" *Proceedings* 1983 *Annual Industrial Engineering Conference*, Louisville, Ky., May 1983.

accelerating. In fact, many industrial engineering curricula are dropping the more traditional areas of workplace measurement and improvement. We feel, however, that it is important to keep this in perspective. Thus, this chapter will briefly cover traditional areas and move on to the more company-wide techniques.

Perhaps Figure 6.1 does the best job of placing all this in perspective. Organizational productivity is simply a measurement of an organization's ability to turn inputs into outputs. Equation (6.1) states this:

$$\text{productivity} = \frac{\text{output}}{\text{input}} \tag{6.1}$$

Outputs and inputs can be partial or composite. For example, units per person-hour is a measure of labor productivity only and is, therefore, partial. Assuming that we desire quality outputs which meet the consumer demand (effectiveness), we can see that productivity is affected by efficiency, effectiveness, and quality, as shown in Figure 6.1. Productivity, together with innovation and quality of working life, determine the total organizational performance, which is usually measured by profitability in the free-enterprise system.

Now, where does work design and measurement (timing of individual operations) fit? Referring to Figure 6.1, we see that *work measurement is the standard by which we measure efficiency. Similarly, work design helps determine efficiency.* Work measurement and design can be distinguished from total organizational performance through considerations of *focus* and *scope*.

Total organizational system performance has a much broader *focus* in that it addresses more performance criteria than just efficiency (i.e., quality, quality of working life, effectiveness, innovation, profitability, productivity). Work measurement/design in the workplace has a narrow focus in that it addresses efficiency, primarily. In regard to *scope*, total organizational system performance is concerned with the collection of individuals organized to accomplish a purpose. Work measure-

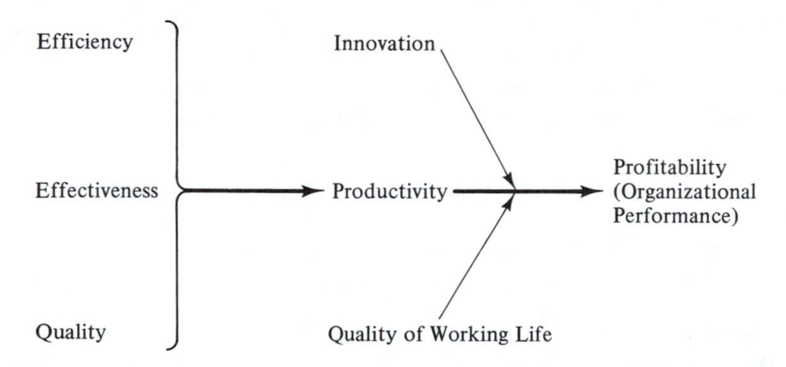

Fig. 6.1. Total organization performance. [Adapted from D.S. Sink, *Productivity Management: Planning Measurement, Evaluation, Control, and Improvement* (New York: John Wiley & Sons, Inc., 1984)].

ment/design concerns itself with the individual person and the person's interaction with his or her environment.

This chapter examines the measurement of organizational performance and the design of individual workstations. It does that by first discussing work simplification (probably the birthplace of industrial engineering) and then work measurement. Then the chapter discusses organizational performance with an emphasis on productivity measurement.

Some of you may never do work methods simplification or work measurement, as many industrial engineers are branching out into different areas. However, you should understand that industrial engineering started here and that more I.E.s are employed in this area today than in any other. It is sometimes very helpful to know your discipline's birthplace.

The following example is presented to demonstrate the need for the topics of this chapter.

Example 6.1

You have been contacted by the president of Gadgets, Inc., in Corvallis, Oregon, and asked to design and lay out the manufacturing processes for production of their new "super-gadget." Furthermore, he wants you to stay on and manage the plant once it is completed.

Since Gadgets, Inc., has done a great deal of research on processing methods, you find that there is little work needed there. Your goal is to produce 10,000 units per year. You must gear up to produce that, *but* to be able to determine how much equipment is necessary, you must know how long each operation takes. Since Gadgets, Inc., does not have that kind of information, you hire a time study analyst to do work measurement. Furthermore, since the equipment is expensive, you want the most efficient methods possible. The analyst must also design the most efficient work procedures for the workers to follow.

With the results of the time study analyst's work, you have the information necessary for determining manpower needs. Again, you find you must know how many super-gadgets each person can produce. You go back to the previously taken time studies to determine your labor needs.

Of course, the people must be trained. Who is going to train them, and what methods should they use? You decide to let your foremen do the training, *and* you find that the time study analyst specified the proper method for each job. The training is then simplified.

Because you are human relations oriented, you want to pay your people on the basis of their work quantity and quality. One way of doing this is to have an *incentive system*[3] in which the people are paid according to how much they do. The previously taken time studies provide you with the time information you need for this incentive system.

Next, it is necessary to start production (construction and layout are complete). One of the first things you must do is schedule the orders for super-gadgets (each order has a unique requirement or design change) throughout the shop. To do this, you must figure how long a production order takes for each operation and the total time necessary in the plant. Since your time studies provide you with this information, you can do the scheduling. Production now begins.

---

[3]Incentive systems are discussed in Chapter 9.

As you sit back in your chair, you reflect, "How can I know if the plant is performing as it should?" Once again, the time studies give you a bench mark against which you can compare performances since you know how long the operations should take.

Finally, you need to decide how much to charge for super-gadgets. There are many considerations, for example, the market supply, but one consideration is the cost of production. Since the plant is just starting, you do not know the cost of production. The time studies tell you what the labor cost *should* be. You then add on material and overhead cost and you are ready.

Or, at least you thought you were. The president recently attended a productivity improvement seminar; so he tells you to design a productivity measurement model for the company. Using the material of this chapter, you find you can, indeed, measure productivity. In fact, you feel the need to inform the president that he also needs to worry about total organizational performance (hopefully, profitability) and that productivity is only a part of total performance.

Although the above example is grossly oversimplified, it does point out *some* of the uses of work simplification and measurement. Those uses were in:

(1) Determining equipment requirements.
(2) Determining labor requirements.
(3) Designing training methods.
(4) Designing scheduling procedures.
(5) Designing incentive systems.
(6) Gauging work performance.
(7) Estimating cost.

3

This is not an exhaustive list by any means, but it should be sufficient to demonstrate the importance of work simplification and measurement.

We have demonstrated that the correct design of work methods and timing of these methods is very important to an organization, but how can these be done? For obvious reasons, work simplification should be performed on a task before that task is timed. Therefore, it is logical to examine some procedures for work simplification first.

6.2. Methods Improvement

Over the evolutionary sequence of industrial engineering, work methods improvement has been called many things, including work simplification, methods improvement, and work design. Whatever it is called (we chose the term *work methods improvement*) a definition might be: *A systematic approach to finding easier and better ways to accomplish a task.*

The basic objective is to avoid waste of any kind (time, human effort, materials, capital, etc.) or, as often quoted

Work smart, not hard.[4]

You have been making attempts at work methods improvement all your life whenever you try to find better ways to cut grass, wash cars, travel to work, and so on, but have you used a systematic approach? In some cases, probably yes, but in others, no. To do work methods improvement correctly, a systematic approach is necessary. We will discuss some systematic approaches in this section. A five-step procedure might be:

(1) *Selecting a Job*—One must select a job to study. This is often done because the job is the most expensive or obviously in need of such a study.
(2) *Getting and Recording the Facts*—The present method of doing the job must be recorded along with other facts such as the frequency of the job.
(3) *Questioning Every Detail*—The present method must be questioned. Some elements may be deleted; most can be changed.
(4) *Developing and Testing a Better Method*—From step 3, better efforts may be obvious. These ideas should be developed and tried.
(5) *Installing and Maintaining Improvements*—For the study to be worthwhile an improved procedure must be installed and maintained.

Charts are a convenient way of applying parts of the foregoing procedure, especially in steps 2 and 3. The operations process chart already covered is one. Some of the other charts will be discussed next in their approximate order of application.

6.2.1 Flow Process Charts

A flow process chart is a chart of all the activities involved in a process. It is similar to an operations process chart, except that more detail is shown by including transportations and delays as well as operations, inspections, and storages. The symbols used are explained in Chapter 4. As in operations process charts, the objective is to list every detail and to examine each detail closely. The following questions should be asked about each activity:

(1) Is this activity necessary, or can it be eliminated?
(2) Can this activity be combined with another or others?
(3) Is this the proper sequence of activities, or should the sequence be changed?
(4) Can this activity be improved?
(5) Is this the proper person to be doing this activity?

The construction is similar to that of an operations process chart in that flow is from the top to the bottom. Only one item can be charted per form, however,

---

[4]Stephan Konz, *Work Design: Industrial Ergonomics* (Columbus, Ohio: Grid Publishing, Inc., 1983), p. 21.

and that item can be either a person or an object. A blank flow process chart is shown in Figure 6.2.[5] An example of its use is demonstrated below.

Example 6.2

Mr. Cafe, a school food cafeteria supervisor, has asked you to prepare a flow process chart for the preparation of spaghetti. You have decided to make a chart of the noodles going through the cafeteria and see how the process can be improved.

The noodles start on the storage shelf. They are picked up and carried to the cooker, which is a large vat in which the noodles are cooked. Next they are placed in a large bowl, which is carried to a sink where they are washed. Then the bowl is carried to a work area where the noodles are placed in six serving pans. The meat and sauce (a separate flow process chart covers their production) are added, and the pans are carried to a heating unit where they are kept warm unitl ready for serving.

You prepare a flow process chart (shown in Figure 6.3). After consultation with Mr. Cafe and the operators and examining the chart carefully, you decide that by adding a water tap adjacent to the cooker (it already has a drain) you could wash the noodles right in the cooker, add the meat and sauce directly, and keep them warm in the cooker. This saves several transportations. The revised procedure is shown in Figure 6.4.

The advantages of the revised procedure over the existing one are obvious and numerous. No heavy lifting is required, only three transportations are required —a saving of four (57%)—and the same number of operations and storages are required.

In the existing procedure, steps 4 and 8 are called *transportations*, but they could have been called *operations*. Either term would have worked very well since the term is not important *in cases like this*. Many times an activity can be called several things; usually all are shaded on the chart, *or* only the most predominant ones are shown, as was done here.

6.2.2 Left-Hand–Right-Hand Charts

Left-hand–right-hand charts are useful in analyzing the work performed by one person at one specific workstation. As the name implies, the chart follows the motion of the left and right hands of one operator. Consequently, it is applicable only when one operator is performing repetitive work—usually at one station. An example is given below.

---

[5]This is a simplified form in that much more information usually is presented. It is slightly different from the form used in Chapter 4; but either form or another can be used, depending on the preference of the user.

Process	Man ________ or Material ________
Begins	Ends
Charter	Date
Present ________ Proposed ______	

Steps	Symbol	Notes
1.		
2.		
3.		
4.		
5.		
⋮		

Fig. 6.2. Flow process chart.

Process: Prepare Spaghetti	Man ___ or Material _X_ Noodles
Begins: In Storage	Ends: Ready to Serve
Charter: Ind. Eng.	Date: 7/24
Present _X_ Proposed ______	

Steps	Symbol	Notes
1. Noodles on Shelf	▽	
2. Carried to Cooker	⇨	
3. Cooked	○	
4. Placed in Bowl	⇨	Wash in Cooker?
5. Carried to Sink	⇨	Very Heavy
6. Washed	○	
7. Bowl Carried to Work Area	⇨	Very Heavy
8. Placed in Serving Pans	⇨	Six Pans–Full
9. Meat and Sauce Added to Pans	○	Repeated Six Times
10. Pans Carried to Heater	⇨	Repeated Six Times
11. Warmed	▽	
12. Carried to Serving Line	⇨	Repeated Six Times

Fig. 6.3. Flow process chart—present method.

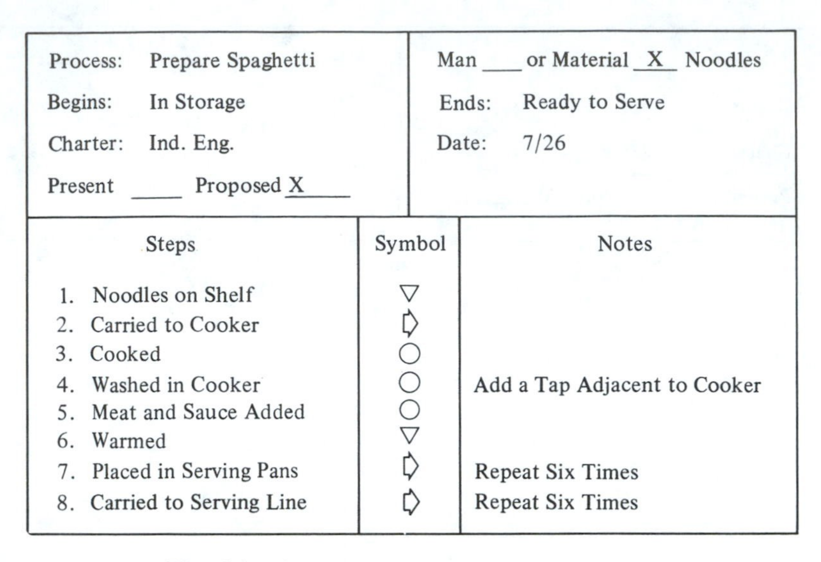

Process: Prepare Spaghetti	Man ___ or Material _X_ Noodles
Begins: In Storage	Ends: Ready to Serve
Charter: Ind. Eng.	Date: 7/26
Present ___ Proposed _X_	

Steps	Symbol	Notes
1. Noodles on Shelf	▽	
2. Carried to Cooker	⇨	
3. Cooked	○	
4. Washed in Cooker	○	Add a Tap Adjacent to Cooker
5. Meat and Sauce Added	○	
6. Warmed	▽	
7. Placed in Serving Pans	⇨	Repeat Six Times
8. Carried to Serving Line	⇨	Repeat Six Times

Fig. 6.4. Flow process chart—proposed method.

Example 6.3

In this example a pot handle is formed by plastic injection molding. This process requires a sophisticated die, and consequently, much time is devoted to the design of the die. Regardless of the amount of time spent on the design, a small burr is left on the part after it is removed from the die. This burr is ground in a small grinder; then it is polished by a buffing wheel to eliminate any grinding marks. The process used is presented below. Plastics, Inc., has asked you to improve the process and you have decided to use a left-hand–right-hand chart. First, you ask for a drawing of a typical burr. That is presented in Figure 6.5. Then you ask for a sketch of the workplace and a description of the process. The workplace sketch is shown in Figure 6.6.

All pot handles are placed on the gravity chute by a material handler. The chute is inclined so that parts are always available to the operator. The operator reaches to the chute with her left hand and picks up three pot handles. In the center of the work station she holds the handles with her left hand. Her right hand picks one pot handle out of the left hand, grinds the pot handle (takes about 2 seconds), and drops it in a cardboard box on the floor marked for completed units. The right hand does all three parts and the left hand reaches for three more pot handles. This is continued approximately 56 times until the cardboard box on the floor is full. The operator stops, shoves the box over to the area marked full boxes, and

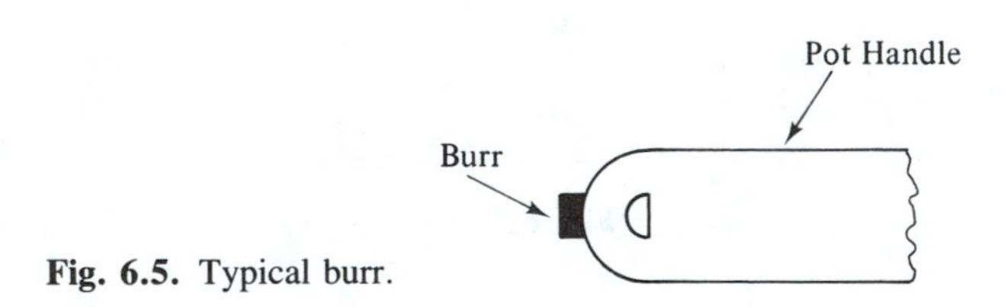

Fig. 6.5. Typical burr.

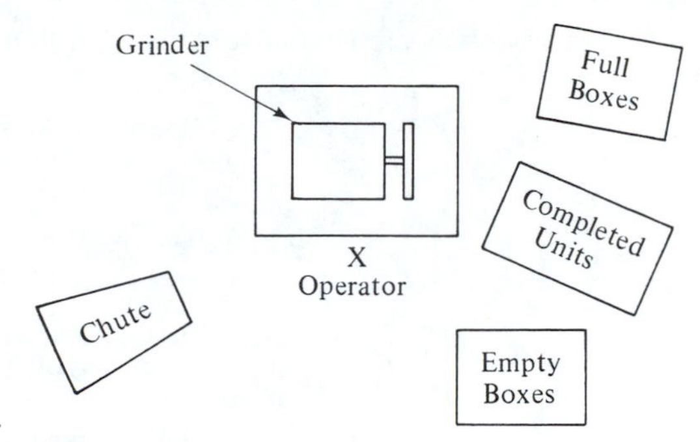

Fig. 6.6. Workstation layout.

slides another empty box into place. The process is repeated. (The material handler keeps a supply of empty boxes in place.)

Figure 6.7 shows a simplified left-hand–right-hand chart. First, the information is filled in on the part number, and so on; then a sketch of the workplace is made. After that, the left and right hands are simultaneously charted so that at any time the movement of the left hand should match that of the right. Accordingly, a description of activities being on the same horizontal line implies that the activities are done at the same time.

Operation ______________ Name ________________ Date __________

Existing ______________ Proposed _______

(Sketch of workplace and part if necessary)

Left-hand Activity	Symbol	Symbol	Right-hand Activity

Fig. 6.7. Left-hand–right-hand chart.

The symbols are similar to those of a flow process chart, for example:

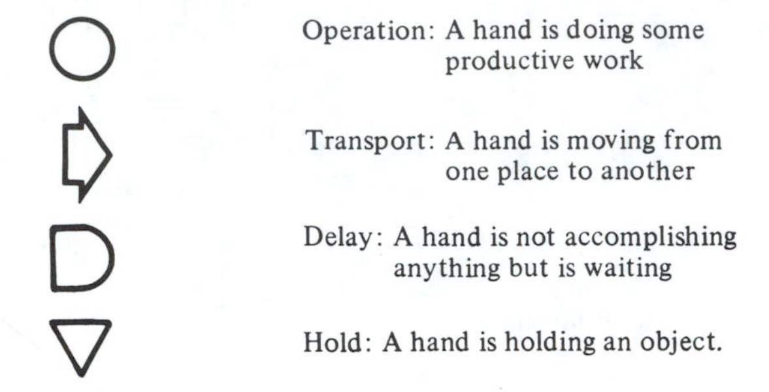

Example 6.3 (Continued)

You construct a left-hand–right-hand chart for the operation. That chart is shown in Figure 6.8. You notice that the right hand is delayed a large portion of the time while it is waiting for the left hand to obtain the parts. Also, the left hand holds the parts while the right hand does the work. (Holding is really not doing any productive work.) Productive work then really only occupies a small percentage of the total cycle.

After working with the operator and the foreman and after trying several new methods, you come up with a proposed procedure. (You also consider the principles of motion economy that will be discussed in Section 6.2.4.) That proposed procedure is shown in Figure 6.9 with a sketch of the new work plan. You time the two procedures and find that you save 37% of the time formerly used, not counting the substantial material handling savings.

Now there are two chutes feeding the operator from the left and the right. There is also a work shelf added to the grinding table. The work shelf has a drop hole in it directly in front of the wheel. The operator reaches simultaneously left and right and slides two parts on the work shelf (one per hand) to the two grinding wheels. After the grinding occurs, the operator slides the parts to the middle where they drop through to a conveyor underneath. The conveyor carries the handles to the polishing operation. The cycle is repeated continuously because boxes never have to be moved.

6.2.3 Other Charts

This discussion only skims the surface of charting for work simplification. There are many other charts that have been used in designing work for many different types of jobs. Some of the other popular charts are described briefly below.

Flow Diagram. A flow diagram is essentially a flow process chart drawn to show the layout of a facility. Only the symbols and the step numbers are shown. Usually, the objective is to look for spatial relationships.

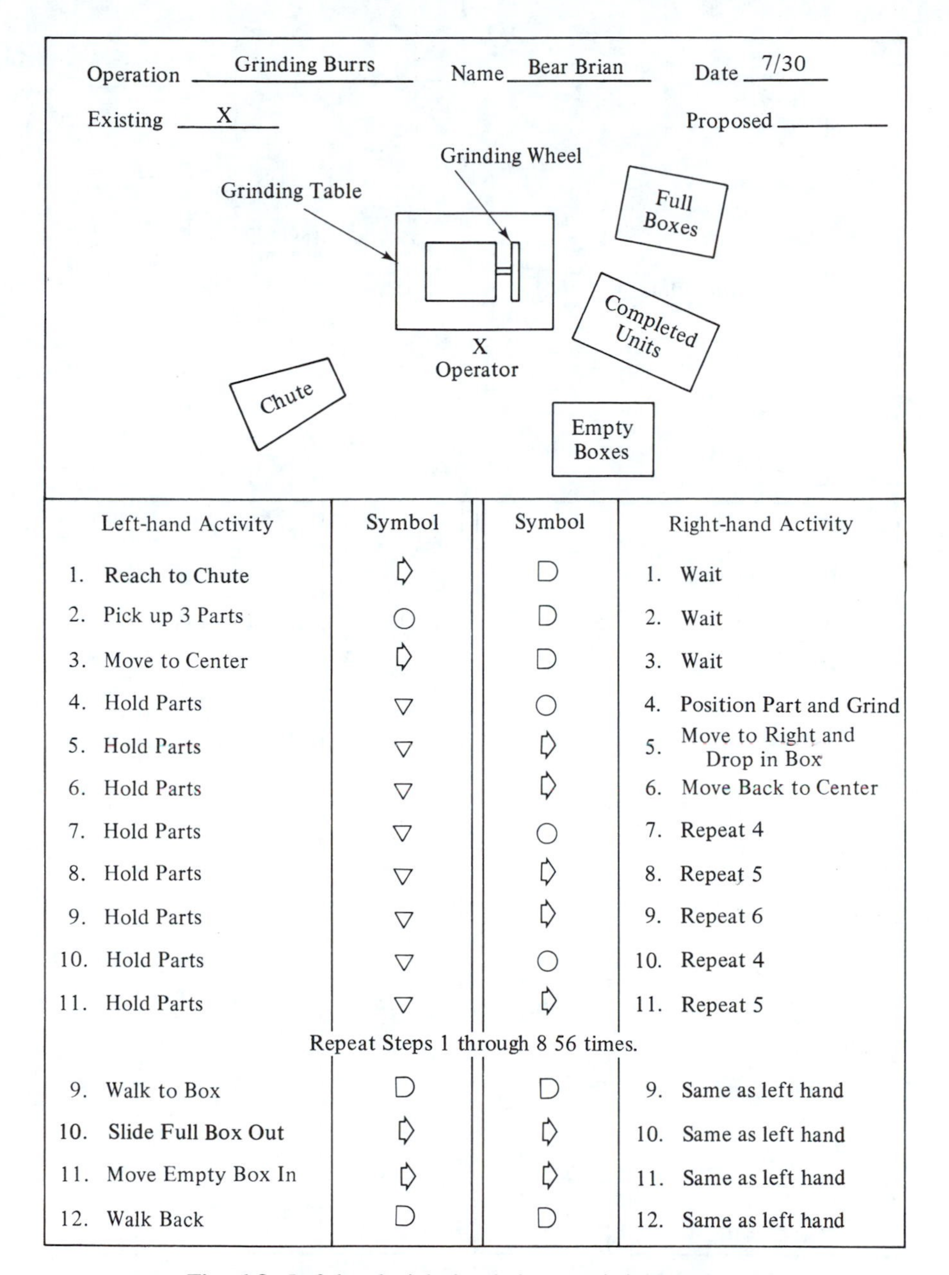
Operation Grinding Burrs Name Bear Brian Date 7/30

Existing X Proposed ____

Left-hand Activity	Symbol	Symbol	Right-hand Activity
1. Reach to Chute	⇨	D	1. Wait
2. Pick up 3 Parts	○	D	2. Wait
3. Move to Center	⇨	D	3. Wait
4. Hold Parts	▽	○	4. Position Part and Grind
5. Hold Parts	▽	⇨	5. Move to Right and Drop in Box
6. Hold Parts	▽	⇨	6. Move Back to Center
7. Hold Parts	▽	○	7. Repeat 4
8. Hold Parts	▽	⇨	8. Repeat 5
9. Hold Parts	▽	⇨	9. Repeat 6
10. Hold Parts	▽	○	10. Repeat 4
11. Hold Parts	▽	⇨	11. Repeat 5
Repeat Steps 1 through 8 56 times.			
9. Walk to Box	D	D	9. Same as left hand
10. Slide Full Box Out	⇨	⇨	10. Same as left hand
11. Move Empty Box In	⇨	⇨	11. Same as left hand
12. Walk Back	D	D	12. Same as left hand

Fig. 6.8. Left-hand–right-hand chart—existing procedure.

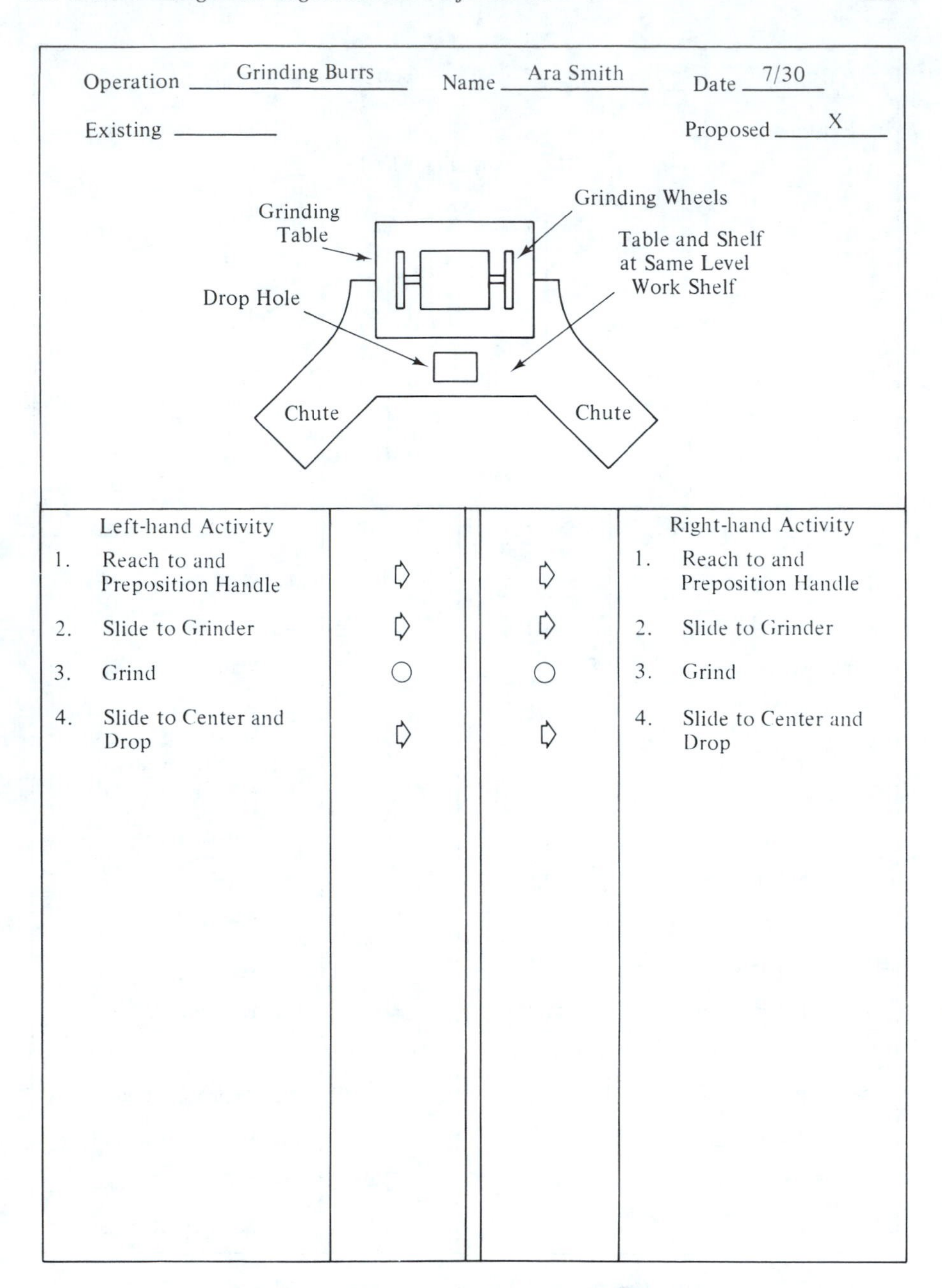

Fig. 6.9. Left-hand–right-hand chart—proposed procedure.

Multiple Activity Chart. A multiple activity chart is similar to a left-hand–right-hand chart in that the simultaneous activities of two or more objects or persons are shown. A chart simultaneously showing the activities of a worker and those of one or more machines he is operating is an example. The simultaneous activity of several workers on an assembly line is another. The objective is to obtain good balance—minimizing the total cycle time and subsequently the waiting time of any of the components.

Work Distribution Chart. A work distribution chart is a listing of all activities or responsibilities of every person in a department or group. It is similar to a multiple activity chart except that the activities are not necessarily simultaneously performed. The objective is to obtain proper balance of assignments and to be certain that the proper level of employment is performing the function (e.g., a department head should not deliver mail or answer telephone calls).

Operation Process Chart (already discussed)—The objective is to eliminate unnecessary operations, inspections, or storages and to obtain the proper sequence of them.

Gantt Chart. A Gantt chart is a horizontal time bar chart that shows relative timing of various activities. It is not often used in work simplification but may be used in scheduling. The objective is usually to perform all activities in the least amount of total time—often subject to some resource availability constraints.

6.2.4 Principles of Motion Economy

The *principles of motion economy* is a list of thoughts or concepts that have proven helpful in designing efficient work methods. There have been many lists generated by different authors; but the lists always carry the same essential concepts.

Many industrial engineers use motion economy in conjunction with left-hand–right-hand charts, since the list is most useful for single workstations. There are other uses, however, and one of them is to use the principles of motion economy as a list of commandments in designing and performing work. This last is perhaps the most rewarding but least utilized approach.

The following list of motion economy principles is taken from Mundel.[6]

A. Elimination:
   1. Eliminate all possible jobs, steps, or motions. (This applies to body, leg, arm, hand, or eye.)
   2. Eliminate irregularities in a job so as to facilitate automaticity. Provide fixed places for things.

---

[6]Reproduced, with permission, from Marvin E. Mundel, *Motion and Time Study: Improving Productivity*, 5th ed. (Englewood Cliffs, N.J.: Prentice Hall, 1978), pp. 173–174.

3. Eliminate the use of the hand as a holding device.
4. Eliminate awkward or abnormal motions.
5. Eliminate the use of muscles to maintain a fixed position.
6. Eliminate muscular force by using power tools, power feeds, etc.
7. Eliminate the overcoming of momentum.
8. Eliminate danger.
9. Eliminate idle time unless needed for rest.

B. Combination:
1. Replace with one continuous curved motion short motions which are connected with sudden changes in direction.
2. With fixed machine cycles, make a maximum of work internal to the machine cycle.
3. Combine tools.
4. Combine controls.
5. Combine motions.

C. Rearrangement:
1. Distribute the work evenly between the two hands. A simultaneous symmetrical motion pattern is most effective. (This frequently involves working on two parts at the same time.) With crew work, distribute the work evenly among members of the crew.
2. Shift work from the hands to the eyes.
3. Arrange for a straightforward order of work.

D. Simplification:
1. Use the smallest muscle group capable of doing the work, providing for intermittent use of muscle groups as needed.
2. Reduce eye travel and the number of fixations.
3. Keep work in the normal work area.
4. Shorten motions.
5. Adapt handles, levers, pedals, buttons, etc., to human dimensions and musculatures.
6. Use momentum to build up energy in place of the intense application of muscular force.
7. Use the simplest possible combination of therbligs. (A *therblig* is an elementary activity such as reach, grasp, pick up.)
8. Reduce the complexity of each therblig, particularly the "terminal" therbligs.

6.2.5 Human Engineering

Human engineering (or ergonomics) may be defined as the design of the human–machine interface so that workers and machines may function more effectively and efficiently as integrated systems. Much of what has been covered in this chapter can be called human engineering, but there is much more to human engineering than work simplification and principles of motion economy.

Formerly, machines were designed with an emphasis on their capabilities for production but with little or no emphasis on the human element. Now, with human engineering, the machine may be lowered, tilted, or completely redesigned so that the worker and the machine may function better as a complete system. Much work in human engineering is done in design and location of dials for maximum effectiveness and efficiency. The design of a cockpit for a large modern jet is an excellent example of human engineering.

Human engineering is also useful when applied to day-to-day operations. For example, at an individual workstation it is important that all work be located close to and in front of the worker, as implied by the principles of motion economy. Using human engineering, we can design the workstation for the "average" person. Human engineers have calculated the reach of the average worker for both easy and strained moves. The workstation can be designed so that as many moves as possible are within easy reach. Figure 6.10 is an example of such measurements.

Another example of human engineering is the design of tools for more effective use. Pliers and scissors with curved handles are results of human engineering studies in which it was found that correctly designed curved handles are easier for human beings to use.

6.3. Work Measurement

The objective of any work measurement system is to determine the time it should take an average, trained person to perform a task if he or she were doing that over an 8-hour day under usual working conditions and working at a normal pace. This time is called *standard time*. Historically, there have been two basic approaches to defining standard time. We shall call these two the bottom-up and top-down approaches. The *bottom-up approach* starts with a basic measurement of time, adjusts for operator pace, and then allows for fatigue, personal needs, and delays. The *top-down approach* is used in many labor contracts, and it normally defines standard time as that time under which a qualified employee working under usual conditions *can* make an incentive pay (specified) percent above base pay. Whichever definition is used, the approach taken to calculate standard time is *almost* always the bottom-up approach. Consequently, we shall explore that approach. First, some definitions are in order.

Normal Time. The time required for an average, trained operator to perform a task under usual working conditions and working at a normal pace. (It does not include allowances for personal needs and delays that would be necessary if the task were done all 8 hours.)

Normal Pace. The pace of an average, trained, and conscientious operator working over an 8-hour day.

Actual Time. The observed time required for an operator to perform a task.

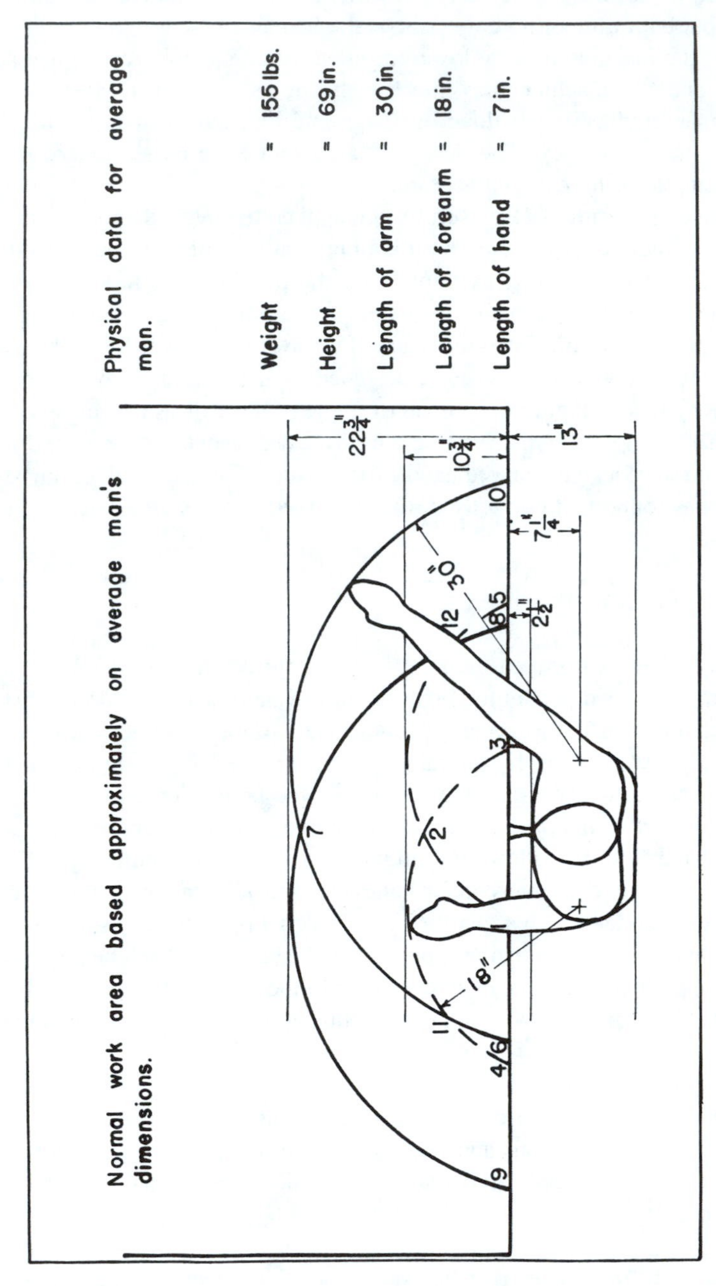

Fig. 6.10. Human work areas.

Allowances. The amount of time added to the *normal time* to provide for personal needs, unavoidable delays, and fatigue.

We can now examine some of the possible methods of determining standard times. Following are the methods to be discussed:

(1) Direct time study.
(2) Time study standard data.
(3) Predetermined time systems.
(4) Predetermined time systems standard data.
(5) Work sampling.

6.3.1 Direct Time Study

In this context, direct time study is a work measurement technique in which a physical measurement is made of the actual time required to do a task by using a watch or some other timing device. This measured time is then modified by considering the operator's pace, and finally, allowances are added.

The form used to record task times varies greatly among companies. A general rule of thumb is to present as much information as might be required to duplicate the operation from the form alone. Therefore, the form should include a good description of the task involved, a sketch of the workplace, perhaps a left-hand–right-hand chart of the operation, and a clear description of the elemental breakdown.

Figure 6.11 shows a typical watch used in direct time studies and Figure 6.12 shows a typical time study board. A time study board is a device for holding the forms so that the analyst can make a recording while standing or even moving.

Figure 6.13 shows a simplified sample time study sheet for a continuous

Fig. 6.11. Time study watch.

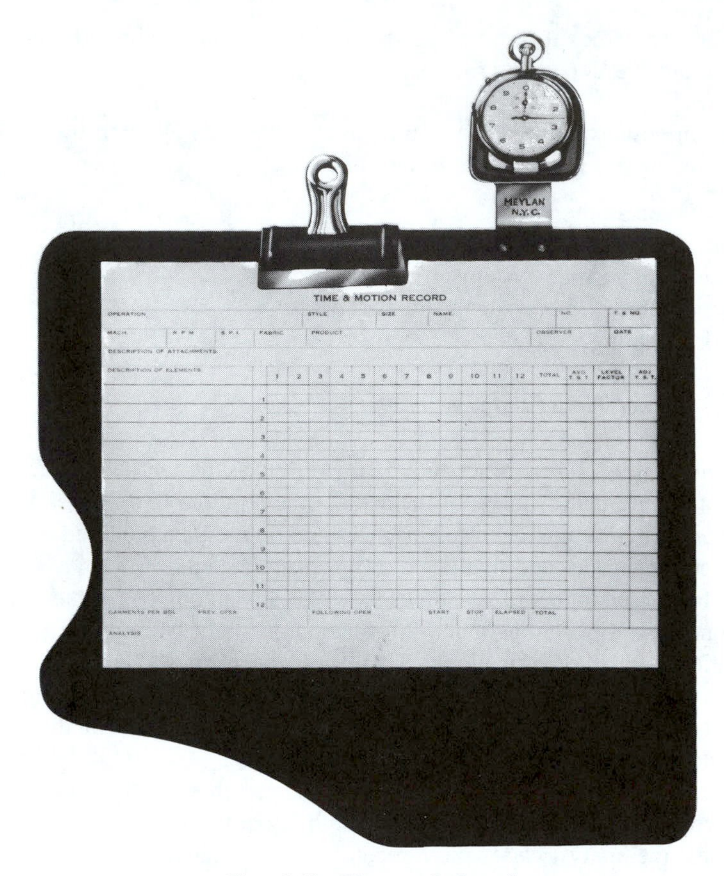

Fig. 6.12. Time study board.

reading time study (the watch is not stopped). The front of the sheet is for recording the task being studied, the existing procedure in detail, and any other pertinent information. The back of the sheet is for recording the actual time study. The task is broken down into easily discernible elements, the elements are recorded, and then the timing is done. The watch reading is recorded in column R, and the time for each element is entered in column T later. For example, the following might be sample results:

	1		*2*		*3*		*4*	
Element	*R*	*T*	*R*	*T*	*R*	*T*	*R*	*T*
1	5		18		27		40	
2	10		22		32		45	
3	12		23		35		47	

	1		*2*		*3*		*4*	
Element	*R*	*T*	*R*	*T*	*R*	*T*	*R*	*T*
1	5	5	18	6	27	4	40	5
2	10	5	22	4	32	5	45	5
3	12	2	23	1	35	3	47	2

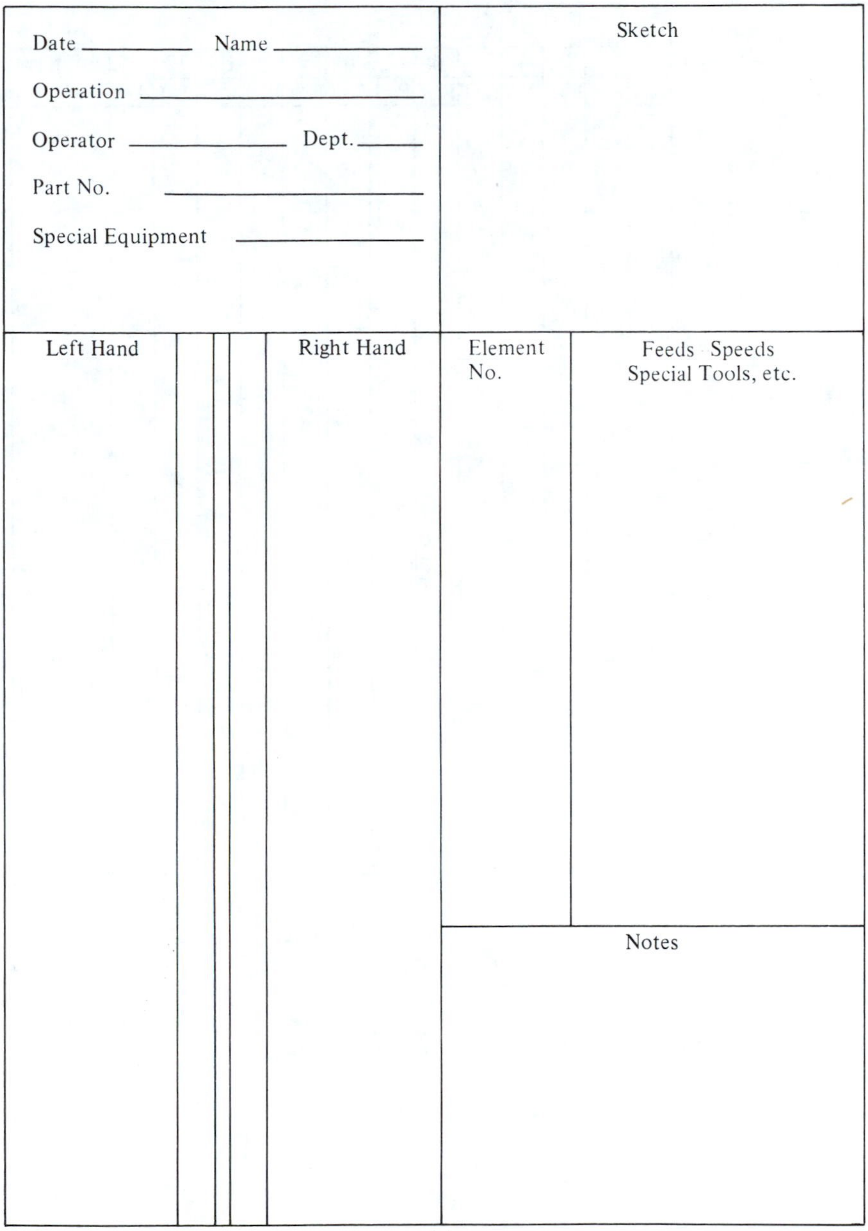

Date ________ Name ____________ Operation ____________________ Operator ____________ Dept. ____ Part No. ____________________ Special Equipment ______________				Sketch	
Left Hand			Right Hand	Element No.	Feeds Speeds Special Tools, etc.
				Notes	

(a)

Fig. 6.13. Time study form.

No.	Element Description	Cycle																									
		1		2		3		4		5		6		7		8		9		10		11		12		13		Average
		R	T	R	T	R	T	R	T	R	T	R	T	R	T	R	T	R	T	R	T	R	T	R	T	R	T	

Symbols Explanation	Calculations
	Performance Rating ____________

(b)

Fig. 6.13. (cont'd)

Next, each element is averaged. Sometimes a performance rating is applied to each element, but in this case, we shall assume that the entire job (not each element) is rated. The calculations for a *different* example would be as follows (the average time for each element is obtained by averaging all the good readings for that element):

Element	*Average*
1	15
2	7
3	5
4	8

Total actual time = 15 + 7 + 5 + 8
= 35
Performance rating = 115%
Normal time = (35)(1.15) = 40.25
Allowances = 12%
Standard time = (40.25)(1.12) = **45.08**

In this case, the analyst believes that the operator is performing at 115% or 15% faster than would normally be expected. Consequently, the normal time should be adjusted to show that *more time is allowed*. To obtain the normal time, the actual time is multiplied by the performance rating. If the operator has been performing at 90%, or 10% less than what normally would be expected, the normal time would be

$$(35)(0.90) = 28.50$$

Now, realizing that there will be other problems, the problems must be "allowed" for. For example, the operator will have personal needs (rest room, water, etc.) and will become fatigued as the day progresses. Finally, there will be unavoidable delays (machine breakdowns, shortage of parts). We make a study and find that personal needs occupy 4% of the workday, fatigue accounts for 5% of the performance, and unavoidable delays occur 3% of the time. The total allowance then is

$$4\% + 5\% + 3\% = 12\%$$

To calculate the standard time, we add the allowances to the normal time and find

$$\text{Standard time} = (40.25)(1.12) = 45.08$$

The form and example above are highly simplified. No foreign elements (unnecessary moves, for example) were encountered, and performance rating was on the entire job, not on each element. Consequently, real-world time studies can often be very complicated. A competent time study analyst is a highly trained technical person who, since he or she is dealing with other employees almost all the time, must also be a human relations expert.

6.3.2 Time Study Standard Data

Often, elements of a task are repeated many times throughout an organization. Instead of a detailed study of each time a task is performed, the task could be studied in depth once and a standard data file developed. Then, whenever that element reappears, the time could be directly applied from the standard data file. *Time study standard data* may be defined as normal time values not obtained from direct time measurement of the *particular* element but obtained from direct time measurement of the element in a similar operation earlier.

For example, in a machine shop that produces many small parts all of which are the same approximate size, the element ''obtain two parts and place in fixture, *X* in. to parts and *Y* in. to fixture'' would probably be repeated very often. A left-hand–right-hand chart of the element is shown in Figure 6.14.

After they have been studied, the elements could be placed in a standard data file, such as that shown in Table 6.1. (The values in this table are fictitious.) Then, any time a time study analyst encounters this element, he or she can obtain the normal time directly without having to record stopwatch readings or performance ratings.

The advantages in using standard data are numerous. It must be realized, however, that standard data are not always applicable and that even when they are, some other measurement must be used for unusual elements. A list of advantages given below.

(1) Individual time studies are quicker and less costly.
(2) Consistency exists between time studies.
(3) There is less chance for error on a study.
(4) Fewer analysts may be needed.
(5) Some work simplification is done automatically as the method is prescribed.
(6) It is good for cost estimating and production planning before a job is run.
(7) There is less disturbance in the shop while the time study is being taken.

The chief disadvantage is that there is a high developmental cost to obtain the necessary data. In some organizations the cost can be completely prohibitive. In

Left-hand Activity	Symbol	Symbol	Right-hand Activity
Reach X ''	⇨	⇨	Reach X ''
Grasp	○	○	Grasp
Move Y ''	⇨	⇨	Move Y ''
Position	○	○	Position
Release	○	○	Release

Fig. 6.14. Left-hand–right-hand chart of the element, ''Obtain two parts and place in fixture, *X* in. to parts and *Y* in. to fixture'' (time in minutes).

Table 6.1. STANDARD DATA FILE FOR "OBTAIN TWO PARTS AND PLACE IN FIXTURE, X IN. TO PARTS AND Y IN. TO FIXTURE"
(Time in minutes)

	Y			
X	*6 in.*	*8 in.*	*10 in.*	*12 in.*
6 in.	0.025	0.026	0.026	0.027
8 in.	0.026	0.026	0.027	0.027
10 in.	0.027	0.028	0.028	0.029
12 in.	0.027	0.028	0.029	0.030

general, however, whenever standard data are applicable and the developmental cost is not prohibitive, the data should be used.

6.3.3 Predetermined Times

If jobs or tasks are broken down into finer and finer elements, there will be a point reached where all tasks are made up of the same set of elements. At this point, time values can be assigned to these elements, and the total time required to perform a task can be determined by adding the appropriate elements. This approach is called *predetermined times*.

Predetermined time systems are very similar to standard data in that the time values can be read directly from charts. They are not unique in their application, however, and can be used by any company.

There are several systems in use today. Among the better known are (1) work factor and (2) methods time measurement (MTM).

The time values applied to the various elements can be obtained in several ways. One approach is to use highly accurate movie cameras and use the films (slowed down) to obtain the times (micromotion analysis). Another approach is to use electronic timing devices. Finally, time studies can be used when appropriate.

As an example, one such element might be "move object to other hand," and the time might be recorded as

Weight > 2 lb

Distance	*Time*
4 in.	0.00412 min
8 in.	0.00591 min

Of course, since the data must be extremely accurate, more decimal places are necessary.

6.3.4 Predetermined Time Standard Data

Predetermined time standard data are standard data exactly as in time study standard data *except* for the source of the information. Here, the repeated elements are timed by using predetermined time systems. For example, Figure 6.14 and Table 6.1 would appear exactly the same, except that now times would be applied to the subelements in the chart of Figure 6.14 from predetermined time systems. Often this distinction is made by saying predetermined systems time elements at a level 1. The level 2 elements (standard data) may be made up of several level 1 elements.

The advantages and disadvantages are exactly the same as those for time study standard data. The only possible difference is that predetermined time standard data may be developed more quickly since all times are already available.

6.3.5 Work Sampling

Work sampling is the estimating of the proportion of time devoted to a given type of activity by the relative frequency occurrence of intermittent randomly spaced instantaneous observations. Suppose, for example, that it is desired to estimate the amount of time that should be set aside for personal needs during a workday or that it is desired to know the amount of time taken up by unavoidable delays such as material shortages or machine breakdowns. This can be done by sampling over an extended period of time. Basically, a series of observations is made randomly (probably through the use of random number tables such as those available in almost all texts on work sampling), and the results are recorded. Sample random numbers are shown below with possible interpretations.

Random Number	*Interpretation (When to Observe)*	
915	5:15 P.M.	(starting with 100 as 8:00 A.M. and skipping lunch)
723	3:23 P.M.	
019	8:19 A.M.	

Once a series of numbers has been generated, the observations can be made and the percent of time occupied by one activity can be estimated as follows:

$$\text{Percent occurrence for activity A} = \frac{\text{number times A occurred}}{\text{total number of observations}}$$

The number of observations necessary can be determined scientifically, but in general, enough observations are taken so that the accumulated data stabilize. For example, Figure 6.15 shows the typical response and the approximate total number of observations that should be taken—N^*.

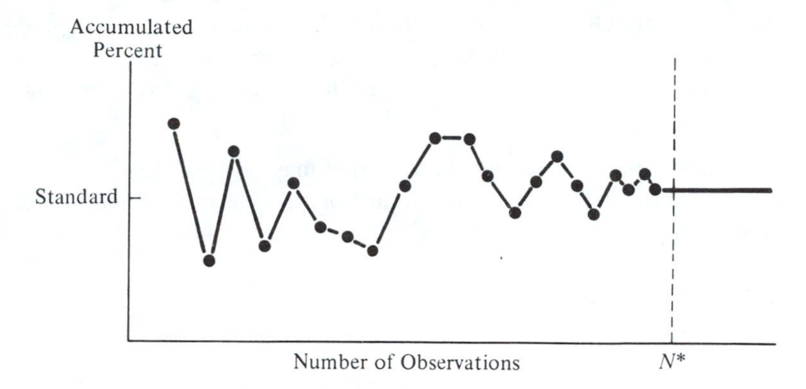

Fig. 6.15. Typical work sampling response.

6.4. Organizational System Performance Measurement

As discussed in Section 6.1, the goal of an organization is usually some composite function of several measures. Profitability, for example, is a composite function bringing together quality, customer satisfaction, sales, cost of labor, material, energy, and others. Naturally, there are other objectives that may not be measured by profitability, but they are usually less important and we will not discuss them here.

A control system designed to optimize profitability is really a series of control subsystems—each with its own subobjectives. For example, a measured day work system controls direct labor by comparing its performance to that determined by standard times (work measurement). This is efficiency measurement in Figure 6.1. Another popular control scheme is quality control.

Recently, the industrial engineering profession has recognized that another control system is needed. *Productivity measurement and control* is the newest control system and it is receiving a tremendous amount of attention in the literature, at conferences, and in day-to-day discussions. The rest of this chapter will discuss productivity measurement (control is exercised as a result of the measurement system, hopefully).

One must question whether or not productivity measurement is necessary. Are our time-proven systems of quality control and work measurement enough, or do we need more? There is increasing evidence that more is directly needed. Once the leading nation in productivity improvement, we (the United States) are finding our rate of productivity improvement very low—so low, in fact, that we are trailing most industrialized nations in the rate of productivity improvement. Perhaps by measuring and improving productivity, we can reverse this trend.

6.4.1 Productivity Measurement Basics

Productivity measurement is simply a ratio of organizational outputs to organizational inputs [Equation (6.1)]. The ratios can be total or partial. For example, consider Figure 6.16. Given this introduction, let's look more deeply into productivity measurement. First, productivity measurement seems exceedingly simple, and, in fact, it can be. However, the following often make this measurement more complex:

(1) The relationship between input and output may be obscure.
(2) It may be difficult to measure outputs and/or inputs (e.g., hospitals and banks).
(3) Product mixes may vary considerably, affecting the ratio in a fashion difficult to understand.
(4) Input and/or output quality may vary.
(5) Data may be hard to obtain.

In addition to being total or partial measures[7] (Figure 6.16), productivity indices can be "static" or "dynamic." *Static measures* are ratios taken as a snapshot with no base year comparison. All the ratios in Figure 6.16 are static. *Dynamic measures* are actually ratios of static productivity indices, as shown in the equation

$$\text{dynamic productivity index} = \frac{\dfrac{\text{outputs this year}}{\text{inputs this year}}}{\dfrac{\text{outputs base year}}{\text{inputs base year}}} \tag{6.2}$$

Therefore, indices greater than 1 show productivity improvement, while indices less than 1 show productivity deterioration. As with static measures, dynamic measures can be partial or total.

There are two popular productivity measurement approaches. The first uses a group-generated model and is called the *normative productivity measurement methodology* (NPMM). The second is less participative in that one model can be modified to fit any organizational scheme. It is called the *multifactor productivity measurement model* (MFPMM). Each will be discussed in some detail.

6.4.2 Normative Productivity Measurement Model

The NPMM relies on experienced managers to develop the productivity measurement ratios or indices applicable to a specific company. It does this through a structured group process known as the *nominal group technique* (NGT). Although not a panacea to productivity measurement, the NPMM can usually lead to group consensus on what is important to monitor.

The procedure is first to develop the group. Normally, 5 to 12 people are chosen. They should all be familiar with the organization and should represent various interest groups. Give the group a charge. For productivity measurement, the charge might be something like: "Develop a list of productivity measures, ratios, or indices important to our organization." Then the group is immersed into

[7] Sink, "Organizational System Performance."

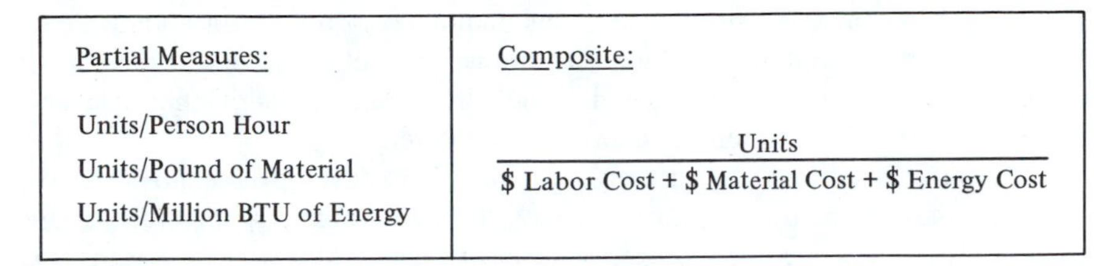

Partial Measures:	Composite:
Units/Person Hour Units/Pound of Material Units/Million BTU of Energy	$\frac{\text{Units}}{\text{\$ Labor Cost + \$ Material Cost + \$ Energy Cost}}$

Fig. 6.16. Productivity indices.

the NGT. The process (which must be carefully adhered to by a chairperson) is as follows:

(1) *Silent Generation*—Each person is asked to generate a list of measures.
(2) *Round Robin*—After 5 to 15 minutes, the group members are asked to present their ideas. Each person gives a succinct description of one measure. The next person does the same; and so on. This continues around the group as many times as necessary to exhaust everyone's list.
(3) *Group Clarification*—First, all ideas are clarified. Then overlap or duplication of ideas is identified and remedied. (Normally, several ratios are the same or very similar and can be combined.)
(4) *Voting*—Each person is given N cards (N is determined by the leader) and asked to list the N best ideas or ratios. After this is done, each person chooses:
   (a) The best one and gives it N points.
   (b) The lowest of their list and gives it 1 point.
   (c) The second best and gives it $N - 1$ points.
   (d) The second worst and gives it 2 points.
   Etc.

The results might be as shown in Figure 6.17. From these results, we can see two things. First and most important, there is a strong consensus on what is important and what should be monitored. Second, there is some confusion on productivity. Measures 1, 4, and 5 are productivity measures, but 2 and 3 really are not productivity measures (they are more measures of effectiveness than ratios of output to input). This confusion is normal and should be expected.

	Individual Votes (5 = most impt., 1 = least)	Number of Votes/Vote Score (Total)
1. Units per Direct Labor Hour	5–4–4–3–4–5	6/25
2. Number Orders Completed on Time	3–2–2–1–1	5/9
3. Number Customers with Repeat Orders	1–1–3–5	4/10
4. Units per Million BTU (Energy)	4–5–5–4–5	5/23
5. Units per Ton of Purchased Material	2–3–1	3/6

Fig. 6.17. Results from NGT on productivity ratios.

At any rate, there is a strong consensus on units per direct labor hour as being the most important; units per million Btu comes in a close second (perhaps the company is very energy intensive); and so on. From this the list of final measures can be developed and the measurement process started.

Given the list of ranked measures, it is possible to develop a composite model that actually combines these identified ratios through weighting techniques. One such process is known as the *multicriteria performance/productivity measurement technique*.[8] Details will not be presented here.

6.4.3 Multifactor Productivity Measurement Model

MFPMM is a more comprehensive and analytical approach to measuring changes in productivity. Basically, MFPMM recognizes that there are two ways profitability can be affected in an organization:

(1) Price recovery.
(2) Productivity changes.

Under price recovery, prices can be raised or lowered as the market allows or demands. Both can dramatically affect profitability. Price increases can usually be tolerated only as long as competition and consumer reaction eventually occur. The desired change is through productivity improvement(s).

The MFPMM uses accounting techniques to break the total variation (change in profit) into price effects and productivity effects, resulting in a valid productivity measurement model. This analysis can be done on both inputs and outputs. For example, profitability is affected by sales price and sales quantity (outputs). It is also, however, affected by price and quantity changes in labor, energy, materials, and so on (inputs). Careful manipulation of the data will demonstrate what effects each change has.[9]

6.5. Computers and Work Measurement and Design

Computers have had a strong impact on the field of work measurement and design, but particularly on work measurement. For example, several firms use microprocessors directly wired to workstations. This way, a long-range time study can be conducted yielding very accurate measurements.

There has been a great deal of work toward computerizing predetermined time

---

[8] D. Scott Sink, "Multi-criteria Performance/Productivity Measurement Technique," *Productivity Management*, Vol. II, No. IV, Stillwater, Okla., Fall 1983.

[9] The details of MFPMM techniques are beyond the scope of this chapter and will not be presented. For more, see J. C. Swaim et al., "Current Developments in Firm or Corporate Level Productivity Measurement and Evaluation," *Proceedings 1983 Fall Industrial Engineering Conference*, Toronto, Canada, Nov. 1983.

systems. For example, the computer can take simple coded activities, look up the predetermined times, and do all the calculations. Another more sophisticated approach is to store some previously determined times for higher levels of activities (several predetermined elements as in standard data) and to use these whenever appropriate.

When similar products are produced, it may be possible to store standard times by product descriptions, such as weight or material used. By adequately describing the product to the computer, the appropriate standard time could be obtained. Interpolation could be used for given characteristics that have not previously been calculated.

Finally, there are numerous routines available on computers that generate random numbers. These random numbers can then be used in the design of a work sampling procedure.

DISCUSSION QUESTIONS

1. Define work measurement, work design, and productivity measurement. Discuss the relationship between work measurement and productivity measurement.
2. Discuss organizational systems performance measures other than "productivity" that might be important.
3. Ajax Manufacturing produces wooden kegs utilized in plants in Kentucky and Tennessee. The only product is wooden kegs, but main inputs include labor, energy, and wood. List potential productivity measures for the plant and state whether they are partial or total.
4. You are asked to study the procedure for issuing and executing back orders. (Back orders occur when inventory is temporarily depleted. An order is issued so that the product will be shipped as soon as it is available.)

   In another situation, you are asked to study the procedure for assemblying fountain pens. Basically, one operator assembles all fountain pens and all parts are brought to her workstation.

   State what tools of work methods simplification you might utilize.
5. Define normal times, normal pace, allowances, standard times, direct time studies, standard data, predetermined time systems, and work sampling. Discuss the uses of the last four.
6. Give the five-step approach to work simplification and show the use of charting in the five steps.
7. Discuss the advantages and disadvantages of predetermined time studies over direct time studies.
8. Discuss the primary difference between a static productivity ratio and a dynamic productivity index.
9. List several performance measures that a university might use.
10. Discuss why you think the nominal group technique (NGT) might work well for group consensus on productivity measurement.
11. Eskimo Mike's is a tremendously popular college town pub. It produces and sells hamburgers, beer, mixed drinks, and other fast foods. Most of the employees are older

students and recent graduates. You are the manager. List one example each of the measures shown in Fig. 6.1 (efficiency, effectiveness, quality, innovation, productivity, quality of working life, profitability).

PROBLEMS

1. Given the following data, compute the production standard (pieces per hour), the standard labor cost per 100 pieces, and the standard hours per 100 pieces. The allowances are 4% for fatigue, 8% for unavoidable delays, and 8% for personal needs. The operator is paid \$5.00 per hour. Performance rating = 105%. (Instructor: Look also at Problem 12.)

	Reading (min)			
Element	*1*	*2*	*3*	*4*
1	*0.020*	*0.010*	*0.009*	*0.012*
2	*0.002*	*0.003*	*0.004*	*0.004*
3	*0.100*	*0.110*	*0.009*	*0.050*
4	*0.205*	*0.195*	*0.200*	*0.210*

2. Take an operation that is performed at one workstation with which you are very familiar and make a left-hand–right-hand chart of it. List the principles of motion economy and show which areas are applicable to that operation. Finally, show a revised left-hand–right-hand chart and calculate labor savings for an appropriate rate.

3. Make a flow process chart for a process with which you are familiar. Examine it and see which activities can be eliminated, combined, changed in sequence, or improved (in that order of preference). Show the revised flow process chart.

4. You are working with Giziwizets, Inc., on productivity measurement. They give you the data below. For those data, calculate:

(a) Partial productivity ratios for this year and the base year.

(b) Productivity ratios for this year.

What can you say about the company's productivity?

	Base Year	*This Year*
Output (number of giziwizits)	110	160
Labor hours	250	275
Energy (10^6 Btu)	10,000	60,000
Material (lb)	500	675

5. You have made time studies of a task that occures frequently within your plant. You find that standard time varies directly with the weight of the product. For the given values, graph time vs. weight and draw a straight line. Fill in your "estimate" for the line (i.e., visually estimate *a* and *b* below).

$$y = a + bx$$
$$a = y \text{ intersect}$$
$$b = \text{slope}$$
$$x = \text{weight}$$
$$y = \text{time}$$

Weight (lb)	*Time (min)*
1.00	0.190
5.75	0.300
3.50	0.230
4.00	0.260

Why is a not a zero value?

6. You are to design the workstation for installing shoelaces in shoes. The shoes are delivered to the workstation in a cardboard box and are taken to the next station in the same box. (Assume that you must use the box even though a conveyor would be better and assume that the operator does not have to move the boxes except to move the full one out of the way and an empty one in its place.) Using the principles of motion economy, design the best method and record it on a left-hand–right-hand chart. Time the method by using a stopwatch obtained from your instructor (ignore any fumbles by dropping elements where fumbles occur). Using allowances of 15%, determine the standard time per pair of shoes. Assume a standard pair of shoes with four holes on each side and 18-in. laces.

7. Repeat Problem 6 for an operation with which you are familiar, for example, changing spark plugs in a specific 4-, 6-, *or* 8-cylinder car and sharpening a box of pencils (assume an infinite supply and an electric sharpener).

8. Make a flow process chart of a person changing a flat tire from the time the person first gets out of the car to check the tire. Follow the person and stop when he gets back in the car (flat tire in the trunk). For the first time, have a person perform the activity as he normally would. Then make a revised flow process chart and show the total time for both. (For each activity, first try to *eliminate the activity*, then try to *combine it* with another, then try to *change the sequence* for better flow, and finally, try to *improve the method* of performing the activity.)

9. To test the validity of your allowance for personal delays (8%), you ran a work sampling study. Out of 638 observations, 51 were personal delays. *Assuming* that this is enough observations, should the allowance be changed? If so, to what value?

10. In conducting a work sampling study, you find the following results. Graph the results and from that visually estimate how many more readings will be required.

Number of Observations	*Accumulated %*
50	34
60	26.5
70	32.5
80	28

11. Form into groups of approximately five. Using the Normative Group Technique, determine a list of seven productivity measures that Eskimo Mike's should use. (See Question 11 of this chapter. Comment on the process.)

12. Redo Problem 1 if the performance rating was 90%. Discuss the change and the role of performance rating.

Home-work:

Dody: First 5 chapt.

CHAPTER 7

Operations Planning and Control[1]

7.1. Introduction

A large portion of modern industry consists of fabrication and assembly processes. Consider, for example, a plant that manufactures television sets, radios, and stereos. This plant must purchase materials, and perhaps parts, convert these materials into specific components, then assemble the components into the several products offered to the consumers. This greatly oversimplified process is represented in Figure 7.1.

Several observations are important at this point. First, there could be many different configurations of each of the three primary products. For example, there are many sizes and styles of television sets. Second, different products could contain several common components. It would not be surprising, for example, to find the same power switch in all three products. Third, different components could contain common raw materials. For example, certain components for each product may be stamped from the same roll of aluminum. The three components shown in Figure 7.2 could be for different products.

---

[1]A significant portion of this chapter has been adapted, with permission, from J.H. Mize, C.R. White, and G.H. Brooks, *Operations Planning and Control* (Englewood Cliffs, N.J.: Prentice Hall, 1971).

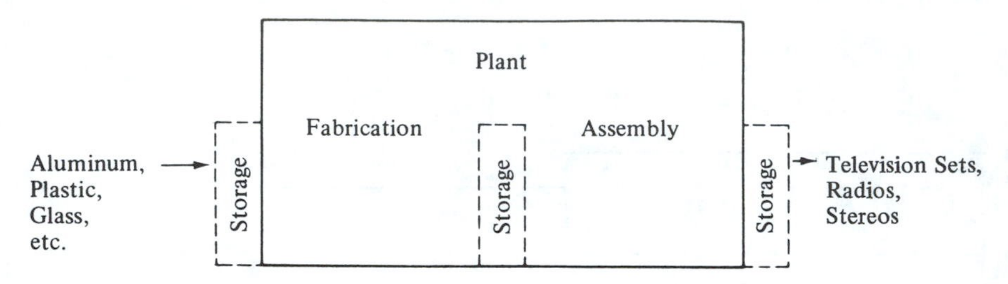

Fig. 7.1. Representation of a typical manufacturing plant.

7.2. Overview of Operations Planning and Control

Figure 7.3 is a schematic overview of an integrated operations planning and control system. The details of performance of the planning and control functions are presented in Sections 7.3 through 7.7. An intricate flow diagram, such as Figure 7.3, is conceptually difficult to understand in its entirety. Many of the component blocks shown in this master flow diagram can be expanded into detailed flow diagrams.

7.2.1 Demand Forecasting (I)

Inputs. The most tangible input to the demand forecasting function comes from product markets in the form of demand histories of existing products and, especially, trade information. Demand trends can be determined quantitatively both for one's own company and for the entire industry. Special trade information can either modify

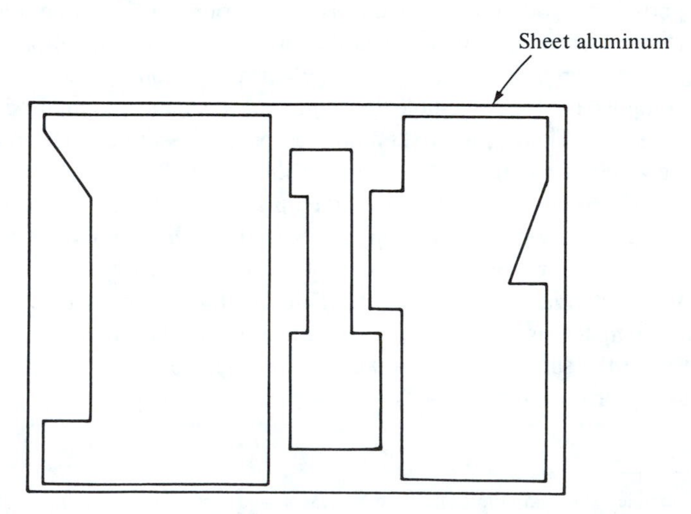

Fig. 7.2. Components for three different products stamped from the same sheet of aluminum.

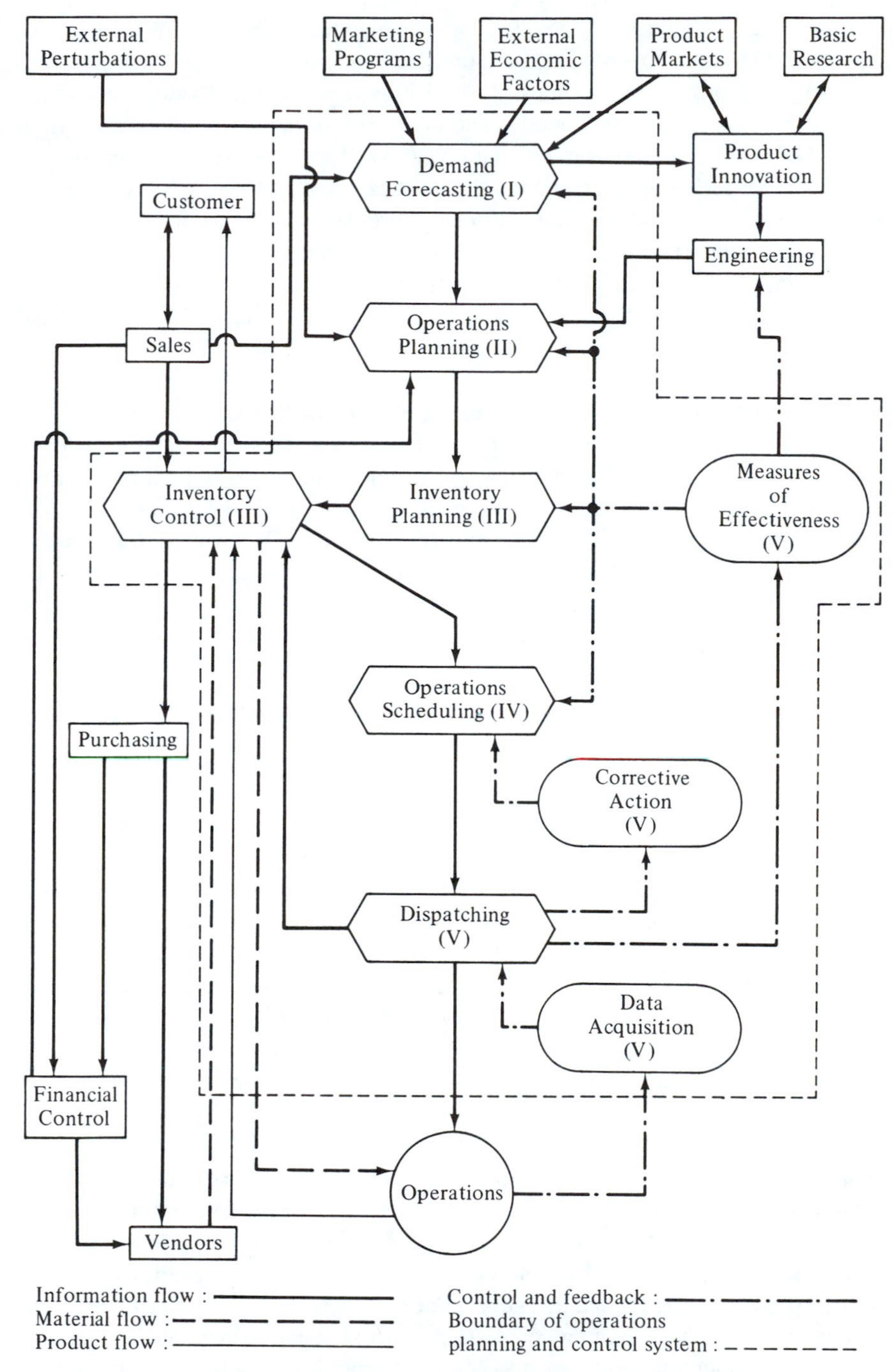

Fig. 7.3. Master flow diagram: operations planning and control system.

or explain the demand trends. Another input comes from marketing programs. An attempt must be made to anticipate variations in the demand pattern that result from advertising and promotion schemes. A third important input is external economic factors, which represent the general economic health of the nation (and possibly of other nations). The state of the national economy at any point in time depends on the collective effects of such factors as the political situation, the military situation, the extent of government involvement in large projects (space, welfare, road building, other construction), labor conditions, stock market trends, and literally millions of other factors. It is very difficult for a company to evaluate quantitatively the total effect of the national economy upon its own operations. Finally, actual demand comes from sales and is compared to forecasted demand.

Outputs. The primary output of the demand forecasting function is a statement of expected demand quantities for the several products over some planning period. This output is conveyed to the operations planning function. Another important but often overlooked output is the recognition of declining demand for a product. Any indication of declining demand is conveyed to product innovation for possible modification of the product. When it becomes evident that a product line will soon cease to be sufficiently profitable, a decision must be made by higher management to phase it out. This decision then has implications in the operations planning function in terms of changes in capacity.

7.2.2 Operations Planning (II)

Inputs. The primary input to the operations planning function is from demand forecasting. The demand forecast provides the basis for determining the activity level at which the company should plan to operate over the planning period. Another large and important set of input data comes from engineering and concerns new products, modifications to existing products, or modifications to the production process. These data include manufacturing sequences, bills of material, standard operation times, setup times, standard process manuals, and the like. A third input, from financial control, concerns monetary constraints and budget limitations. A fourth input, called external perturbations, includes emergency orders, canceled orders, labor strikes, unavailability of resources, and other problems from the world at large.

Outputs. There are two broad categories of outputs from the operations planning function. One deals with long-range planning of plant expansion and new facility acquisition. In this category consideration is given to the total life cycle of a product line and the impact on production and resource capacities of proposed new or modified products. The second category, concerned with relatively short term (usually one year or less) planning, consists primarily of the allocation of available resources to production requirements. Machine groupings, facility layout, and assembly line balancing are performed. Make or buy decisions are made, as are decisions regarding work force size, skill mix, staff leveling, number of shifts, approximate time phasing of production runs, and the like. This information is conveyed to inventory planning and control and operations scheduling.

7.2.3 Inventory Planning and Control (III)

The inventory planning and control function is shown in Figure 7.3 as two blocks: inventory planning and inventory control. Inventory planning determines the material requirements (components, parts, raw materials, assemblies, supplies, etc.) necessary to satisfy the operations plan. Inventory control determines the proper inventory levels, reorder points, safety stocks,[2] and the like.

Input to Inventory Planning. The primary input to inventory planning is the approved operations plan from operations planning. Specifically, the operations plan must provide a time-phased outline of the plant's activities in terms of production for the market.

Outputs of Inventory Planning. The operations plan is extended into the net, time-phased requirements for components, parts, raw materials, assemblies, and supplies. These requirements are forwarded to inventory control.

Inputs to Inventory Control. Primary are the time-phased requirements from the inventory planning function. Other important information inputs are sales orders from sales and material requisitions from dispatching. In addition to information input and output, we are concerned also with the receipt and issue of materials and products. Material is received from the vendor and product is received from operations.

Outputs of Inventory Control. Order quantities, reorder points, safety stock of raw materials and purchased parts, manufacturing batch sizes, and safety stock of assemblies, fabricated parts, and finished product are determined. When stocks on hand of raw materials and purchased parts reach their reorder level, a purchase order is sent to purchasing. When stocks on hand of assemblies, fabricated parts, and finished products reach their reorder level, a production order is sent to operations scheduling. Material and product outputs include finished product sent to customers and materials sent to operations. Note that subassemblies and fabricated components go from operations to material control as products; however, when they are brought back into the production process, they return to operations as materials.

---

[2]Safety stocks are needed as protection against unusually large demand during a short period of time.

7.2.4 Operations Scheduling (IV)

Inputs. Primary input to the scheduling function are requests for fabrication and assembly from inventory control. Another extremely important input comes from corrective action in the form of updated priorities and schedule adjustments of production orders currently in process.

Outputs. Detailed operation sequences for individual work activities, as well as start and stop times for all operations, are released to dispatching. New work is assigned to operating facilities, with proper consideration given to work already in process and to priority assignments. Schedule conflicts on production facilities are resolved. This function is at the heart of the entire planning and control system, since it is here that compromises must be made between economic batch sizes, due dates, resource constraints, manpower leveling, and facility utilization. Here also corrections are made for nonstandard process behavior.

7.2.5 Dispatching and Progress Control (V)

Since this function is shown in Figure 7.3 as four blocks, we discuss each block in turn.

Dispatching. This function is responsible for initializing production, that is, for releasing work orders to operations at the appropriate time. Material is requisitioned from materials control, and arrangements are made for changing the production process over to the new production item.

Data Acquisition. As work progresses, data are acquired from different processing points on the shop floor. Specific data required are the progress of work in production, the status of work awaiting processing, the status of production facilities, and the availability of required workers. Material usage is relayed back to inventory control. It should be noted that much of the data is also useful for accounting functions, quality control, personnel, etc.

Corrective Action. Short-term corrective action is performed here. Specific decisions are made regarding operating problems. These decisions include expediting critical jobs, determining new priorities, balancing workloads between work centers, handling personnel problems, problems with product quality, and equipment breakdowns, etc. These decisions are conveyed to operations scheduling and are included in the next execution of the scheduling function.

Measures of Effectiveness. Long-term corrective action is performed here. Control parameters are maintained to which measures of plant performance are related. Actual performance is compared with planned performance for such measures of effectiveness as production output, standard and premium labor cost, investment in inventory, hiring and layoff cost, amount of scrap, facility utilization, schedule slippage, and so on. Particular attention is given to the identification of cause–effect relationships so that we can anticipate problems and try to prevent them. Information from this function constitutes the principal form of long-range system feedback and is used for recalculation of pertinent governing parameters.

7.2.6 Interfaces

The purposes of several of the interfaces shown in Figure 7.3 were discussed in the preceding sections; others are self-explanatory. Explained below are certain important interfaces whose functions are important to the operations planning and control system but whose purposes may not have been clear from the flow diagram.

Operations. Production orders reach the shop floor through the dispatching function. Materials and components are issued by inventory control. Production is accomplished, products are returned to inventory control, and progress information is fed back through the data acquisition function.

Basic Research. Basic research originates ideas for new products essentially independently of market considerations. An example would be the development of nylon by the du Pont research group. This development opened the way for hundreds of commercial uses. However, many research ideas prove fruitless.

Product Innovation. Ideas for new products and modifications of existing products are consolidated by this function, whether the ideas originate in product markets, basic research, an independent inventor, or a member of the product engineering group.

Engineering. Specific configurations of all models in the product line are finalized here, often in consultation with product innovation. The manufacturing engineering functions, including the determination of manufacturing operations, raw material and purchased parts requirements, standard processing times, setup times, and so on, are also performed here. Standard operating procedures for performing the operations are prepared. The complete file for each new or modified product is conveyed to operations planning. Important feedback, including cost information from the production process, quality control information, labor usage, material usage, scrap rates, and so on, comes from measures of effectiveness.

Purchasing. Purchase orders are received from inventory control. A vendor is selected, and the order is placed. Material costs are conveyed to financial control.

Financial Control. This function determines and reports the effects of the company's activities upon its financial position. Standard costs are developed from production records, material costs, line labor, and overhead rates. Labor charges are accumulated against work orders, accounts, and departments. Reports are produced which cover variances between actual and standard costs of material, labor, and overhead. Budget constraints are transmitted to operations planning. Vendors are paid for raw materials. Production costs are received from operations, and sales dollars are received from sales.

7.2.7 Integrating the Functions

There is, of course, considerable overlap among the five functions discussed in preceding sections. The performance of certain of these functions depends on the previous performance of other functions, which may depend on still others. It is usually necessary, therefore, to perform some functions essentially simultaneously, while performing certain groups of functions iteratively. The concept of an integrated operations planning and control system is explored further in Section 7.8.

7.3. Techniques for Demand Forecasting

Demand forecasting provides the major input to the other functions in the operations planning and control system. The other functions convert this forecast into material requirements, parts lists, labor requirements, schedules, and other decisions.

All planning must begin with an estimate, or forecast, of the amount of business our company can expect during the planning period. The means by which the estimate is arrived at may be completely subjective or unscientific, but the fact remains that all other planning of the company's activities depends on an estimate of business volume.

In this section we examine some of the elementary techniques of demand forecasting.

7.3.1 Moving Average

A commonly used forecasting technique is the *moving average method*. This method generates next period's forecast by averaging the actual demand values for the last n time periods. Mathematically, we have

$$\hat{x}_t = \frac{\sum_{i=1}^{n} x_{t-i}}{n} \tag{7.1}$$

where $\hat{x}_t$ = forecasted demand for period t

x_{t-i} = actual demand for ith period preceding t

n = number of time periods to include in the moving average

The choice of the value for n is arbitrary and should be determined by experimentation (perhaps in a computer simulation model).

A variation of the procedure above is the *weighted moving average*, in which more weight is given to more recent data. The forecasting equation is

$$\hat{x}_t = \frac{\sum_{i=1}^{n} c_i x_{t-i}}{n} \tag{7.2}$$

where c_i = weight given to ith actual value

$$\sum_{i=1}^{n} c_i = n$$

The choice of values for n and the weighting coefficients are arbitrary and can be determined by experimenting with several combinations.

7.3.2 Exponentially Weighted Moving Average

The exponentially weighted moving average (EWMA) is a forecasting method which overcomes many of the disadvantages of the ordinary moving average and weighted moving average methods. Like the weighted moving average method, the EWMA method gives more weight to newer data; however, the EWMA method is much more efficient and is more readily adapted to computer application.

EWMA assigns weights to the demand values of previous periods in inverse proportion to their age. It does this in a very ingenious manner in which only three pieces of data are required to generate next period's forecast: (1) last period's forecast, (2) last period's actual demand, and (3) a smoothing constant, which determines the relative amount of weight given to recent demand values.

The rationale behind EWMA is that we would like our forecasting method to track the demand pattern through its general trends (i.e., its ''ups and downs'') without overreacting to purely random fluctuations. Suppose, for example, that for a particular product we forecasted a demand of 100 units. When actual demand became known, it was only 95 units. We must now forecast demand for the following month. Our question now becomes: How much of the difference between 100 units and 95 units can be attributed to an actual shift in the demand pattern and how much can be attributed to purely random causes? If our next forecast is for 100 units, we would be assuming that all of the difference should be attributed to chance and, therefore, that the demand pattern has not shifted at all. If our next forecast is for 95 units, we would be assuming that the entire difference should be attributed to a shift in the demand pattern and none to chance causes.

EWMA attributes part of the difference between actual demand and the forecasted demand to a trend shift and the remainder to chance causes. In our example, suppose we decide that 20% of the difference should be attributed to a trend shift and the remaining 80% to chance causes. Our new forecast, then, would be determined by reducing the previous forecast by 20% of the observed difference of 5 units. Since our previous forecast was 100 units and 20% of 5 units is 1 unit, the new forecast would be 99 units. After the actual demand for the new period became known, we would determine the forecast for the following period by again computing 20% of actual demand minus forecasted demand and adding that quantity (which may be positive or negative) to the previous forecast, or 99. The 20% value is called the *smoothing constant* and is denoted α.

We can generalize this procedure as follows:

$$\text{Next forecast} = \text{previous forecast} + \text{smoothing constant} \times \text{forecast error}$$

$$\hat{x}_t = \hat{x}_{t-1} + \alpha(x_{t-1} - \hat{x}_{t-1}) \qquad (0 \leq \alpha \leq 1) \tag{7.3}$$

This expression can be rearranged algebraically to obtain an equivalent expression more convenient for computation.

$$\hat{x}_t = \alpha x_{t-1} + (1-\alpha)\hat{x}_{t-1} \tag{7.4}$$

7.3.3 Regression Analysis

When a demand pattern is consistently increasing or decreasing, we can develop our forecasting transfer function by using regression analysis, the details of which may be found in almost all standard statistics texts.[3,4]

The rationale for using regression analysis for forecasting is that the total set of factors that generated the cause system in the past will continue to operate in the future. Thus, forecasting becomes a matter of determining the general trend line and extrapolating this line into the future.

Returning to our hypothetical company, Gadgets, Inc., suppose that during a 12-month period a particular model of our product, super-gadget, had sales volumes as shown in Figure 7.4. These values are plotted (months 1 through 12) in Figure 7.5. The solid line plotted through the 12 points represents the general trend of sales. The dashed line extended from the solid line represents our best guess as to what the future demand will be.

There are two ways of determining the trend line. Often, as in this case, a good trend line can be fitted to a set of points "by eyeball." If more precision is desired, we can use regression analysis. This method fits a line to a set of observations such that the sum of the squared vertical distances of the observations from the line is minimized.

The basic equation for a straight line that expresses demand (x) as a function of time (t) is

$$\hat{x}_t = a + bt \tag{7.5}$$

where a is the intersection of the line with the vertical axis when $t = 0$ and b is the slope of the line. Our problem is to determine the values of a and b such that the

---

[3]Irwin Miller and J.E. Freund, *Probability and Statistics for Engineers, 3rd ed.* (Englewood Cliffs, N.J.: Prentice Hall, 1985).

[4]W.W. Hines and Douglas C. Montgomery, *Probability and Statistics in Engineering and Management Science*, 2nd ed. (New York: John Wiley & Sons, Inc., 1980).

Month	*Demand*
Jan,	500
Feb.	510
Mar.	480
Apr.	600
May	600
June	660
July	590
Aug.	700
Sept.	680
Oct.	740
Nov.	790
Dec.	760

Fig. 7.4. One-year sales records for super-gadget.

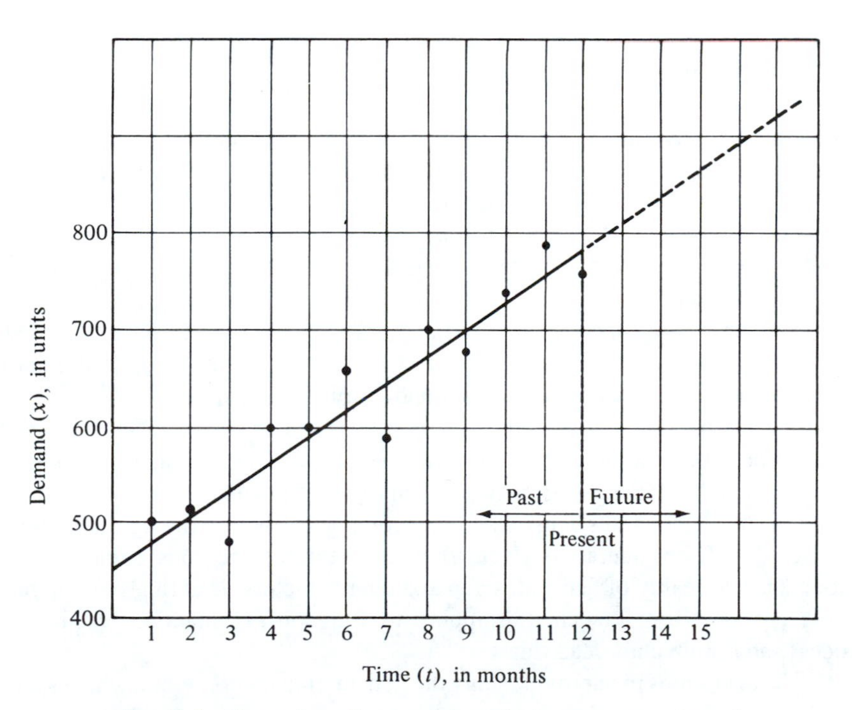

Fig. 7.5. Illustration of regression analysis in an upward trend.

least squares criterion is satisfied. The following equations for a and b are derived in almost all statistics texts:

$$b = \frac{n\sum_{i=1}^{n} t_i x_i - \sum_{i=1}^{n} t_i \sum_{i=1}^{n} x_i}{n\sum_{i=1}^{n} t_i^2 - \left(\sum_{i=1}^{n} t_i\right)^2}$$

$$a = \bar{x} - b\bar{t} \tag{7.6}$$

where n is the number of periods of demand data that we include in the calculations.

Applying these equations to the data in Figure 7.4, we obtain

$$b = 27.8, \qquad a = 452.6$$

Thus, our forecasting equation is

$$\hat{x}_t = 452.6 + 27.8t$$

which says that demand is increasing by an average of 27.8 units per month. Although we realize that our extrapolated line will not agree precisely with future demand values, it is expected that deviations of actual demand from the extrapolated line will be the result of random fluctuations from the line. Remember, it is the trend that we hope to predict accurately, not each individual demand value.

7.4. Techniques for Operations Planning

The demand forecasting function provides us with a refined estimate of how many units of each product are likely to be demanded by our customers and when during the planning period the demands are expected to occur. Operations planning must now convert the demand forecasts into a complete production program in which all resources (persons, machines, and materials) are coordinated for the maximum benefit to the company.

The purpose of operations planning is to assure that all resources needed to produce the required items (determined by the forecast) are at the right place at the right time and in the needed quantities and, furthermore, that waste of resources (idle time and overly large inventories of materials and product, etc.) is minimized. All this must be accomplished within the overall constraints (such as budgetary limitations) and policies (such as steady employment practices) imposed by higher management. These tasks must also be accomplished with existing or obtainable resources, with consideration given to such factors as previous commitment of resources, seasonality of demand, scrap and quality factors, lead times of purchased items, in-transit times (such as shipping to warehouse), quantities required in-process, and production lead times.

The operations plan provides only the general framework within which specific activities are to be performed. The plan contains reasonable flexibility, so that detailed task assignments can be made within the general framework.

It is helpful in operations planning to represent various requirements as a function of time. Suppose that a certain super-gadget model has a seasonal sales pattern and that we have forecasted its demand for the next 12 months to be the values shown in Table 7.1. The cumulative demands are also shown.

The cumulative demand values are plotted as a solid line in Figure 7.6.

The significance of the cumulative demand curve is that through a combination of inventory and production, we must have available at least the cumulative values by each of the corresponding time periods. At any one time our combined inventory and production can exceed the cumulative demand, but it cannot be less (if no backordering is permitted).

Two possible production plans (production plus inventory) are plotted as dashed lines. Plan 1 permits a level production rate of about 70 units per month, but assumes that we begin the year with about 340 units in inventory. Note also that we end the year with almost no inventory. This condition could be undesirable for next year's planning. Plan 2 employs a production rate of 120 units per month for the first 6 months and 60 units per month for the last 6 months. In plan 2 we must start the year with only 100 units in inventory, but this plan also leaves us with almost no inventory at the end of the year. The two plans are summarized in Table 7.2.

The plans discussed here are but two of many different ones that could be used to satisfy the expected demand. It is the purpose of operations planning to devise the best plan, that is, the one that results in lowest total cost to the company.

We can convert units to labor-hours and machine-hours and tabulate these requirements as a function of time. We could then perform the same procedure for all products produced in our company and summarize all requirements for each department.

There are several matrix-based techniques for generating requirement-time profiles and summarizing the requirements by department and machine center. These techniques, mostly computerized, are beyond the scope of this book.

To illustrate the problem of planning production when demand varies from month to month, consider again the monthly demand values for super-gadgets shown in Table 7.1. If we planned to produce exactly 103 units in month 1, 117 units in month 2, and so on, our monthly production rate would vary greatly. If we had a

Table 7.1. MONTHLY AND CUMULATIVE DEMAND FORECASTS IN UNITS

	Month											
	1	*2*	*3*	*4*	*5*	*6*	*7*	*8*	*9*	*10*	*11*	*12*
Demand	103	117	115	121	123	109	89	74	71	73	81	98
Cumulative	103	220	335	456	579	688	777	851	922	995	1,076	1,174

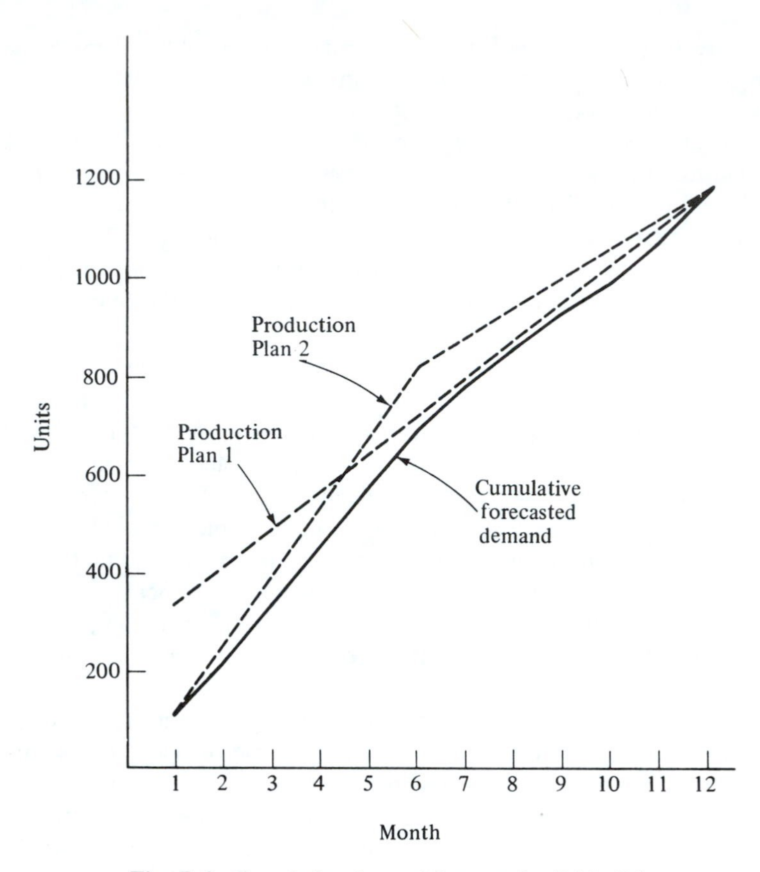

Fig. 7.6. Cumulative demand forecast for Table 7.1.

Table 7.2. PRODUCTION PLANS 1 AND 2 OF FIGURE 7.6

			Production Plan 1			*Production Plan 2*		
Month	*Demand Forecast*	*Cumulative Forecast*	*Begin Inventory*	*Production*	*End Inventory*	*Begin Inventory*	*Production*	*End Inventory*
1	103	103	340	70	307	100	120	117
2	117	220	307	70	260	117	120	120
3	115	335	260	70	215	120	120	125
4	121	456	215	70	164	125	120	124
5	123	579	164	70	111	124	120	121
6	109	688	111	70	72	121	120	132
7	89	777	72	70	53	132	60	103
8	74	851	53	70	49	103	60	89
9	71	922	49	70	48	89	60	78
10	73	995	48	70	45	78	60	65
11	81	1,076	45	70	34	65	60	44
12	98	1,174	34	70	6	44	60	6

constant size work force, we would be forced to work a lot of overtime during the high-demand months, and there would be a lot of idle time during the low-demand months.

If, however, we insisted on a level production rate, we would either have to lose sales during the high-demand months or carry a high inventory of finished goods to meet demand during those periods.

There are extra costs associated with both of the policies described above. The first cost involves overtime and idle time. The second cost involves inventory carrying costs.

Our objective is to smooth production over the planning period (usually one year) such that total cost is minimized.

Fairly good results can be achieved in smoothing production over the year simply by "eyeballing" it. More elaborate procedures are presented in most modern texts on production planning and control.

7.5. Techniques for Inventory Planning and Control

The demand forecasting function (Section 7.3) provides us with an estimate of demand for each of our products. Forecasts are generated for each operating period, and these values are summed to determine cumulative forecasts for the entire planning period. The operations planning function (Section 7.4) converts the forecasts into a time-phased production program. Units of finished product are translated into material requirements of raw materials, purchased parts, fabricated parts, and assemblies. Inventory planning and control must now provide procedures that, while assuring that materials and products will be available when they are needed and in the quantities required, will also guard against the costs of excessive inventories. The purpose of the inventory planning and control function is to determine appropriate inventory policies and to keep all associated costs at a minimum.

Almost all inventory problems are concerned with answering two fundamental questions:

(1) *How much* to order (either from a supplier or from our own production facilities) at one time?
(2) *When* (or *how often*) to place an order?

There are opposing costs in each of these questions. In the first question there is a cost associated with ordering too much at one time and another associated with ordering too little at one time. In the second question there is a cost associated with placing orders too frequently and another associated with not placing orders frequently enough. Our general objective is to determine the course of action that minimizes the sum of all such costs.

The expression for total costs depends on the particular situation. Several models are developed later in this section. It would be instructive at this point, however, to identify the major classes of inventory costs.

- *Procurement Cost*—The procurement cost of a purchased item is the order cost and consists of the clerical cost of making up and processing a purchase requisition. In general, order cost includes any cost whose size or amount is affected by the number of orders processed during a given time period.
- *Carrying Cost*—Carrying cost is the cost of holding inventory and includes several component costs:
  (1) The cost of money invested in inventory which could be used in other ways.
  (2) The cost of storage space—warehousing costs, utilities such as heat and light, etc.
  (3) The cost of obsolescence, spoilage, and pilferage.
  (4) The cost of insurance and taxes on the items in inventory.

The classical inventory model determines the optimal number of units to order on each order placed. This number is derived under the following assumptions:

(1) The usage rate of the item is linear and known with certainty.
(2) The order is received instantaneously when placed (i.e., lead time is zero).

An inventory system operating under these idealized conditions would behave as shown in Figure 7.7. Under the conditions we have assumed, we can wait until the inventory level reaches zero, then place an order for the desired amount. This order is received immediately. Our objective is to determine the number that should be ordered at one time.

We can express the total cost as the sum of procurement cost and carrying cost:

$$\text{total cost} = \text{procurement cost} + \text{carrying cost}$$

We use the following symbols in the development of the model:

$$
\begin{aligned}
TC &= \text{total cost} \\
PC &= \text{procurement cost for each order} \\
CC &= \text{carrying cost per unit per year} \\
D &= \text{annual demand (or usage) of the item} \\
Q &= \text{quantity ordered on each order placed (lot size)} \\
Q_0 &= \text{optimal lot size}
\end{aligned}
$$

To determine the particular lot size that will result in the lowest value for total cost, we develop a total cost expression in which the two opposing costs, procurement cost and carrying cost, are in terms of Q, the quantity ordered.

We can express the procurement cost in terms of Q by recognizing that the total annual procurement is equal to PC (the procurement cost for each order) multiplied by the number of orders placed during the year. (The number of orders placed during the year is just the total annual demand divided by the number of units ordered on each order.) Symbolically, therefore, we have

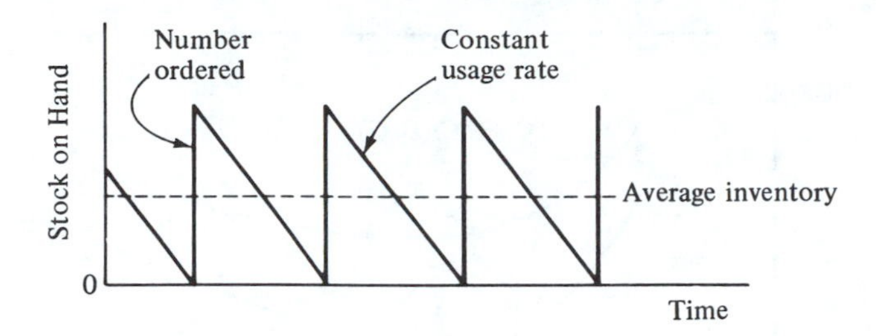

Fig. 7.7. Idealized inventory system.

$$\text{annual procurement cost} = PC\frac{D}{Q}$$

We can express the carrying cost in terms of Q by recognizing that the total annual carrying cost is equal to CC (carrying cost per unit per year) multiplied by the average number of units held in inventory. It is clear from Figure 7.7 that the average number of units held in inventory is just $Q/2$. Therefore,

$$\text{annual carrying cost} = CC\frac{Q}{2}$$

Our total cost equation can now be written as

$$TC = PC\frac{D}{Q} + CC\frac{Q}{2} \tag{7.7}$$

For a particular inventory problem we will know values for PC, CC, and D. With these values fixed, total cost becomes a function only of Q, the lot size. Suppose, for example, that PC = \$10.00 per order, CC = \$0.20 per unit per year, and D = 10,000 units. If we order 2,000 units at a time, then from Equation (7.7) we obtain

$$\begin{aligned} TC &= \$10.00\,\frac{10{,}000}{2{,}000} + \$0.20\,\frac{2{,}000}{2} \\ &= \$50 + \$200 = \$250 \end{aligned}$$

Other total cost values would be obtained by using different lot sizes. In fact, it is instructive to plot several values of Q, as has been done in Figure 7.8, to show its effect on each cost component in Equation (7.7), as well as on TC itself.

Our objective is to determine the particular value of Q that will minimize our total cost equation. In Figure 7.8 we see that the minimum value on the TC curve occurs at a point where the curve has zero slope. We know from differential calculus that the first derivative of a mathematical function gives the slope (instantaneous rate of change) of the function at a particular point. In our case, we are hunting the point where the slope is zero. Our approach, then, is to take the first derivative of

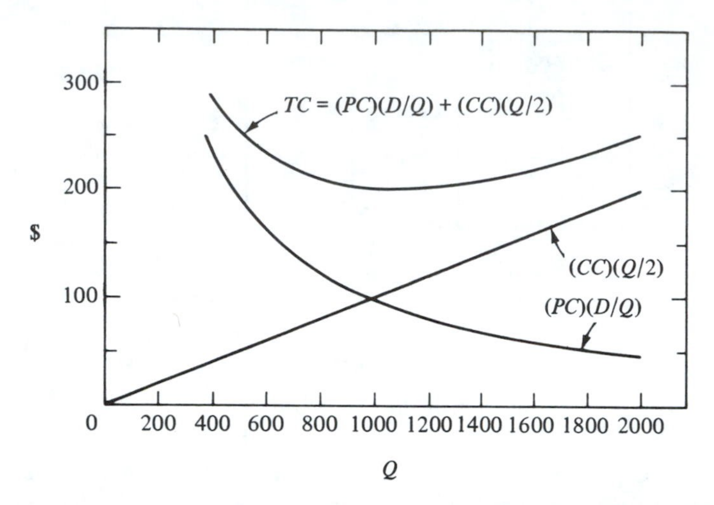

Fig. 7.8. Total cost function of classical inventory model where $D = 10{,}000$ units per year, $CC = \$0.20$ per unit per year, and $PC = \$10{,}000$ per order.

the total cost equation, Equation (7.7), set the derivative equal to zero, and then solve for Q_0, the optimal lot size:

$$\frac{d(TC)}{dQ} = -PC\frac{D}{Q^2} + \frac{CC}{2} \tag{7.8}$$

$$-PC\frac{D}{Q^2} + \frac{CC}{2} = 0 \tag{7.9}$$

$$Q_0 = \sqrt{\frac{2(PC)(D)}{CC}} \tag{7.10}$$

By substituting our example values into Equation (7.10), we obtain

$$Q_0 = \sqrt{\frac{2(\$10.00)(10{,}000)}{\$0.20}} = 1{,}000 \text{ units}$$

We see from Figure 7.7 that a lot size of 1,000 units does indeed result in lowest total cost of $200.

Several other models are presented in other texts which allow for such factors as variable demand, lead time, and safety stocks.

7.6. Techniques for Operations Scheduling

Demand forecasting (Section 7.3) provides an estimate of demand for each product during each operating time period. Operations planning (Section 7.4) develops a time-phased production plan, in which resources are matched to production requirements and are smoothed over the planning period. Inventory planning and control (Section 7.5) regulates the flow of materials, parts, and products into, through, and out of the production system. This regulatory action results in the placement of purchase orders for purchased items and production orders for manufactured items. The operations scheduling function must now fit the production orders into the existing work loads of the various departments or operating facilities.

7.6.1 Purpose of Operations Scheduling

The purpose of operations scheduling is to assign specific operations to specific operating facilities with specific start and end times indicated.

Operations scheduling attempts to assign work to the required facilities in such a way that all the various costs associated with manufacturing are minimized. These costs are caused by factors such as in-process inventory, idle workers and equipment (for whatever reasons), overtime, orders completed behind schedule, and so on.

To fully appreciate the magnitude and importance of the scheduling function, we must first attempt to visualize the complexity of large manufacturing operations. Orders (regular production orders from inventory control and special orders through sales) are received periodically. Individual jobs are assigned to specific facilities. Certain precedence restrictions must be observed. As jobs progress through the facilities, they often must compete with other jobs for the same scarce resources. This competition results in conflicts on various facilities. Machines break down, some workers fail to appear and others perform below or above standard, tools break or wear out, materials are defective, and machines go idle waiting for work from preceding work centers. Orders are canceled, reduced, or increased. Raw materials fail to arrive when expected. Sales drop suddenly or increase sharply. Marketing begins a crash advertising campaign that causes a sudden increase in demand, after which demand will oscillate sharply until the system is allowed to stabilize again. Engineering introduces product design modifications that alter standard processing times, setup times, operation sequences, operator instructions, etc.

It is very difficult to exercise close control over a dynamic environment such as that described above. In spite of the difficulties, however, the scheduling function must be performed. The objectives of operations scheduling are:

(1) High percentage of orders completed on time.
(2) High utilization of facilities and workers.
(3) Low in-process inventory.
(4) Low overtime.
(5) Low stockouts of manufactured items.

These objectives must be accomplished within the overall framework specified by the operations plan.

One of the oldest techniques available for sequencing and scheduling operations to facilities is the *Gantt chart*. The Gantt chart is very useful in showing planned work activities versus actual accomplishments on the same time scale. There are numerous possible variations of the Gantt chart. Many are color coded to indicate certain conditions such as material shortages, machine breakdowns, and so forth.

The basic Gantt scheduling chart is shown in Figure 7.9. Operations assigned to each facility are sequenced into a feasible schedule. Actual progress is indicated on the same chart to show any deviations from planned operation process times. The status of the facilities can be read from the chart. Operations 44*A* and 65*D* have been completed. Operation 65*E* is exactly on schedule as of now. Operation 44*C* is a half-day behind schedule. Operation 86*B* is 2 days ahead of schedule. Operations 86*D* and 103*C* are scheduled but have not yet begun. We should recognize that the time scale may be in any appropriate time units—hours, days, weeks, or months.

When the Gantt chart is used to sequence several operations resulting from new production orders, a good rule to follow is to sequence the most heavily loaded facility first, the second most heavily loaded facility second, and so on.

Although a fairly good schedule can be generated in this way, it will not be optimal under any criterion except by accident.

The Gantt scheduling chart must be updated very frequently in order to reflect the changing status of the shop. Certain jobs will take longer than expected, while others will be completed ahead of schedule. Machines will break down, workers will be absent, and materials will run out. All these factors make it difficult to keep

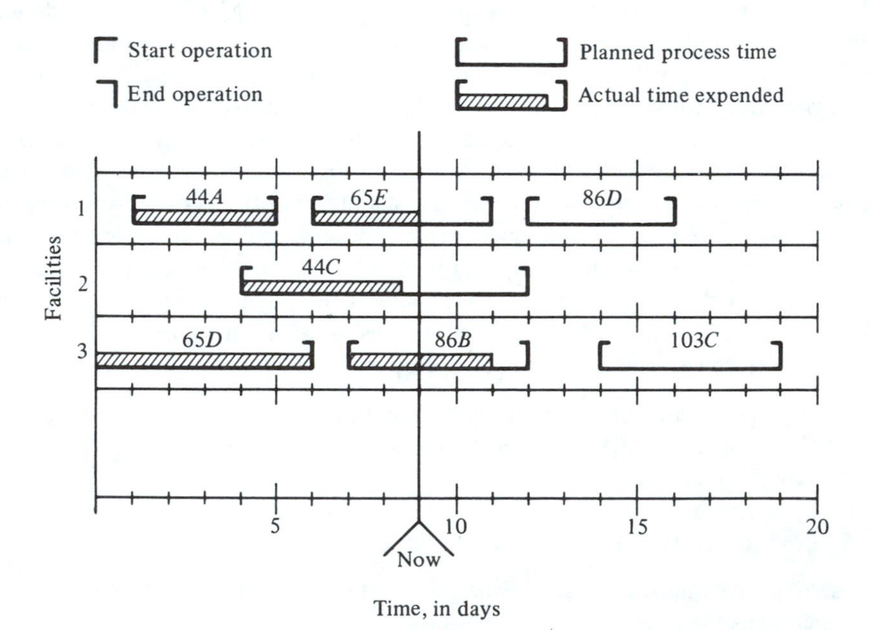

Fig. 7.9. Gantt scheduling chart.

the Gantt chart updated on a timely basis. Nevertheless, the Gantt chart should not be overlooked as an aid for loading, sequencing, and scheduling.

7.7. Dispatching and Progress Control

In the preceding section we concentrated primarily on the way decisions are made in an operations planning and control system. For example, we talked about inventory decisions, forecasting decisions, scheduling decisions, and the like. Now we must consider how all these decisions are implemented, that is, how they are transferred into action. Somehow, the decisions must be relayed from the production control office to the supervisors and workers on the shop floor; material has to be issued; instructions have to be gotten to workers; parts must move from one operation to the next. As work is accomplished, we must have a means of recording progress, comparing it to planned progress, and then taking whatever corrective action that seems appropriate.

We could say that the primary purpose of dispatching and progress control is to close the loop in our operations planning and control system. To accomplish this, we must provide ways to detect when a deviation has occurred, analyze what has happened, decide what corrective action would be appropriate, implement the corrective action by incorporating it into subsequent scheduling decisions, and finally, feed back progress information to the planning section to improve future planning.

Procedures for accomplishing this function are beyond the scope of this text. An essential component of such procedures is obviously a good information system. Consequently, this function is often performed with the aid of modern computer systems. See Chapter 20 for more details.

7.8. MRP Systems

One of the most important recent developments in operations planning and control is that of material requirements planning (MRP) systems, and their derivatives, MRPII (manufacturing resource planning) and closed-loop MRP. Details of MRP may be found in other sources.[5] Our purpose here is to introduce the concepts briefly and illustrate MRP's usefulness in effectively managing an operation.

The basic concepts of operations planning were introduced in Section 7.4. The primary output of operations planning is a *master production schedule*, which is a listing of the specific products to be produced and the number of units of each product that will be produced in each time period. This becomes a primary input to MRP.

---

[5]T. E. Vollman, W. L. Berry, and D. C. Whybark, *Manufacturing Planning and Control Systems*, 2nd ed. (Homewood, Ill.: Richard D. Irwin, Inc., 1988).

Another primary input to MRP is the *bill of materials* (BOM) for each product. The BOM is a detailed breakdown of the various subassemblies, components, and raw materials that make up the product. Typically, this detailed breakdown is shown in the form of a *product structure diagram*, an example of which is shown in Figure 7.10. P1 indicates product 1; SA1 and SA2 indicate subassemblies 1 and 2; C1 through C5 indicate components 1, 2, 3, 4, and 5. The number in parentheses to the right of each box indicates the number of units of that item required to produce 1 unit of the item it goes into (i.e., the item at the next-higher level in the tree). For example, 4 C1s are required to produce 1 unit of SA1, and two SA1s are required to produce 1 unit of P1. Thus a total of 8 units of C1 is required to produce 1 unit of P1. The number just above each box indicates the lead time for that item. Lead time is the number of time periods required to obtain an order of a particular item. The lead time for SA2, for example, is 2 time periods.

The MRP program (usually, a computer software package) calculates the number of each subassembly and component required to produce a specified number of end products. It does this by "exploding" the end product requirements into successively lower levels within the product structure.

Suppose that our master production schedule specifies 200 units of product P1 needed in week 17. Referring to the product structure diagram in Figure 7.10, we can calculate the number of units of each subassembly and each component that will be needed:

SA1	400
SA2	200
C1	1,600
C2	800
C3	400
C4	200
C5	3,200

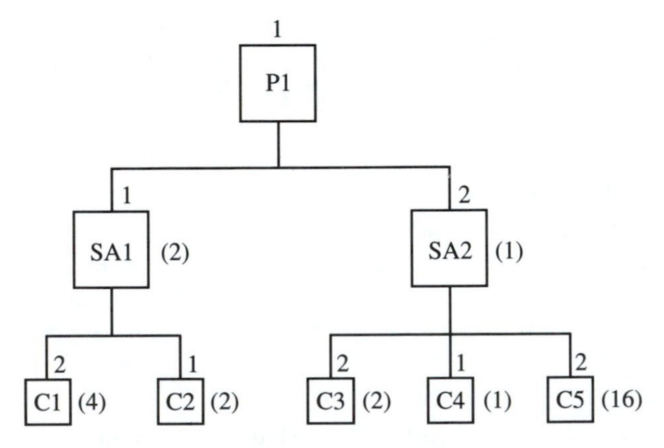

Fig. 7.10. Product structure diagram for product P1.

We must now consider the lead times required for each item. P1 has a lead time of 1 week. (For the remainder of this example, the time period will be considered as one week.) This means that 400 units of SA1 and 200 units of SA2 must be available at the end of week 16. The lead time for SA2 is 2 weeks. This means that 400 units of C3, 200 units of C4, and 3,200 units of C5 must be available at the end of week 14. Therefore, an order for 3,200 units of C5 must be placed no later than the end of week 12.

Notice that the calculated requirements for each item is the gross number of units needed in order to satisfy the master production schedule. If some of these items are already in inventory, or if an order is scheduled for delivery in the near future, these must be subtracted from gross requirements to determine net requirements for satisfying the master production schedule. For example, suppose that we already have 716 units of C5 in inventory. The net requirement for C5 is, therefore, 3,200 − 716 = 2,484 units. Suppose further that C4 is a purchased item and that 75 units of C4 are scheduled to arrive in week 12. Our gross requirements for C4 were calculated earlier as 200 units to be available at the end of week 14. So our net requirements for C4 are 200 − 75 = 125 units.

A systematic means of performing these calculations is needed. Numerous computer software packages are available for performing MRP calculations. For manual calculations, a matrix approach is convenient. For the calculations discussed above, the MRP solution is shown in Figure 7.11. Notice that only those calculations pertaining to the right branch (SA2) in Figure 7.10 are shown. Similar calculations would be required to determine the time-phased requirements for all items in the left branch (SA1).

Figure 7.11 is merely a "slice" of a total MRP solution, in that it illustrates the backward explosion calculations from the demand for P1 in week 17. Similar calculations would be made for the demand for P1 in all other weeks. These demand values, shown as "gross requirements" on the form, come from the master production schedule for end products. However, the "gross requirements" value for all subassemblies and components must be calculated according to the information provided on the product structure diagram (Figure 7.10).

When implemented properly into a manufacturing operation, a MRP system provides tight, accurate control over the many complex, interdependent activities of the plant. All activities are synchronized such that parts are being manufactured (or received from vendors, if purchased) exactly when they are needed for the next stage of production. Such a system avoids the typical problems of excess in-process inventory, unavailability of components when needed, grossly uneven utilization of resources, and so on.

When these concepts are extended to the entire range of manufacturing operations, and when corrective action is incorporated into the system, the result is a "closed-loop MRP system." In a closed-loop MRP system, the various functions of operations planning and control (forecasting, operations planning, inventory management, MRP calculations, dispatching, and progress control) have been integrated into one unified system. It also provides for feedback from vendors

Period		10	11	12	13	14	15	16	17	18	19
Item: P1	LT 1										
Gross Requirements									200		
Scheduled Receipts											
On Hand	0										
Net Requirements									200		
Planned Order Releases								200			
Item: SA 2	LT 2										
Gross Requirements								200			
Scheduled Receipts											
On Hand	0										
Net Requirements								200			
Planned Order Releases						200					
Item: C3	LT 2										
Gross Requirements						400					
Scheduled Receipts											
On Hand	0										
Net Requirements						400					
Planned Order Releases				400							
Item: C4	LT 1										
Gross Requirements						200					
Scheduled Receipts				75							
On Hand	0			75	75	75					
Net Requirements						125					
Planned Order Releases					125						
Item: C5	LT 2										
Gross Requirements						3200					
Scheduled Receipts											
On Hand	716	716	716	716	716	716					
Net Requirements						2484					
Planned Order Releases				2484							
Item:	LT										
Gross Requirements											
Scheduled Receipts											
On Hand											
Net Requirements											
Planned Order Release											

Fig. 7.11. MRP calculations. [Form adapted from M. P. Groover, *Automation, Production Systems, and Computer Integrated Manufacturing* (Englewood Cliffs, N.J.: Prentice Hall, 1987), Fig. 24.10, p. 739.]

(delayed order deliveries, etc.) and from customers (changes in orders, models, etc.).

One further extension to the MRP concept can be made by linking the closed-loop MRP system to the financial function of the organization. This type of structure, called "manufacturing resource planning," is typically indicated by the label MRPII. Commercially available MRPII software usually includes a simulation capability. The simulator allows the manager to explore various "what if" questions. For example, if a particular customer requests that its order for 500 units be delivered 3 weeks earlier than the original contract date, the simulator allows the manager to determine whether this request can be granted without an adverse effect on other customer orders.

7.9. Just-in-Time Manufacturing[6]

A very important recent development in operations planning and control is that of just-in-time (JIT) manufacturing. Sometimes referred to as "stockless production," JIT is more than just a method to reduce inventory. It is also concerned with the total organization of the production process so that defect-free components (both manufactured and purchased) are available for the next stage of production exactly when they are needed—not too late and not too early.

A company implementing the JIT philosophy must attend to several important details:

- *Smooth Material Flow*—Simplify material flow patterns. This sometimes requires a total rearrangement of the production lines. It also requires direct access to and from receiving and shipping docks. The goal is to have material flow uninterrupted from receipt, directly through each production stage, then to delivery. Anything that interrupts this flow is a target for investigation and elimination.
- *Reduction of Setup Times*—Prior to JIT, many discrete part manufacturers had machine setups that sometimes required several hours. This is intolerable in a JIT system. Dramatic reductions in setup times have been achieved by many companies, sometimes from 4 to 7 hours to 3 to 7 minutes. This allows batch sizes to be reduced to very small values, thereby allowing the company to be very flexible and responsive to changes in customer requirements.
- *Reduction of Vendor Lead Times*—Instead of receiving large shipments of purchased parts every 2 or 3 months, under JIT we want to receive parts "just in time" for the needed production operation. Companies sometimes have to negotiate long-term contracts with suppliers in order to achieve this feature.
- *Zero-Defect Components*—A JIT system cannot tolerate defective compo-

[6]See Chapter 7, W. W. Luggen, *Flexible Manufacturing Cells and Systems* (Englewood Cliffs, N.J.: Prentice Hall, 1991).

nents, either manufactured or purchased. For manufactured components, statistical process control techniques must be implemented to assure that the *process* is producing parts within tolerance at all times. For purchased components, vendors are required to guarantee that the components they are supplying have been produced on processes that are in statistical control. A company will usually have a vendor certification program to assure that this occurs.

- *Disciplined Shop Floor Control*—In traditional shop floor control systems, the emphasis was on high machine utilization, relatively long production runs to keep setup costs low, and on low labor idle time. Consequently, production orders were released to the production units with these factors in mind. Under JIT, these traditional performance measures are far outweighed by the desire to keep inventories very low and to maintain an "unclogged," responsive operation. This requires that strict adherence to precise order release times be adhered to at all times. This means that machines and machine operators are sometimes idle. Many production managers who have spent most of their professional lives keeping machines busy and operators working are having a difficult time making the necessary mental adjustments required for a successful JIT operation. For those companies that have successfully implemented the JIT philosophy, the rewards have been significant.

DISCUSSION QUESTIONS

1. What are the elements of a control system? Describe each element individually and discuss how the elements operate together as a system.
2. Discuss the nature of the interdependencies between the five basic functions of operations planning and control.
3. In general, how much money should management be willing to spend in obtaining an accurate forecast of demand?
4. How would you select a particular forecasting technique to use in a given situation?
5. What is the *basic purpose* of operations planning?
6. Discuss the relationship between operations planning and operations scheduling.
7. What are the objectives in operations scheduling?
8. Discuss what would likely happen in a company that *did not* have a dispatching and progress control function.
9. Discuss how the framework presented in this chapter could be used in the management of
   (a) A hospital.
   (b) A bank.
   (c) A church.
   (d) A professional football team.
   (e) A university.

PROBLEMS

1. Suppose that the demand data for super-gadgets, models X and Y, are as follows:

Month	*Model X*	*Model Y*
Jan.	430	110
Feb.	380	160
Mar.	420	150
Apr.	370	100
May	410	130
June	380	70
July	440	120
Aug.	380	80
Sept.	420	80
Oct.	370	120
Nov.	410	40
Dec.	390	90

(a) Estimate the demand for each model for each of the following 12 months.
(b) Of the three forecasting procedures presented in the text, which do you believe is most appropriate for this case? Why?
(c) How reliable do you think your estimates are? Justify your answer.
(d) The Marketing Department informs you that, beginning in April, it will conduct a vigorous sales campaign for model X. Past records show that such campaigns usually increase sales by 20%, 15%, and 10% for the first, second, and third months, respectively, of the sale. During the fourth month of the campaign, sales are usually approximately 10% below what they would have been had there been no sale. Sales are 5% below "normal" during the fifth month and return to normal during the sixth month after the beginning of the campaign. Recalculate the estimated demand for each of the 12 months.

2. Consider the 36 months of historical demand data shown below.

Month	*Demand*	*Month*	*Demand*	*Month*	*Demand*
1	102	13	104	25	103
2	115	14	125	26	112
3	128	15	124	27	126
4	121	16	132	28	112
5	118	17	125	29	126
6	110	18	103	30	112
7	82	19	92	31	94
8	68	20	74	32	81
9	75	21	68	33	69
10	75	22	77	34	66
11	80	23	82	35	82
12	97	24	104	36	93

(a) Estimate the demand for each of the following 12 months.
(b) Discuss this particular sales pattern in terms of how well each of the three forecasting techniques presented in the text would perform.

3. From the modified forecast of Problem 1, determine a requirement-time profile for models X and Y, such as the one shown in Figure 7.6 and Table 7.2. Determine three workable production plans for each product. Show your plans on a graph. Are any of your plans optimal? Discuss the concept of optimality for problems such as this one.
4. From the forecast derived in Problem 2, determine a requirement-time profile. Determine three workable production plans for this product. Construct one of your plans such that there will be 100 units of the product remaining in inventory at the end of the year.
5. Consider again the modified forecast of Problem 1(d). Suppose that five production departments are involved in manufacturing models X and Y. The per unit production times (in labor hours) for each department are given below.

Department	*Model X*	*Model Y*
1	10	5
2		10
3	20	10
4	5	10
5	10	10

(a) Determine a combined requirement-time profile for each department and three workable production plans for each department.
(b) Determine the smoothest overall production plan you can develop "by eyeball." Attempt to minimize overtime and in-process inventory simultaneously.
(c) Discuss how you would adjust your production plan if past records showed the following scrap factors for the five departments:

Department	*Scrap Factor*
1	0.02
2	0.05
3	0.03
4	0.01
5	0.05

6. Annual usage of an item is 1,000 units. Ordering cost is $10.00 per order, and carrying cost is $0.50 per unit per year. Determine the optimal order quantity for this item. How many orders will be placed during the year? How many days will there be between consecutive orders? (Assume 250 working days in a year.)
7. Purchased components C1, C2, C3, and C4 are used on models X and Y of Problem 1(d) in the following quantities:

Component	*Model X*	*Model Y*	*Order Cost*	*Carrying Cost*
C1	4	2	$ 5.00	1.00
C2	0	5	10.00	0.50
C3	1	2	10.00	0.60
C4	5	5	20.00	1.20

Ordering costs (per order) and carrying costs (cost per unit per year) are also shown.

(a) Determine the optimal order quantity for each component by using the classical inventory model presented in the text.

(b) Discuss all of the assumptions pertaining to the classical inventory model as they apply to this problem. Which are unreasonable? Do you have any suggestions regarding these?

8. Consider the product structure diagram shown in Figure P7.8. The numbers on the diagram are explained in Section 7.8. Suppose that the master production schedule for P3 specifies demands for 30, 20, and 25 for time periods 9, 10, and 11. Perform the MRP calculations for all items. Assume that the company already has 16 units of C4 in stock, and that an order for 12 units of C1 will arrive at time period 4.

9. Consider the production structure diagram in Figure 7.10. Suppose that the company's engineering design department was able to make modifications such that component C4 could be replaced with component C2. Discuss how this would affect the MRP calculations.

10. In the example in Section 7.8, MRP calculations were completed only for the right side of the product structure diagram shown in Figure 7.10. Perform the MRP calculations for the left side. Assume that 40 units of SA1 are already in stock, and that an order of 60 units of C1 will arrive in week 12.

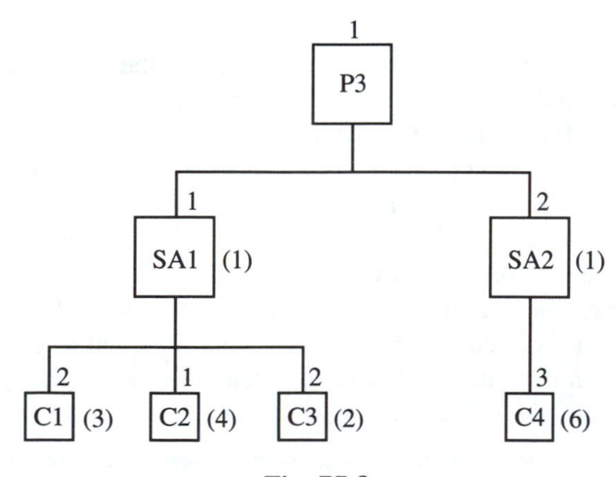

Fig. P7.8.

CHAPTER 8

Quality Control

8.1. Introduction[1]

Quality is judged by the customer. It is meeting and exceeding customer needs. It is delighting the customer. It is customer satisfaction. Quality sounds so simple, but only the best organizations worldwide are making these things happen on a day in, day out basis.

Quality and customer satisfaction have become the key drivers behind some of the most profound changes that have occurred in the world of work in all of recorded history. The focus today is on the customer, recognizing that there are many customers. For example, there are the external customers who buy and use a product or service. There are the internal customers who are the next persons in the process, perhaps working right next to us. We are trying to meet or exceed the needs of *all* of our customers today.

The emphasis on quality has not been an isolated event. It is intertwined with virtually all of industrial engineering. For exmaple, JIT and MRP (Chapter 7) will simply not work unless quality of products, deliveries, record keeping, and so on,

[1]Much of the content of this chapter comes from the materials of Exxon Chemical Company, Associates in Process Improvement, and from the Malcolm Baldrige National Quality Award Office of the National Institute of Standards and Technology, U.S. Department of Commerce.

from suppliers is top notch. Quality and JIT have affected facilities design, requiring that workstations be placed close together or even in cells, placing the internal supplier and internal customer elbow to elbow. Similar examples could go on and on.

8.2. A Bit of History

During the twentieth century, there has been a big flurry of ideas put forth under names such as quality control, statistical process control, designed experimentation, quality planning, quality costs, reliability, quality circles, zero defects, quality audit, quality function deployment, total quality management, benchmarking, and others. Each of these topics and many others have generated enthusiasm, research, and implementation. Each was successful in various places; similarly, each failed in other places.

Companies have evolved. After World War II, those manufacturers not severely affected by the war (like those in the United States) had little global competition. High short-term profitability was achieved by high production levels. Many manufacturing companies neglected quality in the push for production. Quality suffered, but manufacturers saw little reason to change since markets and profits continued.

We all know that the situation has changed. Global competition is intense. Industries and companies are faced with closings, sectors of high unemployment, and the loss of many high-paying jobs. Companies around the world that sacrifice quality improvement for increased production are in deep trouble. This is because a ''world class'' breed of companies now achieve both high quality and high productivity at lower cost. *No industry or company is immune to this new competition.*

Fortunately, we continue to learn what common principles and tools make some ideas work and some companies successful. Some of these common principles and tools are detailed in this chapter.

8.3. The Malcolm Baldrige National Quality Award[2]

The single most dramatic event in the past decade to foster quality in the United States (and even worldwide) has been the offering of the Malcolm Baldrige National Quality Award. The criteria for the award are extensive and they give a de facto definition of the words *total quality management* or (TQM). The specific topics addressed by the criteria fall into the seven categories shown in Figure 8.1. The seven major categories of the Baldrige criteria are often illustrated as shown in

---

[2]Kenneth E. Case has served on the panel of judges and also as a senior examiner in the Baldrige process.

1.0 LEADERSHIP

Senior Executive Leadership
Management for Quality
Public Responsibility

2.0 INFORMATION AND ANALYSIS

Scope and Management of Quality and Performance Data and Information
Competitive Comparisons and Benchmarks
Analysis and Uses of Company-Level Data

3.0 STRATEGIC QUALITY PLANNING

Strategic Quality and Company Performance Planning Process
Quality and Performance Plans

4.0 HUMAN RESOURCE DEVELOPMENT AND MANAGEMENT

Human Resource Management
Employee Involvement
Employee Education and Training
Employee Performance and Recognition
Employee Well-Being and Morale

5.0 MANAGEMENT OF PROCESS QUALITY

Design and Introduction of Quality Products and Services
Process Management-Product and Service Production and Delivery Processes
Process Management-Business Processes and Support Services
Supplier Quality
Quality Assessment

6.0 QUALITY AND OPERATIONAL RESULTS

Product and Service Quality Results
Company Operational Results
Business Process and Support Service Results
Supplier Quality Results

7.0 CUSTOMER FOCUS AND SATISFACTION

Customer Relationship Management
Commitment to Customers
Customer Satisfaction Determination
Customer Satisfaction Results
Customer Satisfaction Comparison
Future Requirements and Expectations of Customers

Fig. 8.1. Malcolm Baldrige National Quality Award Categories/Items.

Figure 8.2. Companies are using these criteria to measure themselves over time and against others, both in their own industry and in other industries. The primary motivation for this is to continuously improve, increase competitiveness, capture the market, survive, and thrive.

There are several ''bedrock quality concepts'' underlying the MBNQA criteria and the worldwide quality movement today. Interestingly, they are all also bedrock concepts underlying the practice of industrial engineering.

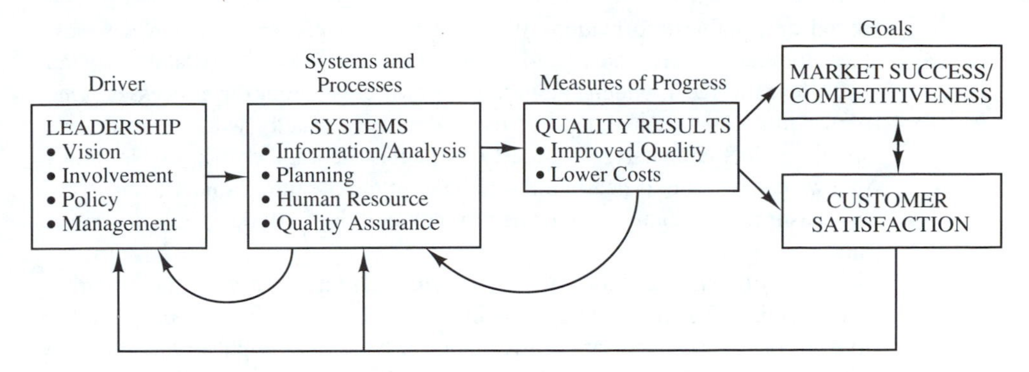

Fig. 8.2. National quality award categories dynamic relationships.

(1) *Quality is customer driven*. Since quality is judged by the customer, all product and service attributes that contribute value to the customer and affect customer preference must be considered in a quality system. Customer-driven quality demands constant awareness of changing customer wants and needs and rapid response to meet these requirements.

(2) *Leadership creates quality values*. A company's senior leaders must create clear quality values, specific goals, and well-defined systems and methods for achieving the goals. These systems and methods need to guide all activities of the company and to encourage participation by all employees. Senior leaders can make or break a company.

(3) *Optimized systems and processes*. Quality excellence comes from well-defined and well-executed work systems and processes (this is exactly what industrial engineers are concerned with). Quality is achieved by focusing on systems and processes and doing the right things right the first time. It is not achieved by focusing on inspection and correction of products or services after they have been made or delivered.

(4) *Response-time reduction*. Meeting changing customer requirements and expectations in competitive markets demands shorter cycles of concept–design–build–deliver. For example, an automaker that can deliver a new concept car in three years will far outdistance a car company that requires four or five years. Reducing response time often occurs when work processes are simplified and shortened, usually also resulting in improved quality.

(5) *Continuous improvement*. Improvements to all operations and work activities may be of several types: (a) increasing value to the customer by improving product features, (b) reducing errors and defects, (c) improving responsiveness and cycle time performance, and (d) improving efficiency and effectiveness in the use of all resources.

(6) *Actions based on facts, data, and analysis*. Real improvement requires that actions be taken to change and improve processes. These changes must be

based on reliable information. We must first get good data. Then we must look at trends, levels, and cause-and-effect relationships. Good data, analyzed well, leads to facts, understanding, and knowledge about our processes. One of the most important things for us to decide is exactly what measures we should collect and analyze. In most companies the number of possible things that we can measure is nearly infinite. Selecting those things that are important to measure can be done with the help of some of the analytical tools described later.

(7) *Strategic planning and well-developed goals*. It has been said that planning is everything. The planning process in an organization should be crystal clear to everyone. Employees at all levels should be able to imput to the planning effort. The major goals should lay out a well-defined direction for the company. The plan should then cascade through the company, with specific activities for individuals and teams at all levels.

(8) *Employees trained, developed, and involved*. Real progress toward world class requires that all employees be committed, well trained, and encouraged to participate in continuous improvement activities. The reward and recognition systems must be aligned with the goals of the company—world class will be impossible to achieve if the company rewards employees for ''business as in the past.'' Safety, health, well-being, and morale of employees are an integral part of getting them involved in continuous improvement.

(9) *Supplier involvement*. Incoming products and services must be of high quality if we are to add value and satisfy our customers. This is why suppliers are being treated as an extension of a company. The customer–supplier relationship is much closer than in the past, with confidential process information being exchanged if it promotes better quality. Partnerships are being formed over long-term periods, and customers are devoting time to ''growing'' their suppliers. As a result of the need to focus on the customer–supplier marriage, customers are vastly reducing the number of suppliers used, and the primary criterion for retaining them is their quality of products and services.

8.4. Deming's Thoughts on Continuous Improvement

One of the most outspoken proponents of continuous quality improvement is Dr. W. Edwards Deming. Dr. Deming has been a great supporter of the worker while being critical of management practices. While controversial, his teachings continue to grow in acceptance as more and more people realize that the management ways of the past have ceased to work in this worldwide competitive workplace. Dr. Deming's 14-Point Special Program may be paraphrased as in Figure 8.3.[3]

---

[3]This is simply an updating from the set of 14 points provided by Dr. Deming's office, Washington, D.C., Nov. 1983.

1. Create constancy of purpose toward improvement of product and service, with the aim to become competitive and to stay in business, and to provide jobs.
2. Adopt the new philosophy. We are in a new economic age. Western management must awaken to the challenge, must learn their responsibilities, and take on leadership for a change.
3. Cease dependence on inspection to achieve quality. Eliminate the need for inspection on a mass basis by building quality into the product in the first place.
4. End the practice of awarding business on the basis of price tag. Instead, minimize total cost. Move toward a single supplier for any one item, on a long-term relationship of loyalty and trust.
5. Improve constantly and forever the system of production and service, to improve quality and productivity, and thus constantly decrease costs.
6. Institute training on the job.
7. Institute leadership. The aim of supervision should be to help people and machines and gadgets to do a better job. Supervision of management is in need of overhaul, as well as supervision of production workers.
8. Drive out fear, so that everyone may work effectively for the company.
9. Break down barriers between departments. People in research, design, sales, and production must work as a team, to foresee problems of production and in use that may be encountered with the product or service.
10. Eliminate slogans, exhortations, and targets for the work force asking for zero defects and new levels of productivity. Such exhortations only create adversarial relationships, as the bulk of the causes of low quality and low productivity belong to the system and thus lie beyond the power of the work force.
11. a. Eliminate work standards (quotas) on the factory floor. Substitute leadership.
    b. Eliminate management by objective. Eliminate management by numbers, numerical goals. Substitute leadership.
12. a. Remove barriers that rob the hourly worker of his right to pride of workmanship. The responsibility of supervisors must be changed from sheer numbers to quality.
    b. Remove barriers that rob people in management and in engineering of their right to pride of workmanship. This means, *inter alia*, abolishment of the annual or merit rating and of management by objective.
13. Institute a vigorous program of education and self-improvement.
14. Put everybody in the company to work to accomplish the transformation. The transformation is everybody's job.

Fig. 8.3. Dr. Deming's elaboration on the 14 points. [From W. Edwards Deming, *Out of the Crisis* (Cambridge, Mass.: MIT Press, 1986), pp. 23–24.]

8.5. Juran's Contributions to Quality Thought

Dr. J. M. Juran has perhaps been the most prolific of the world's quality thought leaders. Dr. Juran, an engineer and lawyer, has leveraged his theories, observations, and knowledge to a worldwide audience through his writings and his consulting organization. Dr. Juran presents pragmatic, up-to-date, and detailed approaches for helping practitioners build an organization within which TQM is a reality.

Dr. Juran often speaks of his Juran Trilogy, which includes (1) quality planning, (2) quality control, and (3) quality improvement. *Quality planning* is determining current and future customer needs and developing products and processes to meet or exceed those needs. *Quality control* is evaluating actual performance, comparing it to goal, and taking action if the two are different. *Quality improvement* is the methodical attacking of waste (often thought of as non-value-added operations). In the remainder of this chapter we focus on some specific tools used in all three: planning, control, and improvement efforts.

8.6. Tools for On-Line vs. Off-Line Quality Control

We often hear people talk about off-line and on-line quality control. Figure 8.4 graphically presents a picture of off-line QC leading into on-line QC. *Off-line QC* refers to design of products and processes to produce them. Generally, these take place in advance, before the production system goes on-line. Off-line, there is a great deal of emphasis on correctly capturing the wants and needs of current and potential customers, then designing products and services to meet their needs, and then designing the necessary production processes. Experts agree that the majority of the deficiencies we realize during the production of goods and services is not because we are careless during the on-line phase, but because we actually do an insufficient job during the off-line phase. Some experts estimate that between 60 and 80% of the deficiencies we see are actually designed into the product or its production process. We can see that there is a great opportunity for the prevention of problems during the off-line phase.

On-line QC is usually synonymous with actual production activities. During production, our intent is to operate our production processes in a way that (1) meets targets, and (2) minimizes variation. That is, we want to operate our processes so stably, consistently, and predictably that our customers know that every time they buy from us, they will receive a consistent product or service. Stable processes are said to be "in control." Unstable processes are referred to as inconsistent, unpredictable, or out of control. If a process is stable and also meets customer requirements, it is said to be a capable process. We want all of our processes to be capable.

There are many tools that we can use during both off-line and on-line QC. Only some of the most successful and popular tools will be presented in the remainder of the chapter. These tools are shown in Figure 8.5. We discuss them in turn.

8.7. Quality Function Deployment

QFD is a concept that helps translate customer requirements into product and production requirements. The idea is to begin by learning and agreeing upon the customers' requirements, and to use that knowledge to drive development or improvement of the process. More time is spent up-front in the design phase, but that time is more than made up in time reductions in production since the product and processes were thought out well in the beginning. As the man says, "pay me now or pay me later." More up-front time spent on getting it right the first time is more than rewarded later as we make and sell product.

Figure 8.6 presents a QFD matrix that translates customer requirements (the voice of the customer) for a connector (plug or socket) into final product control (technical) characteristics. Another matrix might be used to translate the final product control characteristics into needed process characteristics. Still another matrix might be used to translate needed process characteristics into process control points.

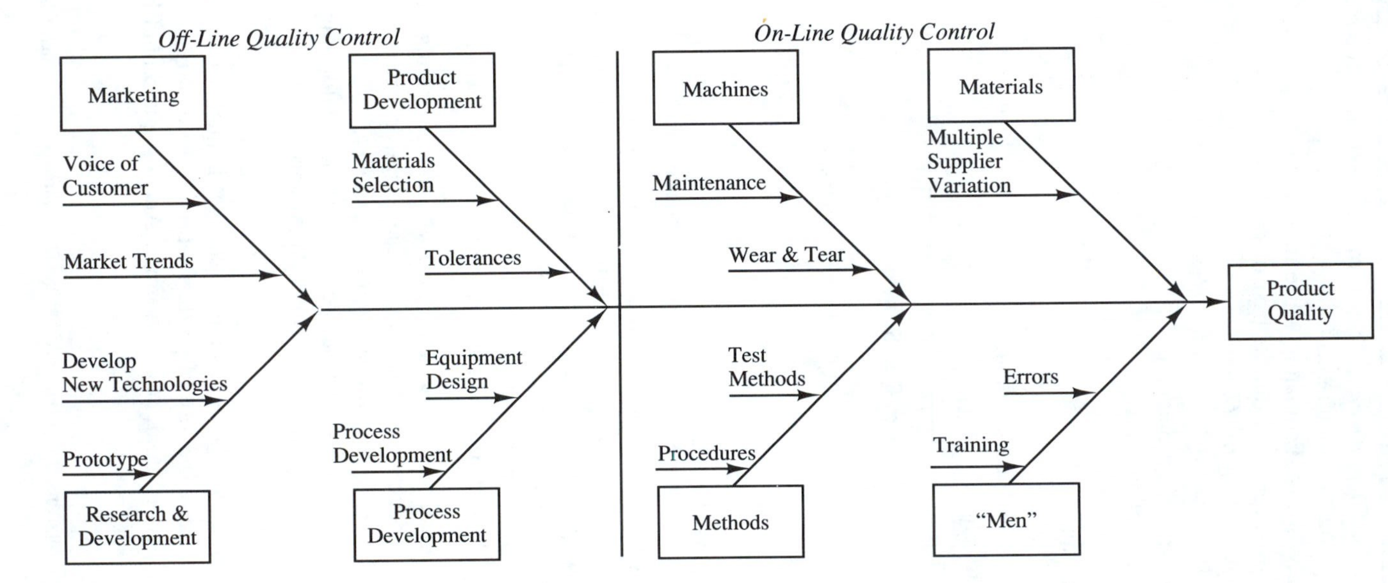

Fig. 8.4. On-line and off-line quality control.

Quality Function Deployment—Transforming customer requirements into product and production requirements.
Quality Cost Systems—Estimating waste in dollars.
Benchmarking—Learning from the best of the best.
Statistical Process Control Tools

- Flowchart.
- Cause-and-effect diagram.
- Data collection form.
- Pareto chart.
- Histogram.
- Scatter plot.
- Designed experimentation.
- Control chart.

Fig. 8.5. Some quality tools.

Typically, industry applications of QFD include at least four matrices. They do take time!

8.8. Quality Cost Systems

Quality costs are often described as "the cost of doing things wrong." In fact, Mr. Phil Crosby[4] speaks of this cost as the *price of nonconformance* (PONC). For example, sorting inspection, rework, repeat testing, time spent resolving customer complaints, material scrapped, downtime, and returns are a few good examples of the wastes we want to avoid.

Two major elements of quality costs are as follows:

(1) *Internal Failure Costs*—Costs associated with nonconforming materials, components, or products that cause losses due to rework, repair, retest, scrap, sorting, and so on, prior to release to the customer.
(2) *External Failure Costs*—Costs associated with nonconforming products that cause losses due to warranties, returns, allowances, and so on, after release to the customer.

A quality cost system helps us monitor changes in quality costs: for example, a recent trend upward in the cost of product returns. The quality cost system also helps us identify pots of gold, ready to mine. An example would be in-process

[4]Philip Crosby and his firm have provided quality awareness and training for thousands of practitioners.

Customer Needs \ Technical Characteristics		Ends			Number of Components	End Finish Technique	Pull-out Force	Roundness	Crimping Method
		ECC	Mat'l	Float					
Good Performance	Low Loss	◎		◎		◎		○	
	Repeatable Loss	○	△	◎		○			
	Low Reflection					◎		◎	
	Wide Temp. Range		○						
Low Cost	Inexpensive Materials								△
	Fast Assembly				◎				
	Inexpensive Tools								○

◎ Strong Relationship
○ Medium Relationship
△ Weak Relationship

Fig. 8.6. QFD matrix-connector.

inventories that are consistently high, eating up capital, space, and morale. The important concept underlying quality cost systems is that they identify areas to prioritize for process improvement. Such investments are usually highly leveraged, yielding a 6:1 return. That is, if we are willing to pay $1 now to prevent problems, we will save $6 in the absence of those problems.

8.9. Benchmarking

Benchmarking is the search for industry best practices that lead to superior performance. It has also been described as the continuous process of measuring products, services, and practices against the toughest competitors or those companies recognized as industry leaders.

The four main features of benchmarking are:

(1) We are striving to learn from the "best of the best."
(2) We are striving to be the "best of the best."

(3) We are *externally oriented*, looking outside our company and even our industry.
(4) We are focusing first on best *practices* (how to do key work processes) and then on *metrics* or measures (to assess performance).

As examples, suppose that you are a specialty electronics manufacturer. You want to benchmark warehouse material handling. You might go completely outside your industry, to a mail-order dry goods business that is well known for their warehouse material handling efficiency. Or, suppose that you have a critical, expensive, and much needed set of precision painting areas that have specialized capabilities. Scheduling them is a fright. You might wish to benchmark a world-renown hospital to determine how they schedule operating rooms.

Example 8.1

As president of Gadgets, Inc., you and your employees are determined to produce a high-quality product. One of your teams has been through a highly successful QFD effort and you believe Gadgets now has a solid fix on customer requirements, product characteristics, and processes necessary to efficiently and effectively deliver the product—the new supergadget. Simultaneously, recent quality cost figures have been assessed and you have determined that there is at least one begging need—for improvement in the consistency of raw materials received from suppliers. Variations in their product cause all sorts of instability and constant readjustment in Gadgets' machining operations. You have benchmarked an outstanding firm known for excellent relations and results with suppliers. You have determined many of the practices they have used, applying the tools of statistical process control to assist their suppliers.

You have also implemented many of Dr. Deming's 14 points. For example, you are no longer using end-of-line inspection to try to inspect quality into the product. Rather, you are focusing on your processes upstream in the process to ensure that quality is built in from the beginning. You no longer submit requests for bid to any and all potential suppliers, awarding business on the basis of first cost alone. Rather, you have selected your overall best supplier for each type of raw material, and have decided to ''grow'' those suppliers as if they were part of your own organization. That is, you plan to nurture those suppliers as if they were an extension of Gadgets, Inc., embracing them in a long-term relationship.

You are focusing on both product and service delivery processes and have implemented rigorous training of all employees. Management and supervisors are now thrilled when employees surface problems to be worked, rather than hiding those problems out of fear. Different parts of the company are working together as a team. For example, purchasing, materials management, design engineering, process engineering, and marketing plan to work together on the team to help improve the key suppliers. Some members of the team will be hourly workers; others will be salaried personnel.

Even though reducing the supplier base got rid of a lot of variation and quality problems, there is still room for improvement. You think there is great potential to work closely with suppliers, using the tools of SPC, to help them become a better supplier to you. First, however, it is time to review those tools of SPC.

8.10. Tools of Statistical Process Control[5]

The tools of SPC are absolutely necessary to help us understand and improve processes. These tools help teams communicate, share and document ideas, understand variation, and measure results of process changes. The eight common tools of SPC are pictured in Figure 8.7. Seven of them will be addressed briefly in this section. Control charts will be treated separately and more extensively.

8.10.1 Flowchart

A *flowchart* is a picture of the activities that take place in a process. Drawing a flowchart is usually one of the first things a team does. It gives everyone on the team a chance to communicate about, understand, and agree on how the process is accomplished. It defines customer–supplier relationships and helps standardize procedures. It is a simple but powerful tool. An invoice-processing flowchart is illustrated in Figure 8.8.

Several guidelines are helpful to us in drawing a flowchart:

(1) Establish boundaries for the process you want to flowchart. Exactly where does it start and where does it stop?
(2) Try to hit a level that shows the truly key steps in the process, without becoming too detailed. If the level is too shallow, you can go back in and draw in more detail. If the level is too deep, you may get bogged down in detail that is not truly helpful.
(3) Keep the symbols simple. There is little to be gained by using all of the common "computer program" flowchart symbols.
(4) Be sure that there are no dead ends (other than at "stop") or loops that continue forever.
(5) Show those feedback loops where something goes back for rework, checking, and so on.

After drawing the flowchart, keep a few things in mind:

(6) After drawing a chart of the existing process, we might want to "imagineer" a flowchart of the process if everything worked right. Then the two can be compared and important areas for improvement may be identified.
(7) Feedback loops are a likely sign of waste and opportunity for improvement.
(8) Consider collecting data immediately following a "decision" diamond. This is where you can gain some real insight as to what is really going on in the process.
(9) Between every pair of "action" or "step" boxes is a customer–supplier relationship.

---

[5]Much of this material is taken directly from the Exxon Chemical Company Quality Principles Course manual.

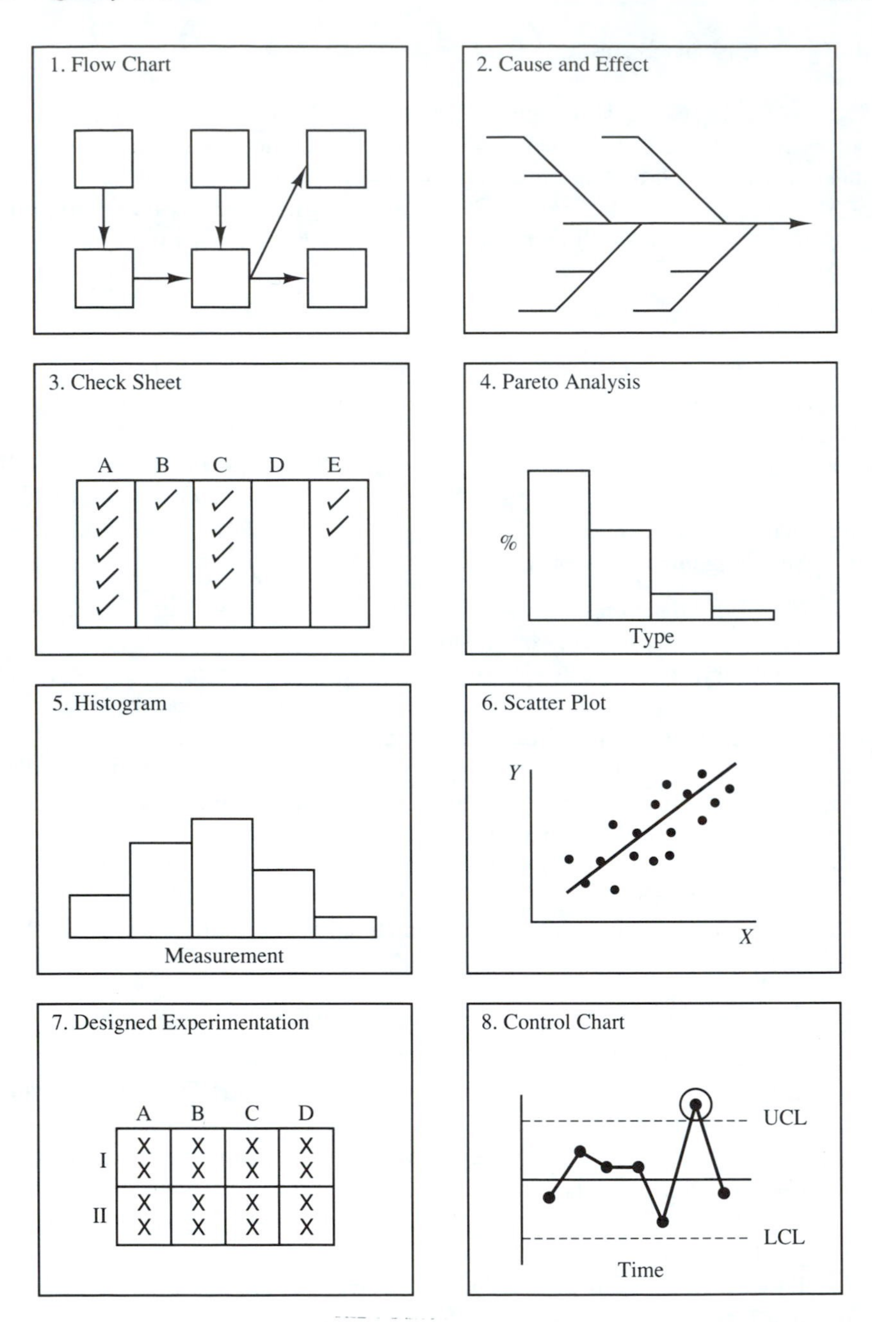

Fig. 8.7. Eight basic SPC tools.

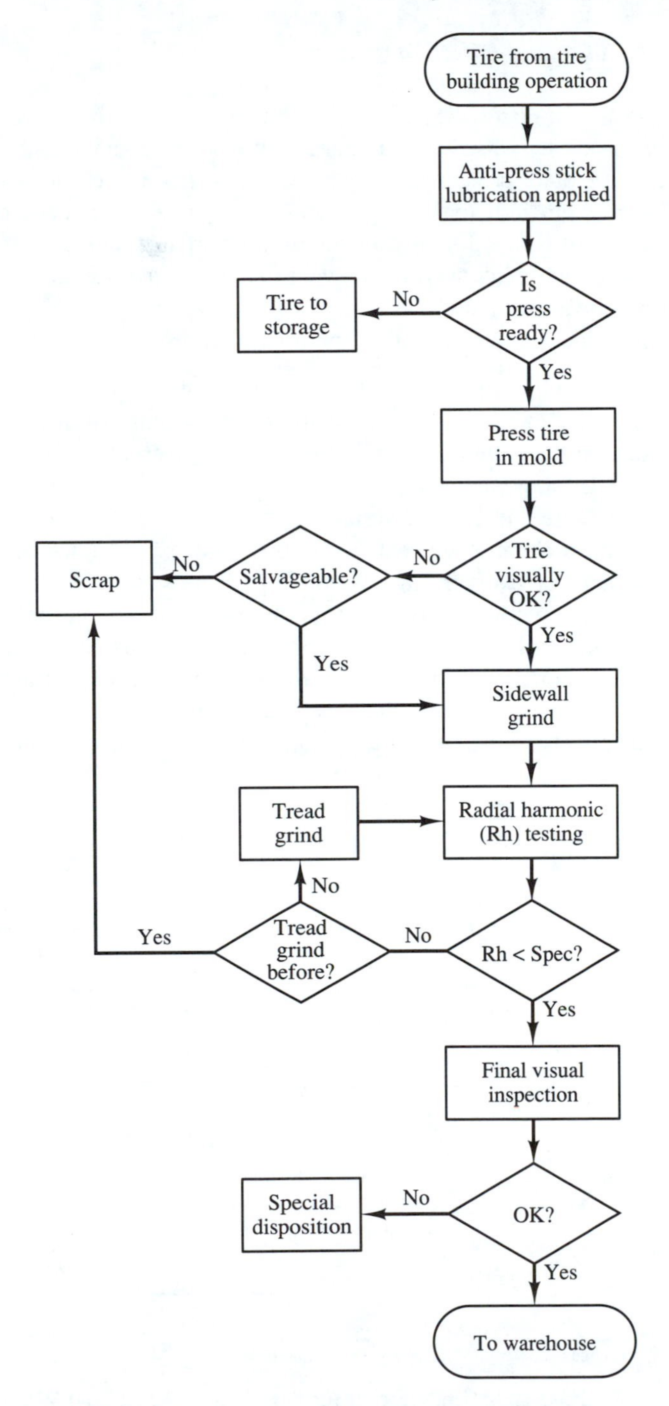

Fig. 8.8. Tire cure and finish process flowchart.

8.10.2 Cause-and-Effect Diagram

The *cause-and-effect diagram* is also known as a *fishbone diagram* or an *Ishikawa diagram*. It is used to summarize knowledge about possible causes of variation or some other problem. It organizes the causes of variation or the causes of a quality problem into logical categories. It helps a team focus on different possible causes and is therefore a valuable tool for organizing efforts to improve a process. A cause-and-effect diagram relevant to the ham radio station of K5KC and WA5DSH is shown in Figure 8.9.

Several guidelines are helpful in building a cause-and-effect diagram:

(1) Use a team to build the cause-and-effect diagram.
(2) Generate a list of potential causes using brainstorming. Brainstorming allows everyone on the team to speak and encourages everyone to listen. All ideas are evaluated at a later time.
(3) Build the cause-and-effect diagram:
   - Place the problem statement in the box on the right (at the fish's head). Make sure that everyone agrees on the problem statement and that it is well defined and measurable (if you cannot measure it, you cannot improve it).
   - Draw and name three to six major ''bones'' or cause categories. Naming the major bones may be easier if the brainstormed causes are written on ''stick-um'' notes so that they may be moved and grouped by the team. When all else fails, use the old standby major bones: men, machines,

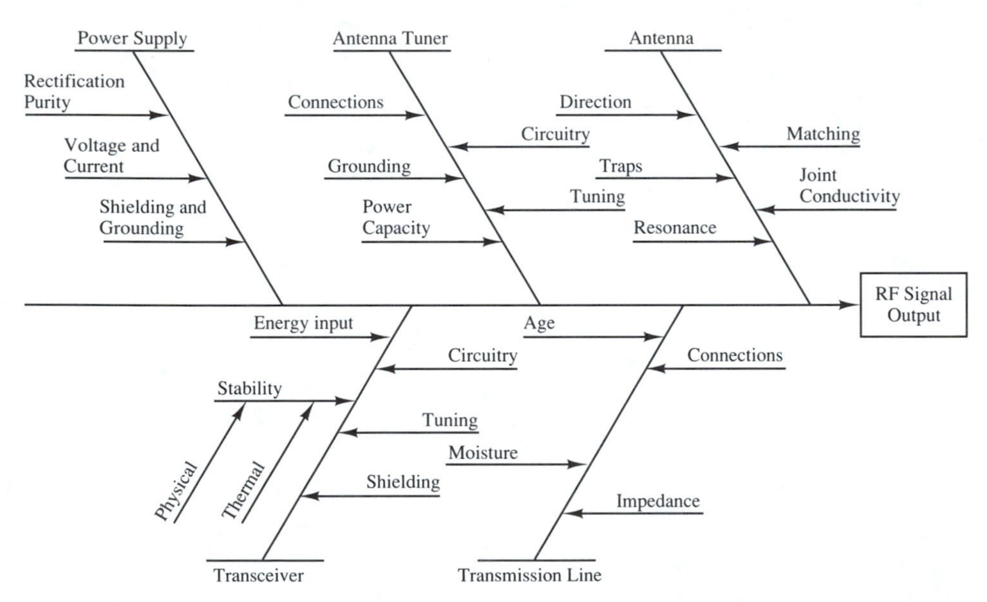

Fig. 8.9. Cause-and-effect diagram for ham stations K5KC and WA5DSH.

methods, materials, measurement, and environment, but these are seldom the best breakouts.

- Place the brainstormed ideas on the major, most related bones.
- For each cause, ask "Why does it happen?" and list the responses as subbones or sub-subbones.

After drawing the cause-and-effect diagram, focus on finding the most important causes of the problem by doing the following:

(4) Look for causes that appear repeatedly. Data may be needed to identify these.
(5) Discuss each of the causes listed, as desired by the team. Use "expert knowledge" from the team, plus the question "why" to identify the most basic causes.
(6) Reach a team consensus as to which of the possible causes deserves the most attention. Focus on these for greater understanding and knowledge of the process, with the intent of eliminating or reducing those causes that are clearly affecting the problem.
(7) Keep the cause-and-effect diagram updated as knowledge of the process improves. It is a working document that should capture the "brains" of the team on paper.

8.10.3 Data Collection Form

Part of any effort to improve a process is to gain knowledge about the process. This usually requires us to collect and analyze data. Unfortunately, most of the data we *routinely* collect alerts us to problems but is useless for analysis and helping us take the right action that leads to real process improvement. When we decide to define and collect new data, we need some type of a *data collection form*.

Several guidelines are helpful in designing a data collection form:

(1) Decide exactly what data is to be collected. For each characteristic, state its units of measure (dollars of scrap/month; hours overtime/week, etc.)
(2) Clearly define each characteristic so that everyone has a common understanding of it and its units. For example, if the characteristic is "time lost" in hours per week, there should be a clear definition of exactly what "time lost" means.
(3) Decide if you wish to stratify (separate) the data according to some factor. For example, different strata might be by plant, shift, continent of customer, type of equipment, and so on.
(4) Decide when, where, how, and in what order the data should be collected. The idea is to "visualize" the data collection process and to make it as easy and failproof as possible.
(5) Determine and build the most suitable type of data collection form. Three types receive a lot of use.

- First, a *traditional data collection form* may be designed. It should have a clear place for all measures to be recorded, and its layout should conform to the order in which data are collected.
- Or a simple *checksheet* might be designed to collect data in the form of "check marks." The categories into which check marks are placed should be well defined so that there is no overlap between them. A defect identification check sheet is shown in Figure 8.10 for bales of butyl rubber just produced.

Some hints on using the data collection form include:

(6) After designing and building a data collection form, give it a trial under fire. Let those who will be collecting the data try it for a day or two. Be sure that they are people who have *not* been involved in the design of the form.
(7) Take their comments and make design changes as needed.
(8) Nearly all data must be converted into information. Plotting the data on a Pareto chart, frequency plot, scatter plot, control chart, or other device are common methods of making sense out of data.

8.10.4 Pareto Analysis

"Pareto" is "prioritization." A *Pareto analysis* orders occurrence data by type, category, or other classification. It helps us focus on the important things. It identifies the "vital few" problems or types of defects from the "useful many." This helps us pinpoint the right problem or effect to study. The "Pareto principle," also known as the "80/20 rule," says, in effect, that 80% of our "heartburn" is caused by 20% of the things with which we deal. A Pareto diagram showing reasons for a chemical unit's downtime is shown in Figure 8.11.

Several guidelines are helpful in building a Pareto diagram and doing Pareto analysis:

(1) Decide the data that are to be collected. This includes deciding on the measurement axis of the diagram. It is usually a measure of "pain," such as dollars per year, number of occurrences per month, downtime per week, or number errors per 100 invoices.It includes deciding on the classification axis of the diagram. This is usually a type, category, or classification of defect. Examples include type of injury, type of defect, area of plant, day of week, country from which order placed, and so on.
(2) Clearly define each classification that will be along the classification axis. These should cover all possible classifications and not overlap in definition.
(3) Design a data collection form. Begin to collect data. After collecting at least 30 occurrences, set up the Pareto diagram and look for the most frequently occurring classification. Pareto diagrams order the classifications on the axis from most frequently occurring to least frequently occurring.

Date 12/14/9X Shift AM Line #2

<table>
<tr><td>REASON FOR INSP. (a)'</td><td colspan="5">0600 – 0800</td><td colspan="5">0800 – 1000</td><td colspan="5">1000 – 1200</td><td colspan="5">1200 – 1400</td><td colspan="5">1400 – 1600</td><td colspan="5">1600 – 1800</td></tr>
<tr><td>BALERS</td><td>A</td><td>B</td><td>C</td><td>D</td><td>E</td><td>A</td><td>B</td><td>C</td><td>D</td><td>E</td><td>A</td><td>B</td><td>C</td><td>D</td><td>E</td><td>A</td><td>B</td><td>C</td><td>D</td><td>E</td><td>A</td><td>B</td><td>C</td><td>D</td><td>E</td><td>A</td><td>B</td><td>C</td><td>D</td><td>E</td></tr>
<tr><td>WEARSTRIP WORMS (Blacks, stringy, soft piece of rubber)</td><td></td><td></td><td></td><td></td><td></td><td></td><td>||||/ |||</td><td></td><td></td><td></td><td></td><td>||||/ |</td><td></td><td></td><td></td><td></td><td>|||</td><td></td><td></td><td></td><td></td><td>||||/ ||||/ |</td><td></td><td></td><td></td><td></td><td></td><td></td><td></td><td></td></tr>
<tr><td>BLACK SPECKS (Unidentifiable, soft black spot)</td><td></td><td></td><td></td><td></td><td></td><td></td><td></td><td></td><td></td><td></td><td></td><td></td><td></td><td></td><td></td><td></td><td></td><td>|</td><td></td><td></td><td></td><td></td><td></td><td></td><td></td><td></td><td></td><td></td><td></td><td></td></tr>
<tr><td>GREASE (Black or greenish spot, may rub off on hand or may be mixed into rubber)</td><td></td><td></td><td></td><td></td><td></td><td></td><td></td><td></td><td>||||</td><td></td><td></td><td></td><td></td><td>||</td><td></td><td></td><td></td><td></td><td>||||/ ||</td><td></td><td></td><td></td><td></td><td>|||</td><td></td><td></td><td></td><td></td><td></td><td></td></tr>
<tr><td>RUST (Reddish hard magnetic)</td><td></td><td></td><td></td><td></td><td></td><td></td><td></td><td></td><td></td><td></td><td></td><td></td><td></td><td></td><td></td><td></td><td></td><td></td><td></td><td></td><td></td><td></td><td></td><td></td><td></td><td></td><td></td><td></td><td></td><td></td></tr>
<tr><td>COLOR (Noticable change in rubber color, usually caused by poly upset)</td><td></td><td></td><td></td><td></td><td></td><td>||</td><td></td><td></td><td></td><td>|</td><td></td><td></td><td></td><td></td><td></td><td></td><td></td><td></td><td></td><td></td><td></td><td></td><td></td><td></td><td></td><td></td><td></td><td></td><td></td><td></td></tr>
<tr><td></td><td colspan="5">0</td><td colspan="5">15</td><td colspan="5">8</td><td colspan="5">11</td><td colspan="5">14</td><td colspan="5">0</td></tr>
</table>

Fig. 8.10. Check sheet for defects in butyl rubber.

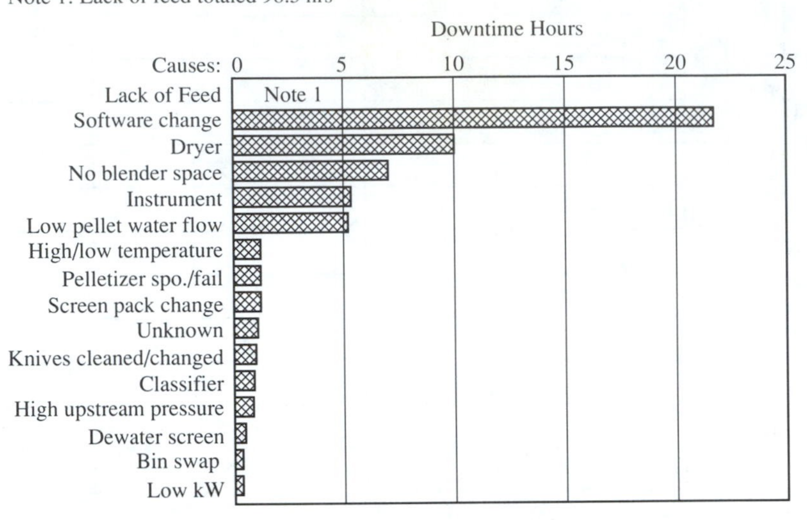

Fig. 8.11. Pareto chart of downtime causes.

Some helpful hints in using Pareto analysis include:

(4) Be sure that the measurement axis is really a measure of "pain." If there are multiple measures that are equally meaningful (e.g., percent downtime and number of occurrences), do a Pareto diagram for each.

(5) A Pareto diagram may be broken down further. That is, Pareto analysis may be done on data from the biggest bar of the Pareto diagram. Repeating this several times often leads to the right aspects of the process to study. This is often called "macro-to-micro" or "Pareto within Pareto" analysis. For example, data from the most frequent invoice error may again be broken down by type of invoice.

(6) Spend the most time on defining the classifications along the classification axis. These should be crystal clear, or the data will be misleading due to improper classification.

8.10.5 Histogram

The histogram presents a snapshot in time of a set of data. It is particularly useful for seeing the shape, centering, and spread of a set of data from some process. If the process is stable (see control charts later in this chapter), the histogram is a picture of what we might expect to see in the future. If the process is not stable, the histogram is only a picture of what has happened in the past. A histogram of student test scores is presented in Figure 8.12.

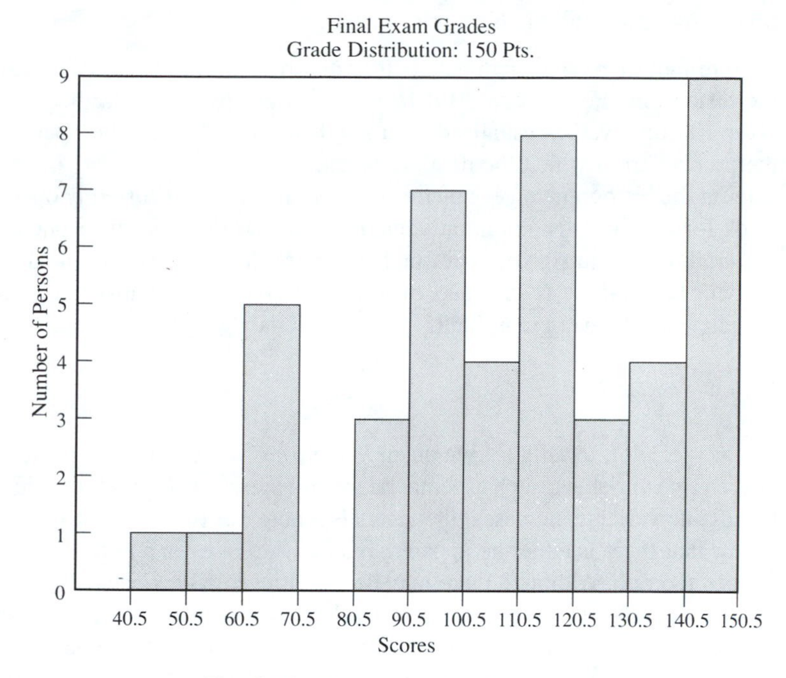

Fig. 8.12. Histogram of the test scores.

Several guidelines are helpful in building a histogram:

(1) The characteristic of interest (e.g., height, weight, viscosity, melt point, time, etc.) is scaled on the horizontal axis.

(2) The scale of the characteristic of interest is usually broken into cells of equal width.

(3) It is best if cell boundaries are such that data must fall inside a cell and cannot fall on a cell boundary. It is common to set up cell boundaries to one decimal place past the data, with the last place ending in a 5. For example, cell boundaries of 7.5 to 11.5 would include data numbered 8, 9, 10, and 11.

(4) The number of cells is often from 5 to 20, with 10 being a good starting number. To get the cell width, begin by taking the difference between the maximum and minimum number and dividing that difference by the approximate number of histogram cells desired. Then round that to a workable cell width. For example, if the largest and smallest data values are 77 and 53, respectively, and about 10 cells are desired, $(77 - 53)/10 = 2.4$. Use cells of either width 2 or 3.

(5) Either the frequency or the percent of occurrences is scaled on the vertical axis.

Some hints on using the histogram include:

(6) Be sure that there is enough data before putting a lot of stock in the shape of the histogram. Fewer than 30 data values often give a misleading picture. Over 100 data values usually does a good job of reflecting the operation of the process from which the data are taken.
(7) Look at the histogram to get a picture of how the process is currently operating. Look for (a) the most common value (the mode), (b) more than one clearly discernible peak (a sign of an unstable process), (c) symmetry of the data, (d) outlying data values, (e) any data over- or underrepresented (unexpected peaks or valleys in the histogram), and (f) percent of data within specification limits.

8.10.6 Scatter Plot

The scatter plot, usually shown on an *X-Y* plot, gives a picture of the relationship between two variables, such as temperature and pressure. That relationship may show that as one variable increases, the other also increases (positive correlation); or it may show that there is either no apparent relationship or even a negative relationship between the two. Although there may be a relationship between two variables, there is not necessarily a cause-and-effect relationship between them. A scatter diagram showing business calls answered versus calls received (some persons hung up before someone answered the phone—all call data recorded electronically) is shown in Figure 8.13.

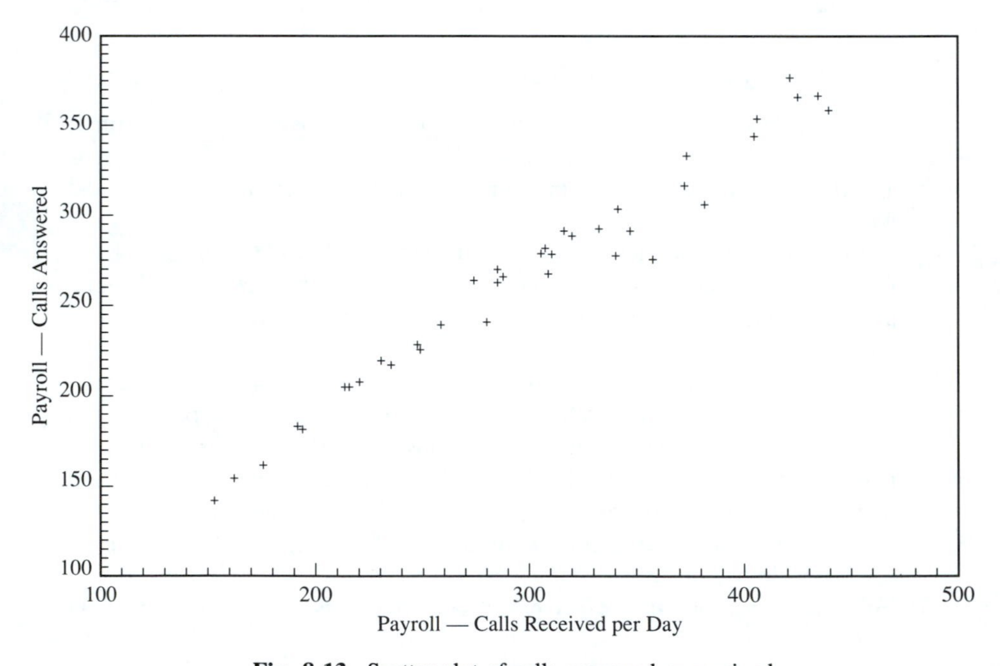

Fig. 8.13. Scatter plot of calls answered or received.

Several guidelines are helpful in building a scatter plot:

(1) Draw two perpendicular axes of approximately the same length.
(2) Scale the axes so that the data collected use about 90% of both axes. Be sure that both scales are increasing as you go up the vertical scale or to the right on the horizontal scale.
(3) Plot the data pairs on the *X-Y* plot, look at the data, and attempt to understand the process and how it is operating.

Some hints for using and interpreting the scatter plot include:

(4) If there is a relationship, use your knowledge of the process to determine if there is a cause-and-effect relationship between the two variables, or if there is some other common variable influencing both.
(5) If there is a straight-line relationship in the plot, eyeball-fit a line through the data (this can be done more formally using least-squares regression). Do not extend the line beyond the data. That is, do not extrapolate, assuming the process will continue to operate in a straight-line relationship well beyond the data collected.

8.10.7 Designed Experimentation

Designed experiments are one of the most powerful tools we have in our toolkit to improve processes. Basically, we are attempting to determine economically which of many factors have the most significant effect on process centering and spread. We would like not only to be on target, but also to minimize variation.

Everyone has done designed experiments. For example, in high school or in a college science lab, you probably varied a factor such as temperature or pressure or slope or time to see its effect. Typically, such experiments are 1FAT designs (one factor at a time). They are particularly inefficient at best, and misleading at worst. They are not nearly as powerful as if multiple factors are varied simultaneously in a particular pattern as desigated by one of many possible experimental designs. Common designs used effectively in industry include factorials, fractional factorials, Taguchi designs, Plackett–Burman screening designs, central composite designs, and others.

Experimental designs can help us study quantitative effects (e.g., temperatures of 100 and 140 degrees) and qualitative effects (e.g., machine A vs. machine B). We design the experiment in advance to ensure that we will be able to discern if the effects of interest are important factors in our process. Proper designs let us test for interaction effects. An interaction exists, for example, if process yield is better at high temperature when raw material A is used, but yield is better at low temperature if raw material B is used.

Designed experimentation is one of the most needed and important bodies of knowledge you can have today in industry. If you do not get a course in this in industrial engineering, take it as an elective in the statistics department. Tell your other engineering friends (non-I.E.s) that they need it, too. An example of data and marginal mean plots is shown in Figure 8.14.

In this section we have had a very practical introduction to many of the SPC tools. Next we cover one of the most important and useful tools, control charts, in more depth.

8.11. Background on Control Charts

Control charts are a tool for studying variability. They tell us if a process is showing stable (or consistent) variation. A stable process is often called an in-control process, a predictable process, or a "common cause" process. It is said to be in a state of statistical control (SOSC). An unstable process is also known as out-of-control (OOC), unpredictable, or a "common plus special cause" process. A control chart tells us whether or not a process is stable.

Everything we do in life is a process. Everything in life varies. Brushing your teeth is a process. You have never brushed your teeth twice in exactly the same way. We want to learn about variation and how it affects our work processes. More specifically, we are trying to determine whether or not all of our variation is due to

Tire Pressure	Gasoline Octane	Miles per Gallon Trial				Mean
		y_1	y_2	y_3	y_4	$\bar{y}$
25	87	16.5	18.3	15.7	16.6	16.78
25	91	18.2	16.6	17.3	17.8	17.48
32	87	21.3	22.6	19.5	21.8	21.30
22	91	21.5	20.9	23.1	21.7	21.80

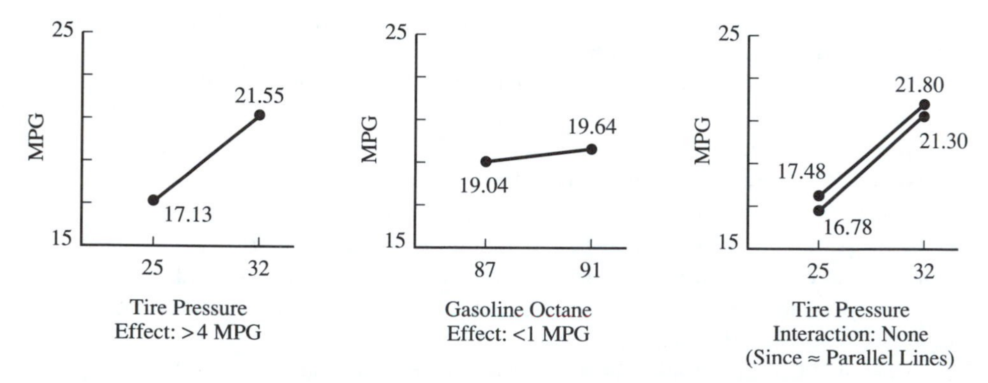

Fig. 8.14. Designed experimentation on car gasoline mileage.

"common causes" or if "special causes" are also present. All processes have variation due to common causes.

Common causes are those that are inherent in the process over time, affect everyone working in the process, and affect all outcomes of the process. Learning about industrial engineering is a process. Common causes to all students in an introduction to industrial engineering course include (1) your textbook, (2) the classroom, (3) the lighting in the room, (4) the teacher, (5) the number of credit hours, and so on.

Special causes are those that are not part of the process all the time, do not affect everyone, but arise due to specific circumstances. Examples would be an individual's mathematics background, their experience working in industry, whether or not they are sick, and so on. If special causes are at work, they can be eliminated by attacking their underlying specific causes (e.g., remedial math, summer industry experience, medical treatment followed by preventive medicine). If common causes are at work, the only way to reduce or remove them is to make fundamental changes in the system or the process (e.g., change classrooms, improve lighting, modify the number of credit hours, etc.). *Note:* This book and your instructor probably cannot be improved!

Control charts have been used as a diagnostic and maintenance tool in the control of production processes since first introduced by Shewhart, the father of control charts. By far the most popular control charts are the $\overline{X}$, R, p, and c charts. The $\overline{X}$ and R charts are used together on variables data (e.g., numerical measurements) to analyze the central tendency and spread of a process on a single measurable characteristic. The p chart is used to analyze attributes data (e.g., good/bad classification), such as the fraction nonconforming of a process. The c chart is used when attributes data, such as a count of the number of nonconformities per sample unit, is the measure of interest.

Control charts can be used as a management tool to help (1) bring a process into control, (2) keep a process in control, and (3) determine process capability to meet specifications. Output from a process is sampled and a statistic is calculated which bears the name of the control chart on which it is to be plotted (e.g., the sample average $\overline{X}$ is plotted on the $\overline{X}$ chart). When process variability is due only to *common* causes, the process is in a *state of statistical control* (SOSC) and the plotted pattern of statistics over time is reasonably well behaved, falling within predictable limits and according to predictable patterns. When process variability is due to common *and special* causes, such as people, equipment, and/or materials, the process is *out of control* (OOC) and nonrandom patterns appear on the control chart.

Control charts have a center line plus upper and lower control limits. These control limits are usually spaced 3 standard deviations (of the sample statistic being plotted) above and below the center line. As such, it is highly unlikely that a plotted point will fall outside control limits as long as an SOSC exists. In addition to points outside control limits, the *pattern* of plotted points within control limits is equally important in identifying an OOC condition and its cause. Therefore, sample statistics are almost always plotted in time order.

8.12. Control Charts for Variables

The $\overline{X}$ and R control charts are used together in analyzing a single measurable characteristic. Approximately $m = 20$ or 30 subgroups of size n each are selected. Typical subgroup sizes are $n = 4$ or 5, selected consecutively from the process or produced under as nearly identical conditions as possible. Time between subgroups is dependent on judgment and may be once per hour, twice per day, once per shift, and so on. The idea is to have *within*-subgroup variation be as small as possible, and represent common-cause variation only. Production process differences, if any, then showing up over time *between* subgroups, and represent special cause variation. For each subgroup i, the average $\overline{X}_i$ and range R_i statistics are calculated.

Once the desired number of subgroups m has been inspected, the next step is to calculate the average range, $\overline{R}$, of the subgroups. This will be the center line on the R chart. Factors for the upper and lower control limits are given as D_4 and D_3, respectively, in Table 8.1. The equations pertinent to an R chart based on m subgroups are given as follows:

$$\text{center line} = \overline{R} = \frac{\sum_{i=1}^{m} R_i}{m} \tag{8.1}$$

$$\text{upper control limit} = \text{UCL}_R = D_4\overline{R} \tag{8.2}$$

$$\text{lower control limit} = \text{LCL}_R = D_3\overline{R} \tag{8.3}$$

Note that for small subgroup sizes, three standard deviations below $\overline{R}$ would yield a negative range control limit, which is impossible. Therefore, when the subgroup size is six or less, D_3 is set to zero, resulting in $\text{LCL}_R = 0$.

The $\overline{X}$ chart control limits depend on $\overline{R}$; therefore, it is important to establish the R chart first. The $\overline{X}$ chart assumes an underlying normal distribution in accordance with the central limit theorem. Also, on the $\overline{X}$ chart we are dealing with a distribution of subgroup averages with variance

$$\sigma_{\overline{X}}^2 = \frac{\sigma^2}{n} \tag{8.4}$$

Our control limits will be set at a point equivalent to three subgroup average standard deviations above and below $\overline{\overline{X}}$. The equations for the $\overline{X}$ chart, assuming m subgroups, are given as follows:

$$\text{center line} = \overline{\overline{X}} = \frac{\sum_{i=1}^{m} \overline{X}_i}{m} \tag{8.5}$$

Table 8.1. CONTROL CHART FACTORS

Subgroup Size, n	*LCL_R Factor,* D_3	*UCL_R Factor,* D_4	*$CL_{\overline{X}}$ Factor,* A_2	*$\overline{R}/\sigma$ Ratio,* d_2
2	0	3.267	1.880	1.128
3	0	2.575	1.023	1.693
4	0	2.282	0.729	2.059
5	0	2.115	0.577	2.326
6	0	2.004	0.483	2.534
7	0.076	1.924	0.419	2.704
8	0.136	1.864	0.373	2.847
9	0.184	1.816	0.337	2.970
10	0.223	1.777	0.308	3.078
11	0.256	1.744	0.285	3.173
12	0.284	1.716	0.266	3.258
13	0.308	1.692	0.249	3.336
14	0.329	1.671	0.235	3.407
15	0.348	1.652	0.223	3.472
16	0.364	1.636	0.212	3.532

$$\text{upper control limit} = \text{UCL}_{\overline{X}} = \overline{\overline{X}} + A_2\overline{R} \qquad (8.6)$$
$$\text{lower control limit} = \text{LCL}_{\overline{X}} = \overline{\overline{X}} - A_2\overline{R} \qquad (8.7)$$

The factor A_2, values of which are given in Table 8.1, takes account of the relation between the range R, the process variance σ^2, the subgroup average variance $\sigma^2_{\overline{X}}$, and our desire to have $\pm 3\sigma_{\overline{X}}$ control limits based on only common-cause or within-subgroup variation. For subtle reasons beyond the scope of this text, it is important that the control limits for both the $\overline{X}$ and R charts be calculated using the formulas given above rather than trying to take some alternative approach to the calculation of control limits.

Example 8.2

You have now embarked on a course of assisting one of your suppliers. Your quality engineer (Q.E.), while at the vendor's location, collects 30 subgroups of size $n = 4$ over several days, testing and recording the strengths of each unit of raw material. The data, averages, and ranges are given in Table 8.2. The center lines and control limits are calculated as follows:

$$\overline{R} = \frac{\sum_{i=1}^{m} R_i}{m} = \frac{6{,}445}{30} = 214.83$$
$$\text{UCL}_R = D_4\overline{R} = 2.282(214.83) = 490.24$$
$$\text{LCL}_R = D_3\overline{R} = 0.0$$

Table 8.2. STRENGTH TEST DATA OF RAW MATERIAL

Subgroup	Sample Data X_1	X_2	X_3	X_4	Subgroup Average, $\overline{X}$	Subgroup Range, R
1	908	1,095	1,159	1,215	1,094.25	307
2	1,337	1,080	1,086	1,242	1,186.25	257
3	1,212	1,326	1,175	1,380	1,273.25	205
4	1,286	1,278	1,151	1,359	1,268.50	208
5	1,401	1,313	1,384	1,466	1,391.00	153
6	1,114	1,096	1,060	1,204	1,118.50	144
7	1,119	974	1,007	1,212	1,078.00	238
8	1,001	1,030	1,063	1,037	1,032.75	62
9	1,152	1,161	1,159	1,046	1,129.50	115
10	1,000	882	870	823	893.75	177
11	835	667	787	803	773.00	168
12	707	904	961	918	872.50	254
13	949	809	842	859	864.75	140
14	828	1,061	918	964	942.75	233
15	1,043	753	813	861	867.50	290
16	784	847	856	945	858.00	161
17	1,086	925	855	840	926.50	246
18	1,117	1,201	1,034	929	1,070.25	272
19	1,115	1,114	1,038	1,227	1,123.50	189
20	1,140	928	1,332	1,061	1,115.25	404
21	1,229	1,127	1,030	1,253	1,159.75	223
22	1,314	1,242	1,087	1,140	1,195.75	227
23	1,305	1,219	1,353	1,207	1,271.00	146
24	1,181	1,326	1,245	1,548	1,325.00	367
25	1,312	1,285	1,444	1,176	1,304.25	268
26	993	1,052	1,018	1,074	1,034.25	81
27	1,196	980	1,116	1,161	1,113.25	216
28	948	1,139	1,222	1,062	1,092.75	274
29	993	1,283	1,231	1,105	1,153.00	290
30	1,050	928	1,009	1,058	1,011.25	130
					32,540.00	6,445.00

$$\overline{\overline{X}} = \sum_{i=1}^{m} \frac{\overline{X}_i}{m} = \frac{32,540}{30} = 1084.67$$

$$\text{UCL}_{\bar{x}} = \overline{\overline{X}} + A_2\overline{R} = 1,084.67 + 0.729(214.83) = 1,241.28$$

$$\text{LCL}_{\bar{x}} = \overline{\overline{X}} - A_2\overline{R} = 1,084.67 - 0.729(214.83) = 928.06$$

The control charts for the data given in Table 8.2 are drawn in Figures 8.15 and 8.16.

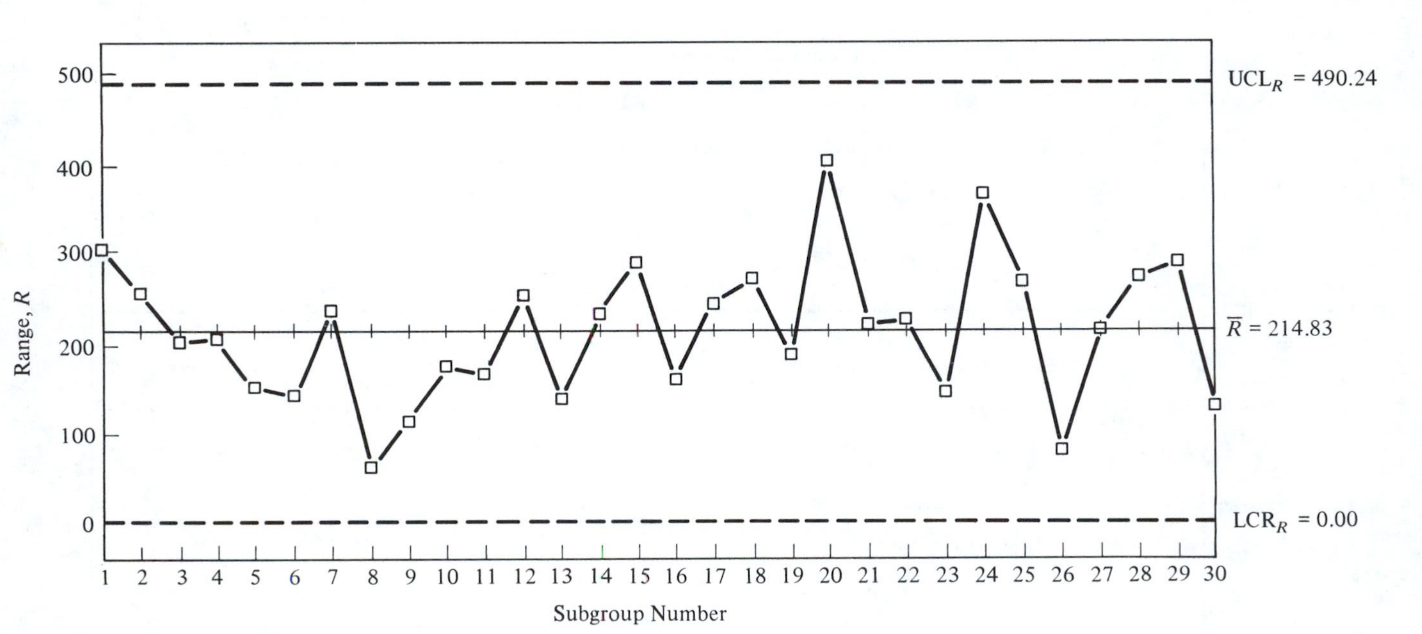

Fig. 8.15. *R* chart for strength test data.

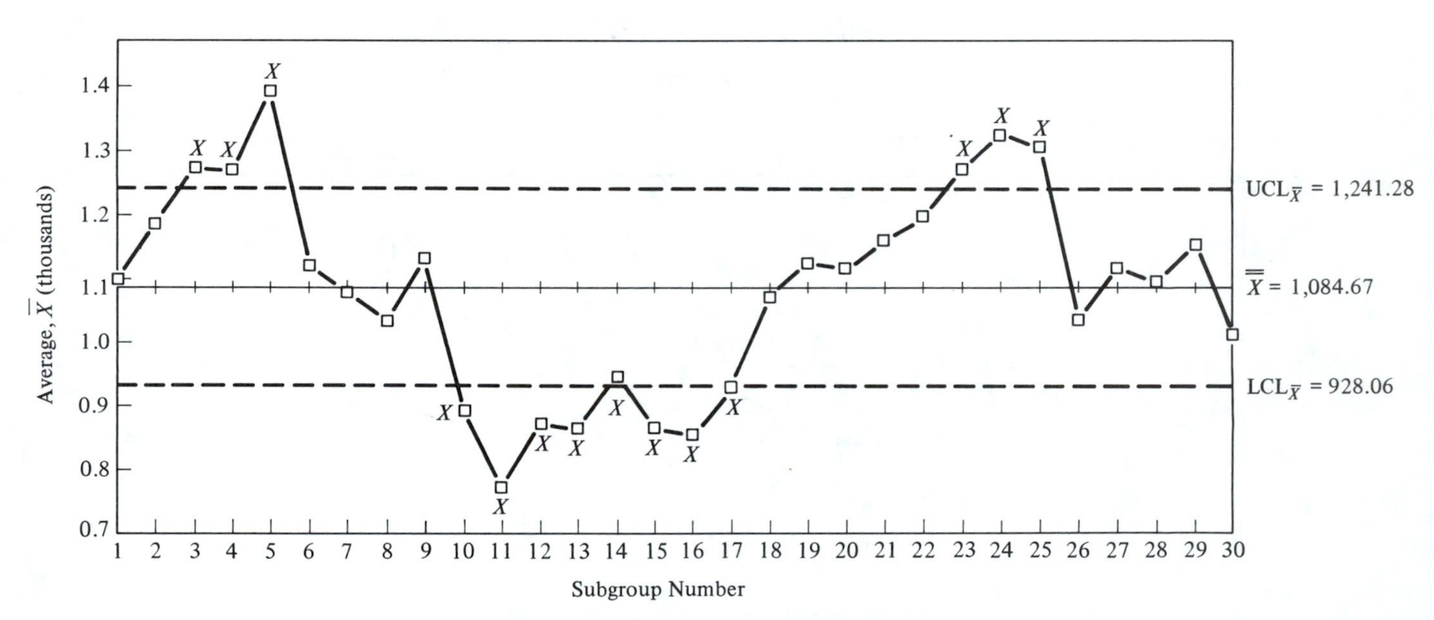

Fig. 8.16. $\overline{X}$ chart for strength test data.

8.13. Sensitivity Checks for Control Charts

When used alone, Shewhart control charts are often not sensitive to relatively small process shifts. A scheme consisting of four checks is recommended for making a control chart ($\overline{X}$, R, p, etc.) much more sensitive to OOC conditions.[6] The steps of the scheme are listed below and illustrated in Figure 8.17.

(1) Divide the area on either side of the control chart center line into three equal zones: A, B, and C.
(2) *Check 1*—If one point falls outside 3-sigma control limits (beyond zone A), mark each such point with an X slightly above or below the point, away from the center line.
(3) *Check 2*—If two out of any three successive points fall in or beyond zone A of the same side, mark the second of the two points with an X. The "other" point may be anywhere.
(4) *Check 3*—If four out of any five successive points fall in or beyond zone B of the same side, mark the fourth of the four points with an X. The "other" point may be anywhere.
(5) *Check 4*—If eight successive points fall in or beyond zone C of the same side, mark the eighth such point with an X. These checks were applied to the $\overline{X}$ and R charts of Figures 8.15 and 8.16.

Example 8.3

Your Q.E. concluded that the process spread as indicated by the R chart was in a SOSC, but that process centering was OOC. The cause for the out-of-control state was due to a faulty thermal isolation network on the metering equipment controlling the process. This was not the only problem, however. One of the critical process input valves was incorrectly set, causing the process centering to be far too low (at 1,084.67 when it should have been approximately 1,232.60). Correcting these difficulties corrected the process for both centering and control. This changed the operation of the $\overline{X}$ chart centering and control limits, requiring the Q.E. to update the control chart based on recent data.

Thirty new subgroups of $n = 4$, taken after correction of the process, resulted in $\overline{R} = 208.70$, $UCL_R = 476.25$, $LCL_R = 0.00$, $\overline{\overline{X}} = 1{,}228.45$, $UCL_{\overline{X}} = 1{,}380.59$, and $LCL_{\overline{X}} = 1{,}076.31$. The highest R value was 432 and the highest and lowest $\overline{X}$ values, respectively, were 1,354.75 and 1,099.50. No "X's" were triggered by any of the four sensitivity checks. The process is now in a SOSC.

8.14. Process Capability Analysis

It is important for us to note that a process in control does not necessarily mean that all of the product will be good—or that *any* of it will be good. When a process is

[6] Adapted from *AT&T Statistical Quality Control Handbook*, 2nd ed. (Easton, Pa.: Mack Printing Co., 1984).

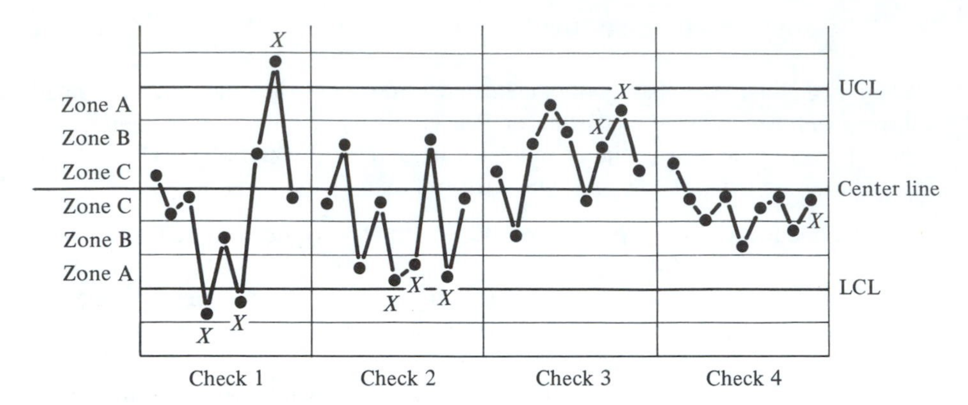

Fig. 8.17. Checks for sensitizing control charts.

in control, it simply means that a common-cause system is operating. It is possible, for example, to have a process in control far away from the desired specification limits. We must, therefore, assure ourselves that the process is not only in control, but that it is also *in control at the correct process centering or mean dimension.*

It is also possible for us to have an in-control process with perfect centering and still have an excessive fraction nonconforming. This can occur when we have both upper and lower specifications (U and L) that are too close when compared with the inherent variability of the process. This means that the process is operating as well as possible, but it is just not capable of producing all of the items within specifications. In this situation, we must either change processes, widen the design specifications, or be prepared to sort good items from bad.

From our control chart data, we can easily determine if the specifications U and L will be too narrow for our process. We know that nearly all (99.73% in the case of a normal distribution) of the items produced by a distribution will be within three standard deviations of its mean; that is, the natural variability of the process is $\overline{\overline{X}} \pm 3\sigma$. We estimate the process standard deviation for an in-control process as follows:

$$\sigma = \frac{\overline{R}}{d_2} \tag{8.8}$$

where d_2 is a factor that depends on the subgroup size and is given in Table 8.1. Now, if 6σ exceeds U—L when the process is in control, we know that the inherent variability of the process is simply too great.

Example 8.4

Your Q.E. can now estimate the inherent variability of the strength of super-gadgets based on recent, in-control data:

$$\sigma = \frac{\overline{R}}{d_2} = \frac{208.71}{2.059} = 101.36$$
$$\overline{\overline{X}} \pm 3\sigma = 1{,}228.45 \pm 3(101.36) = 924.37,\ 1{,}532.53$$

Since only a lower specification is applicable, the Q.E. cannot compare 6σ to U–L, but he can see that some fraction of product will still be nonconforming. Assuming a normal distribution, calculating $Z = (1{,}000 - 1{,}228.45)/101.36 = -2.25$, and using the standard normal table in Appendix B, it can be seen that the fraction nonconforming below a strength of 1,000 is now about 1.22%.

8.15. Control Charts for Attributes

8.15.1 The p Control Chart

The *p chart* is an attributes data chart based on the fraction of product nonconforming to specs. A subgroup of n items is selected from a process and each is inspected for nonconformities on multiple characteristics. Any item found to have one or more nonconformities is considered a nonconforming item. The number of nonconforming items x divided by the sample size n is the fraction nonconforming.

The p chart, like $\overline{X}$ and R charts, is very popular in industry. It requires only attributes data, so go–no go gauging, visual inspections, and the like yield appropriate data. The p chart is particularly attractive where combining several characteristics into one chart is desired, unlike $\overline{X}$ and R charts. Unfortunately, common subgroup sizes are often in the range of 50 to 300 items.

Approximately $m = 20$ or 30 subgroups of size n are selected from the process, keeping in mind the desire for within-subgroup homogeneity, with production process differences, if any, showing up over time between subgroups. For each subgroup i, the statistic $p_i = x_i/n_i$ is calculated.

We recall that the number of nonconforming items in a sample of size n from an infinite process is described by the binomial distribution with mean $\mu = np$ and variance $\sigma^2 = npq$, where $q = 1 - p$ and where p is the process fraction defective. Since we are interested in a p chart, the appropriate measure for each subgroup is x/n, or the subgroup fraction nonconforming which has mean $\mu = p$ and variance $\sigma^2 = pq/n$. Thus, 3σ control limits are established from m subgroups by using the following equations:

$$\text{center line} = \overline{p} = \frac{\sum_{i=1}^{m} x_i}{\sum_{i=1}^{m} n_i} \tag{8.9}$$

$$\text{upper control limit} = \text{UCL}_p = \overline{p} + 3\sqrt{\frac{\overline{p}\,\overline{q}}{n_i}} \tag{8.10}$$

$$\text{lower control limit} = \text{LCL}_p = \overline{p} - 3\sqrt{\frac{\overline{p}\,\overline{q}}{n_i}} \tag{8.11}$$

The center line and control limits are plotted on graph paper, and the individual statistics p_i are plotted in chronological order. The same four-check sensitivity scheme presented earlier may be applied.

Two additional items are important to note. First, the subgroup size n_i may not be constant from subgroup to subgroup on the p chart. In this case, individual control limits for each point may be calculated using Equations (8.10) and (8.11) with the appropriate subgroup sizes inserted for n_i. Second, it may be desired for later troubleshooting to keep a record of which characteristic nonconformities are observed during inspection, even though a knowledge of specific nonconformities is not needed for the p chart.

Example 8.5

Your Q.E. is now in the habit of achieving successes using SPC tools to diagnose processes. A decision is made to utilize a p chart in-house on the good-gadget line. The data in Table 8.3 are collected over a 3-week period.

The center line and control limits are calculated using Equations (8.9), (8.10), and (8.11) as follows:

$$\bar{p} = \frac{263}{2,885} = 0.0912$$

$$\begin{aligned} \text{UCL}_p &= 0.0912 + 3\sqrt{\frac{(0.0912)(0.9088)}{n_i}} \\ &= 0.1775 \text{ at } n = 100 \\ &= 0.2026 \text{ at } n = 60 \\ &= 0.2639 \text{ at } n = 25 \\ \text{LCL}_p &= 0.0912 - 3\sqrt{\frac{(0.0912)(0.9088)}{n_i}} \\ &= 0.0048 \text{ at } n = 100 \\ &= 0.0 \text{ at } n = 60 \text{ and } n = 25 \end{aligned}$$

A p chart for the data is shown in Figure 8.18. The three X's indicate that the good-gadgets line was not operating in a SOSC. The Q.E. finds that subgroup 7 was a recording error and subgroups 13 and 19 were due to an untrained operator. Also, a separate list of nonconformities clearly showed that most nonconforming items were due to scratches caused by improper material handling. The Q.E. quickly instituted a training program on material handling for all operators, and a comprehensive training program for occasional substitute operators.

8.15.2 The c Control Chart

The *c chart* is based on the number of nonconformities observed per subgroup. The size of each subgroup must first be established and can, for example, consist of an airplane skin section, five stereo receivers, three forklift trucks, 10 pairs of overalls, and so on. The size of the subgroup must remain fixed from subgroup to subgroup in order to maintain constant the area of opportunity for the occurrence of nonconformities. The c chart, like the p chart, requires only attributes data. It

Table 8.3. SAMPLE DATA FROM GOOD-GADGET LINE

Subgroup Number	*Subgroup Size*	*Number Nonconforming*	*Fraction Nonconforming*
1	100	9	0.09
2	100	4	0.04
3	100	6	0.06
4	100	11	0.11
5	100	13	0.13
6	60	12	0.20
7	100	0	0.00
8	100	8	0.08
9	100	15	0.15
10	100	12	0.12
11	100	8	0.08
12	100	6	0.06
13	100	18	0.18
14	100	9	0.09
15	100	7	0.07
16	100	11	0.11
17	100	8	0.08
18	100	8	0.08
19	25	9	0.36
20	100	10	0.10
21	100	7	0.07
22	100	3	0.03
23	100	12	0.12
24	100	4	0.04
25	100	6	0.06
26	100	7	0.07
27	100	10	0.10
28	100	4	0.04
29	100	11	0.11
30	100	15	0.15
	2,885	263	

is attractive where a simple *count* of nonconformities (deviant visuals, speeds, frequencies, dimensions, etc.) is the performance measure of interest.

Approximately $m = 20$ or 30 subgroups are selected over time. The items selected for a subgroup should be homogeneous in the sense that they are produced together or under similar conditions. Once again, production process differences, if any, should show up over time between subgroups. For each subgroup, the statistic c_i is kept, representing the total count of nonconformities observed over all units in the subgroup. The c chart center line and control limits are given as follows:

$$\text{center line} = \bar{c} = \frac{\sum_{i=1}^{m} c_i}{m} \tag{8.12}$$

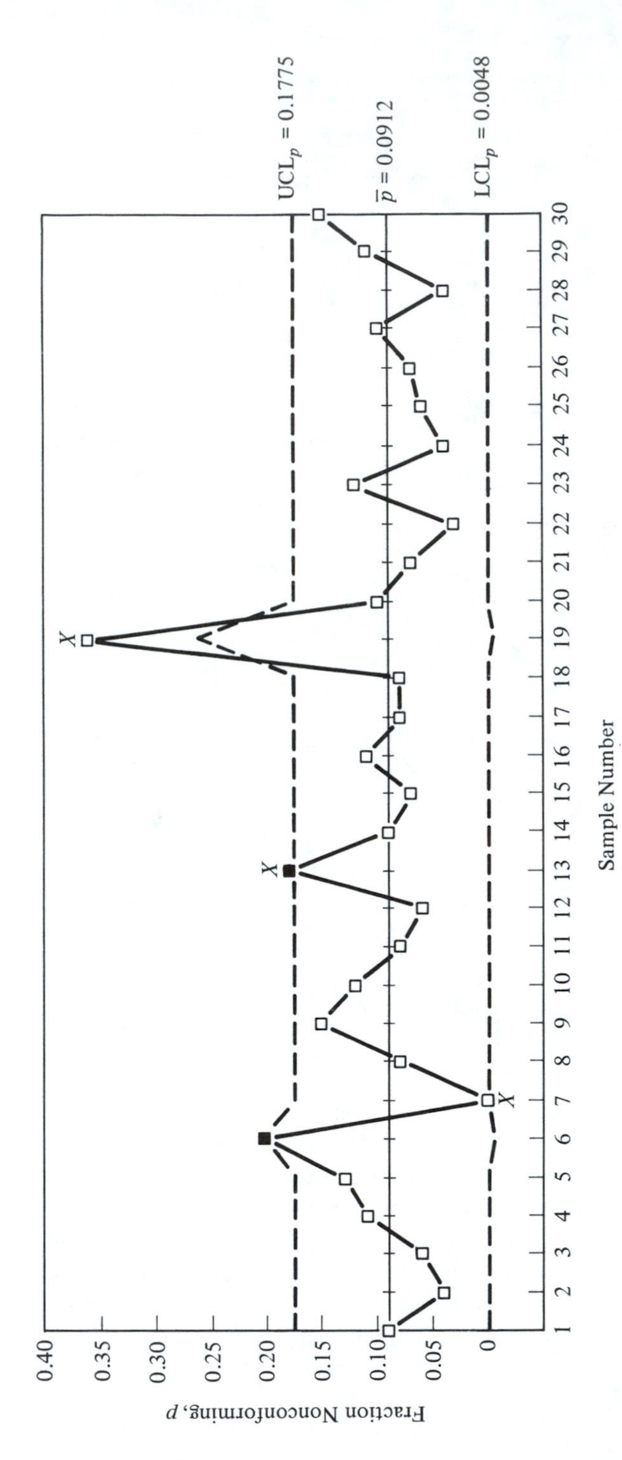

Fig. 8.18. *p* chart for good-gadget line.

$$\text{upper control limit} = \text{UCL}_c = \bar{c} + 3\sqrt{\bar{c}} \tag{8.13}$$

$$\text{lower control limit} = \text{LCL}_c = \bar{c} - 3\sqrt{\bar{c}} \tag{8.14}$$

The center line and control limits are plotted on graph paper, and the individual statistics c_i are plotted in chronological order. The same four-check scheme discussed previously is applicable.

Example 8.6

You now have a new dilemma. Your Q.E. has learned that an industrial engineering degree coupled with a knowledge of SPC is in high demand. The vendor has just made the Q.E. an offer to come to work as quality manager at a 30% increase in salary. What are you going to do as president of Gadgets, Inc.?

DISCUSSION QUESTIONS

1. Define quality. Why has the emphasis on quality changed from the producer to the customer?

2. Many areas must be coordinated to provide a quality product or service. Why wouldn't the only really important area be "production"?

3. Why does reduced quality often result in decreasing market share and productivity?

4. Consider the enrollment process at your university. Identify and quantify (in terms of time) typical sources of waste (e.g., time lost waiting) from your perspective as a student.

5. Identify your "customers" and your "suppliers" on your college campus. Are some customers also suppliers?

6. What is the difference between quality planning, quality control, and quality improvement? Who refers to these as the "trilogy?"

7. Identify two examples each of off-line and on-line quality control.

8. Give two examples each of internal and external failure costs, based on your own work or school experience.

9. Name some benefits of a team taking the time up-front to develop both a flowchart and a cause-and-effect diagram of the process they are setting out to improve.

10. Give examples of variables that would be expected to have a positive relationship on a scatter plot. A negative relationship. No relationship.

11. If we desire to minimize quality costs, why not just fire everyone associated with quality control?

12. What is the difference between attributes data and variables data?

13. What is meant by a "state of statistical control"? Does an SOSC imply that all product is good? That a majority is good?

14. For a basketball game, state three factors that would be considered common causes and three that are special causes.

15. Is it possible to produce 100% in-spec product with a process that is "out of control"? Explain.

16. Why would we sometimes prefer to use a p chart instead of an $\overline{X}$ and R chart combination?

17. Give an original example of data that would be appropriate for a c chart.

PROBLEMS

1. Develop a flowchart of your enrollment process at school. Clearly show on it any areas of needed improvement.

2. Develop a cause-and-effect diagram for why you and your boyfriend/girlfriend/spouse have spats.

3. Develop a Pareto chart for the data presented on the checksheet in Figure 8.10.
(a) Use baler number as your classifications.
(b) Use type of defect (e.g., wearstrip worms) as your classifications.
(c) Use time of day as your classifications.

4. (a) Develop a histogram for the first 80 raw data values appearing below in the first 20 subgroups in Problem 7.
(b) Develop a histogram for the first 20 averages appearing below in Problem 7.
(c) Describe the relationship between the centering and spread of your histograms.

5. Draw a scatter plot showing the relationship (if any) between the first 20 subgroup averages and the first 20 subgroup ranges given below in Problem 7. Do you see a relationship? Did you expect to?

6. In a plating operation, we think the thickness is related to the time the metal surface being plated remains in the plating solution. Following are some data:

Hours	*Thickness*
2	60
3	80
6	70
8	110
12	110

Show graphically using the appropriate tool if there is a relationship between time and thickness.

7. Twenty subgroups of $n = 4$ each are taken from a bottle filling process. Weights are expressed in hundredths of an ounce over 16 ounces. The coded data are shown below. Using only the first 20 subgroups:

Subgroup, Number, i	*Data*				*Subgroup Average,* $\overline{X}_i$	*Subgroup Range,* R_i
	X_{i1}	X_{i2}	X_{i3}	X_{i4}		
1	20	45	46	10	30.25	36
2	20	50	38	30	34.50	30
3	29	18	36	29	28.00	18

Subgroup, Number, i	Data				Subgroup Average, $\overline{X}_i$	Subgroup Range, R_i
	X_{i1}	X_{i2}	X_{i3}	X_{i4}		
4	38	38	44	19	34.75	25
5	35	36	25	28	31.00	11
6	37	32	19	49	34.25	30
7	49	40	16	7	28.00	42
8	36	7	40	32	28.75	33
9	29	18	30	44	30.25	26
10	34	20	31	26	27.75	14
11	25	26	38	15	26.00	23
12	36	35	20	24	25.75	16
13	33	21	22	22	24.50	12
14	18	41	38	40	34.25	22
15	36	36	25	25	30.50	11
16	21	31	22	41	28.75	20
17	32	20	38	46	34.00	26
18	21	37	21	25	26.00	16
19	20	26	30	30	26.50	10
20	21	28	33	32	28.50	12
				Sum	595.25	433
21	46	43	33	36	39.50	13
22	33	49	37	41	40.00	16
23	42	42	65	26	43.75	39
24	42	43	56	48	47.25	14
25	30	47	41	35	38.25	17
26	45	44	41	44	43.50	4
27	45	35	21	25	31.50	24
28	33	24	35	27	29.75	11
29	31	43	19	15	27.00	28
30	34	48	48	35	41.25	14

(a) Calculate $\overline{X}$ and R chart center lines and control limits.
(b) Plot the $\overline{X}$ and R charts.
(c) Mark X's on the charts as required.
(d) What is your conclusion about the process?
(e) Now, plot the next 10 $\overline{X}$'s and R's.
(f) Mark X's on the next 10 subgroups on both charts.
(g) What is your conclusion about the process?

8. Reconsider the first 20 subgroups of data from Problem 7.
(a) Make all necessary calculations, plot $\overline{X}$ and R control charts, and state your conclusions about the process as in parts (a) to (d) of Problem 7.
(b) Suppose that the original 20 subgroups were modified such that odd subgroup data are all increased by 10 and even subgroup data are all decreased by 10. Make all necessary calculations, plot $\overline{X}$ and R control charts, and state your conclusions about the process.

9. Subgroups of size $n = 5$ are taken from a process. After 40 such subgroups, $\sum_{i=1}^{40} R_i = 930$ and $\sum_{i=1}^{40} \overline{X}_i = 3{,}000$. The maximum and minimum values of R_i are 45 and 3, respectively. The maximum and minimum values of $\overline{X}_i$ are 86.8 and 63.0, respectively.
(a) Based on all that you know, is the process in control? Why or why not?
(b) If specification limits are at 55 and 95, what can you conclude about the ability of the process to produce items within specifications?
(c) If the specs are 55 and 125, what do you recommend? Why?

10. An in-control process has $\overline{\overline{X}} = 200$ and $\overline{R} = 46.52$ using subgroups of size $n = 5$.
(a) What are the $\overline{\overline{X}} \pm 3\sigma$ values which indicate the natural variability of the process?
(b) If you would like to have specs at least 5σ from the center of the process, where should they be?

11. Consider a process which, when in control, has $\overline{\overline{X}} = 500$ and $\sigma = 25$.
(a) Determine the $\overline{X}$ and R control chart limits if $n = 4$.
(b) Determine the $\overline{X}$ and R control chart limits if $n = 9$.
(c) Determine the $\overline{X}$ and R control chart limits if $n = 2$.
(d) Suppose that a shift in process centering occurred from $\overline{\overline{X}} = 500$ to 530. Which subgroup size above yields an $\overline{X}$ chart that is best capable of detecting the shift? Why? What is the major disadvantage of using that sample size?

12. The following numbers of nonconforming items are observed in 25 subgroups of (usually) 200 items each: 24, 0, 12, 8, 8, 15, 8 (subgroup size $n = 30$), 3, 9, 16, 7, 12, 1, 21, 28, 32, 23, 9, 8, 4, 4, 7, 3, 1, 1, 1. Develop a p chart.

13. Consider the following real data from a production process involving several presses in the electronics industry. A subgroup of size $n = 200$ is used throughout.

Subgroup Number	*Press Number*	*Number Nonconforming*
1	3	0
2	5	0
3	10	1
4	12	5
5	13	2
6	7	0
7	5	0
8	4	0
9	3	0
10	1	0
11	1	0
12	3	0
13	4	0
14	10	4
15	3	0
16	12	7
17	3	0
18	13	0
19	15	0

Subgroup Number	*Press Number*	*Number Nonconforming*
20	11	4
21	10	5
22	4	0
23	12	6
24	4	0
25	3	1
26	1	0

(a) Calculate control limits for a p chart.

(b) Plot the p chart and mark the X's on the top half of the chart.

(c) Assuming that you correctly find some points indicating an out-of-control condition, state specifically what you would recommend be done next.

14. Overalls are processed in bundles of 18 pairs each. One bundle will be treated as a subgroup for purposes of collecting data to be analyzed using a c chart. The following numbers of nonconformities (string hanging, stitch missed, etc.) are observed during inspection of 20 bundles: 2, 0, 0, 6, 4, 4, 10, 2, 2, 4, 0, 2, 0, 22, 20, 42, 38, 6, 2, 0.

(a) Calculate the center line and control chart limits for the c chart.

(b) Plot the c chart and mark X's in the top half of the chart as required.

(c) What do you conclude about the sewing process?

CHAPTER 9

Financial Compensation

9.1. Introduction

Financial compensation encompasses many areas having to do with pay for the accomplishment of work. In this chapter we cover such areas as job evaluation, job descriptions, job specifications, wage surveys, and incentive plans. To set the stage, let us consider an example to demonstrate the need for sound financial compensation management.

Example 9.1

You are still plant manager of Gadgets, Inc., in Corvallis, Oregon, and are trying to set up a profitable operation. One of the first problems you face is in determining labor needs. Once they have been determined, you must go to the labor market and fill these needs. To do this, the personnel manager must know what the work is and how it is done (*job description*), and then he or she must know what requirements are needed in the individual who will be doing the work (*job specification*). For example, does the person need a high school degree? Is there a height or weight requirement? The industrial engineer, using the material in this chapter, writes the *job description* and *specification*. (You will find these useful later in the training programs.)

Since you must pay a competitive salary, you feel that you must find out how much industry in your local area is paying employees and how much competing industries are paying elsewhere. This information is obtained for you by a *wage survey* conducted within your area and industry. Furthermore, you must pay fairly "within" your own organization. This is difficult to do since few jobs are alike. You find that the methods of *job evaluation* help you there. You decide to set up an *incentive program* that has been designed using sound principles of wage and salary payments presented in this chapter.

One of the exciting trends in industry today is paying incentives according to the organization's *performance*. These "group" plans have been around many years; but lately, industrial engineers have been finding total system performance is often dramatically enhanced through group incentive plans. You decide to consider these group plans in designing your incentive system.

Once you have designed all these, you have completed the financial compensation system design and are ready to proceed. (To be continued.)

Again, the example is oversimplified, but it does point out the need for the following areas, which are covered in this chapter:

(1) Job analysis.
  (a) Description.
  (b) Specification.
(2) Job evaluation.
(3) Wage surveys.
(4) Wage payment (incentive vs. daywork).
(5) Group incentive plans.

9.2. Job Analysis

Job analysis is a procedure by which not only a description of the job can be obtained but also a description of the traits needed in the person who is to do the job. The first is called a *job description*. The second is called a *job specification*. We shall assume that the reason for conducting a job analysis is to obtain a job description and a job specification for later use in job evaluation and wage determination. But there are other uses. For example, the job description can be very useful in a training program, and a job specification is needed in an active recruitment/screening personnel program (see Chapter 21).

Before proceeding further, some definitions are in order:

Task. A task is human effort exerted for a specific purpose (sharpening pencils, sweeping floors).

Position. A collection of all the tasks performed by or requiring one person (janitor).

Job. A group of positions highly similar or identical (janitors).

As an example, consider an office that has one supervisor, four typists, and two clerks. There are 1 + 4 + 2, or 7, positions; but there are only three jobs (supervisor, typist, clerk). Each job has its own collection of tasks.

There is always a question of who should do the job analysis. The employee himself could do it by answering a questionnaire, the supervisor could do it, or a trained analyst could perform the study. Probably, the best method is to use a trained analyst who works with the supervisor and the employee.

Also, the method must be determined. Basically, there are three methods that could be used: questionnaire, observation, and interview. The best approach varies between companies, but a good one is to use all three or at least an observation-interview combination.

Finally, you must decide on how much information should be obtained. The ultimate use will determine this. For example, if the analysis is to be used for hiring, training, and job evaluation, a great deal of information is required.

Once the analysis has been completed and all the information has been recorded, the next step is to prepare the job description. A job description shows the following information about the job:

(1) Job title.
(2) Job summary.
(3) List of tasks.
(4) Miscellaneous information.

The job title is a short descriptive title that identifies the job. The job summary is a short description of the job that summarizes the tasks involved. The list of tasks is a list of all tasks constituting the job. Miscellaneous information is any other information necessary, for example, the date, the grade number of the job (from job evaluation), the analyst, equipment necessary, and so on. Figure 9.1 is a sample job description.

Once the job description has been completed, the next step is to prepare a job specification stating the qualifications required by the person performing the job. The information basically comes from the job analysis study, but feedback to the employees for correction is usually very helpful.

Many factors can be involved in job specifications, as reflected in Figure 9.2, a blank form showing some of the more often used factors. The writing is done in concise, usually incomplete sentences to save space and avoid unnecessary detail. When the job specification is complete, a person meeting the requirements as shown in the job specifications should be able to perform the job well.

9.3. Job Evaluation

After the job analysis is complete and the job description and specifications have been written, the next step is usually to determine "within house" ranking and rating of the jobs. The purpose is to establish and maintain a fair distribution of wages within the plant. This ranking or rating of jobs is called *job evaluation*.

Job Title Secretary Analyst Jim Brown

Date 8/25 Job No. 750 Grade 4

Job Summary: Perform normal secretarial duties including taking dictation, typing, using word processor, distributing mail, answering telephone, greet visitors, and other miscellaneous duties.

Task List

1. Take dictation in person or from machine, transcribe and type.
2. Answer all incoming telephone calls and take messages when necessary.
3. Receive and distribute all incoming mail and distribute all internal correspondence.
4. Greet all visitors and schedule appointments.
5. Keep coffee and tea available for all visitors.
6. Keep and maintain a petty cash fund.
7. Maintain all files for office supervisor.
8. Make airline, hotel, and rental car reservations.
9. Type on word-processing equipment.

Special Equipment

Typewriter, dictation machine, adding machine, microprocessor

Fig. 9.1. Job description example.

Job Title ________ Analyst ________

Date ________ Job ________ Grade ________

Education

Training and Experience
Resourcefulness
Physical
Mental-Visual
Personality
Supervision
Safety to Others
Equipment-Process

Fig. 9.2. Job specification form.

There are four basic methods of job evaluation: ranking, classification or grade description, factor comparison, and point rating. The first two are nonquantitative and are, therefore, highly relative. The last two are quantitative and yield direct comparisons and ratings.

9.3.1 Ranking Method of Job Evaluation

The ranking method involves a listing of all jobs in descending or ascending order. The supervisor, analyst, or someone who is very familiar with all the jobs must rank all the jobs and then assign wages according to that ranking.

The method has one chief advantage—it is simple to understand if not to implement. Another advantage is that it takes less time than any of the other methods.

The disadvantages include the following:

(1) No rating is applied; only a listing or order is known. There are no distinctions between levels.
(2) It is highly subject to error and thus probably less accurate (e.g., a "halo" effect can occur because a very capable employee may hold the job).

9.3.2 Classification or Grade Description

This nonquantitative approach to job evaluation is very good for organizations that have a large number of jobs, for example, the U.S. Civil Service. The method may be compared to a bookcase in which each cubbyhole is given a certain classification (grade). Each job falling into that classification is assigned that cubbyhole (grade), and the employee is paid accordingly. Of course, the major problem lies in determining the number and proper description of the grades.

The basic advantage of this method is that a large number of jobs can be handled easily once the grade descriptions are written. Also, it is easy for people to think in terms of classifications or grades, which makes it relatively easy to "sell" the procedure. The disadvantage is that the method takes more time and costs more than simple ranking and does not provide the accuracy or detail of factor comparison or point rating.

The procedure usually begins by creating the grades with descriptions for each grade. This can be done several ways, but one is to choose jobs with sharply distinguishing characteristics and wages that cover the range involved. The grade descriptions can be written accordingly. Care must be taken to avoid writing grade descriptions that simply match current wage structures, unless the current structure is appropriate. Once the grade descriptions have been written and accepted by all concerned, the job description and specifications of an unclassified job can be compared to the grade descriptions, and the job can be assigned a grade. Figure 9.3 gives examples of possible grade descriptions.

GRADE	
A	Operation of very simple machines. Requires little concentration or creativity. No hazardous environmental conditions nor safety problems. No previous experience or training required. Ability to read and write helpful but not necessary.
.	.
.	.
.	.
G	Position requires technical training either through a two-year (or more) program or apprenticeship. Requires substantial creative ability and works with little supervision. May involve some supervision or training of other employees.

Fig. 9.3. Grade description examples.

9.3.3 Factor Comparison

This method is one of the two quantitative approaches to job evaluation. It is similar to the grade description method in that levels or grades are used. These levels or grades are for each of several key factors. A composite score is obtained for all factors. The advantages are that it is quantitative yet relatively easy to apply once the factors and levels have been chosen. The main disadvantage is that it is costly and time consuming to set up. Also, the plan has to be revised each time wage rates change.

The procedure usually begins by selecting a number of "key" jobs (usually 15 to 25). Since these jobs are used to establish the guidelines, their choice is critical. They should be chosen such that they adequately cover the entire wage range, vary as to type of work, and should be familiar to all personnel affected by the evaluations.

Next, the list of "key," or "critical," factors are chosen. Usually, the following factors are used because they are almost universal:

(1) Mental requirements.
(2) Physical requirements.
(3) Skill requirements.
(4) Working conditions.
(5) Responsibility.

Each of the key jobs is ranked within *each* factor (the rank will vary between factors), and wages are assigned according to each factor. The total wage rate for a job is calculated by adding the wage rates for each factor. Figure 9.4 is an example for only five key jobs.

Using this structure, we see that the wage rate for a machinist is $18.50, calculated as follows:

Factors

Wages	Mental	Physical	Skill	Working Conditions	Responsibility
$6.00	Tool and Die Maker				
5.40			Tool and Die Maker	Painter	
4.80	Special Assembler		Machinist Painter		Tool and Die Maker
4.00	Machinist	Material Handler		Machinist Tool and Die Maker	Special Assembler
3.60		Special Assembler		Material Handler	Machinist
3.00					
2.70	Painter		Special Assembler		
2.40					
2.10		Machinist			
1.80		Tool and Die Maker			Painter
1.50					
1.20					
0.90				Special Assembler	
0.60	Material Handler	Painter			Material Handler
0.30			Material Handler		

Fig. 9.4. Factor comparison example.

Mental	$ 4.00
Physical	2.10
Skill	4.80
Working condition	4.00
Responsibility	3.60
	$18.50

Now let us use another job as an example, that of a punch press operator. The analyst, the supervisor, perhaps the union steward and employee, and perhaps a job evaluation committee agree that being a punch press operator "fits in," as shown in Figure 9.5. The wage rate can thus be calculated as

$$0.90 + 2.40 + 0.90 + 4.80 + 2.40 = \$11.40$$

All other new jobs would be handled in the same manner.

9.3.4 Point Rating

Point rating is perhaps the most popular of all the methods and is the most detailed. It is similar to factor comparison in that factors are involved (although

Factors

Wages	Mental	Physical	Skill	Working Conditions	Responsibility
$6.00	Tool and Die Maker				
5.40			Tool and Die Maker	Painter	
4.80	Special Assembler		Machinist Painter	"Punch Press Operator" Machinist	Tool and Die Maker
4.00	Machinist	Material Handler		Tool and Die Maker	Special Assembler
3.60		Special Assembler		Material Handler	Machinist
3.00					
2.70	Painter		Special Assembler		
2.40		"Punch Press Operator" Machinist			"Punch Press Operator" Painter
2.10					
1.80					
1.50		Tool and Die Maker			
1.20					
0.90	"Punch Press Operator"		"Punch Press Operator"	Special Assembler	
0.60	Material Handler	Painter			Material Handler
0.30			Material Handler		

Fig. 9.5. Factor comparison—new job.

there are usually more than five factors), and it is similar to the classification method in that jobs are compared with and inserted into certain levels. Point plans can be developed for individual companies or there are several available commercially.

The procedure begins by breaking the jobs down into component factors similar to those in factor comparison. Figure 9.6 presents a list of some of the more often used factors.

The next step is to break the factors into degrees and assign point values to each degree. In this manner, a maximum point value for each factor is established. The total points or value of a job can be calculated by examining each factor, assigning degrees and appropriate point values, and adding overall factors. Figure 9.7 is an example of a point rating plan. Suppose that one of the jobs in this example must be evaluated. By working with the supervisor, the employee, the union steward, and the job evaluation committee, the analyst decides that degree 3 is appropriate for factor 1 (mental demands), and so on, as shown below.

Factor	*Degree*	*Points*
Mental demands	3	150
Training and experience	7	42
Effect of error	4	24
Personal contacts	2	35
Job conditions	3	22
		273

The total points for the job (273) can then be compared to other jobs, and wage rates are assigned accordingly.

To assign wage rates, a scatter gram can be prepared as shown in Figure 9.8 for certain key jobs, and the best straight line constructed. Wage rates can be calculated from Figure 9.8 and the total point values for new jobs. For example, the new job above with 273 points should pay approximately $10.75.

The advantages to point rating are numerous. First, it is the most quantitative and thus perhaps the most accurate. Second, it does not have to be revised each time the base rates are changed (the line in Figure 9.8 would simply move up). Finally, it is very flexible and can be used in many different types of organizations. The chief disadvantage is that it is time consuming and costly to develop (unless it is a purchased procedure).

Education	Responsibility for Safety of Others
Knowledge	Hazards
Mental Requirements	Responsibility for Tools and Equipment
Accuracy	Effect of Error
Effort	Personal Contacts
Training and Experience	Job Conditions

Fig. 9.6. Popular factors in point rating plans.

TECHNICAL AND OFFICE JOBS

Element	Total Points (Weight)	Per Cent
Mental Demands	200	48.2
Training and Experience	84	20.2
Effect of Error	48	11.6
Personal Contacts	45	10.8
Job Conditions	38	9.2

MENTAL DEMANDS

Mental demands is the mental capacity required of an individual to perform a given job efficiently. Factors considered are judgment, analytical ability, initiative, and originality The training and experience acquired by an individual are not considered.

Degree	Factor	Point Value
1	(a) Independent judgment in making decisions, from many diversified alternatives, that are subject to general review in final stages only. (b) Analysis and solution of complex problems affecting production, sales, or company policy. (c) The establishment of procedures in a field in which pioneer work has been negligible and with no reference of detail to higher supervision.	200
2	(a) Independent judgment in making decisions from various alternatives, with general guidance only from higher supervision. (b) Analysis and solution of nonroutine problems involving evaluation of a wide variety of data. (c) The establishment of procedures in conformance with administrative policies and general instructions from supervision.	175
3	(a) Independent judgment in making decisions involving nonroutine problems under general supervision. (b) Analysis and evaluation of a variety of data pertaining to nonroutine problems for solution in conjunction with others. (c) The carrying out of nonroutine procedures, under constantly changing conditions, in conformance with general instructions from supervision.	150

Fig. 9.7. Point rating plan example. [Reproduced, with permission, from David W. Belcher, *Compensation Administration* (Englewood Cliffs, N.J.: Prentice Hall, 1974), pp. 84–187.]

Degree	Factor	Point Value
4	(a) Independent judgment in planning sequence of operations and making minor decisions in a complex technical or professional field. (b) Research and analysis of data pertaining to problems of a generally routine nature. (c) The carrying out of nonroutine procedures in conformance with instructions from supervision.	125
5	(a) Independent judgment in making minor decisions where alternatives are limited and standard policies have been established. (b) Analysis of standardized data for information of, or use by, others. (c) Performance of semiroutine operations with guidance by supervision, but where detailed instructions are lacking.	100
6	(a) Independent judgment is negligible; however, minor decisions sometimes must be made. Work is checked by others. (b) Analysis of noncomplicated data by established routine. (c) Performance of semiroutine operations from detailed instructions.	75
7	(a) Independent judgment is negligible but must be able to receive and transmit simple information obtained from written and verbal sources. (b) Analysis of data is negligible but must be accurate in recording information for use by others (c) Performance of routine, standardized operations under direct supervision.	50
8	(a) Independent judgment is not involved. (b) Analysis not required. (c) Performance of simple, repetitive tasks under close supervision.	25

Fig. 9.7. (cont'd)

9.4. Wage Surveys

The job evaluation techniques discussed in Section 9.3 handle the problem of distributing wages fairly within an organization, but the organization is competing in the labor market for its personnel. Consequently, it must pay wages comparable to those paid by other organizations in the geographical area and by other organizations in the same industry throughout the country. This is normally handled by the use of wage surveys.

A wage survey is a concentrated effort or study by one organization to deter-

TRAINING AND EXPERIENCE

Training and experience are the length of time required for an individual of average mental capacity to acquire the training and knowledge needed to perform a given job efficiently.

Degree	Factor	Point Value
1	7 years or more	84
2	5 years but less than 7 years	77
3	4 years but less than 5 years	70
4	3 years but less than 4 years	63
5	$2\frac{1}{2}$ years but less than 3 years	56
6	2 years but less than $2\frac{1}{2}$ years	49
7	$1\frac{1}{2}$ years but less than 2 years	42
8	1 year but less than $1\frac{1}{2}$ years	35
9	8 to 11 months	28
10	5 to 7 months	21
11	2 to 4 months	14
12	Less than 2 months	7

EFFECT OF ERROR

Effect of error is the extent to which an employee's decisions may affect the company from a cost standpoint. The loss may involve loss of own or others, time and/or loss of company funds. (Loss is considered on the basis of an average top figure for a single occurrence, not an extreme maximum.)

Degree	Factor	Point Value
1	Probable error in basic information developed independently and upon which important management decisions are based could cause substantial loss of company funds, serious delays in schedule, or loss of customers' accounts. Work is not generally subject to check.	48

Fig. 9.7. (cont'd)

Degree	Factor	Point Value
2	Probable error not easily detected and may adversely affect outside relationships. Work is subject to general review only and requires considerable accuracy and responsibility.	40
3	Probable error may create a serious loss of production, waste material, or damage to equipment, but loss is usually confined within the company. Most of work is not subject to verification or check.	32
4	Probable error may result from decisions based on recommendations, but error is usually confined to a single department or phase of company activities. Most of work is subject to verification or check.	24
5	Probable error would normally result in loss of own and others' time to correct error. Practically all work is subject to verification or check.	16
6	Probable error would normally result in loss of own time to correct error. Practically all work is subject to verification or check.	8

PERSONAL CONTACTS

Personal Contacts is the nature, purpose, and level of company and public contacts required of an individual to perform a given job efficiently. Factors considered are ability to orally express ideas, maintain poise, and exercise tact and persuasiveness.

Degree	Factor	Point Value
1	Contacts are with company supervision, supervision of other companies, customers or other person in all types of positions requiring the ability of the employee to influence and to establish and maintain company good will.	45
2	Contacts with persons of substantially higher rank within the company and concerning matters requiring explanation, discussion, and obtaining approvals. Contacts with persons outside the company, involving carrying out of company policy and programs, the improper handling of which will affect operating results.	35

Fig. 9.7. (cont'd)

Degree	Factor	Point Value
3	Contacts with persons in other departments within the company and concerning matters of a general nature. Contacts with persons outside the company but where the primary responsibility rests with supervision.	25
4	Contacts are normally with persons within the company and outside contacts are infrequent. Contacts usually concern routine reporting and exchange or giving of information requiring little or no interpretation or discussion.	17
5	Contacts generally with immediate associates and own supervision.	9
	JOB CONDITIONS Job conditions are the environmental working conditions and unavoidable hazards involved in the performance of the job. These conditions are beyond the employees' control.	
Degree	Factor	
1	Continuous exposure to one or more disagreeable factors such as fatiguing physical exertion, burns, sprains, noise, glare, weather, ventilation and so forth.	38
2	Frequent daily exposure to one or more disagreeable factors such as fatiguing physical exertion, burns, sprains, noise, glare, weather, ventilation, and so forth.	30
3	Somewhat disagreeable conditions due to occasional exposure to noises, open drafty places, possibility of damage to clothing, and occasional exposure to minor accident hazards.	22
4	Good working conditions, but infrequently exposed to extremes in temperatures, loud noises, disagreeable odors, and so forth. Practically no exposure to any accident hazard.	14
5	Usual office working conditions.	6

Fig. 9.7. (cont'd)

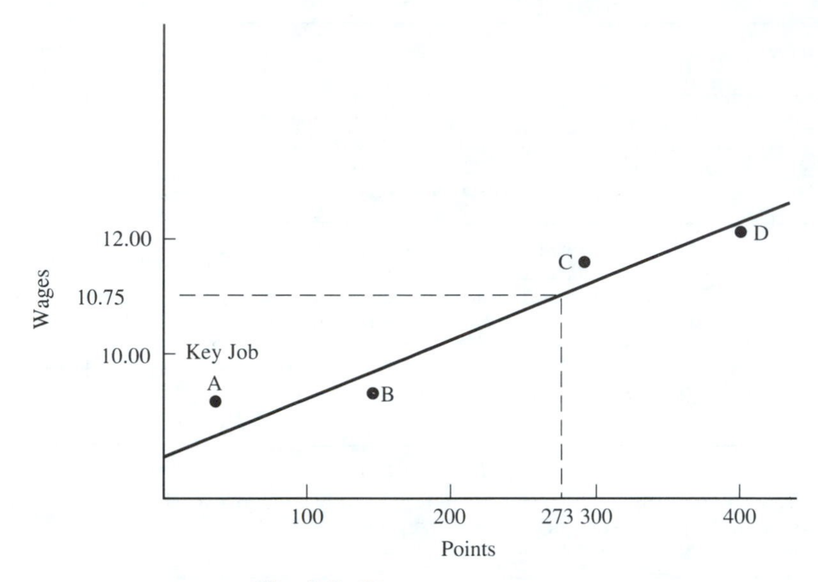

Fig. 9.8. Wage rate vs. points.

mine the wage and fringe benefit structure in other organizations. The study is usually conducted by the industrial engineering and personnel departments and can be as formal or informal as the organization desires.

One of the least costly and most popular methods is to use questionnaires. The questionnaire is designed very carefully by the organization and is mailed to other organizations in the area and sometimes to similar industries outside the area. This method is not expensive. If it is done correctly, much information can be obtained. Sometimes an interviewer goes to the various companies to make the survey. Usually, the interviewer takes the questionnaire and fills it in during the interview. This approach yields more complete information, but it is very expensive.

Telephone calls are a popular means for making spot checks or getting ''emergency''-type information. The data obtained are incomplete, but the speed of the response sometimes makes it worth the effort.

There are other sources of wage survey information, and the astute industrial engineer or personnel manager is always looking for them. For example, some are available commercially, particularly in large metropolitan areas. Some information is usually given in the various trade magazines. Local chambers of commerce and local governments usually have wage and salary information for their areas.

9.5. Wage Payment

The *primary* use of the preceding material is in developing fair wages and methods of payment. This section discusses the various methods of actually paying wages.

Through the use of job evaluation and wage surveys, the hourly base rate of all the jobs has been determined; now the people must be paid. Many methods are available, but those given below are among the most popular:

(1) Daywork.
(2) Measured daywork.
(3) Piecework incentive.
(4) Standard hour incentive.
(5) Group incentive.

9.5.1 Daywork

In this method the worker is paid a wage based solely on the number of hours worked. The base rate is determined by the job evaluation. Usually, there are no enforced standards on the work performed, and no direct attempt is made to have the employee work harder for a monetary reward.

The advantage is that without enforced standards employees are less likely to be unhappy (since almost all grievances are on the standards when the standards are enforced or used as a basis for pay). Also, the method is simple, straightforward, and easy to apply.

Following are the disadvantages:

(1) Production rate tends to be slow.
(2) Production rate may be erratic.
(3) Schedules and cost estimates are difficult to obtain with any degree of confidence.

9.5.2 Measured Daywork

Measured daywork is similar to daywork in that wages are paid on the basis of the number of hours worked and base rates. In measured daywork, however, standards are used, and periodic reports are issued to management. In this manner, management knows how well employees are performing, but no incentives are paid. Sometimes, merit increases or bonuses may result from consistently good reports.

The method is still relatively simple. The production rate is probably higher and less erratic than in daywork, and schedules and cost estimates are more reliable.

9.5.3 Piecework Incentive

Under a piecework incentive plan, an employee is paid according to his or her output. Almost always the employee is guaranteed a base rate. Wages are based strictly on performance above the base rate.

As an example, suppose that an employee has a base rate of $8.00 per hour and is working on a job that the time study shows should require .300 min/piece. The *expected* output can be calculated as

$$\left(\frac{60 \text{ min}}{\text{hr}}\right)\left(\frac{1 \text{ piece}}{0.300 \text{ min}}\right) = 200 \text{ pieces/hr}$$

The piecework rate would then be

$$\left(\frac{\$8.00}{\text{hr}}\right)\left(\frac{1 \text{ hr}}{200 \text{ pieces}}\right) = \$0.04/\text{piece}$$

The employee's wages (for various levels of output) would then be as shown in Table 9.1.

This method results in individual performance rates being maximized, good supervisory information made available (performance reports), and reliable cost estimates and schedules. The method is costly to operate, however, because standards have to be maintained and much clerical work is involved in calculating wages. Also, the piecework rate must be changed each time the base rates are changed.

9.5.4 Standard Hour

The standard hour method of payment is very similar to the piecework method in that base rates are guaranteed to a certain level (standard output) and incentives are paid after that. The only difference is that with the standard hour method, wages are paid on the basis of "earned time."

Let us use the same examples as in Section 9.5.3. The standard time per piece is 0.300 min, or 0.005 hr. The earned time is calculated as the number of pieces produced times the standard time per piece. Table 9.2 shows the calculations for the same output as in Table 9.1. The hourly rates paid are the same, but they are calculated in a different way. In the standard hour method the base rates are not used except at the very end to multiply by the earned time. Therefore, it is not

Table 9.1. WAGE RATE VS. OUTPUT FOR PIECEWORK INCENTIVE

Output/Hour	*Base Rate or Incentive*	*Wage Rate/Hour*
150	Base rate	$ 8.00
175	Base rate	8.00
200	Base rate	8.00
220	Incentive	8.80
240	Incentive	9.60
260	Incentive	10.40
280	Incentive	11.20
•	•	•
•	•	•
•	•	•

Table 9.2. WAGE RATES VS. OUTPUT FOR STANDARD HOUR INCENTIVE

Output/Hour	*Base Rate or Incentive*	*Earned Time (Hours)*	*Wage Rate*
150	Base rate	0.750	$ 8.00
175	Base rate	0.875	8.00
200	Base rate	1.000	8.00
220	Incentive	1.100	8.80
240	Incentive	1.200	9.60
260	Incentive	1.300	10.40
280	Incentive	1.400	11.20
•	•	•	•
•	•	•	•
•	•	•	•

necesary to change the standard each time base rates are changed as must be done in the piecework method. This is the main advantage of the standard hour method; all other advantages and disadvantages are the same as for the piecework method.

9.5.5 Group Plans

Individual incentive plans do help. Some employees are highly motivated and reach high levels of efficiency (and consequently pay rates), but there are problems. Maintaining the necessary standards is a real concern, as changes in equipment or prescribed procedures probably necessitate changes in the standard. Also, some employees are satisfied with base rates and simply do not respond to individual incentives.

The main problem, however, is more subtle. Under individual incentive plans, there is no guarantee that the employee's goals and that of the organization are aligned. As an example, an employee may be able to reach high levels of earnings by producing inferior-quality products (not necessarily "bad," just inferior). Obviously, this does not help the organization.

Group plans offer an alternative approach. Under group plans, each person earns an incentive based on the output (or performance) of the group as a whole. This group output may be measured in many different ways. For example, group incentives may be paid on the basis of the safety record of the group, the quality record, the amount of production, or just about any criteria that management wishes to choose.

The advantages are that less clerical work may be required to maintain the system (only one group calculation as opposed to many) and that group spirit may be encouraged. Performance of the top workers is usually less than that of an individual plan and many of them may become discouraged because the group pulls their wages down. The total group performance, however, may be increased and the group may pull as a whole. Finally, peer pressure is often exerted on the low producers, yielding higher performance from them.

Many group plans are in existence today, but some of the more popular are profit sharing, Scanlon, Rucker, and Improshare. Two of these are briefly discussed below (profit sharing and Scanlon). The reader is referred to the references for information on Rucker and Improshare. This does not imply that Scanlon plans are more popular than Rucker or Improshare plans; it is simply not possible to cover them all.

Profit Sharing. Profit sharing is the most widely used type of group plan. In profit sharing, a predetermined share of profits is paid to the employees in some form. They may occur as cash, investments in retirements, or partial ownership (stock) in the company.

The advantages are numerous, some of which are listed below:

Alignment of organizational and individual goals can occur.
Comprehensiveness and breadth are ensured.
Administration is relatively simple.
Motivation of management is usually good (they are used to dealing with longer-term issues such as year-end profits).
Compatibility with other incentive plans is good (e.g., shorter-term group plans or even individual incentive plans).

There are some major disadvantages, however, as follows:

Individual employees often have difficulty seeing how their single performance affects the total company profitability.
Profitability is complex, involving accounting maneuvers, business cycles and actual performance. Consequently, performance could be high and profits low, and vice versa.
Bonuses are usually paid at year's end, diluting the impact.

Basically, profit-sharing plans tend to work well in smaller organizations and/or where payouts can occur on a more frequent basis (e.g., monthly). Also, they tend to complement other incentive plans that can provide the necessary feedback and tie into individual performance.

Scanlon Plan. A Scanlon plan is as much a philosophy of management as an incentive plan. Following the often-quoted "Theory Y," the Scanlon philosophy is that the average worker likes to work, seeks responsibility, and is a tremendous source of profit-improvement ideas.

Therefore, the key ingredients to a Scanlon plan are:

(1) A management that supports this philosophy.
(2) A system to enlist *and act on* workers' ideas.

(3) A reward scheme that encourages the flow of their ideas and increased productivity.

The philosophy and the organization necessary for all the above is really beyond the scope of this chapter, but a review of the reward scheme is necessary.

First, the increased productivity is shared with the company (often 25% to the company and 75% to the employees). This is to reward and encourage management and to recognize the importance of management functions (e.g., production scheduling, inventory control, etc.).

All companies experience "ups and downs." To recognize this and to help smooth the peaks and valleys, a part (often 25%) of the bonus is placed in a reserve fund. The rest is paid to the employees (usually as a percent of their pay). If the bonus is negative, no additional pay occurs and the negative amount is made up by the reserve fund. At the end of the year, the reserve fund is usually distributed and the system starts over.

Table 9.3 presents a sample calculation of the Scanlon plan. Actually the calculation is more complex, as new employees typically do not participate (nor is their payroll counted) and changes in inventory must be accounted for carefully. However, the example does point out the basic simplicity of the Scanlon plan. The literature shows numerous examples of Scanlon plans that work well and (unfortunately) those that do not. Reasons for failure usually involve the organization and management philosophy for soliciting ideas and the calculation used to determine the bonus pool.

Highlights of Rucker[1] and Improshare.[2] The Rucker plan is unique in that it uses value added as a measure of productivity. Value added is simply the sales value minus production costs (excluding payroll). A historical average is compared to the present value added and the bonus determined accordingly. Company employee splits and bonus reserve funds are used as in Scanlon plans.

Improshare calculates the performance (and the bonus) by comparing actual production to a standard production and develops a base productivity factor. This is then used to calculate Improshare hours (similar to "earned hours") and the bonus is calculated accordingly. Improshare establishes a 30% ceiling on bonuses and "banks" hours above that. The banked hours are used for low-productivity periods. If banked hours consistently build, the company and the employees agree on a buy-back rate and the company pays a lump sum to each employee. This buys the banked hours and raises the standard. The standard is raised such that the employees are still expected to earn a 30% bonus. These ceiling and buy-back provisions are unique to Improshare.

---

[1] A registered trademark of Eddy–Rucker–Nickles Co., Cambridge, Mass.

[2] A registered trademark of Mitchell Fein, Inc., Hillsdale, N.J.

Table 9.3. EXAMPLE OF A SCANLON PLAN CALCULATION

1. Sales	$2,000,000
2. Less sales returns and allowances	50,000
3. Net sales (1 − 2)	1,950,000
4. Change in inventory	150,000
5. Value of production (3 + 4)	2,100,000
6. Allowed payroll*	525,000
7. Actual payroll	465,000
8. Bonus pool (6 − 7)	60,000
9. Company share (25% 8)	15,000
10. Employee share (75% 8 or 8 − 9)	45,000
11. Employee share to reserve (25% 10)	11,250
12. Employee share to be paid immediately (75% 10 or 10 − 11)	33,750
13. Percent bonus paid immediately (12/7)	7.3%
14. Percent total bonus (12/7)	12.9%

* Calculated as a base (standard) percentage of line 5, in this case, 25%.

DISCUSSION QUESTIONS

1. Distinguish among a job, a task, and a position.
2. List three uses for job descriptions and job specifications.
3. Why isn't a questionnaire approach to job analysis as accurate as observation or interview?
4. Why do you think it is necessary to write both a job description and a job specification?
5. Which two job evaluation procedures involve comparing something to a norm or scale? Which two are quantitative in nature?
6. Discuss situations in which ranking might be a preferred method of job evaluation. Do the same for classification, factor comparison, and point rating.
7. Discuss the motivational impacts of group incentive plans vs. individual incentive plans.
8. Discuss situations in which telephone calls might be the best approach for wage surveys.
9. Discuss the primary differences between profit sharing and Scanlon plans.
10. Considering only job satisfaction (how well the employee likes the job), which of the pay plans are best (day work, piece rate, group)?

PROBLEMS

1. Write a job description and a job specification for the position of hall monitor or counselor in a dormitory.
2. A new job, machinist, is to be added to the payroll. Using the point rating plan shown in Figure 9.7 and the following information, determine the approximate wage rate for the new job.

New Job, Machinist	*Old Jobs*	*Points*	*Wages*
Mental demands—degree 4	A	75	$ 6.00
Training and experience—degree 3	B	150	8.30
Effect of error—degree 4	C	225	12.20
Personal contact—degree 3	D	310	15.20
Job conditions—degree 3	E	395	19.00

3. Given that a standard 60 pieces per hour and base rate is guaranteed, how much would an employee get paid under a piecework system if he made 520 pieces in an 8-hour day and his base rate is $8.00 per hour? What would be the earned time under a standard hour system? Repeat both questions for a total production of 400 pieces over 8 hours.

4. Using the factor comparison chart of Figure 9.5 and the information given below, calculate the range of possible wages for the new job, heliarc welder.

Factor	*Location of New Job*
Mental	Between Special Assembler and Machinist
Physical	Same as Special Assembler
Skill	Between Machinist and Tool and Die Maker
Working conditions	Same as Material Handler
Responsibility	Between Machinist and Special Assembler

5. Given the time study results given below, allowances of 12%, and a base rate of $12.00 per hour: What is the earned time for each unit of production, and what is the piecework rate? Graph the pay for production rates of 30, 40, 50, 60, and 70 units per hour.

Time Study Element	*Average Elemental Time (min)*
A	0.600
B	0.712
C	0.305
D	0.207
Performance rating 110%	

6. Using the point rating plan given in Figure 9.8, perform a job evaluation (calculate the point rating) on the hall monitor or counselor's job in Problem 1. Using the points and wages given below for existing jobs, calculate the approximate base rate.

Job	*Points*	*Rate*
A	75	$2.00
B	100	2.20
C	250	3.25
D	300	3.85

7. Using the questionnaire shown in Figure 9.1 and any other information form, conduct a job analysis interview for a college department head and write a job description and a job evaluation for that job.
8. Write a job description and specification for a major college football coach.
9. Given the information below, calculate the percentage bonus (total), bonus to reserve, and bonus paid immediately for a Scanlon group plan.

Sales	1,000,000	Sales returns 100,000	
Changes in inventory	200,000	Payroll percentage standard	30%
		Employee share	70%
Bonus split		Reserve fund percentage	30%
Company	30%		
Employee	70%		
Actual payroll	165,000		

10. Repeat Problem 9 if the actual payroll was $200,000.
11. Your company has several machinists and so has created a position entitled "lead machinist." This person reports directly to the foreman. He or she is responsible for all machinists and so must keep track of each one's performance. Using Figure 9.4 (factor comparison) and your knowledge, determine a pay scale.

CHAPTER 10

CAD/CAM, Robotics, and Automation

10.1. The Second Industrial Revolution

Since about 1980, numerous business magazines have published many articles proclaiming that a *second industrial revolution* is under way. Whether we believe this to be the case or not depends on our perception of what constitutes a "revolution" as contrasted to "evolution."

Industry has "evolved" in a generally orderly fashion for two centuries. Why, then, do some authors claim that a revolution is under way? Are the changes and innovations now being experienced in industry so significant that they are more properly classified revolutionary rather than evolutionary? We will examine several aspects of these questions in the following sections.

10.1.1 A Brief History of Manufacturing

Manufacturing as such did not exist until approximately the middle of the eighteenth century. Before that time, most people lived in rural areas and produced their own clothing and farm implements. A degree of specialization occurred when certain craftsmen began to produce items for sale to others. The village blacksmith produced farm implements, tools, kitchen utensils, and similar items; other specialists produced shoes, candles, clocks, and so on.

Even after craftsmen began to specialize in the production of certain items, the location of this production was usually in or near the homes of the craftsmen. This was the beginning of *cottage industries*. As the work load increased, the craftsman had to employ others to assist in accomplishing the work. Only the most elementary form of planning and organization was followed. Essentially every item produced was one of a kind. Most of the energy used in the production processes was human and animal energy. Essentially all communication was verbal. Much of the work was performed out-of-doors.

Certain innovations occurred which led to remarkable progress. Flowing water was harnessed to provide power to large-scale machinery. Water power was used to grind wheat and other grains, to run textile equipment, and to turn lathes and other equipment. Wind power was also used, but not as extensively.

The widespread application of steam power in the nineteenth century resulted in the concentration of large numbers of machines in a central location. Workers who operated these machines had to leave their homes, travel to the central location, and work closely with dozens (even hundreds) of other people. The cottage industries disappeared rather quickly, replaced by the *factory system*.

With so many people involved, the factories were forced into more conscious forms of planning and organization. Specialization of labor became commonplace. With the introduction of the concept of *interchangeable parts*, a large number of almost identical items could be produced. The production cost per unit decreased, which led to more profit for the owners, higher salaries for the workers, and lower prices for the customers. Everybody won!

The factory system of production swept much of Western Europe and the United States. It has eventually found its way into essentially every part of the world. But there are other developments within the factory system that are important to mention in our brief chronology. Electricity replaced steam as the predominant power source. Machinery employed in the factory became increasingly sophisticated, accurate, and reliable. Materials improved dramatically.

A very important development was the concept of *mass production*, usually involving an assembly line. Fundamental principles of factory management began to emerge. By the early part of the twentieth century, hundreds of thousands of factories in many countries were rapidly and efficiently transforming enormous quantities of iron ore and other materials into a bewildering array of consumer products at unbelievably low prices.

The first 60 years of the twentieth century saw even greater advances in manufacturing. The entire chemical industry was developed with profound and far-reaching impacts. Plastics became a fundamental feedstock for a wide variety of consumer products. Nuclear energy was developed for commercial purposes. Communications became an industry within itself.

As astounding as the aforementioned developments have been, the advancements in electronics and digital computers have had the greatest impact on manufacturing. These impacts are so dramatic and so pervasive that we must conclude that we are now in the initial stages of a second industrial revolution. The changes we are seeing are too rapid and fundamentally different to be called evolutionary. They are, indeed, revolutionary.

10.1.2 Impact of Computers and Electronics

Many people believe that the parallel developments in electronics and digital computers constitute the most profound innovations in the history of human progress. Certainly, the development of computer technology was a major milestone whose impact is still unfolding. The first generation of digital computers, even though relatively clumsy and unreliable, permitted scientists and engineers to perform computations at speeds that were orders of magnitude greater than was possible with mechanical calculators. As vacuum tubes gave way to transistors, speed and reliability increased significantly. Interactive, on-line programming became possible through the development of greater memory sizes, remote communication capabilities, and sophisticated operating systems that could accommodate multiple users simultaneously.

Computer technology continued its progress through the development of sophisticated data-base management systems, mass storage systems, and ever-improving input/output capabilities. Of particular significance was the development of interactive graphics and supporting software. These developments, together with the development of workstations that operate in either a stand-alone mode or driven by a central host computer, made possible the development of today's CAD/CAM systems.

Supporting the developments in computer technology were significant advancements in electronics. Some of the important developments were integrated circuits, printed circuit boards, LSI (large-scale integrated circuits), VLSI (very large scale integrated circuits), and programmable controllers. These advancements have culminated in the microelectronics revolution, which has literally made "computer chips" a household term. Computer chip technology pervades practically every aspect of our lives. Numerous chips are found in household appliances, automobiles, calculators powered by any light source, watches, credit cards, and even in greeting cards!

Modern computer technology also pervades our factories. In later sections of this chapter, we will see some of these applications. There is no question that computers and electronics are changing dramatically the way products are designed, manufactured, distributed, and serviced. This is forcing companies to rethink the way work is organized and performed and the way people relate to each other in our manufacturing facilities. A second industrial revolution is, indeed, well under way.

10.1.3 Other Recent Developments

Together with, and partly because of, the developments in computers and electronics, there have been other recent developments which are contributing to the way the second industrial revolution is unfolding.

Numerically controlled (NC) machine tools became common in manufacturing during the 1960s. This was an important development that helped pave the way toward digital control of manufacturing processes. Prior to this development, machine tools were under manual control. Accuracy, precision, and repeatability in the production of parts were greatly enhanced through this development.

The concepts of *group technology* (GT) began to emerge during the 1960s and 1970s. These concepts contributed to greater design simplicity (through the emphasis on standard design features) and to much greater manufacturing efficiencies (through the grouping of production parts into "families" with similar characteristics).

The first industrial *robot* was developed and employed in the 1950s. However, it was not until the late 1970s and early 1980s that robots were widely used in industry. Many questions and issues remain unanswered regarding the field of robotics, but there is no doubt that they will be a significant force in future industrial enterprises.

The concept of *"flexible manufacturing systems"* (FMSs) began to emerge in the 1970s. This movement was spurred by the desire of companies to have manufacturing facilities that could quickly be modified to meet rapidly changing customer demands. As is true for the robotics movement, the FMS movement is in its infancy, but is expected to be a major consideration in future manufacturing plants.

Another set of significant developments is concerned with *factory management* concepts. MRP (materials requirements planning) evolved into MRP II (manufacturing resource planning). Methods for master scheduling, detailed shop floor scheduling, and shop floor corrective action are continuing to evolve. Some of the lessons we have learned from the "Japanese management style" are being incorporated into the newer factory management systems. These include "just-in-time" production, zero or near-zero inventory levels, very small batch sizes, and quality circles in which workers identify and solve production problems.

10.1.4 The Factory of the Future

Authors are beginning to use the term *factory of the future* to characterize manufacturing facilities that embody all the modern concepts we have just discussed. The term is probably misleading, but it serves as a mechanism to focus management's attention on the need to continually modernize a company's manufacturing capability.

There can be no precise definition of the factory of the future because the concept itself is vague and the "future" consists of all time from this point forward. Using 1990, say, as a reference point, we can project ahead to a particular point in time, say 2000, and attempt to characterize the typical factory of 2000 relative to the typical factory of 1990. We defer a discussion of this to Section 10.6.

10.2. Computer-Aided Design

Just a few short years ago, essentially all engineering design activities consisted of reducing design concepts to a scaled drawing, showing the geometry of the part, dimensions, tolerances, and other specifications. A separate engineering drawing was required for each major component of a product. Other drawings showed groupings of components into subassemblies.

Tens of thousands of such drawings were required in the design of a Boeing 727 airliner. These had to be cataloged, retrieved, routed to production areas, and returned to storage each time a component was produced. Anytime that a change occurred in an engineering design, a new drawing had to be produced and inserted into the active drawing file. The drawing that was superseded had to be removed. Hundreds of thousands of engineering labor hours were expended each year in the generation and maintenance of engineering drawings.

10.2.1 Computers in Product Design

With the developments in interactive graphics in the late 1960s and during the 1970s, great strides were made toward improving the productivity of the design engineer. Using the tools of computer-aided design (CAD) at a graphics terminal, an engineer defines a part shape, analyzes stresses and other factors, simulates mechanical performance, and (if desired) automatically generates an engineering drawing. This process results in a design data base, consisting of geometric (shape) data and nongeometric data such as bills of materials, tooling requirements, and other data useful to the user of the design data base.

The categories of computer-aided design are as follows:

(1) *Design Development*—The design image is created on a graphics terminal from a library of basic geometric elements, such as lines, points, cones, and spheres that are added, subtracted, intersected, or otherwise transformed to construct the geometric shape desired.

(2) *Design Analysis*—On the design just created, the engineer now performs a series of analyses. Software packages accessible from the graphics workstation are used to calculate properties of the design (weight, volume, center of gravity, surface area, etc.) and to analyze stresses, heat transfer properties, and other factors of interest. Finite-element analysis software is one of the more common packages available to perform design analysis. Using this technique, the component is subdivided into a network of simple elements that the computer uses to calculate stresses, deflections, and other structural characteristics. In this manner, the engineer can observe how the proposed component would behave and modify it if necessary to obtain the desired behavior. Thus the engineer avoids the building of costly physical models and prototypes.

(3) *Design Simulation*—The design analysis procedure described above can be extended to a complete system model and the performance of a total product can be evaluated. Simulation and animation software routines are employed to examine the paths of moving parts and to analyze more complete mechanisms.

(4) *Design Review and Evaluation*—After the design has been analyzed and simulated, several aspects of design accuracy can be checked on the CAD graphics terminal. Interference checking is one useful procedure for design review that helps reduce the risk that two or more components of an assembled system are trying to occupy the same space at the same time. Dimension and tolerance checking routines are available to help reduce the possibility of dimensioning errors. Software packages are becoming available that assist in assessing producibility (or manufacturability) of a particular design.

(5) *Automated Drafting*—Most CAD systems will automatically generate, upon command, hard-copy engineering drawings for use in process planning and in manufacturing. Some of the features available in automated drafting systems are automatic dimensioning (mentioned earlier), generation of specific sectional views, crosshatching, scaling of the drawing, and angle views.

(6) *Design Retrieval and Modification*—Engineering designs undergo frequent modification. CAD data bases allow specific designs to be retrieved for modification, improvement, further analysis, and so on. If a group technology type of parts classification and coding system has been incorporated into the CAD/CAM system, designers can use the classification and coding system to retrieve an existing part design for possible consideration for a new part design. Perhaps the existing design can be modified slightly to serve the function required of the new part.

The CAD system consists of both hardware and software. The CAD hardware will normally include a computer, one or more graphics design terminals, keyboards (alphanumeric, numerical keypad, and function keypad), a digital tablet, a light pen, a large disk storage device, and a plotter. The computer may be a large mainframe serving as a host to many design terminals, it may be a powerful minicomputer serving four to six design terminals, or it may be a stand-alone CAD design workstation. The CAD software typically consists of special computer graphics packages and a wide variety of application software programs, depending on the type of manufacturing involved, the nature of the product lines, and customer requirements.

10.2.2 Computers in Process Design

Computers have been used for many years in the design of continuous processes, such as oil refineries and chemical plants. The basic characteristics of the system variables involved in specific unit processes are expressed as differential equations. The parameters of these equations are ''tweaked'' until the response variables behave as desired. The individual unit processes are optimized in this manner. The next step is to connect the unit processes in a stagewise manner, and then adjust the parameters until the enlarged system is behaving as desired (recognizing that trade-offs are always necessary between competing objectives). This process is continued until the total system has been incorporated into the model.

Analogous progress has not been achieved in discrete-part manufacturing. CAD systems are now being developed that will enhance the abilities of the design engineer whose responsibility is to design discrete-part manufacturing systems.

Computer-aided plant-layout packages have been available for over a decade. One of these packages was illustrated in Chapter 4.

CAD systems make it possible to develop detailed plant layouts directly on the graphics terminal. A library of machine shapes/geometries can be employed. Aisles, storage locations, inspection stations, tool cribs, receiving and shipping docks, conveyors, office areas, and so on, can all be incorporated into the library. A three-dimensional data base can be created which is essentially a scale model of the factory. Wiring, piping, and all other utilities can be included.

As is true for the product design case, the computerized process design can be rotated, certain sections enlarged, and so on. If the CRT is a color monitor, any desired color coding may be employed.

Software is becoming available that allows the facility design engineer to create realistic animation of the factory. This software makes it possible to see parts moving from work center to work center, building up in queues, being recycled for rework, and so on. The operation of automated storage and retrieval systems, automated material handling systems, and the movement of robots are all included in the animation. The industrial engineer and the plant manager now have a true "management laboratory" to evaluate the outcomes of alternative policies and system changes, based on the performance of the model.

10.2.3 Computers in Electronics Design

In principle, ECAD (electronic computer-aided design) is basically the same as CAD (mechanical computer-aided design). There are certain differences, of course.

The major steps involved in designing electronic devices are as follows:

(1) *Concept Design*—The first stage of designing an electronics product involves conceptual design and feasibility analysis. The result of this step is documentation forming the foundation for the final product. At this stage, the design is usually expressed in terms of functional requirements, cost and quality (reliability) constraints, and other pertinent operational parameters.

(2) *Product Design*—The conceptual design is translated into a product description. Schematics are developed on ECAD graphics terminals using software packages and input devices (digitizers, etc.) especially for this function.

(3) *Design Verification*—At this step, special software packages are used which emulate the functionality of the circuits being tested without building a prototype. Each gate array is thoroughly simulated to assure full functionality when the actual component is manufactured. Software packages are also becoming available to perform automatic test generation and detailed timing analysis. These packages will allow potential design changes to be tried and verified interactively.

(4) *Design Layout and Wire Routing*—Once the electrical design for a particular component has been completed, the physical design of a printed circuit board or integrated circuit follows. This is accomplished at a graphics terminal using a stored library of templates. A wire-routing software package is then employed (a small percentage of the wires still must be routed manually). Satisfactory trace separation is assured by programmed "design rule" checks. The ECAD system produces drawing documentation and parts lists. The layout is checked automatically against the schematic diagram, and an NC drill tape is generated automatically.

10.3. Computer-Aided Manufacturing

Computers were first employed in manufacturing companies in the early 1960s. The first applications were concerned primarily with computerizing the accounting function and other routine financial transactions. Gradually, computers were applied to other tasks, such as inventory management, labor reporting, production scheduling, and production routing.

Another major development in the use of computers in the manufacturing environment was that of numerical control (NC) technology. The initial efforts on NC were conducted in the early 1950s.

We will explore several current aspects of computer-aided manufacturing. This exciting field is undergoing rapid change and development, and is considered a key weapon in the second industrial revolution.

10.3.1 Computer-Aided Process Planning

Ultimately, we would like to remove the slash from CAD/CAM to symbolize the integration of the design and manufacturing functions. A key element in our efforts to remove the slash is the development of an automated process planning methodology.

The CAD function results in a design data base which consists of a complete description of part geometry, dimensions, tolerances, and parts lists (including material specifications). The design data base may remain totally electronic or it may be used to produce hard-copy engineering drawings of the work parts. In either case, the design data base is the input to the next function, which is to interpret the design in terms of specific manufacturing operations required to produce the work part.

This function is commonly called *process planning*. The output of process planning is a detailed routing sheet which shows the sequence of manufacturing operations to be performed, the specific machines or work centers involved, the tools and fixtures required, material requirements, and standard processing times.

Significant progress has been made toward automating the process planning function. Software packages are being developed which determine the sequence of manufacturing operations for a given part, based on a detailed set of specific characteristics for that part.

Two basic approaches have been followed in developing automated process planning techniques. These are known as *variant systems* and *generative systems*.

(1) *Variant Process Planning Systems*—This approach, also known as *retrieval process planning*, is based on the concepts of group technology and parts classification (discussed in Section 10.3.3). The entire set of parts manufactured in the plant are categorized into part "families," which are established according to common manufacturing characteristics and requirements. A standard process plan is created for each part family, and stored in the manufacturing data base.

When a new part is introduced it is analyzed and classified into one of the existing part families. The process plan for that part family is retrieved. If necessary, modifications are made to the standard process plan for that family to accommodate any specific requirements of the new part. Both the standard plan and the modified plan are stored in the manufacturing data base. All of these steps can be performed at a graphics terminal that has access to both the design data base (from CAD) and the manufacturing data base. Software programs are available to assist the process planner in determining appropriate cutting conditions for machining operations (speeds, feeds, lubricants, etc.), work standards, and standard costs. If marketing has access to this system, the standard cost programs can be used as a basis for determining total product costs for pricing purposes. Sales could use the system for preparing price quotes when bidding on jobs.

(2) *Generative Process Planning Systems*—In its purest form a generative process planning system automatically creates a unique process plan for each new part, based on comprehensive information about the characteristics of the part. Even though no such system is currently available, extensive development work is under way which will likely lead to a working system for generative process planning.

10.3.2 Numerical Control[1]

The first successful numerical control (NC) system was developed in the early 1950s at the Massachusetts Institute of Technology under contract to the U.S. Air Force. The first prototype consisted of retrofitting a milling machine with position controllers using servomechanisms for each of the three axes of the machine tool. Machine tool companies quickly picked up on this innovation and developed their own NC machine tools.

An NC system on a machine tool activates the tool in response to a prepro-

---

[1]M. P. Groover, *Automation, Production Systems, and Computer-Integrated Manufacturing* (Englewood Cliffs, N.J.: Prentice Hall, 1987), Chap. 8.

grammed series of commands stored as digital data on punched tape, magnetic tape, or some other medium. The NC program sends digital signals to activators which position the tool point in three dimensions relative to a workpiece (which may be turning or moving laterally). The program may also control secondary functions such as tool speed, material feed rate, coolant flow, gauging, tool selection, and tool changing. The components of an NC system are as follows:[2]

- *Program of Instructions*—The detailed set of directions which tell the machine tool what to do, step by step. It is coded on some type of input medium which can be interpreted by the machine control unit (below). The most common input medium on early NC systems was punched tape. A more advanced method is through a direct link with a computer. This is called *direct numerical control* (DNC). With DNC, a group of NC or DNC machines can be controlled simultaneously by a host computer.
- *Machine Control Unit*—This is an electromechanical device that reads and interprets the program of instructions and converts them into mechanical actions of the machine tool. Most modern NC systems use a microprocessor as the controller unit. This type of numerical control is called *computer numerical control* (CNC).
- *Processing Equipment*—Usually a machine tool, this is the part of the system that actually performs the work. A typical machine tool consists of a spindle to turn either the cutting tool or the workpiece; a worktable, chuck, or some other means of holding and positioning the workpiece; and the motors and controls that drive the spindle and the table. The cutting tools themselves, work fixtures, and other associated equipment are also part of the controlled process.

Let us back off for a moment to get our perspective. The machining operations are for the purpose of creating the desired geometry of the workpart. Data describing the part geometry are created by design engineers working through a CAD system (see Section 10.2). Geometrical and other design data are entered into the design data base.

It would seem logical for the engineers concerned with various aspects of CAM (computer-aided manufacturing) to be using as their primary input the CAD-generated design data base. Indeed, considerable effort is being directed toward just such an integrated system. The ultimate aim is to be able to generate NC programs automatically (through appropriate software) directly from the design data base. Currently, these programs are generated through one of several NC part programming languages,[3] the most common of which is APT. Under development are NC programming languages based on interactive graphics and others based on voice

---

[2]Groover, pp. 200–201.
[3]Groover, pp. 247–260.

input of NC commands. Further into the future, it may be possible to generate process plans directly from the design data base.

10.3.3 The Concepts of Group Technology

We have referred to "group technology" several times in previous sections. Indeed, much of the advancement in flexible manufacturing and other manufacturing technologies may be attributed to the fundamental concepts of group technology.

The following definition of group technology will motivate our discussion:[4]

> Group technology is a technique for manufacturing small to medium lot size batches of parts of similar process, of somewhat dissimilar materials, geometry and size, which are produced in a committed small cell of machines which have been grouped together physically, specifically tooled, and scheduled as a unit.

This lengthy definition needs further clarification. "Similar process" means that processing of the parts can be completed in the machine cell, except perhaps for heat treatment, final polishing, and so on. "Somewhat dissimilar materials, geometry and size" means that the various parts are not identical, but must be within the general size and shape constraints of the clamps, holding fixtures, chucks, and so on. "Produced in a committed small cell of machines which have been grouped together physically" implies that the cell is committed to processing only parts from that particular family, irrespective of machine utilization. "Specifically tooled" implies that most toolholders and tool stations remain fixed. They may be adjusted, but never removed and replaced. "Scheduled as a unit" implies that the machine cell is scheduled as if it were a single machine using a single cycle in the shop floor scheduling system.

A key to the successful implementation of the group technology concept is the establishment of a workable parts classification and coding system. Many such systems have been developed throughout the world, and several packages are available from consultants and other vendors.

A particular part is classified and coded according to two sets of attributes: design attributes and manufacturing attributes. The reason for classifying parts by their *design attributes* is to facilitate design standardization and design retrieval capabilities in a CAD system. Parts are classified by their *manufacturing attributes* to facilitate computer-aided process planning (see Section 10.3.1), tool design, materials requirement planning, and so on.

Design attributes include *geometric shape* (both internal and external), *size* (length, diameter, length/diameter ratio), *material type*, *environment in which it must function*, *tolerances*, *surface finish*, and so on.

---

[4]W. F. Hyde, *Classification, Coding, and Data Base Organization* (New York: Marcel Dekker, Inc., 1981), pp. 152–153.

Manufacturing attributes include *sequence of operations* (particular parts may skip certain operations in the basic sequence for that family), *treatment similarities* (surface treatment, laser hardening, chemical baths, etc.), *machine operation time*, *similarity of cutting tools and fixtures*, *similarity of batch sizes*, and so on.

Group technology has had a major impact on the arrangement of equipment for discrete part manufacturing. Traditionally, equipment in such companies have been arranged by functional grouping. All the lathes were located together, all the milling machines together, and so on. Such an arrangement is shown in Figure 10.1. Notice that this arrangement results in a lot of complex material flow, including backtracking. Equipment arrangement based on group technology concepts, on the other hand, results in relatively smooth material flow, as shown in Figure 10.2. The specific equipment needed for processing a particular *family of parts* is arranged in a logical sequence to permit smooth material flow. The book by Hyde presents a thorough treatment of classification and coding systems.

10.3.4 Automated Storage, Retrieval, and Handling

Much progress has been made in recent years in the development of automated storage and retrieval systems (AS/RS). These systems are used to store raw materials, purchased parts, in-process fabricated parts and subassemblies, and final products. A series of storage racks (often reaching to the ceiling of a 60-foot-high warehouse) provides a variety of bin sizes for storing pallets, boxes, or trays of

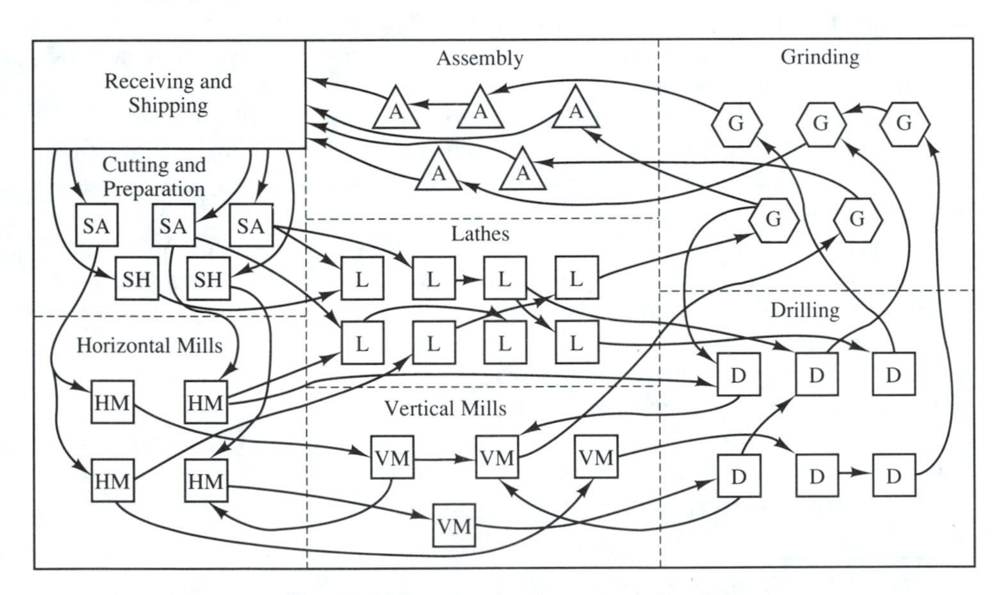

Fig. 10.1. Functional arrangement of equipment.

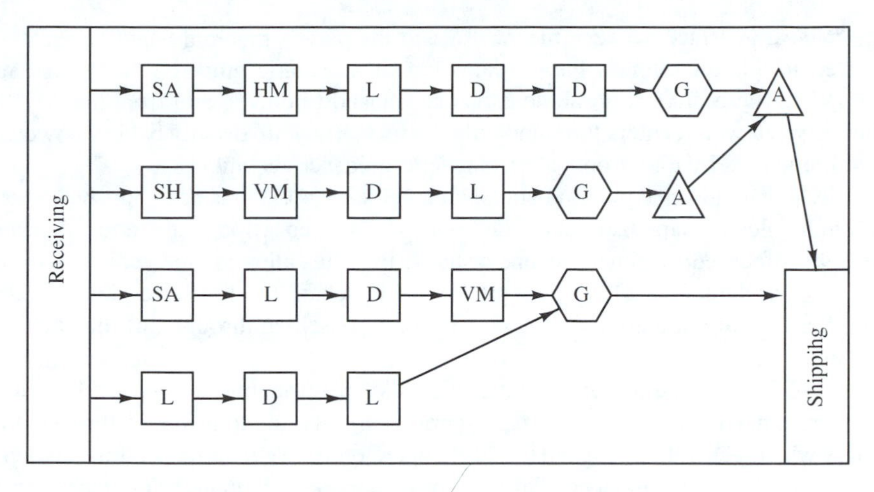

Fig. 10.2. Equipment arranged via group technology concepts.

materials. An aisle between each pair of storage racks allows access to every storage location. A cranelike vehicle operates under computer control to store incoming materials and to retrieve materials needed for manufacturing or for fulfilling a customer order.

The receiving and shipping docks are normally adjacent to the "entry" side of the systems. The "exit" side usually opens to the manufacturing work centers.

When storing materials or parts, the system delivers the items to an open random location appropriate for the characteristics (size, weight, etc.) of the items, and records the location for future reference so that the items may be retrieved. Computer algorithms in the AS/RS control computer determine storage locations such that total distance traveled is minimized.

Miniature AS/RS systems are frequently installed in the manufacturing facility as a temporary storage location for parts and subassemblies that will soon be routed to subsequent machine operations or to final assembly.

Materials and parts are automatically retrieved from the AS/RS in preparation for movement to the various manufacturing work centers in the plant. The items retrieved are usually accumulated at a staging area, where they are transferred to various material handling devices for delivery to the specific work centers.

There is a great degree of diversity from plant to plant at this stage of the manufacturing process, that is, at the interface between the AS/RS and the plant's material handling system. Many plants employ some form of circulating conveyor system. It is not uncommon to see parts being manually transferred to such a conveyor at the staging area of the AS/RS. In other cases, parts are loaded onto pallets or placed in trays and delivered to work centers via forklift trucks, overhead cranes, and even manually pushed part-carts.

A much needed advancement in manufacturing systems is the elimination of the manual interface between the AS/RS and the plant's material handling system. A step in this direction is the development of automatic guided vehicle systems (AGVS). These driverless shuttle carts can transport items to and from the AS/RS and between work centers throughout the factory. They are usually battery powered and are guided by either optical or magnetic guidance techniques.

Optical guidance of automatic guided vehicles (AGVs) is accomplished either through reflective tape laid out on the floor, or a painted stripe on the floor. A beam of light is focused on either the tape or the stripe, thus allowing the vehicle to track the path as laid out. The guide path is easily changed if needed, but has the disadvantage of unreliability if the path becomes obscured through dirt, damage, or for other causes.

AGVs that operate on the magnetic guidance principle are guided by wires that are embedded in the floor. Low-current signals are transmitted through the wires which generate a magnetic field. Sensors on the vehicle respond to this field in providing vehicle guidance. This system is more reliable than the optical system, but is more expensive to install and much more expensive to modify.

Significant advancements in automated material movement are needed to realize the full potential of factory automation.

10.3.5 Computer-Aided Testing and Inspection

As each manufacturing operation is performed on a work part, the part should be inspected to assure that the part conforms to the specifications established for it in the design phase. This function has traditionally been performed using manual inspection methods after the part has been removed from the machine. Testing and inspection are very time consuming and costly activities.

Some of the characteristics subject to inspection among machined parts are dimensional measures (length, width, depth, diameter), geometric shape (roundness, concentricity, fillet curvature, angles), distance from a reference point or surface, surface smoothness, and so on. In general, inspection procedures are used to examine a part in relation to the design specifications established for it.

Testing is distinguished from inspection in that testing is concerned with assessing the functionality of the manufactured product. We are all familiar with such terms as *test cells*, *test tracks*, and *test chambers*. The complete product or major subassemblies are operated under a range of all possible conditions that the item may be expected to encounter. Its performance is observed and critical system variables are measured.

Test cells are becoming increasingly automated with extensive use of computer control of cell operation. During testing, the computer monitors the data using real-time readouts and analyzes the results. If any problem occurs, the computer assists in diagnosing the problem, pinpoints the faulty component, and recommends appropriate corrective action.

Computer-aided inspection is on the verge of becoming one of the most critical components of automated manufacturing systems. Computer-aided inspection may be performed either on-line, while the part is being machined (sometimes called ''on-the-fly'' inspection) or off-line.

On-line inspection involves the measurement of specific variables during the actual performance of an operation. As an example, procedures are being developed to use the cutting tool tip on a lathe as the measuring instrument to continuously measure the diameter of a turned part.

Another motivation for performing on-line inspection is to obtain input data for adaptive control functions. In-process sensors detect what is happening, at the time it is happening, and provides these data to a supervisory controller to make appropriate adjustments in the process. For example, as tool wear is experienced, the depth of cut is adjusted automatically to compensate for the wear.

Whenever possible, inspection should be done on-line, while the machine tool is performing its operations. The advantages are obvious. It is not possible to perform all the necessary inspection steps while the machine is operating. The next best procedure is to perform the inspection with the machine stopped, but with the work part still loaded in the machine. If tolerances are not met, machining can continue without the necessity of unloading and reloading the part, as would be necessary with most of today's manual inspection methods.

Computer-aided inspection procedures are dependent on a variety of sensors for measuring the critical attributes of machined parts. These procedures fall into two broad categories, contact inspection and noncontact inspection.

(1) *Contact Inspection*—The most common type of equipment used for contact inspection is the coordinate measuring machine (CMM). These systems access data from the design data base and automatically measure the part to see if it has been manufactured to design specifications. The CMM consists of a table which holds the work part in a registered position and a movable sensing probe. The probe is automatically moved to a series of preprogrammed points on the surface of the part. These measurements provide the basis for accepting or rejecting the part for subsequent processing. The CMM also stores the data in a manufacturing data base for use in a variety of shop floor management functions.

(2) *Noncontact Inspection*—For some inspection functions, it is preferable to use noncontact inspection methods. Machine vision systems, for example, can perform noncontact gauging of dimensions and can also perform inspection based on pattern recognition of specific object features. Scanning laser beam systems have also been used extensively in measurement applications.

Ultrasonic techniques are another category of noncontact inspection devices for measuring the dimensional features of manufactured parts. X-ray radiation techniques are used to measure thickness and also to detect flaws in castings and welds.

Electrical field techniques are also being used in noncontact inspection.

For example, inductance devices generate eddy currents in the work part. These eddy currents can be used to inspect an object below its surface to detect voids, cracks, and other flaws.

Computer-aided testing and inspection will normally permit 100% inspection of parts and completed products. This will essentially eliminate a fraction of bad parts moving on to the next manufacturing process and bad products getting into the hands of customers. What is more, the testing and inspection processes will be accomplished in less time and at less cost than is now realized with conventional methods.

10.3.6 Computer-Aided Factory Management

The purpose of a factory management system is to coordinate, review, and control the activities of the entire plant. The factory management system ties together all the concepts discussed in the preceding sections into an integrated framework for managing and controlling available resources.

Preceding portions of the text have dealt with this aspect of computer-aided manufacturing. The concepts of feedback control, discussed in Section 2.8, are fundamental to computer-aided factory management systems. Chapter 7 dealt with this subject in greater detail.

Much progress will be made in this area over the next two decades. Eventually, factory managers will have a "control room," showing the on-line, real-time status of essentially all production system variables. Simulation programs will be available to allow the manager to test the probable effect of a particular decision *before* that decision is implemented.

10.3.7 The Concepts of Flexible Manufacturing Systems

In the past, manufacturing systems have been classified as either a flow line or a batch manufacturing system. *Flow lines* are appropriate for products having a very large annual volume, such as television sets and automobiles. Special-purpose equipment is designed for flow lines. *Batch systems* are appropriate for companies that must produce a relatively small number of a great variety of products. General-purpose equipment is required for batch systems (also called *job shops*).

Beginning in the early 1970s, machine tool manufacturers have been developing equipment which retains its flexibility for the production of a variety of components and at the same time achieves production economies approaching that of flow lines. Three major developments have made this possible. The first is the application of group technology, discussed in Section 10.3.3. By properly grouping parts into families, several batches of similar parts can be processed together, with only a minimal amount of setup required between the different parts. The second major development has been the dramatic improvement in machine tools regarding tool changeover time. One large machine can perform several dozen operations on a workpiece, with the tools changed automatically. Significant advancements have also been made on the tables to which the workpiece is attached. These tables are now automatically indexed such that the workpiece is presented to the cutting tool at the correct position and angle for each of the several machining steps required. The third major development has been the advances in electronics and computer control, which have made possible the achievement of dramatically improved production economies for small to medium batch sizes.

Equipment that can perform the type of machining operations just described are generally called *flexible manufacturing systems* (FMSs). There is not general agreement on the definition of an FMS, and the general concept will continue to evolve for some time in the future. It is useful, however, to identify four general types of FMS systems:

(1) *Flexible Modules*—A machine with an automatic tool changer, a tool magazine capable of holding a variety of tools, and an automatic loading/unloading system for the work parts.
(2) *Stand-Alone FMS*—Usually consists of a single machining center or CNC lathe. It is equipped with automatic carousels or rotary tables for loading a small variety of workpieces. It may also have a built-in robot arm for loading and unloading the machine.
(3) *Classical FMS*—An automated production system for producing midvolume, midvariety workpieces. It consists of several machine tools, tied together by an automated workpiece handling system, all controlled by a central computer. The central computer downloads NC programs to individual machine tools in the system, controls workpiece flow, and generates performance reports. The system may or may not include automated inspection capability.
(4) *Robotized FMS*—Same as "classical FMS," except that robots are used for handling the workpieces, thus providing an additional measure of flexibility. The robots would also be controlled by the central computer system.

For a more complete discussion of FMS, the reader is referred to Groover[5] and Luggen.[6] A schematic of an FMS is shown in Figure 10.3.

10.4. Robotics

The general public has little awareness of the technical details associated with the second industrial revolution. However, because of popular movies such as *Star Wars*, and because of several television specials concerning the productivity war between the United States and other countries (principally Japan), the robot has

[5]Groover, Chap. 17.

[6]W. W. Luggen, *Flexible Manufacturing Cells and Systems* (Englewood Cliffs, N.J.: Prentice Hall, 1991).

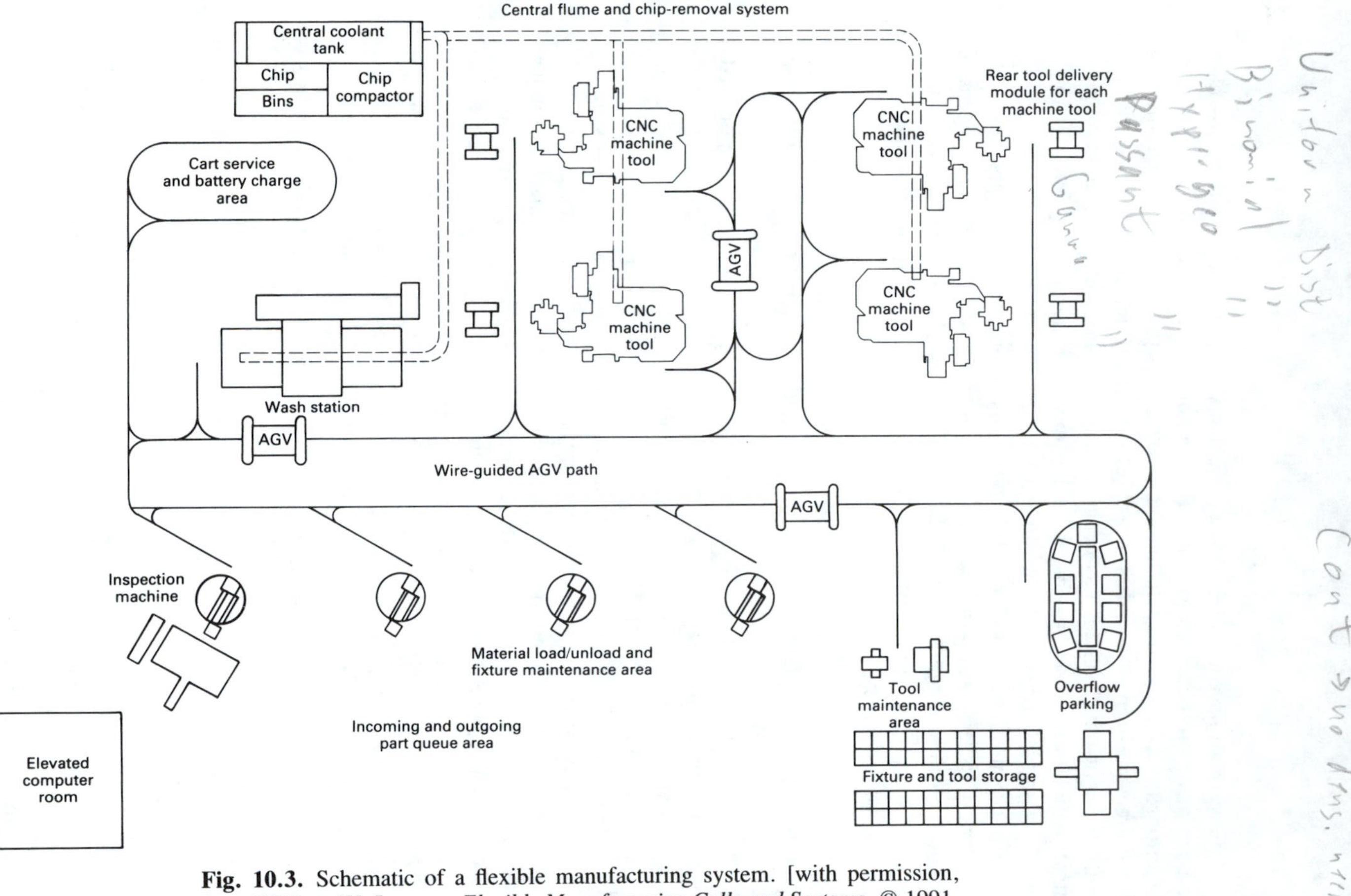

Fig. 10.3. Schematic of a flexible manufacturing system. [with permission, from William W. Luggen, *Flexible Manufacturing Cells and Systems*, © 1991, p. 407. Reprinted by permission Prentice Hall, Englewood Cliffs, N.J.]

become a symbol of all that is "good" and "bad" about modern manufacturing methods.

On the one hand, we are led to believe that the only hope for U.S. productivity is to convert to factories staffed with robots rather than human beings. At the same time, workers fear that robots will replace them, leading to widespread unemployment.

Both perceptions, of course, are highly inaccurate. Within the foreseeable future, robots will not be capable of simply stepping in and doing all the production activities now performed by human beings. The transition to the type of production system that can take advantage of the capabilities of robots (and many of the other developments discussed in this chapter) will require many years of effort. Although there will be specific instances of robots replacing human workers, the specter of massive unemployment resulting from factory modernization is just not going to happen.

It is the responsibility of the industrial engineer to understand the capabilities of robots and how they should be incorporated into the design of productive work systems, just as other equipment is considered. In this section we go through an overview of the field of robotics. More details may be found in other books.[7,8]

10.4.1 Definition and Basic Concepts

Perhaps the most widely accepted definition of a robot is that adopted by the Robot Institute of America:

> A robot is a programmable, multifunction manipulator designed to move material, parts, tools, or special devices through variable programmed motions for the performance of a variety of tasks.

Some robots are stationary, others are mobile. Robots are being developed which will have the capability of walking over uneven terrain, while others will be capable of swimming under water. Some robots have elementary vision capabilities, while others have a sense of feel. We will see a great variety of developments in the future aimed at specific capabilities.

Our interest in robots is primarily in their application in manufacturing systems. Hence, the definition above is adequate for our purposes. The emphasis in industrial robots is on the handling of physical objects in a flexible manner with the capability of modifying the robot's job content through reprogramming.

10.4.2 Physical and Technical Aspects of Robots

Essentially all industrial robots consist of the following basic components:

---

[7]J. F. Engelberger, *Robotics in Practice* (New York: AMACOM Book Division, 1980).

[8]E. Kafrissen and M. Stephans, *Industrial Robots and Robotics* (Reston, Va.: Reston Publishing Co., Inc., 1984).

- *Manipulator(s)*—Usually involves an arm and some type of end-effector (gripper, etc.).
- *Actuator*—Electric, hydraulic, or pneumatic units that supply the power to the manipulator.
- *Control Unit*—Unit that controls the motion and actions of the robot. This may consist of no more than a set of adjustable limit switches, or it could be a microprocessor with a complex control program directing the activities of the robot.

In addition, a particular robot may include one or more optional features, such as mobility units (wheels, legs, etc.), a wide variety of sensors (audio, visual, voice recognition, tactile, etc.), and a measuring system for position and velocity measurement.

Most industrial robots allow six basic motions (sometimes called *degrees of freedom*). The arm itself can move up and down about a horizontal axis. The arm can usually extend in and out relative to the "body" of the robot. The body can rotate about the vertical axis, allowing the arm to swivel right or left. The wrist at the end of the arm can rotate about an axis through the longitudinal center of the arm. The wrist can also bend up or down relative to the fixed position of the arm. Finally, the wrist can swivel right or left relative to the normal vertical plane through the center of the arm.

Although all of these motions may seem confusing, the important thing to recognize at this point is that a typical industrial robot has considerable flexibility of motion, but the motions are limited. A robot's motion flexibility does not approach that of a human being.

Other important technical characteristics of robots include the following:

- *Work Envelope*—Sometimes called the *work volume*, this is the physical space that can be reached by the robot's end-effector.
- *Payload*—Refers to the size of workpiece or tool the robot is capable of carrying within its work envelope. The range of payloads for most industrial robots is from 1 pound to over 1,000 pounds.
- *Precision of Movement*—Refers to the ability of the robot to move its end-effector to a particular point in the workspace to within a specified degree of accuracy.
- *Speed of Movement*—Determined by weight of payload, type of actuator, distance moved, and precision with which the end-effector must be positioned.
- *Stability*—Amount of "clatter" or mechanical oscillations resulting from robot movement.

These are several of the technical aspects of a robot that must be considered when determining which robot to apply in a given application. Robot manufacturers provide this type of technical information for each of their robots.

10.4.3 Robotic Applications

Robots are being applied in an ever-growing number of situations. This growth will continue and accelerate as robots acquire additional capabilities, such as better sensors and greater capability for adaptive learning. Some of the more common industrial applicatons are presented briefly below.

- *Machine Loading*—The robot takes a workpiece from a rack or conveyor, positions it on the machine, waits for the machining to take place, then unloads the finished part from the machine and places it on an exit conveyor. In some cases, the robot also changes tools on the machine.
- *Material Transfer*—The robot simply moves materials from one point to another, such as from one conveyor to another. Other examples include loading a pallet, packing a box, and stacking.
- *Painting/Coating*—The robot contains a spray gun as its end-effector. It is programmed to move through a series of very precise paths, depositing an exact amount of paint on the workpiece. In other cases, the robot grips the workpiece and dips it in a coating material.
- *Welding*—The end-effector is a welding device. For spot welding, two or more metal sheets are fused together at programmed points. Other robots are designed to perform continuous arc welding operations.
- *Assembly*—Robots are being applied in fairly simple, straightforward assembly operations. In the future, product designers will have to design components to accommodate robotic assembly.
- *Processing Operations*—With specially designed end-effectors, robots can be used for operations such as riveting, grinding, drilling, and deburring.
- *Inspection*—Robots containing advanced sensing devices, such as optical sensors, mechanical probes, and so on, can be programmed to inspect workpieces for conformance to dimensional specifications, surface smoothness, "roundness" of a hole or cylinder, and so on. They can also be programmed to remove the workpiece from a machine, test it on go/no go gauges, and then take appropriate action.
- *Packaging*—The robot can insert material into a box or carton, compress it if appropriate, seal the box, place it on a scale, record the weight on the box, and then place it on a conveyor or pallet.
- *Component Insertion*—The robotic end effector is fed a series of sequenced electronic components for insertion into specific locations on a printed circuit board. The same robot is capable of "stuffing" several different board types, simply by following the proper programmed instructions for each type.

Although there are many other applications of robots, these examples illustrate their versatility and potential usefulness in production environments.

To the industrial engineer responsible for designing or improving a productive work system, a robot is simply another device that must be considered in the overall system design. They have great capabilities, but they also have critical limitations, as does every other device or component.

10.5. Automation

The dictionary[9] definition of *automation* is as follows:

> Automatically controlled operation of an apparatus, process, or system by mechanical or electronic devices that take the place of human organs of observation, effort, and decision.

Dorf[10] offers a simpler definition of *factory automation*:

> A process without direct human activity in the process.

In a true automated system, human beings would be involved only in designing the system, monitoring its operation, and maintaining the system. Although no system today qualifies for these definitions of automation, significant progress has been made in removing direct human involvement from many portions of production systems. We will briefly review some of the developments in automation and refer the interested reader elsewhere[11] for more details.

In continous-flow processes, such as oil refineries and chemical plants, automation has reached a high state of development. Sensors measure critical process variables in real time and feed these measurements back to the control computer. The computer exercises an on-line optimization to determine any needed adjustments in the controllable process components. Thus the process is essentially fully automated, with human beings monitoring the process in a control room containing readouts of the process variables and parameter settings.

Progress has also been made in the mass production of discrete items, such as automobile components. These systems are usually called *flow lines* or *transfer lines*. Such systems consist of many special-purpose machines which are arranged sequentially. Work parts are transferred automatically from one machine to the next and the machines perform their operations automatically. In a partially automated flow line, certain operations that cannot be automated are performed manually. Robots may also be employed in a flow line for welding, painting, parts handling, and so on. Inspection may be automated or manual.

In the past, flow-line automation has been relatively rigid and nonflexible. It has been called ''Detroit automation'' because of its widespread use in the production of automobile components, such as engine blocks. In the future, automated

---

[9] *Webster's New Collegiate Dictionary* (Springfield, Mass; G. & C. Merriam Co., 1974).

[10] R.C. Dorf, *Robotics and Automated Manufacturing* (Reston, Va.: Reston Publishing Co., Inc., 1983), p. 28.

[11] Groover, Chaps. 4 through 7.

flow lines can be much more flexible through the use of flexible manufacturing systems and computer control.

In batch production and job shop production, less progress has been made toward automation. The introduction of numerical control (NC), discussed earlier, was a step in this direction. Another step forward is now occurring through the development of flexible manufacturing systems, discussed in Section 10.3.7. In fact, each of the seven subsections of Section 10.3 represents progress toward the automation of batch production systems.

Automation of batch production may never be fully achieved. Nevertheless, progress will continue. The next level of achievement in this direction will come from the concepts embedded in computer-integrated manufacturing (CIM), discussed in the following section.

10.6. The Promise of CIM

Although significant improvements are attainable through the use of any one of the concepts discussed in previous sections of this chapter, the greatest benefits will come when we learn how to put all these concepts into an integrated framework.

The concepts of computer-integrated manufacturing (CIM) are now emerging. CIM is the embodiment of all the previously discussed individual concepts into a whole.

CIM can be achieved in a company only when two conditions are both present. First, the company must have made significant progress in the implementation of the concepts discussed previously in this chapter: computer-aided design, computer-aided process planning, group technology, NC/CNC/DNC, AS/RS, computer-aided testing and inspection, FMS, and computer-aided factory management. The second requirement is that the numerous computerized data bases throughout the company must be designed such that they can be controlled and accessed for optimum information sharing.

The term *data-driven automation* is often used to characterize the CIM environment. Within such a system, the output of one system component (e.g., CAD) becomes the input to another component (e.g., CAM). CIM provides a closed-loop system for product-related information flow for the complete design-through-manufacture process.

The beginning of the information cycle is the preliminary specification of product requirements as agreed upon between the marketing department and the customer. This information is expanded by the design engineer, resulting in a specific, detailed product design. The product design consists of both geometric and textual data in an electronic data base. The product design information is translated to process plans, work instructions, operation sequences, bills of materials, tooling requirements, NC programs, and robot control programs. The next step is to make this information available to the factory management system for detailed scheduling of orders through the production facilities. In a CIM system, the control computer would trigger material release orders, tooling release orders, and would activate an automated material retrieval and handling system. As the production orders move through the manufacturing facilities, production information is generated automatically and stored in a data base. This information includes quality information, material utilization, labor utilization, downtime (with reasons), tool wear, rework, scrap, and so on. The CIM system would also include an inventory management module, covering raw materials, work-in-process, and finished goods inventory. Replenishment orders would be placed automatically when reorder levels are reached and these orders would be forwarded electronically to the appropriate vendors through electronic data interchange (EDI). Vendor performance information (incoming quality, lead time, and price) would be recorded and analyzed automatically.

Our discussion to this point covers the design and manufacturing components of CIM. A full CIM implementation would also include the business-related functions of the company, such as cost accounting, finance, payroll, forecasting, and management performance reporting.

Figure 10.4 shows the relationships among a large number of manufacturing technologies, many of which have been discussed in this chapter. The acronyms included in Figure 10.4 are defined in the following table:

Acronym	*Technology*
CAD	Computer-aided design
CAE	Computer-aided engineering
GT	Group technology
CAPP	Computer-aided process planning
MRP	Materials requirements planning
JIT	Just-in-time
CNC	Computer numerical control
FMS	Flexible manufacturing system
CAI	Computer-aided inspection
AS/RS	Automated storage and retrieval system
CAM	Computer-aided manufacturing
CIM	Computer integrated manufacturing
CIE	Computer integrated enterprise

At the present time, most large companies have portions of this system in place. Large data bases exist throughout the company, but they are frequently isolated from each other and are not in compatible formats. Much effort will be required to create a truly integrated information system.

10.7. Opportunities for I.E.s

We have touched lightly on a large number of topics in this chapter. The topics discussed are those that are currently being pursued worldwide as manufacturing

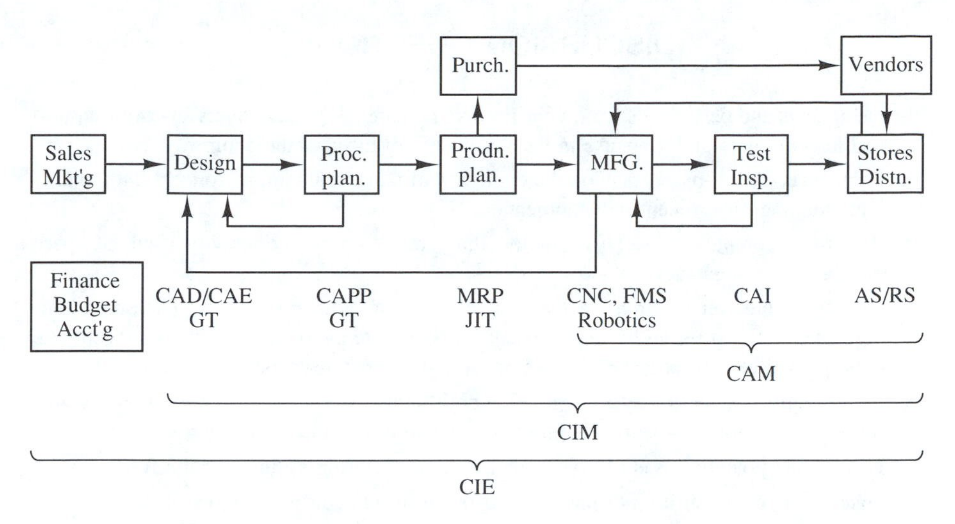

Fig. 10.4. Relationships of various manufacturing technologies.

companies strive to modernize their facilities and their management practices. Factory modernization will be necessary for companies that wish to remain competitive in the world market.

A complete book could be written about each of the many topics we have discussed. Some engineers spend their entire professional career pursuing one or two of these topics in depth. The industrial engineer has a particularly crucial, but difficult, role to play in factory modernization. The I.E. must be familiar with the general concepts associated with each of the topics we have discussed, and must also be able to pull them all together into an integrated framework. The I.E. of the future will be a *system integrator*. No other profession is as well prepared to fulfill this role.

Significant, dramatic, and often traumatic change must occur within companies in order to introduce the innovations we have discussed. Traditional roles will be modified. Traditional organizational structures that emphasize barriers must be replaced with structures that facilitate free and open interchange. Reward systems must be changed to encourage cooperation, so that the full benefits of integration can be attained. Finally, the *culture* of a company must change, that is, its fundamental set of values must be reexamined and brought into line with the new realities.

Change does not come easy in any organization. In the future, a major role for I.E.s will be to *manage change constructively*.

As we look ahead to the next 20 years or so of the second industrial revolution, the industrial engineering profession has its greatest opportunity. We must play two significant roles in the factory of the future. We must be *system integrators*, and we must be *change agents*. These will be exciting challenges for all I.E.s.

DISCUSSION QUESTIONS

1. Using texts and periodicals from your library, prepare a five-page overview of programmable controllers and their use in the design of advanced manufacturing facilities.
2. Write a six- to ten-page essay on the evolution of numerical control concepts, up to and including the most recent developments.
3. Perform a comparative analysis of the three or four most commonly used NC part programming languages.
4. Perform a literature search on the various uses of industrial robots. Determine the percentage of past usage by major category. Speculate as to whether you believe these patterns will change or remain the same over the next ten years.
5. Increasingly, production systems involve both human workers and robots. Identify and discuss the major issues that can arise in such an environment.
6. Discuss the potential uses of voice input in manufacturing systems of the future.
7. Discuss the potential uses of machine vision in manufacturing systems of the future.
8. Do you believe that robots will be widely used outside manufacturing operations within the next five years? Justify your answer.
9. What are "flexible manufacturing systems"? Why are companies purchasing FMS systems?
10. Discuss the relationship between automation and CAD/CAM.
11. Is office automation an important consideration of industrial engineers? Justify your answer.
12. What is the basic rationale underlying group technology?
13. Consider the charting methods in Chapter 6 used for outlining the detailed steps involved in accomplishing a job. These were designed for use with human workers. Would they be applicable for describing the work to be performed by a robot? Discuss thoroughly.
14. Visit a local hospital and determine the feasibility of utilizing a robot for some function within the hospital. Possible applications include the loading of dirty dishes into the dishwasher, dispersement of food items onto trays being prepared for specific patients, and delivery of food trays to patient rooms.

CHAPTER 11

Human Factors

11.1. Perspective

In designing productive work systems, I.E.s put together various components that will operate in such a way as to achieve the outputs desired from the system. The components making up a productive work system include:

- *Equipment*—Machines, material handling devices, gauges, instruments, etc.
- *Processes*—Chemical treatments, thermal treatments, laser hardening, etc.
- *Materials*—Raw materials, purchased parts, lubricants, chemicals, supplies, etc.
- *Capital*—Cash, credits, debits, investments, etc.
- *Energy*—Electrical, steam, mechanical.
- *Facilities*—Buildings, other structures.
- *Information*—Forecasts, sales records, costs, production orders, vendor histories, shop floor status, etc.
- *People*—Workers, supervisors, design engineers, quality control engineers, salespeople, managers, etc.

When the I.E. specifies a specific machine to include in the system, he does so knowing the performance characteristics of that machine. Among the characteristics known about a typical machine are:

- *Range of Operations*—Operations it is capable of performing.
- *Size of Workpiece*—The machine can accommodate a workpiece up to a certain size and shape.
- *Speed of Operation*—Spindle speed, etc.
- *Tolerance Capabilities*—The tolerances the machine is capable of holding.
- *Energy Consumption*—Energy consumed in operation.
- *Maintenance Requirements*—Frequency of overhaul, other maintenance operations.

When the I.E. specifies that a human being should be included as a specific component in the system, it is equally important that proper consideration be given to the performance characteristics of that person. People have performance characteristics just as machines do.

Figure 11.1 illustrates some of the performance characteristics of a "typical" human being compared to a "typical" machine tool. It should be clear from these comparisons that human beings are ideally suited for many functions in a system, but woefully inadequate for others.

The field of study that is concerned with quantifying and measuring the performance characteristics of human beings is called *human factors engineering* or simply *human factors*. Most of the research and teaching associated with human factors is conducted within industrial engineering departments, usually at the graduate level. Human factors is an interdisciplinary field that includes professionals from the behavioral sciences (e.g., psychologists and sociologists) and the biological sciences (e.g., physiologists and bioengineers) as well as industrial engineers.

Human performance characteristics fall into two broad categories, physiological and psychological. *Physiological performance characteristics* have to do with the physical aspects of human activities (we are concerned primarily with work activities), such as lifting, reaching, carrying, hearing, seeing, and speaking. *Psychological performance characteristics* are those dealing with the mental aspects of human activities, such as stress, boredom, and motivation.

11.2. Physiological Aspects of Human Performance[1]

Human beings, by virtue of the biological characteristics of their bodies, are limited in what they can do in a work system. Specifically, the abilities of people to perform various types of motor activities depend on the physical structure of the skeleton, the skeletal muscles, the nervous system, and the metabolic processes.

We are concerned primarily with the bones comprising the upper and lower extremities and the bones of the spine, since these are concerned with the execution

[1]Our coverage of human factors will necessarily be brief. For more complete coverage, see, for example, M. S. Sanders and E. J. McCormick, *Human Factors in Engineering and Design*, 6th ed. (New York: McGraw-Hill Book Company, 1987).

Performance Characteristic	*"Typical" Machine Tool*	*"Typical" Human Being*
Range of Operations	Most M.T.'s perform one basic operation; some perform up to 10.	Extremely broad range.
Size of Workpiece	Some machines perform operations on microscopic-sized workpieces; others machine workpieces weighing serveral tons.	Very limited range. Cannot perform operations on extremely small items. Can perform repetitive operations on items weighing up to 50 pounds, where physical handling is required.
Speed of Operation	Can be extremely fast.	Usually very slow.
Tolerance Capabilities	Machines routinely hold tolerances to 0.001 inch.	Very poor at accuracy and repeatability.
Energy Consumption	Very efficient in converting energy into useful output.	Very low level of energy output; cannot sustain output on even level.
Maintenance Requirements	Regular maintenance required; complete overhaul needed periodically.	Needs minor "maintenance" frequently, such as water, food, oxygen, rest room breaks, bandaids, etc. Major overhauls are performed continuously, as the body mends itself.
Response to Unexpected Occurrences	Very limited capability.	Extremely resourceful and creative.

Fig. 11.1. Performance characteristics of a typical machine tool vs. a human being.

of physical activities. These skeletal bones are connected at body joints, such as the fingers, wrists, elbows, shoulders, hips, knees, ankles, and feet. The manner in which they are shaped and connected puts limitations on our ability to perform physical activities. The wrist will rotate only so far. Our knees bend in one direction but not the other. Our thumbs oppose the four fingers, giving us considerably more dexterity in our hands than any other animal. (Some anthropologists claim that this characteristic is at the root of human mental superiority in the animal world.)

The bones are held together at their joints by ligaments. The skeletal muscles are connected to different bones through tendons, such that when the muscles are activated, the bones apply mechanical leverage. The fibers of the muscles convert chemical energy (from food) into mechanical work, such as lifting, walking, and so on.

The actions of the muscles are controlled by motor nerves. The nerves convey conscious and subconscious messages from the brain to the muscles, causing physical activities to occur. A machine operator reads a work instruction from a blueprint, then executes that instruction by sending the appropriate message over the nervous system to the appropriate muscles, which move bones so as to accomplish the specified task.

Muscles contract to move bones. Contraction requires energy. As it is consumed, it must be replenished through the circulatory system. The heart continuously pumps blood through the circulatory system. The blood picks up nutrients from the digestive process and it replenishes its supply of oxygen from the lungs. The nutrients and the oxygen are carried by the blood to the muscles throughout the body.

Researchers in human factors have carefully studied the process described above and have determined the energy requirements for various types of jobs. Based on these data, jobs can be designed which can be performed by a large percentage of people likely to be involved in such work. Also, optimal rest periods can be specified.

Substantial amounts of detailed information have been generated concerning the biomechanics of motion of the human body. For example, tables of data exist which characterize the movements of the arms, legs, and other body members. As a specific example, the average person's elbow can bend 142 degrees. People at the 5th percentile can bend their elbows only 126 degrees, while people at the 95th percentile can bend their elbows 158 degrees.

Data such as these are extremely useful in designing workplaces for optimum human performance. They are also very useful in designing tools. Better work methods can also be determined, such as having people lift with the large muscles of their legs rather than with the much weaker muscles of their back.

11.3. Psychological Aspects of Human Activities

The physiological aspects of human activities discussed in the preceding section view the human being as a mechanism, much as one would view a machine. While the human body is, indeed, a mechanism governed by physical laws, it also contains a thinking, reasoning, feeling brain. Human beings experience pain, joy, sadness, depression, anger, boredom, frustration, fear, outrage, jealousy, love, hate, and (occasionally) schizophrenia.

We have not made very much progress in quantifying, measuring, or understanding the psychological aspects of human performance. We can predict with much greater accuracy the limit of how much a person can lift than the limit of how much boredom that same person can withstand.

Do people respond better in large open offices or smaller walled ones? With windows or without? Does soft music stimulate thinking, or slow down the work pace? Does fast music result in more output per hour? Unfortunately, the research that has been performed on these questions is inconclusive. Individual differences on such questions are great. Also, people tend to be very adaptable and seem to quickly become oblivious to many such factors.

We can only make some broad, sweeping statements about the psychological aspects of human activity:

- *People are still motivated by money.* Even among people in higher income brackets, the size of their salary is a quantitative measure of their relative worth to the organization.
- *People would rather be doing productive work than be idle.* This has less to do with the "work ethic" than with basic human nature. People are bored when they have nothing to do.
- *Many American workers do not necessarily want "job enrichment."* Behavioral scientists have long proclaimed that workers want more variety and more challenge in their work. Recent attempts to implement "job enrichment" programs have met with surprising resistance.
- *Workers themselves often have the best ideas on how to increase productivity, if they were only asked.* It is astounding that most American managers have not yet learned that their employees have brains as well as hands.
- *Most people respond positively when they know the objectives of their organization, what their own jobs are, and that they will be treated fairly, consistently, and with respect.*

11.4. Human Interface with the World of Work

An important dimension of human performance is the various ways in which human beings interact with their world. Some of the important interfaces are with the following entities:

- *Work Environment*—The physical environment in which the productive work system exists.
- *Social Structures*—Basic values, creeds, laws, traditions, community arrangements, legal structures, educational systems.

- *Machines*—The physical components of equipment and processes in the workplace.
- *Information/Communication Systems*—The total set of systems and procedures by which work system components exchange information.
- *Organizational/Supervisory Structure*—The total set of authority/responsibility/accountability relationships within the work system.
- *Robots/''Intelligent'' Machines*—The rapidly emerging set of electromechanical devices which, via embedded computer-based ''intelligence,'' are assuming certain humanlike characteristics.

As industrial engineers, we are particularly interested in human interfaces with the work environment, machines, information/communication system, organizational/supervisory structure, and robot/''intelligent'' machines. Each of these will be explored briefly in following sections.

11.4.1 Human Interface with the Work Environment

Industrial engineers design workplaces. These may be some combination of tables, desks, seats, chairs, and so on. Machines, equipment, telephones, computer terminals, and other devices are incorporated into the workplace according to the needs of the job.

Most workplaces are such that several different people will work there over a time interval. If the company is working multiple shifts, two or three different people may use the same workplace in one day. These factors must be considered when the I.E. specifies work-space dimensions. Where possible, adjustable chair heights and other dimensions are specified.

Considerable effort is also required to arrange various components of the workplace. Consider the workplace of an airline pilot. Where is the optimal location of the dozens of dials, gauges, levers, and buttons? Some of the components that must be arranged are visual displays, hand controls, foot controls, audio inputs, and audio outputs.

How much light (illumination) is needed for specific jobs? Considerable data exist that provide illumination standards for a variety of jobs. The widespread use of computer terminals has required modification of many lighting systems to reduce glare on the screens. Safety relative to lengthy exposure to CRTs is a growing concern.

Atmospheric conditions are also important in designing work environments. Temperature, humidity, and air cleanliness are the factors considered most frequently. If, for example, dust cannot be controlled adequately in the work environment, respirators should be furnished to all employees affected. Noise is still another atmospheric factor that must be considered when designing work environments. Excessive noise is not only annoying and detrimental to good work, but can be dangerous to the health of the employees.

Finally, the work environment should be safe and attractive. It should be designed with good maintenance in mind. Good housekeeping practices are essential. Guards, carefully marked walkways, warning lights, and so on, are all important in designing work areas with employee safety in mind.

11.4.2 Human Interface with Machines

Virtually all human beings interface with machines of one type or another. The human/machine interface is obvious for truck drivers, lathe operators, crane operators, typists, and computer operators. In most productive work systems involving people and machines, the human being serves a control function. The person adjusts the workpiece in the machine, pushes the start button, watches as the machine operates, stops the machine in emergencies, decides when to shut the machine off, and so on.

Human factors data are extremely helpful in selecting the size, shape, and location of various controls. Consider the human–machine interface involved in driving an automobile. All the displays need to be designed carefully to be immediately recognizable. They also need to be located such that they are easily visible, even when the auto is in motion. The steering wheel needs to be correctly sized so that the driver has the proper ''feel'' of the road as he makes steering corrections. The brake pedal must be located properly and the correct amount of resistance built in. Warning lights and signals indicate ''out-of-norm'' conditions. If the oil in the crankcase is too low, a red light comes on. If the cooling system loses its coolant, the temperature light comes on. Recent automobiles provide verbal warnings.

One of the most important tasks of the I.E. is to allocate functions among machines and human beings in the system. In making this allocation, some general guidelines can serve to delineate between people and machines.

In general, human beings are better in performing the following types of tasks:

(1) Recognizing complex patterns, which may vary considerably over time.
(2) Recognizing unusual and unexpected events.
(3) Storing large amounts of information over long time periods, then assimilating this information into general principles and strategies. (This is the basis of intuition.)
(4) Quickly retrieving pertinent information; using associative abilities to retrieve many related items of information and to project these to inferences involving complex associations between related information.
(5) Adapting the decision process to differing situations.
(6) Reasoning inductively, generalizing from specific observations.
(7) Making subjective evaluations and decisions.
(8) Discerning the most important activities to concentrate on when overload exists.

In general, machines are better in performing the following types of tasks:

(1) Monitoring for prespecified events.
(2) Storing quantitative data quickly and in great volume.

(3) Retrieving and processing quantitative data quickly.
(4) Performing repetitive activities reliably over extended periods of time.
(5) Responding to inputs rapidly and consistently.
(6) Ignoring distractions.

These and other factors are considered when allocating functions among machines and human beings in a complex work system.

11.4.3 Human Interface with Information/Communication Systems

The human being is an essential component in essentially all information/communication systems. These systems are becoming very complex in most modern organizations.

People receive information person to person and from written reports, information displays, telephone calls, buzzers, lights, signals, dials, sounds, and so on. People transmit information in writing, through keyboards, verbally, and by actuating certain input devices.

This is a complex subject. For our purposes, we will examine one small portion of this topic by considering when to use auditory versus visual presentation of information. This is summarized in Figure 11.2.

11.4.4 Human Interface with Organizational/Supervisory Structure

This particular interface has to do with the entire set of issues involving job design, reporting relationships, performance appraisal, evaluation, and training. Details of many of these issues are covered in Chapter 21. These issues are also closely interwoven with the concepts discussed in Section 11.3.

Use Auditory Presentation if:	*Use Visual Presentation if:*
• Message is simple	• Message is complex
• Message is short	• Message is long
• Message not referred to later	• Message will be referred to later
• Message deals with events in time	• Message deals with location in space
• Message calls for immediate action	• Message does not call for immediate action
• Visual systems of the person is overburdened	• Auditory system of the person is overburdened
• Receiving location is too bright	• Receiving location is too noisy
• Person's job requires worker to move about continually	• Person's job allows worker to remain in one position

Fig. 11.2. When to use auditory or visual presentation [From B. H. Deatherage, "Auditory and Other Sensory Forms of Information Presentation," in H. P. Van Cott and R. G. Kinkade (eds.), *Human Engineering Guide to Equipment Design*, rev. ed. (Washington, D.C., U.S. Government Printing Office, 1972).]

We made the point in Section 11.3 that most people respond positively when they know the objectives of their organization, what their own jobs are, and that they will be treated fairly, consistently, and with respect. People want to be treated as individuals and not as numbers or as blocks on an organization chart. They also want to have some input into decisions that affect them directly, such as the arrangement of the workplace and the inspection of the work they perform.

During the early days of the American industrial revolution (roughly 1850–1920), the work force in our factories came primarily from farms and from the masses of people who were immigrating to the United States. These people were largely uneducated and unskilled. As the industries grew from small numbers of workers (say 30) to very large work forces (several thousand) the company managers were faced with the problem of how to group (organize) the workers for maximum efficiency. The early efforts at organizing were largely trial and error.

Organizing and managing large work forces was a new endeavor. The only large organizations in the past had been military organizations. Consequently, company managers adopted many of the practices of military organizations, this being the only experience base from which to draw.

The result has been the rigid, formal-looking organization charts that we see on company walls that are supposed to represent authority/responsibility relationships. We hear a lot about "the chain of command," or "we must go through channels." Yet that is not how real organizations actually work.

The military model also gave managers their supervisory style, which basically amounts to "do what you are told, don't ask questions, and have unswerving loyalty to the company." Unfortunately, this *is* the way the supervisory style actually works in many organizations.

There are no clear-cut solutions to these problems at the present time, but we are beginning to see some alternative styles of organization and supervision. These are largely experimental, and we have a long way to go.

A great amount has been written in recent years about "quality circles," "Japanese management style," "participatory management," and so on. The industrial engineer must play a central role in finding more effective means of organizing and supervising the work force of organizations.

11.4.5 Human Interface with Robots and Intelligent Machines

If any one thing symbolizes the second industrial revolution that is now under way, it is the robot. Although robots are certainly an important aspect of the new manufacturing technology, they are by no means the dominant driving force.

Robots and robotic devices have been sensationalized by the popular press and by science fiction movies. In addition, there is a concern among many blue-collar workers that robots will displace them. Thus, to many uninformed people, robots are mysterious things that appear to be threatening and to have humanlike qualities.

One thing we will *not* see is a bunch of two-legged robots marching into a factory having today's typical equipment, and operating the system in the same manner as would human operators. In those situations where robots can play an important role, the total set of manufacturing equipment with which the robot must interact is specifically designed with the robot in mind. It would be foolish (or at best poor design) to consider robots as one-to-one replacements of human beings with all other equipment unchanged. Such an approach would not take advantage of the robot's superiority to the human being in certain areas, nor would it consider the inferiority of the robot in other areas.

Many early applications of robots did not take advantage of their capabilities, leading one manager to remark that his industrial engineers should design a job enrichment program for the robot! In this particular case, a large, expensive robot was simply grasping items coming from a conveyor, raising them 2 feet, and then loading them onto a take-away conveyor that ran at right angles to the first conveyor. The robot possessed accuracy and repeatability capabilities on the order of 10 times greater than were needed. Furthermore, a fairly standard transfer mechanism using hard automation concepts would have been far superior and much less expensive.

Many factories of the future will contain numerous robots. Within the foreseeable future, there will also be many human beings in these factories. Industrial engineers must develop methods for designing productive work systems which include a varied mix of human beings, robots, and other "intelligent" machines.[2]

There are at least three sets of issues involving the interface between human beings and intelligent machines. The first set of issues is concerned with how to allocate work functions and control functions optimally among the human and machine elements comprising the system. Figure 11.1 compared certain performance characteristics of human beings and "unintelligent" machines. A comparable analysis would have to be conducted for each different type of intelligent machine.

The second set of issues concerning the human being and intelligent machine interface is that of safety. A robot or any machine will activate its components (such as swinging an arm) in response to specific commands. It will not know to stop its movement if a human being happens to be in its path. There have been several confirmed deaths of human beings caused by robot arm movement. Considerable effort is being directed toward the provision of safety features on robotic systems. However, this is expected to remain a critical issue for many years to come.

The third set of issues concerning the interface between human beings and intelligent machines has to do with the question of "who has the final say" when an intelligent machine wants to do one thing and the person wants to do another. At what point, and under what conditions, should a human being be permitted to override a decision made by an intelligent machine? What about the reverse situation? Should a machine ever be permitted to override the decision of a human being?

---

[2]The term *machine intelligence* is frequently used to characterize such capabilities as machine vision (image processing), pattern recognition, audio input/output, tactile sensing, and adaptive control (learning, or modifying its own behavior).

Research into these issues will be pursued for many years. Industrial and systems engineers have an opportunity to contribute to this important effort.

DISCUSSION QUESTIONS

1. Is it proper to compare ''performance characteristics'' of human beings and machines? Discuss thoroughly and justify your answer.
2. Distinguish between physiological and psychological human performance characteristics.
3. Study carefully the furniture in your apartment or dormitory room. Pay especially careful attention to the physical dimensions of the furniture. Speculate on how each major piece of furniture was given its dimensions. Do you believe that these dimensions are proper for the 1990s and the 2000s?
4. Apply the general concepts discussed in this chapter to the design of an ideal engineering student workstation for a dormitory room.
5. Conduct a library research project on the topic ''quality circles.'' Trace the development of this movement and present your conclusions and recommendations.
6. Conduct a library research project on ''human–machine interface'' in industrial organizations. State your conclusions.
7. From literature in your library, speculate on the degree to which robots will have humanlike characteristics by 2000.
8. With human beings having less and less direct involvement with the manufacturing process, per se, do you believe that the study of human factors will diminish in importance in the future?
9. Conduct a library research project on recent innovations in organizational structures. Is it necessarily true that hierarchical organizations are a thing of the past? Justify your response with an appropriate comparative analysis.

CHAPTER 12

Resource Management

12.1. Introduction

Actually, all of industrial engineering involves resource management. Why, then, title a chapter ''Resource Management''? The answer is simple: Certain critical resources have become so scarce and expensive or so difficult to handle that they are taking a disproportionate amount of management time. A separate set of courses is emerging covering the management of these resources.

The three resources discussed in this chapter are *energy, water*, and *hazardous materials* (actually hazardous ''materials,'' ''substances,'' ''chemicals,'' and ''waste'' are involved and each term carries a special connotation and unique set of problems). Every industrial engineer will some day be impacted by these three mammoth problem areas. The judicious I.E. prepares by studying the problem now.

12.1.1 Energy Management

Fortunately, most problems encountered by I.E.s carry a set of opportunities. This is very true for energy, water, and hazardous material management problems. For example, energy management is one of the best profit-improvement tools available to management today. Past rates of increase in energy prices have opened many new opportunity doors. Cogeneration, waste heat recovery, sophisticated control systems, and unique lighting methods are only a few of these prime new opportunities available. Many are now very cost-effective, whereas in the days of cheap energy costs, they were not.

Much can be done to reduce energy costs. Savings of 5 to 10% in the first year are common and two-to three-year program savings of 15% are not unusual. Most energetic programs eventually reduce energy costs by 30% or more. Some extreme cases have involved savings of 50 to 60%. Also, the technology is just now starting to mature. The potential for the future is staggering.

12.1.2 Water Management

Although not yet a major national concern, water management is definitely a concern in some localities and will be nationally before long. Many companies have seen their water and sewerage costs double or even triple in one year. As water resources "dry up," planners are forced to look further and to develop more expensive resources.

Resource management programs have found that simple cost-effective changes can reduce industrial water costs (and consumption) by 20 to 70%. Usually, water recovery or recycling is involved with only simple treatment or filtering. New technologies such as reverse osmosis, waste heat evaporation, and electrodialysis have demonstrated that we have just begun to see results.

12.1.3 Hazardous Material Management

Energy and water problems have placed quite a strain on industry, but hazardous materials problems are rapidly becoming one of the biggest concerns in many industries today. Occupational Safety and Health Administration (OSHA) officials have enacted stringent regulations on hazardous *chemicals* encountered in the workplace. The U.S. Department of Transportation (DOT) has enacted complex regulations on hazardous *materials* (chemicals, wastes, etc.) being transported. Environmental Protection Agency (EPA) personnel have enacted strong controls on the storage, treatment, and disposal of hazardous *wastes* and have issued detailed steps to be followed when hazardous materials are spilled or leak. (Hazardous materials released like this are called hazardous *substances* in the literature.) All these regulations are being revised constantly, requiring management to spend a great deal of time simply understanding the regulations. All this means that the cost of hazardous material, chemical, waste, and substance management is skyrocketing.

As in energy and water, though, these skyrocketing costs have opened new opportunities. It is now often cost-effective to practice hazardous waste recycling (e.g., waste oil and solvent recovery), hazardous material substitution (find a nonhazardous chemical or material that can do the job almost as well), and process changeover to minimize or perhaps eliminate hazardous materials. The industrial engineer can have the biggest impact here as a dramatic set of opportunities is unfolding. The rest of this chapter is divided into three areas: energy management, water management, and hazardous material management.

12.1.4 This Chapter

The remainder of this chapter is divided into three areas: energy management, water management, and hazardous materials management. Obviously, the treatment is very superficial and is only intended to give a broad perspective or overview. Many I.E. departments now have courses designed to make I.E.s better managers of energy, water, and hazardous materials. Thus, in this chapter we introduce those courses.

12.2. Energy Management

Energy management is the judicious and effective use of energy to maximize profits (reduce costs) and enhance competitive positions. Therefore, any activity that affects energy utilization and profits (costs) is part of energy management. If profits are increased (costs decreased), it is *good* energy management. Otherwise, it is *bad* energy management.

Often, we associate energy management and energy conservation as being the same activity. Energy conservation is a large part of energy management, but energy management covers much more, as the following list shows.

Changing utility rate schedules to seek lower costs.
Changing fuel sources or fuel types for lower cost.
''Demand'' leveling or power factor improvement to minimize the cost of electricity.[1]
''Curtailment planning'' to minimize cost if a fuel source fails.

All the activities in the list above (and there are many others) can be real profit improvers for a company, but they do not conserve energy in the truest sense. The following list shows activities that save money by conserving energy.

Controlling thermostats to reduce heating/air conditioning demand when the plant is not operating.
Controlling air and fuel supply to a combustion unit (e.g., boilers or furnaces) to minimize the fuel cost.
Insulating to reduce heat loss (gain).
Repairing steam and compressed air leaks.
Eliminating unnecessary ventilation.

Both sets of activities can save tremendous amounts of money, but the second does it by conserving energy and the first by managing energy utilization better. Both

---

[1]We will discuss demand and power factor later in this section.

sets of activities are vital to a successful energy management program and must be included.

12.2.1 Why Bother?

In a nutshell: Energy management is one of the best profit-improving (cost-reduction) programs available to management today. First-year savings of 5 to 10% usually occur with another 10% coming along rather quickly. Eventual savings of 30% are common and we have seen programs yield reductions of 60 to 70%. Proper planning for new facilities has yielded savings of as much as 80%. This is big money and it gives a definite competitive advantage to the company. Best of all, most companies can accomplish much of these savings without large acquisitions of staff and equipment.

Also, energy management is good for the country. Our balance of payments is improved (as we import less energy) and we are less dependent on foreign energy suppliers. The demand on our environment is reduced, as we are able to slow development of new resources. Finally, energy management helps buy time for the development of new energy resources.

12.2.2 Why Industrial Engineering?

Proper energy management requires a systems approach. A good energy manager knows the workings of the entire plant and how a change in one sector can affect others. He or she also has a good appreciation of economics (especially engineering economy), including utility rate schedules and how to analyze projects for cost effectiveness. Since energy management crosses many organizational lines, the good energy manager knows how a modern organization works and how to accomplish results through people. Finally, the good energy manager *must* be technically sound, understanding engineering science principles.

The industrial engineer is as qualified for energy management as a graduate of any other discipline. Any traditional discipline training does not meet all the requirements. The I.E. may have to receive additional training in subjects such as boilers; heating, ventilating, and air conditioning; lighting; and so on, or at least recognize their limitations. Many energy management positions today are held successfully by industrial engineers.

12.2.3 Required Ingredients

There are certain essentials that successful energy management programs must incorporate. They include:

Management Commitment—Probably the single most important factor is that management must be truly committed. This can be demonstrated through letters to employees, plant-wide talks, pay-envelope stuffers, and other means, but it must be there. Adequate staffing and time must be made available.

Energy Management Coordinator—One person must be responsible for the success/failure of the energy management program. That person, often called the energy management coordinator, should also have the necessary authority to be able to make the program work. That authority can come in the form of budgetary control or organizational structure, but it must be there. The energy management coordinator must be trained in the systems approach, engineering economy, management, and the engineering sciences.

Funding—Energy management can be extremely cost-effective and can yield very large profit improvement, but it cannot do this without adequate funding. If an energy management proposal is cost-effective, it should be funded; if not, it shouldn't. Although simple sounding, too many energy management coordinators see cost-effective proposals turned down for less cost-effective ideas that have more popular appeal.

Objectives and Goals—As with all endeavors, energy management performs best when goals and objectives are set and vigorously pursued. A successful energy management program will set goals, monitor progress, and reset goals.

Persistence—Almost all energy management programs have quick success—so easy are the early pickings. Then a plateau seems to occur where the energy management team is maturing and seeking new technology. Too many programs never go beyond this stage. Persistence is necessary to reach the next level with even larger savings.

Miscellaneous—Certain other items are helpful to a program. They include *incentives* for people to cooperate, *life-cycle costing philosophies* (engineering economy), *monitoring systems* to track actual savings, *energy accountability* for first-line supervision, and *preventative maintenance* philosophies.

12.2.4 Understanding Rate Schedules

Few industrial managers bother to examine their utilities bills at all and very few fully understand how they are billed. This is disappointing and very surprising, for energy bills often constitute a large portion of total production costs. Management should understand utility billing procedures, but it is absolutely necessary that the energy manager fully understand all aspects of his or her rate schedules. This section presents a very brief exposure to utility rate schedules, but you must realize that utility billing procedures are quite complex and that it requires considerable time to understand them.

Oil, Gas, Coal, and Gasoline. Most nonelectricity sources of energy have relatively simple rate schedules. Usually, billing is for consumption only [measured in gallons (fuel oil), tons (coal), or cubic feet (natural gas)]. For some sources, such as fuel oil and especially coal, other characteristics affect the price. For example, sulfur is a strong pollutant, so low-sulfur coal is popular and therefore expensive. If water is present, that water must be evaporated causing a large loss of heat. Consequently, moisture content in the fuel can be important.

Two typical natural gas schedules for an industrial customer are given in Figure 12.1.

Note that the manager has a choice of two rate schedules. The first (and more expensive) has a high priority, meaning that delivery will not be curtailed, in periods of short supply, until all low-priority users have been curtailed. The second, with its lower cost and lower priority, probably means curtailment in times of short supplies (severely cold weather or supply disruptions). Also note that the low-priority schedule has a minimum monthly bill of $4,500, so it is intended for large users.

Management must decide if the company is large enough to go with the cheaper schedule and if curtailment problems might negate savings. Large companies often go with the cheaper schedule and store alternative fuels (such as propane or fuel oil) on site for emergencies.

Electricity. Electrical rate schedules can be just as complex as gas schedules are simple. There are literally hundreds of variations of electricity rate schedules and the energy manager must study his or hers for full understanding. What we present below is a simplified example of only one type of rate schedule. There are many others.

First, it is important to understand that electrical energy is more complex. For example, the following characteristics are important and affect bills.

Voltage Level—To save line losses, electrical utilities transmit at very high voltages. If a company is willing to accept these higher voltages and do its own transforming, the utility is usually willing to reduce the price. If, however, the utility must supply and maintain transformers, the price is higher.

Demand Level—You are not a good customer if you draw large amounts of electricity for short periods of time and lesser amounts in other times. The utility would prefer that you consume the same amount time after time. To encourage you to flatten these "peaks," the utility will bill you for your peak demand in addition to your consumption. This demand is measured in kilowatts (kW).

Power Factor—Inductive loads such as electric motors and fluorescent lights throw voltage and current out of phase, creating problems for the utility. This

High Priority		*Low Priority*	
First 100 ft^3/month	$2.00	First 1,000 MCF/month	$4, 500
Next 9.9 MCF/month	$5.00/MCF	All over 1,000 MCF/month	$4.35/MCF
All over 10 MCF/month	$4.75/MCF		

MCF = thousand cubic feet

Fig. 12.1. Typical industrial gas rate schedule.

is measured by a quantity called "power factor." To encourage you to help them manage this problem, electricity rate schedules often carry penalties for poor power factors.

Consumption—Electrical energy consumption is usually measured in kilowatt hours (kWh). All rate schedules charge per kilowatt hour. Sometimes this charge is on a decreasing block—that is the more electricity you use, the cheaper (per kilowatt hour) it is.

Ratchet Clauses—A customer that draws the same amount of demand month after month is preferred over one that draws a large amount one month and much less thereafter. The utility must have the equipment to supply that higher level and that equipment may sit idle most of the time. To remedy this, utilities often say that demand cannot be lower than 75% (or some other fraction) of the previous 12-month high demand. If it is lower, demand is billed at that minimum level.

A simplified rate schedule for an industrial client is presented in Figure 12.2. Usually, there are three or more voltage levels available, but only two are given in the example to simplify the presentation. Also, note that voltage level 1 is higher than level 2, so demand and energy charges are less.

As an example, assume that a company has just completed one month of operation. In that month their electricity characteristics were:

Consumption (energy)	75,000 kWh
Demand	600 kW
Previous high demand (over last year)	1,000 kW
Voltage Service Level 2	

The company's minimum level of demand billing is

$$(0.75)(1{,}000 \text{ kW}) = 750 \text{ kW}$$

(since the minimum is 750 kW and actual is 600 kW)

Voltage Service Level 1 (4,000 V or higher)		*Voltage Service Level 2 (less than 4,000 V)**	
Customer Charge	$500/month	Customer Charge	$200/month
Demand Charge	$5.00/kW	Demand Charge	$6.00/kW
Energy Charge	$0.03/kWh	Energy Charge	$0.04/kWh
Ratchet Clause	75%	Ratchet Clause	75%
No Power Factor Penalty		No Power Factor Penalty	

*Through utility-owned transformers.

Fig. 12.2. Typical industrial rate schedule.

Therefore, the bill would be

Consumption		
(75,000 kWh)($0.04/kWh)	=	$3,000
Demand		
(750 kW)($6.00/kW)	=	4,500
Customer charge	=	200
Total bill		$7,700

An actual bill would have other charges, such as special assessments to pay for equipment, sales taxes, and fuel cost adjustments (allowed by state utility commissions so that the utility can pass on to the customer exceptionally high fuel costs). The example above is for demonstration purposes only.

12.2.5 Alternative Rate Schedules

Industrial rate schedules have become much more complex recently as the utilities seek ways to flatten demands, lower electricity costs, and boost their energy sales. For example, the following rate schedules are available in most localities. When they are appropriate, many thousands of dollars can often be saved. In a strategic sense, the industrial engineer is in a good position to analyze the possible savings.

Interruptible—An interruptible schedule offers large dollar savings if the industrial firm can save large power draws (often 500 Kw or more) when the utility requests it. Often a company finds that it can move these large loads to off-peak times when necessary.

Curtailable—A curtailable schedule offers large savings to companies that can schedule operations to be out of production at certain (peaking) times. Several companies have found they can start early and shut down during the peaking times. Savings can be very large.

Time of Use—A sophisticated meter is required but time of use schedules allow different costs at different times of the day. Then companies can move large power users away from the high-cost periods but still maintain some operations during the high-cost times.

12.2.6 Energy Management Opportunities

Energy management opportunities (EMOs) are ideas that help a company improve profits or enhance competitive positions. Many universities offer two courses that do nothing but teach EMOs, so it is impossible to do anything but present a cursory list. It is intended to give the reader an idea of what can be done but not to present a comprehensive listing.

EMOs That Do Not Require Large Investments

A. Space heating
   1. Lower the thermostats during the heating season and raise thermostats during the cooling season.
   2. Use night setback on heating systems with zone controls.
   3. Check the boiler stack temperature. If it is too high (more than 150°F above steam or water temperature), clean the tubes and adjust the fuel burner.
   4. Check the flue gas analysis on a periodic basis: the efficient combustion of fuel in a boiler requires burner adjustment to achieve proper stack temperature.
   5. Examine operating procedures when more than one boiler is involved. It is far better to operate one boiler at 90% capacity than two at 45% capacity each.
   6. Lower steam pressure to the minimum pressure that will just satisfy needs.
   7. Reduce blowdown losses.

B. Ventilation
   1. Reduce fresh air to legal and healthy limits.
   2. Reduce exhaust air quantities to practical limits.
   3. Establish a ventilation operation schedule so that the exhaust system operates only when it is needed.
   4. In summer when outdoor air temperature at night is lower than indoor temperature, use full outdoor air ventilation to remove excess heat and precool the structure to reduce the air-conditioning load.

C. Air conditioning
   1. Set the room temperature at 78°F in the summer.
   2. Use water-cooled refrigeration units rather than air-cooled ones since the former are up to 20% more efficient.
   3. Turn off the cooling system during the night.
   4. Maintain equipment to retain "as new" efficiency.
   5. Schedule precooling startup in the morning in accordance with outdoor temperatures in order that the building interior will be at 78°F when occupants arrive.

D. Central air-handling equipment
   1. Whenever possible, use outdoor air for cooling rather than using mechanical refrigeration. Use the economizer cycle, where installed.
   2. Shut off air conditioning, ventilating, and exhaust systems for function rooms and similar rooms when they are not occupied.
   3. When function rooms are to be put in service, use all recirculated air rather than outdoor air to precool or preheat the rooms.

E. Hot water
   1. Reduce generating and storage temperature levels to the minimum required for washing hands, usually about 110°F. Boost hot water temperature locally for kitchens and other areas where it is needed.

2. Consider replacing existing hot water faucets with spray-type faucets with flow restrictors where practical.
3. If the water pressure exceeds 40 to 50 pounds, install a pressure-reducing valve on the main service to restrict the amount of hot water that flows from the tap.
4. Inspect insulation on storage tanks and piping. Repair or replace as needed.

F. Lighting
1. Remove unnecessary lamps and/or replace present lamps with higher-efficiency units.
2. Use energy-conserving fluorescent lamps.
3. Remove lamps or fixtures.
4. Control exterior lighting. If a photocell is used to turn on the lamps and a timer to turn them off, there may be savings of as much as a third of present consumption.
5. Shut off lights in unoccupied rooms.
6. Reduce area lighting in all intense lighting areas and substitute task lighting.
7. Use color-coded light switches to avoid nonessential lights being put on during nonselling hours; designate those needed for cleaning, register reading, and security.

G. General buildings
1. Prepare an energy profile of building in as much detail as possible.
2. Improve the power factor.
3. Make the monthly energy consumption and cost data available to the manager and chief operating engineer so that they can evaluate and compare against previous months and normal budget.
4. Examine the entire building for air leaks around windows, doors, and anywhere else that leaks might occur. Seal up the leaks.
5. Provide proper insulation for all equipment that is heated or refrigerated. This includes tanks, ovens, dryers, washers, steam lines, boilers, refrigerators, etc.
6. Add insulation to the roof whenever needed.
7. Utilize solid-state motor drives instead of motor-generator sets for elevators.

H. Laundry department
1. Use cold-water detergents where permitted and satisfactory for your purposes.
2. Repair all steam leaks promptly.
3. Set schedules to reduce peak use and demand charges.

EMOs That May Require Large Investments

I. Space heating
1. Preheat oil to increase efficiency.
2. Replace existing boilers with modular boilers.

3. Preheat combustion air to increase boiler efficiency.
4. Reduce blowdown losses.
5. Insulate hot, bare heating pipes.
6. Install a flue gas analyzer.

J. Ventilation
1. Consider installing economizer/enthalpy controls on air-handling units in noncritical areas to minimize cooling energy required by using proper amounts of outdoor and return air to permit ''free cooling'' by outside air when possible.
2. When more than 10,000 cubic feet per minute (cfm) is involved, and when building configuration permits, consider installation of heat recovery devices such as a rotary heat exchanger.
3. Consider utilizing revolving doors for main access, in addition to swinging doors needed by those in wheelchairs or on crutches.
4. In locations where strong winds occur for a long duration, consider installing wind screens to protect external doors from direct blast of prevailing winds.

K. Air conditioning
1. Replace inefficient air conditioners.
2. Consider converting systems serving interior zones to variable volume.

L. Domestic hot water
1. Install a small domestic water heater to maintain the desired temperature in the water storage tank to eliminate the need for running a large space heater boiler at a very low efficiency during the summer months.
2. Use a single system to meet handwashing needs in toilets.

M. Lighting
1. Consider replacing present lamps with those of lower wattage which provide the same amount of illumination or (if acceptable in light of tasks involved) a lower level of illumination.
2. Replace all incandescent parking lighting with mercury-vapor HID or sodium-vapor lamps.
3. When natural light is available in a building, consider the use of photocell switching to turn off banks of lighting in areas where the natural light is sufficient for the task.

N. General building
1. Insulate all roofs, walls, and floors having exterior exposures.
2. Use heat pumps in place of electrical resistance heating and take advantage of the favorable coefficients of performance.

12.3. Water Management

Water management is the judicious and effective use of water to maximize profits (or accomplish other objectives). Unlike most other countries, water has been very inexpensive in the United States and there have been ample water supplies *until very recently*. Let's examine some facts.

Each of us uses about 87 gallons of water per day.[2] A steak takes about 3,500 gallons of water from pasture to plate and your car required about 60,000 gallons of water in steel manufacturing. New York City uses 1,500,000,000 gallons per day. We use lots of water!

Yet water supplies are disappearing or becoming more difficult to find. Cities are importing water over hundreds of miles. Droughts now seem to affect water supplies quickly as reserves dwindle. Many underground wells are being polluted daily. The result: Industry is facing rapidly rising water prices and increasing public pressure to use less. We have seen *rate increases of 200% and more in one year*. The problem is just starting to emerge.

Fortunately, industry is finding that it can do much to reduce these costs by the efficient use of water in all phases of manufacturing. Dramatic cost savings are occurring, sometimes as high as 70%. Most of the time, these savings occur because of tightened management controls rather than expensive changes.

As an example, many air compressors in industry are water cooled. A large percentage are "once-through" cooled (i.e., city water passes through the cooling chamber and is then discharged). Often, this water can be used as make up water to rinse tanks, boiler-water makeup, and so on. At a usage rate of 1 to 5×10^6 gallons of water per compressor per year, the savings can be large.

Suppose that a plant operated 24 hours per day 365 days per year and had a 200-hp reciprocating air compressor. Suppose further that the air compressor used 2×10^6 gallons of water per year (once through) and the water was heated from 60°F entering to 110°F leaving. If this water can be reused in a heated water tank, the savings would be[3]

$$(2 \times 10^6 \text{ gal/yr})(\$2.00/10^3 \text{ gal}) = \mathbf{\$4,000}$$

In addition, there would be a large energy savings[4]:

$$(2 \times 10^6 \text{ gal/yr})(8.4 \text{ lb/gal})(110°\text{F} - 60°\text{F})(1/0.8)(\$7.00/10^6 \text{ Btu}) = \mathbf{\$7,350}$$

So the savings can be quite nice for some very simple changes.

Some other ideas are shown in Table 12.1.

Some exciting things are happening in water management. Advanced water recovery techniques such as reverse osmosis, waste heat evaporation, and electrodialysis are available commercially and can be quite cost-effective. Sometimes both pure water and reconstituted chemicals are the result and both may be reusable. The future looks very exciting.

---

[2]These statistics were taken from a special issue of *National Geographics*, Aug. 1980.

[3]Typical water costs.

[4]Conventional water heater efficiency = 0.8, fuel cost = $7.00/10^6 Btu.

Table 12.1. SIMPLE WATER MANAGEMENT IDEAS

Process water	Boilers—steam
Reduce flow	Control blowdown
Recycle	Return condensate
Use effluent elsewhere	
	Domestic water
Cooling towers	Use low-flow devices
Reduce load (thereby evaporating less)	Check for leaks
Monitor bleed	
Reuse bleed	
Management	
Graph usage	
Forecast and compare	
Develop own supply	

12.4. Hazardous Material Management

Hazardous material management is a real problem in industry today that is growing by leaps and bounds. Old disposal methods are now sometimes illegal or too expensive. New regulations require expensive methods of handling and storage. Other regulations require expensive training and contingency planning. The list could go on—but it should be obvious that hazardous material management will be one of the most pervasive problems facing industry in the future. All I.E.s need to understand what's happening.

Consider the following:

The United States generates about 80,000,000,000 pounds of hazardous materials annually, which is enough to fill the Superdome 1,500 times.
It is estimated that 90% of this waste is disposed of improperly.
Forty-eight of our 50 states have at least one EPA priority cleanup site.
Groundwater supplies are being polluted at an increasingly alarming rate.
The trend in industry will be to minimize or eliminate hazardous materials as costs and regulations continue to increase.

Thus, as bad as the picture is for U.S. industry, it will get worse. However, industrial engineers, with their plantwide perspective, can help solve many of the problems.

12.4.1 Government Regulations

At least three government agencies are involved in the life of a hazardous material. When a substance is being offered for transportation, in any mode, the Department of Transportation regulations have precedence. When the hazardous material reaches the plant and is being used or stored for use, the Occupational Safety and Health Administration has jurisdiction. When a hazardous waste has been created, the Environmental Protection Agency steps in and controls the treatment and disposal techniques. Finally, if a spill occurs, three sets of regulations must be referenced: the Clean Water Act (EPA), the Hazardous Material Transportation Act (DOT), and the Comprehensive Environmental Response, Compensation and Liability Act (CERCLA) (EPA). CERCLA, better known as the "Superfund," is a powerful act passed by Congress in 1980 that has wide-reaching authority.

To add more complexity to understanding the requirements, the terms used in the regulations change even though they may be referring to the same chemical. For example, if a company is shipping a drum of sulfuric acid to a plant for use in a process, the acid would be referred to as a *hazardous material* and is under DOT control. When it reaches the plant, it is now referenced as a *hazardous chemical* and is under OSHA control. After the acid has been used and is ready for disposal, the term used is *hazardous waste*, which is under EPA control. If that waste has to be shipped off-site for treatment on disposal, it once again comes under DOT control. See Figure 12.3 for an overall layout of this scenario.

It is beyond the scope of this book to dig into these regulations, but the requirements are rigid and expensive. Fines of up to $100,000 per day and jail terms are possible. In fact, all agencies are relying more and more on criminal sanctions to enforce regulations. I.E.s will be involved—thus they must be informed.

12.4.2 The Role of Industrial Engineering

Industrial engineers are trained to understand the total system and are able to coordinate the responsibilities of many different disciplines. These qualities are

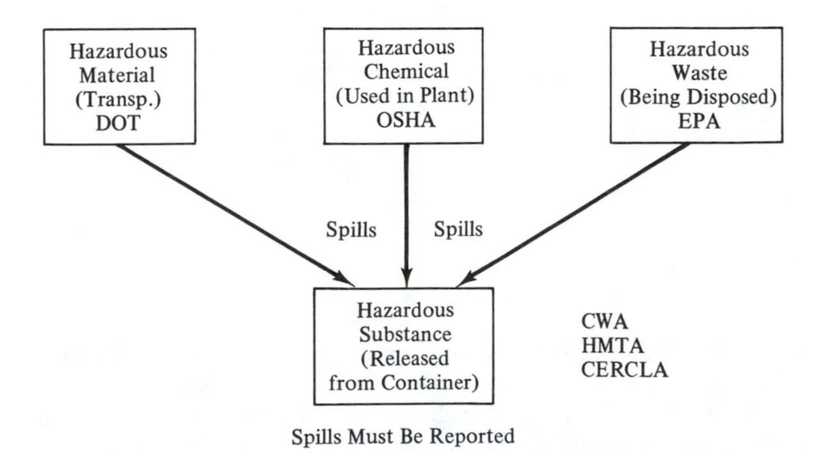

Fig. 12.3. Hazardous materials, chemicals, wastes, and substances relationship.

essential for an effective hazardous materials management program. For example, the industrial engineer's responsibilities might include at least the following:

Supervise shipping operations.
Supervise in-plant handling of hazardous chemicals.
Develop and supervise training programs.
Develop a contingency plan.
Supervise a record-keeping and monitoring program.
Coordinate process changes to minimize quantity and toxicity of hazardous waste.
Develop resource recovery economies.

With the federal mandate to minimize landfills and reduce the amount of hazardous waste being generated at each site, the last two responsibilities listed are perhaps the most crucial. Industrial engineers, by combining cost evaluation and process design, must work with other process engineers to redesign current processes to use nonhazardous materials or to reduce the impact of the waste streams.

In other words, the ever-increasing amount of regulations and associated cost, together with the phasing out of presently popular alternatives (such as landfilling) means that other changes may be cost-effective.

The cost-saving potential for the reduction of hazardous wastes is significant, so the industrial engineer should and must be ready for the challenge. I.E.s are trained in cost-reduction techniques. Calling upon this training, I.E.s can develop a generalized procedure as follows (in order of desirability):

(1) *Eliminate* hazardous chemicals, wastes, and substances.
   (a) Use nonhazardous chemicals.
   (b) Recover and reuse hazardous chemicals.
(2) *Combine* processes to reduce or reuse chemicals.
   (a) Cascade uses.
   (b) Do two or more operations in one process.
(3) *Change* sequence place or thing.
   (a) Use different equipment.
   (b) Move location so waste can be better handled.
   (c) Change sequence to reduce waste.
(4) *Improve* or modify process to reduce amount or toxicity of hazardous waste.
   (a) Improve operations to reduce waste.
   (b) Value-engineer product to reduce need.

Consider a typical spray paint line used to paint punched metal pieces. The paint line process has spent solvent waste that must be disposed of. Industrial engineers have modified paint lines in the past to reduce paint costs. Now, the I.E. should also consider the impact of hazardous waste in the identification of cost. Applying the four approaches above might result in the following:

(1) *Eliminate*—Change to a water-based paint.
(2) *Combine*—Use spent solvent in another process.
(3) *Change*—Paint coil instead of individual pieces and concentrate stream to a single location.
(4) *Modify*—Change painting process to electrostatic, airless, or dip to reduce overspray and solvent use. Use solvent recovery.

Companies are applying these techniques with astounding results.[5] Industrial engineers need to remember to include total system costs in the analyses. For example, eliminating a hazardous material might save money in:

Receiving and shipping requirements.
Handling and storage.
Record keeping.
Contingency planning.
Disposal.
Protective garments.
Insurance costs.
Possible fines.
Litigation costs.

12.5. Summary

Resource management is an important field of industrial engineering that is growing rapidly. Courses are appearing in many I.E. curriculums, but whether or not a course is taught, all I.E.s in the future will be affected by energy, water, and hazardous material management. The prudent I.E. will be prepared.

DISCUSSION QUESTIONS

1. Energy, water, and hazardous material problems seem to involve mechanical, chemical, electrical, and civil engineering. Why should industrial engineers be involved?

2. Discuss why OSHA, DOT, and EPA are all involved in the general area we called "hazardous material management."

3. In Figure 12.2, Service Level 1 is cheaper but it implies the company must have its own transformer. Discuss any disadvantages of this.

4. Your company sets a very high demand peak every July—so high, in fact, that November through April is ratcheted (demand falls below ratchet minimum and is then billed at that

[5]Many companies have reduced their hazardous waste stream (and costs) by 50%. Some have totally eliminated hazardous waste from their operations.

minimum). Discuss the advantages and disadvantages of demand shedding both in the winter and in the summer.

5. Some have said that the energy crisis is over. Does that mean that energy management is no longer needed?

PROBLEMS

1. Ajax Manufacturing consumes 1,000 thousand cubic feet (MCF) of gas per month in its Bozeman, Montana, facility. Using the rate schedules shown in Figure 12.1, calculate the monthly energy cost for both priority schedules. Which would you recommend? Repeat for consumption of 3,000 MCF.
2. Ajax experienced the following characteristics for its monthly electric bill. Assuming that all 12 months in a year are the same (a ridiculous assumption), how much can Ajax afford to pay for transformation equipment to go from Service Level 2 to Service Level 1 in Figure 12.2? The company desires a 3-year payback on its investment. Maintenance of transformers costs $300 per month.

Consumption	200,000	kWh
Demand	2,000	kW
Previous high demand	2,200	kW

3. Ace Manufacturing has a compressor that is "once-through" water cooled. You find that it is using 3×10^6 gal per year and the temperature is raised from 65°F to 105°F. Assuming that you can reuse the water in a heater rinse tank, what is the savings (in dollars) to the company? The company is paying $\$2.00/10^3$ gal for water and $\$5.00/10^6$ Btu for natural gas. The water heater normally used to generate hot water is 70% efficient. Assume that the water cools to 100°F before it can be reused.
4. Zebra Stripes, Inc. has asked you to help them organize an energy management program. They annually spend $\$1 \times 10^6$ per year on natural gas and $\$0.50 \times 10^6$ per year on electricity. The company is organized around three manufacturing centers, each of which is separately metered for energy. Prepare a short (one-page) paper describing the organization of a potential energy management program together with the potential savings.
5. Inden, Inc. presently light their manufacturing space with 100 mercury vapor bulbs of 1000 watts each. They find they can replace those with 100 high-pressure sodium bulbs of 400 watts each. Electricity costs $6.00 per kilowatt for demand and $0.08 per kilowatt hour for consumption. The lights burn 400 hours per month. What is the dollar savings per month? Ignore any factors not stated.
6. One plant generates and disposes of 10,000 kg of spent solvent each month. The solvent is sent to a cement kiln, where it is burned as fuel at a cost of $0.07 per pound. Instead, the company decides to recycle at a cost of $0.05 per input pound. New solvent costs about $0.20 per pound. In the proposed system 90% of the solvent is recycled and 10% must still be disposed of the old way. Ignoring any transportation, what is the savings?

CHAPTER 13

Financial Management and Engineering Economy

13.1. Introduction

Financial management (accounting—cost accounting) and engineering economy are presented in this chapter. Although these two areas are quite different, there is a substantial relationship between the two. For example, the accounting function is responsible for recording, summarizing, and presenting historical financial data, while engineering economy is used to make current and future economic decisions. Also, engineering economy usually requires data obtained from accounting records. Before we continue the chapter, Example 13.1 is given to show the need for both areas.

Example 13.1

As president of Gadgets, Inc., you find that one of your most pressing problems is to understand how the plant is doing financially. The owners of the company have hired you to run the business and it is your responsibility to make sure that the business is financially sound. *Accounting* assists in this endeavor by recording the financial transactions and summarizing them periodically through *balance sheets* and *income statements*.

There is an additional problem. You find that knowing the overall profit or loss of Gadgets, Inc., is not enough. You need to know how the individual product lines perform. Is each line profitable or not? Also, you need to know the cost of individual ''special'' gadgets that you make in order to determine if you should continue making them as you do now, raise the price, or discontinue them. This information is given by a *cost accounting* system which is an extension of the general accounting system.

Also, you are continually faced with decisions that require sound financial analysis, for example:

(1) Should this equipment be replaced?
(2) Is machine A more economical than machine B?
(3) Should gadget X be produced via the existing method or a new method that would cost a lot to install but would be cheaper than the existing method after installation?
(4) How long should machine C be kept?

All these questions and others can be answered by using a set of techniques called *engineering economy*.

Once you have used the material presented in this chapter to design an effective and efficient accounting system and are prepared to analyze economic decisions using engineering economy, you are ready to proceed. (To be continued.)

13.2. Accounting

Accounting is the function of recording, summarizing, and presenting the historical financial data of a business. In accounting, the amounts of and the changes in the company's *assets, liabilities*, and *net worth* are recorded and reported. These terms are defined below.

Asset. Anything of monetary value a company owns (e.g., cash, supplies on hand, inventories, prepaid expenses, land, buildings, equipment).

Liability. Debt a company owes to its creditors (e.g., accounts payable, notes payable, taxes payable, mortgages payable).

Net Worth. Ownership interest in the company. It is made up of *capital stock* (the amount of money originally invested by the owners) and *retained earnings* (the net total of all previous earnings not paid out, but reinvested).

Since net worth is the ownership interest in a business and since the owners have rights to any surplus made by the company, it is logical that

$$\text{net worth} = \text{assets} - \text{liabilities} \tag{13.1}$$

Equation (13.1) is often referred to as the *accounting equation*, although it is most often shown as follows:

$$\text{assets} = \text{liabilities} + \text{net worth} \tag{13.2}$$

A formal presentation of the accounting equation is called a *balance sheet*, which is a statement of the financial position of a company at a point in time. Figure 13.1 is an example of an accounting equation (balance sheet) as it might be for Gadgets,

GADGETS, INC.
Balance Sheet
December 31, 19x2

ASSETS		LIABILITIES	
Cash	$ 30,000	Accounts Payable	$ 20,000
Accounts Receivable	20,000	Bank Loan	50,000
Supplies on hand	30,000	Mortgage	100,000
Land	60,000	Total Liabilities	$170,000
Buildings	90,000		
Total Assets	$230,000	NET WORTH	
		Capital Stock	$ 45,000
		Retained Earnings	15,000
		Total Net Worth	$ 60,000
		Total Liabilities Plus Net Worth	$230,000

Fig. 13.1. Balance sheet.

Inc., as of December 31, 19x2. As can be seen in Figure 13.1, assets do equal liabilities plus net worth.

If the business were static (if the balance sheet did not change), one statement of the accounting equation would always suffice. But business is not static. New items are always being added, old ones are being dropped, amounts are being changed, and so on. Therefore, as soon as the next change occurs (which is almost immediately), the balance sheet is outdated.

Some of the changes could be shown by using the accounting equation [Equation (13.2)]. Some, however, would require a different format. Any change affecting only assets, liabilities, or net worths would be no problem. But what about money received as payment for a product or a service? Cash (an asset) is one item that is changed, but what else is changed? Another item is called a *revenue*. A revenue is the income due to the sale of a product or service eventually resulting in an increase in net assets or net worth (the increase may be delayed). Another item is called an *expense*. This is the expenditure for materials, labor, or a service consumed in the production process. An expense is a cost of production or service eventually resulting in monetary outflow.

The accounting equation must be revised to show these new items. Obviously, a revenue, by itself, increases net worth and an expense, by itself, decreases net worth. Knowing this, we can rewrite the equation as

$$\underbrace{\text{assets} = \text{liabilities} + \text{net worth}}_{\text{balance sheet}} + \underbrace{\text{revenues} - \text{expenses}}_{\text{income statement}} \qquad (13.3)$$

Figure 13.1 has already shown that a balance sheet is a formal statement of the first three items of the revised accounting equation. A formal statement of the other two items is called an *income statement*. (Figure 13.3 presents an income statement.) The following three observations can be made:

(1) An income statement will not balance in general. If revenues are greater than expenses, the company has made a *profit* over the last period. If revenues are less than expenses, there has been a *loss*.
(2) The balance sheet will not balance until the resultant *profit* or *loss* from the income statement has been added. (In Figure 13.1 no revenues and no expenses were given. Therefore, the equation did balance.)
(3) Since revenues and expenses reflect performance over a period of time, periodically the amounts should be set equal to zero, so that a fresh start can be made.

To examine this, consider the balance sheet shown in Figure 13.1 and the following occurrences:

(1) January 5, 19x3. A check for $1,000 is issued to TuFLUK Supply Company for office supplies. The effect on the accounting equation is to reduce cash by $1,000 and increase supplies on hand by $1,000.

Cash	from $30,000 to $29,000
Supplies on hand	from $30,000 to $31,000

(2) January 10, 19x3. A sale of gadgets is made to Blue Wagon, Incorporated ($12,000). If it is assumed that payment will be made later, the effect is to increase accounts receivable by $12,000 and increase a revenue account, sales, for example, by the same amount.

Accounts receivable	from $20,000 to $32,000
Sales	from $ 0 to $12,000

(3) January 18, 19x3. The $2,000 rent expense for the month of January is incurred; but by previous agreement it is paid only every 2 months. The effect is to increase accounts payable by $2,000 and increase an expense account (rent expense) also by $2,000.

Accounts payable	from $20,000 to $22,000
Rent expense	from $ 0 to $ 2,000

(4) January 30, 19x3. $1,000 worth of office supplies are used in the month of January. Supplies on hand are reduced by $1,000 and a new item (supplies expense) is increased by $1,000.

Supplies on hand	from $31,000 to $30,000 [see item (1)]
Supplies expense	from $ 0 to $ 1,000

(5) January 30, 19x3. Salaries totaling $8,000 are paid for the month. Salaries expense must be increased by $8,000 and cash decreased by $8,000.

Salaries expense	from $ 0 to $ 8,000
Cash	from $29,000 to $21,000 [see item (1)]

Figure 13.2 shows the accounting equation after all the foregoing changes have been made.

Usually, assets equal liabilities plus net worth is shown separately as a *balance sheet*. In order for the balance sheet to balance, however, the net difference of revenues minus expenses ($1,000 in this case) must be added to (or subtracted from in case of a loss) net worth. Figure 13.3 presents this. If the revenue and expenses are to be wiped out so that the next period's performance can be shown (called *closing*), the difference is added to retained earnings. Then at the beginning of the next period (February, in this case) there would be no revenues and no expenses, and the cycle would start all over again. Revenues minus expenses is shown separately on an *income statement*. This is presented in Figure 13.4.

13.3. Cost Accounting

General accounting, as discussed in Example 13.1, only provides information on total or overall performance. General accounting procedures do not show which products are profitable and which products are unprofitable or where the major

GADGETS, INC.
Accounting Equation
January 31, 19x3

ASSETS		LIABILITIES	
Cash	$ 21,000	Accounts Payable	$ 22,000
Accounts Receivable	32,000	Bank Loan	50,000
Supplies on hand	30,000	Mortgage	100,000
Land	60,000	Total Liabilities	$172,000
Buildings	90,000		
Total Assets	$233,000	NET WORTH	
		Capital Stock	$ 45,000
		Retained Earnings	15,000
		Total Net Worth	$ 60,000
		REVENUES	
		Sales	$ 12,000
		Total Revenues	$ 12,000
		EXPENSES	
		Rent	$ 2,000
		Supplies	1,000
		Salaries	8,000
		Total Expenses	$ 11,000

Assets = liabilities + net worth + revenues − expenses
233,000 = 172,000 + 60,000 + 12,000 − 11,000 = $233,000

Fig. 13.2. Accounting equation.

GADGETS, INC.
Balance Sheet
January 31, 19x3

ASSETS		LIABILITIES	
Cash	$ 21,000	Accounts Payable	$ 22,000
Accounts Receivable	32,000	Bank Loan	50,000
Supplies on hand	30,000	Mortgage	100,000
Land	60,000	Total Liabilities	$172,000
Buildings	90,000		
Total Assets	$233,000	NET WORTH	
		Capital Stock	$ 45,000
		Retained Earnings $15,000 + $1000 profit for January	16,000
		Total Net Worth	$ 61,000
		Total Liabilities Plus Net Worth	$233,000

Fig. 13.3. Balance sheet.

costs are generated in the machine shop, the paint room, and other departments. Accounting procedures can be modified to show this through the use of cost accounting.

The cost accounting system can generate many different reports showing the performances of individual products, individual departments or shops, and so on. Management then has its fingers on the pulse of the entire organization as a whole, as well as on individual entities.

GADGETS, INC.
Income Statement
January 31, 19x3

REVENUES			
Sales		$12,000	
EXPENSES			
Rent	$2,000		
Supplies	1,000		
Salaries	8,000		
Total Expenses		$11,000	
Net Profit to Retained Earnings			$1,000

Fig. 13.4. Income statement.

Cost accounting develops additional statements that permit us to determine the *cost of goods made* and the *cost of goods sold*. Knowing the cost of material, labor, and overhead applicable to the goods made and sold is essential. These terms are defined below.

Direct Material. Any material whose cost is directly and conveniently allocable to a product (e.g., the wood that goes into a desk and the paper that went into the pages of this book). All other material is called *indirect material* (e.g., sweeping compound used on floors and all lubricants used in the plant). The premise here is that it is easier and more accurate to charge all indirect material (and indirect labor) to a general account and later allocate it to specific products.

Direct Labor. Any labor whose cost is directly and conveniently allocable to a product (e.g., the operator who runs a lathe that produces the part and the painter who actually paints the product). All other production labor cost is called *indirect labor* (e.g., material handling personnel who move in-process parts and materials as needed, machine setters for press operations, etc.).

Overhead. All production cost that is neither direct material nor direct labor. This includes all indirect material and indirect labor as well as such items as depreciation of equipment, taxes on buildings and equipment, maintenance of buildings and equipment, and factory supervision.

The cost of goods made and sold can be determined as shown in Figure 13.5. This is done by taking for each category what is in inventory at the beginning of the month and adding what was applied during the month. The result is the total that is available. The inventory at the end of the month is subtracted, leaving what was finished or completed during the month. If this last entry is added for direct material ($77,400), direct labor ($101,000), and overhead ($117,600), the result is the *cost of goods made* ($296,000). This is what went into finished goods inventory and should be added to the amount in finished goods inventory at the beginning of the month ($32,400), yielding the total amount of finished goods available for sale ($328,400). Subtracting the finished goods inventory at the end of the month ($42,800) yields the *cost of goods sold* ($285,600).

One of the major problems encountered in cost accounting is deciding how to charge overhead to production. By its very nature, overhead is ''neither directly nor conveniently'' chargeable to a specific product. For example, if a material handler helps with 50 different products during a period and if the volume for each ranges from very small to very large, how is the cost to be broken up? Cost accounting does this in several different ways, but three of the most popular are listed below:

(1) Overhead is charged as a percentage of the direct labor cost in the products.
(2) Overhead is charged as a percentage of the direct material cost in the products.
(3) Overhead is charged on a rate per direct labor hour in the products where the rate is estimated beforehand.

Direct Material		
In process Oct. 1, 19xx	$ 6,800	
Applied during the month	$ 79,000	
Total	$ 85,800	
In process Oct. 31, 19xx	$ 8,400	$ 77,400
Direct Labor		
In process Oct. 1, 19xx	$ 8,600	
Applied during the month	$103,800	
Total	$112,400	
In process Oct. 31, 19xx	$ 11,400	$101,000
Overhead		
In process Oct. 1, 19xx	$ 11,600	
Applied during the month	$120,200	
Total	$131,800	
In process Oct. 31, 19xx	$ 14,200	$117,600
Cost of Goods Made		$296,000
Finished goods Oct. 1, 19xx		$ 32,400
Total		$328,400
Finished goods Oct. 31, 19xx		$ 42,800
Cost of Goods Sold		$285,600

Fig. 13.5. Calculation of cost of goods made and cost of goods sold.

13.4. Engineering Economy

Engineering is the application of scientific knowledge to practical problems. One definition of engineering economy, then, is applying "scientific economics to the solution of practical problems." As an example of these practical problems, consider the following:

Example 13.2

As president of Gadgets, Inc., one of your major decisions is to decide which one if either of two new automated machines should be purchased to replace a manual method now used on a product line. The product line will be discontinued entirely in 5 years. The manual method costs $10,000 per year. The new automated machine A costs $30,000 now (purchase price) and will cost $2,000 per year to operate over the next 5 years, at which time it will be worthless. Another new automated machine B is available that costs $40,000 new and will cost $1,000 per year to operate for the next 5 years, at which time it will be worth $9,000. Gadgets, Inc., requires a 12% return on all investments.

To solve problems like these, the first step is usually to draw a *time diagram* showing the flow of money for the various alternatives. Figure 13.6 presents the time diagrams for the three alternatives under consideration. In constructing these

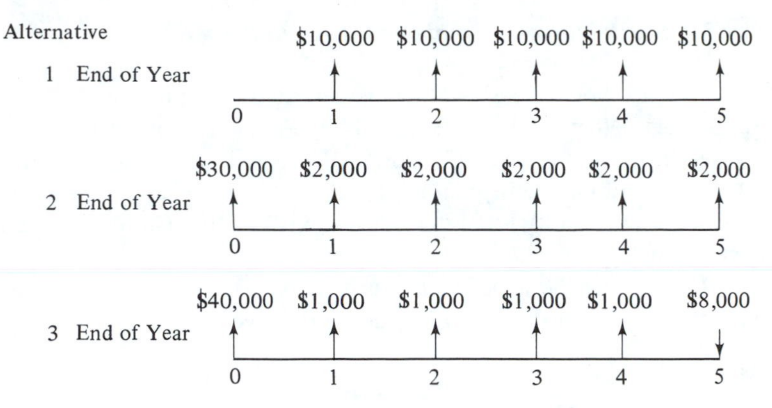

Fig. 13.6. Time diagrams for Gadgets, Inc.

charts, all money flowing during a year is interpreted as flowing at the *end of the year*, all cash outflows are shown as arrows pointing upward, all cash inflows are shown as arrows pointing downward, and all cash flows shown are net cash flows (e.g., year 5 of alternative 3 has a cash outflow of $1,000, a cash inflow of $9,000, yielding a net cash inflow of $8,000).

It might seem that a simple solution would be to add up the cash flows and choose which is cheapest. The results would be:

Alternative 1: 10,000 + 10,000 + 10,000 + 10,000 + 10,000 = $50,000
Alternative 2: 30,000 + 2,000 + 2,000 + 2,000 + 2,000 + 2,000 = $40,000
Alternative 3: 40,000 + 1,000 + 1,000 + 1,000 + 1,000 − 8,000 = $36,000

Alternative 3 seems to be better than alternative 2 and certainly better than alternative 1, but wait! Alternative 3 invests $40,000 for a period of 5 years, while alternative 2 invests only $30,000, and alternative 1 has no initial investment. If the investment is *not* made, then Gadgets' management has $30,000 or $40,000 that it could invest in other activities. Since money is limited, this could be a very important consideration. Some way needs to be found to consider the *time value* of the investment required.

Suppose that in the past Gadgets, Inc. found that it could average a 12% return on all its investments (one way of determining the time value of money). Then it is logical to assign a cost to alternatives 2 and 3 to show that the money invested, or tied up, *could have been* invested in other places. To do this in a methodical way, interest factors are developed.

13.5. Interest Factors

The purpose of the interest factors developed here is to be able to explicitly evaluate the desirability of tying money up in one place versus investing in another place. The first factor is called the *single payment compound amount factor*.

13.5.1 Single-Payment Compound Amount Factor

Suppose that you have $1,000 today and could invest that at 12% interest per year for 5 years. What would the $1,000 be worth then? Obviously, at the end of the first year it would be worth $1,120 ($1,000 + $120 interest). For the second year, there is $1,120 at the beginning of the year; so there would be $1,254.40 at the end of the year [$1,120 + 0.12 (1,120), or $1,254.40]. This could be continued as shown in Table 13.1.

Now, this could be done for all such problems, but there is a simpler way involving the use of factors. But before doing that, we shall define some symbols.[1]

i = annual interest rate

n = number of annual interest periods

P = present principal sum (e.g., $1,500 at the end of year 0, P = $1,500)

A = single payment in a series of n equal payments made at the end of each annual interest period (e.g., $100 at the end of each year for 5 years, A = $100)

F = future sum, n annual periods hence, equal to the compound amount of a present principal sum P, or equal to the sum of the compound amounts of payments, A, in a series (e.g., $2,000 10 years from now, F = $2,000)

In Table 13.1, $1,000 is a present principal sum (P), $1,762.34 is a future sum (F), 12% is the annual interest rate (i), and 5 is the number of annual interest periods (n). Table 13.1 may be rewritten, substituting in the symbols and generalizing to n years, as shown in Table 13.2.

Therefore, for any amount P, any annual interest rate i, and any number of years n, the amount F that it will be equivalent to at the end of n years is given by

$$F = P(1 + i)^n \tag{13.4}$$

The factor $(1 + i)^n$ is called the *single-payment compound amount factor*. Since this is a lengthy title, perhaps a shorthand designation would be more appropriate. The designation used is $F|P\ i,n$, which is interpreted as follows:

What is Given — F (What is needed)
P — What is Given
i — Annual interest rate
n — Number of annual interest periods

$$F \mid P\ i,n$$

In other words, the factor will "find F given that P is known as well as the annual interest rate and the number of annual interest periods."

Finally, since the factor $(1 + i)^n$ is independent of the amount P, the factor

[1]The symbols and the basic approach are taken, with permission, from G. J. Thuesen and W. J. Fabrycky, *Engineering Economy* (Englewood Cliffs, N.J.: Prentice Hall, 1989), p. 39.

Table 13.1. INTEREST EARNED ON $1,000 AT 12% FOR 5 YEARS

A Year	B Investment at Beginning of Year	C Interest Earned (12% Col. B)	D Amount at End of Year
1	$1,000.00	1,000.00 (0.12) = 120.00	1,000 (1.12) = 1,120.00
2	1,120.00	1,120.00 (0.12) = 134.40	$1,000\ (1.12)^2$ = 1,254.40
3	1,254.40	1,254.40 (0.12) = 150.53	$1,000\ (1.12)^3$ = 1,404.93
4	1,404.93	1,404.93 (0.12) = 168.59	$1,000\ (1.12)^4$ = 1,573.52
5	1,573.52	1,573.52 (0.12) = 188.82	$1,000\ (1.12)^5$ = 1,762.34

$(1 + i)^n$ can be tabulated for various interest rates i and annual interest periods n. A sample table will be presented shortly.

13.5.2 Other Interest Factors

As might be expected, factors can be derived in a similar manner for all combinations of P, F, and A. Those not covered so far are

(1) $(^{P\,|\,Fi,n})$ Find P given F.
(2) $(^{F\,|\,Ai,n})$ Find F given A.
(3) $(^{A\,|\,Fi,n})$ Find A given F.
(4) $(^{P\,|\,Ai,n})$ Find P given A.
(5) $(^{A\,|\,Pi,n})$ Find A given P.

All these factors can be tabulated for selected annual interest rates and annual interest periods. Table 13.3 presents all these factors for an interest rate of 12% and various values of n.

Table 13.2. INTEREST FACTOR DERIVATION: SINGLE-PAYMENT COMPOUND AMOUNT FACTOR

(A) Year	(B) Investment at Beginning of Year	(C) Interest Earned (12% Col. B)	(D) Amount at End of Year
1	P	Pi	$P(1 + i)$
2	$P(1 + i)$	$P(1 + i)i$	$P(1 + i)^2$
3	$P(1 + i)^2$	$P(1 + i)^2 i$	$P(1 + i)^3$
•	•	•	•
•	•	•	•
•	•	•	•
n	$P(1 + i)^{n-1}$	$P(1 + i)^{n-1} i$	$P(1 + i)^n = F$

Table 13.3. INTEREST FACTORS FOR 12% INTEREST*

n	Single Payment		Equal Payment Series				Uniform gradient-series factor
	Compound-amount factor	Present-worth factor	Compound-amount factor	Sinking-fund factor	Present-worth factor	Capital-recovery factor	
	To find F Given P F/P, i, n	To find P Given F P/F, i, n	To find F Given A F/A, i, n	To find A Given F A/F, i, n	To find P Given A P/A, i, n	To find A Given P A/P, i, n	To find A Given G A/G, i, n
1	1.120	0.8929	1.000	1.0000	0.8929	1.1200	0.0000
2	1.254	0.7972	2.120	0.4717	1.6901	0.5917	0.4717
3	1.405	0.7118	3.374	0.2964	2.4018	0.4164	0.9246
4	1.574	0.6355	4.779	0.2092	3.0374	0.3292	1.3589
5	1.762	0.5674	6.353	0.1574	3.6048	0.2774	1.7746
6	1.974	0.5066	8.115	0.1232	4.1114	0.2432	2.1721
7	2.211	0.4524	10.089	0.0991	4.5638	0.2191	2.5515
8	2.476	0.4039	12.300	0.0813	4.9676	0.2013	2.9132
9	2.773	0.3606	14.776	0.0677	5.3283	0.1877	3.2574
10	3.106	0.3220	17.549	0.0570	5.6502	0.1770	3.5847
11	3.479	0.2875	20.655	0.0484	5.9377	0.1684	3.8953
12	3.896	0.2567	24.133	0.0414	6.1944	0.1614	4.1897
13	4.364	0.2292	28.029	0.0357	6.4236	0.1557	4.4683
14	4.887	0.2046	32.393	0.0309	6.6282	0.1509	4.7317
15	5.474	0.1827	37.280	0.0268	6.8109	0.1468	4.9803
16	6.130	0.1631	42.753	0.0234	6.9740	0.1434	5.2147
17	6.866	0.1457	48.884	0.0205	7.1196	0.1405	5.4353
18	7.690	0.1300	55.750	0.0179	7.2497	0.1379	5.6427
19	8.613	0.1161	63.440	0.0158	7.3658	0.1358	5.8375
20	9.646	0.1037	72.052	0.0139	7.4695	0.1339	6.0202
21	10.804	0.0926	81.699	0.0123	7.5620	0.1323	6.1913
22	12.100	0.0827	92.503	0.0108	7.6447	0.1308	6.3514
23	13.552	0.0738	104.603	0.0096	7.7184	0.1296	6.5010
24	15.179	0.0659	118.155	0.0085	7.7843	0.1285	6.6407
25	17.000	0.0588	133.334	0.0075	7.8431	0.1275	6.7708
26	19.040	0.0525	150.334	0.0067	7.8957	0.1267	6.8921
27	21.325	0.0469	169.374	0.0059	7.9426	0.1259	7.0049
28	23.884	0.0419	190.699	0.0053	7.9844	0.1253	7.1098
29	26.750	0.0374	214.583	0.0047	8.0218	0.1247	7.2071
30	29.960	0.0334	241.333	0.0042	8.0552	0.1242	7.2974
31	33.555	0.0298	271.293	0.0037	8.0850	0.1237	7.3811
32	37.582	0.0266	304.848	0.0033	8.1116	0.1233	7.4586
33	42.092	0.0238	342.429	0.0029	8.1354	0.1229	7.5303
34	47.143	0.0212	384.521	0.0026	8.1566	0.1226	7.5965
35	52.800	0.0189	431.664	0.0023	8.1755	0.1223	7.6577
40	93.051	0.0108	767.091	0.0013	8.2438	0.1213	7.8988
45	163.988	0.0061	1358.230	0.0007	8.2825	0.1207	8.0572
50	289.002	0.0035	2400.018	0.0004	8.3045	0.1204	8.1597

* Extracted with permission, from G. J. Thuesen and W. J. Fabrycky, *Engineering Economy* (Englewood Cliffs, N. J.: Prentice-Hall, Inc., 1989), p. 659.

13.5.3 Examples

In the preceding section $1,000 was invested at 12% for 5 years. The amount F at the end of the fifth year should be given by

$$\begin{aligned} F &= (^{F\,|\,P i,n}) \\ &\underset{}{=} \overset{F\,|\,P\ 12,5}{1{,}000\ (1.762)} \\ &= \mathbf{\$1{,}762} \end{aligned}$$

which agrees with Table 13.1 except for minor round-off error.

Following are other examples:

(1) How much will $150 be worth 10 years from now at an annual interest rate of 12%?

$$\begin{aligned} F &= P(^{F\,|\,P\ i,n}) \\ &= \overset{F\,|\,P\ 12,10}{150\ (3.106)} \\ &= \mathbf{\$465.90} \end{aligned}$$

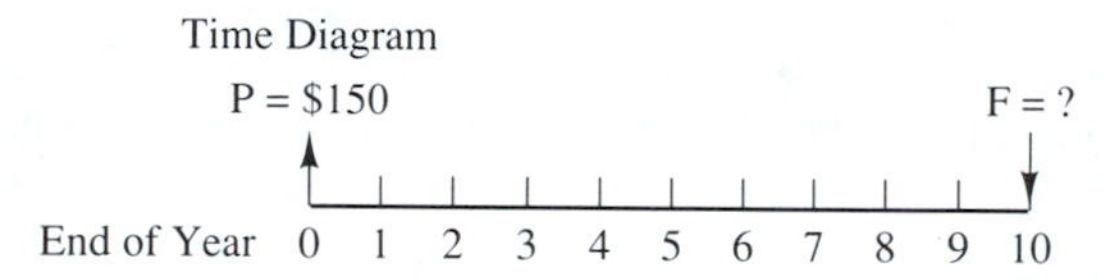

(2) How much would have to be set aside now to provide $10,000 fifteen years from now at an annual interest rate of 12%? Now we are seeking an amount P—given we know an amount F 15 years from now.

$$\begin{aligned} P &= F(^{P\,|\,Fi,n}) \\ &= \overset{P\,|\,F\ 12,15}{10{,}000(0.1827)} \\ &= \mathbf{\$1{,}827} \end{aligned}$$

Time Diagram

P = ? F = $10,000

End of Year 0 1 2 3 4 5 . . . 14 15

(3) If $100 is set aside at the end of each year for 8 years at an annual interest rate of 12%, what would it be worth at the end of the eighth year?
Now we have $100 at the end of every year for 8 years. Therefore,

$$\begin{aligned} A &= \$100 \text{ (equal annual payments)} \\ \mathrm{F} &= \mathrm{A}(^{F\,|\,Ai,n}) \\ &= \overset{F\,|\,A\ 12,8}{100(12.300)} \\ &= \mathbf{\$1{,}230} \end{aligned}$$

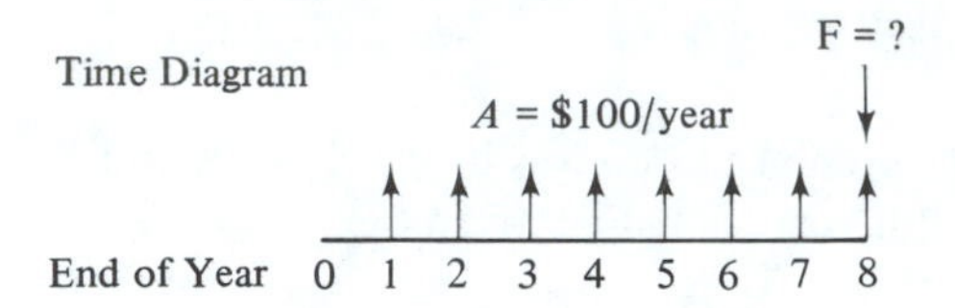

(4) How much would be required at the end of each year for 10 years to repay a loan of $1,500 now if the interest rate is 12% per year?

$$\begin{aligned} A &= P(^{A\,|\,Pi,n}) \\ &= 1500(\overset{A\,|\,P\ 12,10}{0.1770}) \\ &= \mathbf{\$265.50} \end{aligned}$$

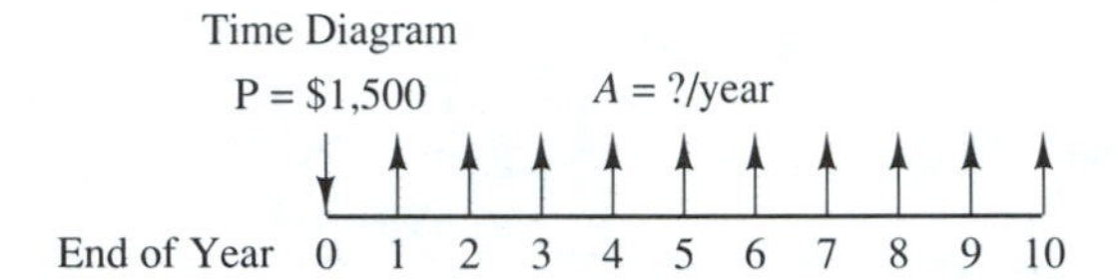

(5) How much would be required at the end of each year for 9 years to accumulate $2,000 at the end of the ninth year if the annual interest rate is 12%?

$$\begin{aligned} A &= F(^{A\,|\,Fi,n}) \\ &= 2{,}000(\overset{A\,|\,F\ 12,9}{0.0677}) \\ &= \mathbf{\$135.40} \end{aligned}$$

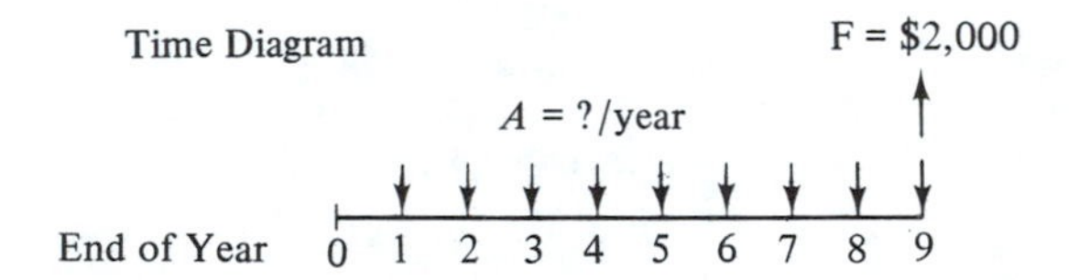

(6) How much can be borrowed now, if it can be repaid by five equal end-of-year payments of $100 each? The annual interest rate is 12%.

$$\begin{aligned} P &= A\ (^{P\,|\,Ai,n}) \\ &= 100(\overset{P\,|\,A\ 12,5}{3.6048}) \\ &= \mathbf{\$360.48} \end{aligned}$$

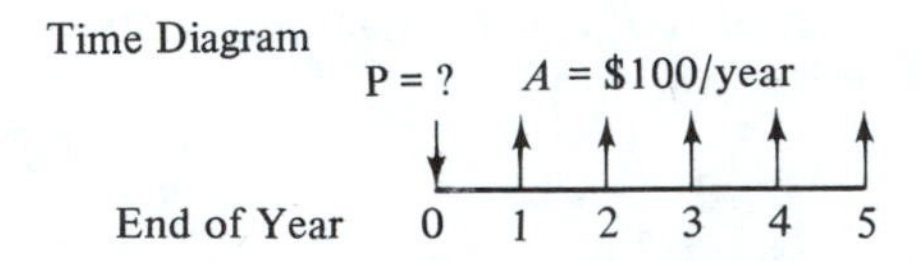

13.6. Back to Gadgets—Present Worth Calculation

In the machine determination problem for Gadgets, Inc., the problem is in deciding which of three alternatives is best, given varying cash flows. One way of doing this is to find the amounts at the end of year 0 that are equivalent to each of the cash flows when interest is considered, and then choosing the best. This equivalent amount at the end of year 0 is called the *present worth amount*.

For example, alternative 1 has the following time diagram:

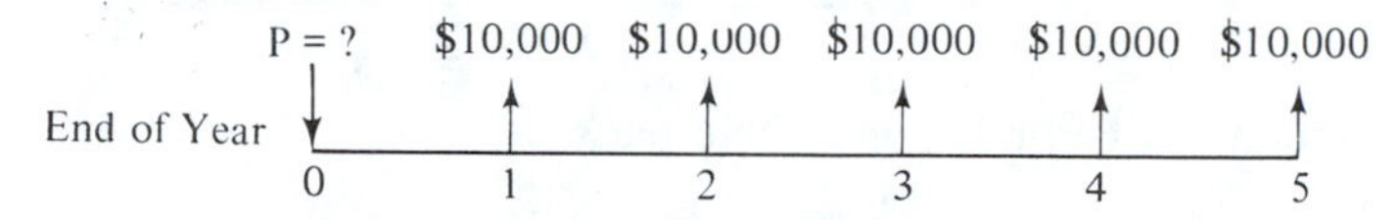

What is needed is the amount P as shown. This is simply

$$P = A(^{P|Ai,n})$$
$$= 10{,}000 \underset{P|A12,5}{(3.6048)}$$
$$= \mathbf{\$36{,}048}$$

An interest rate of 12% is used since the company requires a 12% return (interest) on all investments.

Alternative 2 is a little trickier since its time diagram is more complex. If we break it into two time diagrams, however, it does not seem so difficult.

First, if the two diagrams are "added," the result is the time diagram shown in Figure 13.6. Therefore, if an amount P is found for each diagram and if the two P's are added, the total should be a P for the entire diagram, or the present worth amount needed.

$$P_A = \$30{,}000$$

Time Diagram A

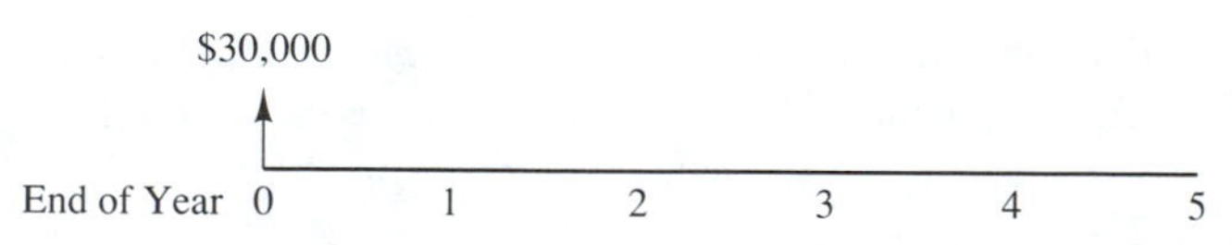

Time Diagram B

$2,000 $2,000 $2,000 $2,000 $2,000

End of Year 0 1 2 3 4 5

since it is already at the end of year 0.

$$P_B = A(^{P|Ai,n})$$
$$= \$2{,}000 \underset{P|A\ 12,5}{(3.6048)}$$
$$= \$7{,}209.60$$

$$\begin{aligned} P &= P_A + P_B \\ &= 30{,}000 + 7{,}209.60 \\ &= \mathbf{\$37{,}209.60} \end{aligned}$$

The present worth amount for alternative 3 can be calculated in a similar manner.

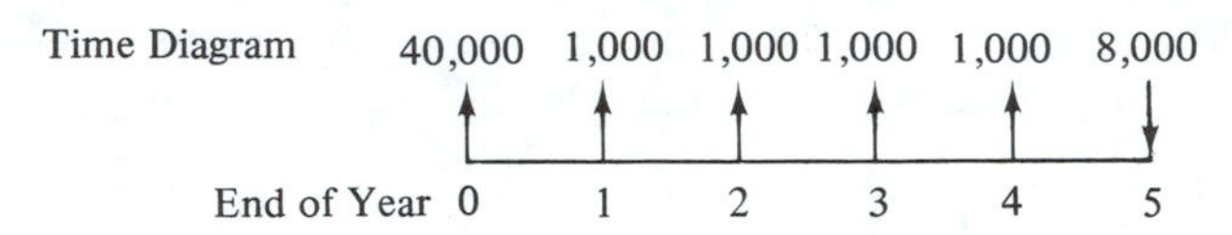

First the diagram can be broken into three parts.

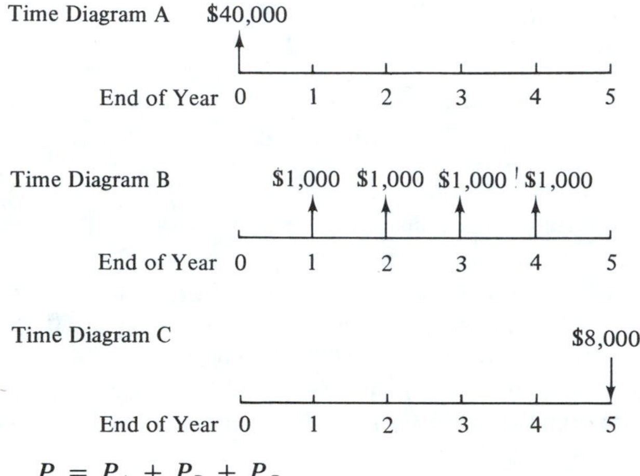

$$\begin{aligned} P &= P_A + P_B + P_C \\ &= 40{,}000 + 1{,}000 \overset{P|A\ 12,4}{(3.0374)} - 8{,}000 \overset{P|F\ 12,5}{(0.5674)} \\ &= \mathbf{\$38{,}498.20} \end{aligned}$$

P_c has a negative sign since the $8,000 is a net inflow and all others are net outflows.

The results for the three alternatives are presented in Figure 13.7. From Figure 13.7 it can be deduced that alternative 1 (stay with the manual method) is best, followed by 2 and then by 3. This is completely reversed from what was decided without considering interest. *Therefore, the time value of money should always be considered.*

Alternative	*Present Worth Amount*
1	$36,048.00
2	$37,209.60
3	$38,498.20

Fig. 13.7. Present worth amounts for three alternatives.

13.7. Impact of the Computer on Accounting and Engineering Economy

Virtually all accounting and engineering economy work is now done on the computer. In accounting, the number of records that must be kept and the amount of labor required would be overwhelming without the computer. Software for data storage and file manipulation, with rapid access to any subset of data, permits reports to be done rapidly and accurately. Also, integrated menu-driven graphics packages allow a large number of data presentation alternatives. Nearly all companies today have most or all of their accounting functions performed on computers, from inexpensive personal computers right on up to large mainframes. Much of the data collection for cost accounting systems is done in real time, through shop-floor terminals or other data transmission devices. The impact of computers in this area will continue to increase.

In engineering economy, certain techniques such as rate of return analyses can only be solved through trial-and-error manipulations or searches. Without computers, the application of these techniques is limited to relatively small classroom type problems. With computers, however, the applications are almost limitless in size. Replacement of equipment is a major engineering economy problem that can be solved much better now that computers are available. Various replacement strategies can be simulated and/or modeled very rapidly through the use of computers. Certain capital budgeting problems can be modeled as mathematical programming problems with unusual structures. The only way these mathematical programming models can be optimized is with sophisticated algorithms that, once again, would be limited to only small classroom type problems without the aid of computers. Again, the computers being used for engineering economy analyses range from personal computers to mainframes.

DISCUSSION QUESTIONS

1. Discuss why cost accounting is usually of more interest than general accounting to industrial engineers.
2. At what point in their careers do industrial engineers need to take a seminar or other continuing education course to become very knowledgeable about general accounting?
3. Give examples of assets, liabilities, net worth, revenues, expenses, direct material, direct labor, and overhead.
4. Classify each of the following as direct labor, direct material, indirect labor, or indirect material.
   (a) A worker's time to machine a part for later assembly into the finished product.
   (b) Sheet metal from which aircraft support structure parts are stamped or sheared.
   (c) A maintenance person's time which is easily allocable to different products operating on different production lines.

(d) A maintenance person's time devoted to a single process where multiple different products are run.
(e) A bank teller's time.
(f) A grocery store stockperson's time.
(g) Pallets used for moving grocery stock.

5. Find a balance sheet and an income statement for some company (financial reports are often in annual reports which are usually kept in the library). Discuss how the two statements are related and show that the income statement is usually needed to enable the balance sheet to match.

6. Identify three sources of overhead:
(a) In a manufacturing plant.
(b) In a service sector operation (e.g., Disneyland).

7. Discuss why overhead must be allocated, and describe three common ways of doing so. Pick one of the ways and generate a small example to illustrate the allocation of overhead.

8. Find an engineering economy book and determine the formulas for the $P|F$, $F|A$, $A|F$, $P|A$, and $A|P$ factors.

9. Describe how you would use the four A interest factors ($F|A$, etc.) if *beginning* of year cash flows were given for A instead of end-of-year values.

10. Describe how you would calculate a future worth if the annual amounts are mixed—some given as beginning-of-year values and others given as end-of-year values.

11. Discuss why the time value of money is important and must be considered in economic analyses that involve cash flows over several years.

12. Give an example that illustrates why compounding of interest is important.

PROBLEMS

1. Given the information shown below for Jacks, Inc., show the accounting equation. Then present the balance sheet and income statement.

Buildings	$22,500	Cash	$10,000
Loans Payable	2,500	Retained Earnings	17,500
Supplies Expense	5,000	Utility Expenses	750
Miscellaneous Expenses	7,000	Accounts Payable	5,000
Supplies on Hand	3,750	Sales	25,000
Accounts Receivable	12,500	Capital Stock	61,500
Salaries	12,500	Land	37,500

2. Given the information shown in Problem 1, show how the following transactions affect the accounting equation:
(a) Loans payable are paid off out of cash.
(b) Ace company buys $1,500 worth of products but does not pay immediately.
(c) Deuce Corporation owes $5,000 to Jacks, Inc. They pay $4,000 of that debt.
(d) Queens Co. sends a bill for $1,500 to Jacks, Inc., for entertainment expenses.

3. Present the balance sheet and the accounting equation for the Mizer Company, using the following information:

Accounts receivable	$ 7,000
Cash	43,300
Land	11,000
Capital stock	100,000
Profit for March	56,100
Notes payable	4,700
Accounts payable	22,000
Dividends declared	15,000
Retained earnings	25,000
Raw materials on hand	9,000
Factory building	82,000
Equipment	35,300
Finished goods	17,000
Work in progress	21,400
Tax accrued and payable	3,200

4. Overhead of $100,000 is to be allocated. Following are direct labor and direct material figures:

Product	*DL (Hours)*	*DL*	*DM*
A	4,000	$36,000	$15,000
B	2,000	21,000	25,000
C	1,500	22,000	30,000

Allocate the overhead on the basis of (a) direct labor hours, (b) direct labor cost, and (c) direct material cost.

5. How much is $30,000 today worth 5 years from now if the annual interest rate is 12%?

6. Suppose that you put $10,000 into an investment at age 25 and leave it there until your first retirement at age 55. The investment earns 12% compounded annually. How much can you withdraw at retirement?

7. How much must be invested now to have $100,000 in 15 years at 2%?

8. How much would have to be invested at the end of each year for 15 years to have $100,000 at the end of the fifteenth year if the annual interest rate is 12%?

9. How long does it take for money to triple if the annual interest rate is 8%? 12?

10. If you wanted to set up a college fund for your child by investing $2,000 each year for 18 years, how much would be available at the end of the eighteenth year? The annual interest rate is 12%.

11. How much money must be invested now at 12% to be able to withdraw $3,000 at the end of each year for 10 years?

12. If you invest $10,000 now at 12%, what equal annual amount will you be able to withdraw at the end of years 6, 7, 8, 9, and 10?

13. You invest $1,000 at the end of each year for 10 years (beginning with the end of year 1) and then stop adding to the investment. Your friend then starts investing $1,000 at the end of each year (beginning with the end of year 11) forever. Both investments earn 12%. In what year will your friend's investment be worth more than yours?

14. If a new machine is available that costs $60,000 and will last 5 years, at which time it is worthless, how much would that machine have to save each year in order for the purchase of it to be economically advisable? The annual interest rate is 12%. (*Hint:* You are trying to find the amount at the end of each year that causes the present worths to be equal.)

15. If a production process were available commercially that would cost $200,000 to purchase, would incur expenses of $20,000 per year to operate, and would generate $70,000 worth of profit per year, should this process be purchased if the annual interest rate is 12%, the process would last 8 years and the salvage value is 0? (*Hint:* Calculate *net* cash flows and draw a time diagram.)

16. Given the two following methods for doing a job, which method should be followed? The annual interest rate is 12%.

Method A—Initial expenditure is $17,500; annual cost for each of 4 years is $3,500.

Method B—Initial expenditure is $7,000; annual cost for each of 4 years is $7,000.

17. Write a simple computer program or use a spreadsheet to calculate the future worth of an investment today for arbitrary interest rates and number of years (both should be "read in" by the program).

CHAPTER 14

Deterministic Operations Research

14.1. Introduction—Definition

Operations Research has been defined by the Operational Research Society of the United Kingdom as follows:

> The attack of modern science on complex problems arising in the direction and *management of large systems of men, machines, materials, and money in industry, business, government, and defense*. The distinctive approach is to develop a *scientific model* of the system, incorporating measurement of factors such as chance and risk, with which to *predict and compare the outcomes of alternative decisions, strategies, or controls*. The purpose is to *help management determine its policies and actions scientifically*.

Let's examine the italicized terms more closely. First, the definition says that for operations research to be applicable, there must be systems of men, machines, materials, and money and that the techniques of O.R. are applicable to systems in industry, business, government, and defense. We can say, then, that O.R. is applicable just about anywhere there are systems that need to be managed.

The next italicized phrase, a "scientific model," is perhaps the key phrase in the definition. This implies that unless a scientific model (usually mathematical) is developed, it cannot be called operations research. Notice that many of the techniques we have discussed in previous chapters definitely use scientific models.

The next italicized phrase says that the objectives are to "predict and compare the outcomes of alternative decisions, strategies, or controls." This implies that any scientific model that predicts or evaluates the results of decisions or policies is operations research.

Finally, the overall objective of operations research is well stated by the last italicized phrase, "help management determine its policies and actions scientifically." It is significant that we say to *help* management. No O.R. tool "makes" decisions. It is only an aid to the decision maker.

14.2. Similarity to Industrial Engineering

The definition of industrial engineering, as adopted by the Institute of Industrial Engineers, is as follows:

> Industrial engineering is concerned with the design, improvement, and installation of *integrated systems of people, materials, information, equipment, and energy*. It draws upon specialized knowledge and skill in the *mathematical, physical, and social sciences* together with the principles and methods of engineering analysis and design to *specify, predict, and evaluate* the results to be obtained from such systems.

It is obvious that I.E. and O.R. have commonalities, as signified by the similarity in the definitions (note especially the italicized terms). It is the authors' opinion that operations research and industrial engineering indeed do have many of the same objectives and work on many of the same problems. The primary difference is that operations research has a higher level of theoretical and mathematical orientation, providing a major portion of the science base of industrial engineering.

Operations research carries a connotation of mathematical orientation, but industrial engineering does not restrict itself to any specific approach. Figure 14.1 illustrates the relationship between the two disciplines in terms of mathematical sophistication.

Many industrial engineers work in the area of operations research, as do

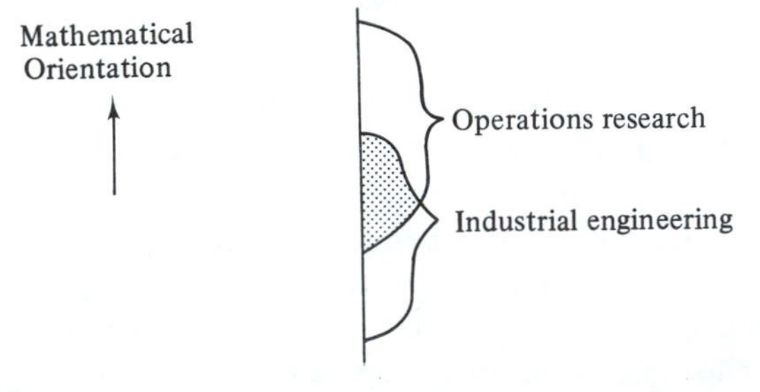

Fig. 14.1. Industrial engineering and operations research.

mathematicians, statisticians, physicists, sociologists, and others. It is significant to observe that many O.R. programs in universities are taught by industrial engineering faculty and many O.R. programs are taught by mathematics and statistics faculty. Some universities have individual programs called Operations Research that are taught by Operations Research faculty.

14.3. Nature of Operations Research

Since we have already touched upon this subject, let's explore the concept of mathematical orientation in some depth. We have said that a chief distinguishing characteristic of O.R. is the mathematical involvement. Let's look at some of the models we have already studied under industrial engineering.

14.3.1 Economic Order Quantity

In Chapter 7 we developed a mathematical model depicting the total variable cost of an inventory situation. That equation is repeated below:

$$TC = PC\frac{D}{Q} + CC\frac{Q}{2} \tag{14.1}$$

We took a first derivative, set it equal to 0, and solved for the optimum order quantity Q_0

$$Q_0 = \sqrt{\frac{(2)(PC)(D)}{CC}} \tag{14.2}$$

14.3.2 Plant Location

In Section 4.3 we developed a mathematical programming formulation of a plant location problem and stated that it could be solved by using a technique consisting of linear programming and branch and bound search (although we did not study them). That formulation is repeated below:

$$\text{Minimize } TC = \sum_{i=1}^{m}\sum_{j=1}^{n} c_{ij}x_{ij} + \sum_{j=1}^{n} k_j y_j \tag{14.3}$$

subject to

$$\sum_{i=1}^{m} x_{ij} \leqslant my_j \qquad (j = 1, \ldots, n) \tag{14.4}$$

$$\sum_{j=1}^{n} x_{ij} = 1 \qquad (i = 1), \ldots, m) \tag{14.5}$$

$$y_j = 0, 1 \qquad (j = 1, \ldots, n) \tag{14.6}$$

$$x_{ij} \geqslant 0 \qquad (i = 1, \ldots, m \text{ and } j = 1, \ldots, n) \tag{14.7}$$

14.3.3 Job Evaluation

In both point rating and factor comparison methods of job evaluation we assumed an underlying mathematical model that says that a measure of a job's total complexity can be obtained by *summing* the measures of complexity for each subfactor.

14.3.4 Quality Control

In the construction of $\bar{x}$ charts we assumed the probability distribution of sample means can be represented by the normal distribution where the normal distribution is defined (in Appendix A) as

$$f(x) = \frac{1}{\sqrt{2\pi}\,\sigma} e^{-(x-\mu)^2/2\sigma^2}$$

14.3.5 Others

This discussion could be extensively continued. The transportation programming problem used in Chapters 4 and 5, the assignment problem used in Chapter 5, the computerized plant layout packages (scoring) of Chapter 4, and the traveling salesman problem and routing problems of Chapter 5 are just a few other examples of industrial engineering techniques with mathematical orientations.

Given all these formulations, would you say that they are industrial engineering or that they are operations research? Actually, they are both and would appear in the shaded area of Figure 14.1. A flow process chart might be an example of a technique that could be called industrial engineering, but not operations research (although we could argue that a flow process chart is a *scientific model* of a real-world situation). Research into new mathematical techniques could be called O.R. but not I.E., although the research is often done by industrial engineers.

14.4. Categorization of Operations Research

Let's explore the nature of operations research in an attempt to categorize the techniques. In every situation that we wish to model using an O.R. approach we encounter the problem of estimating the values of the parameters (factors such as the price of raw material or time to produce a part). Those parameters may not be constant over time. That is, they may behave as random variables or perhaps change in some predictable fashion. One approach is to forget that they are random variables and use a mathematical model that does not recognize variation. This approach is called the *deterministic* approach and is often used. Actually, all models we have seen thus far have been deterministic. If the model recognizes this random variation, the approach is called the *probabilistic* approach. We can categorize O.R. techniques then as deterministic or probabilistic, and this is done in many I.E. and O.R. curricula in which the first course(s) explores deterministic models.

As an example (grossly oversimplified) of these two approaches, consider the inventory problem of Chapter 7. In that chapter we used a deterministic model to determine the optimum order quantity Q_0. Now let's determine when to order if we assume that there is a lead time (delay from the time the order is issued until the material is received) involved. We shall use both a deterministic and a probabilistic approach.

14.4.1 Deterministic Approach

The model in Chapter 7 represented the inventory situation when there was zero lead time and no variation in demand. If we assume that the lead time can be specified by l and that it is deterministic (does not change), we can draw the figure for stock on hand as shown in Figure 14.2. The figure simply shows that the time to issue an order must be exactly l time units before the stock on hand runs out. This also specifies the inventory at that reorder time as

$$\text{RQ} = lD$$

where l is the lead time in units of time and D is the demand per unit of time.

14.4.2 Probabilistic Approach

Now let's assume that the demand is still constant and deterministic but that the lead time is a random variable with a mean of l. This means that we can't be sure how long after we issue an order the order will be received, but we do know that it averages l time units.

Figure 14.3 depicts the situation that might occur if we calculated the reorder quantity and time as before. We can see from Figure 14.3 that many things can occur when random variation is allowed. In the inventory situation the results usually include excess inventory at times and stockouts at other times (Figure 14.3 assumes that the customer waits and his order is filled as soon as the material arrives). This inventory model will be revisited and expanded in Chapter 15.

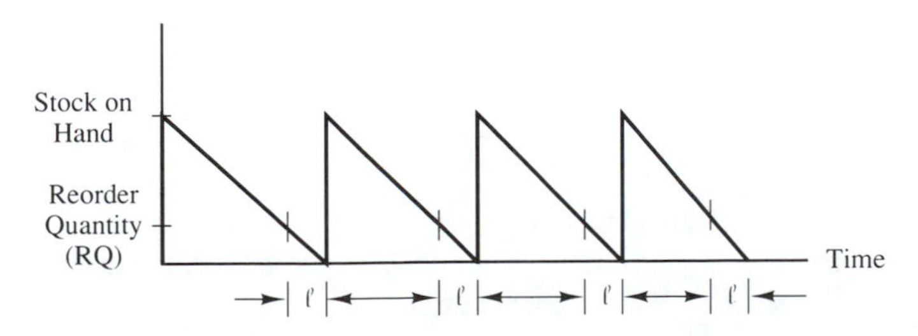

Fig. 14.2. Deterministic lead time.

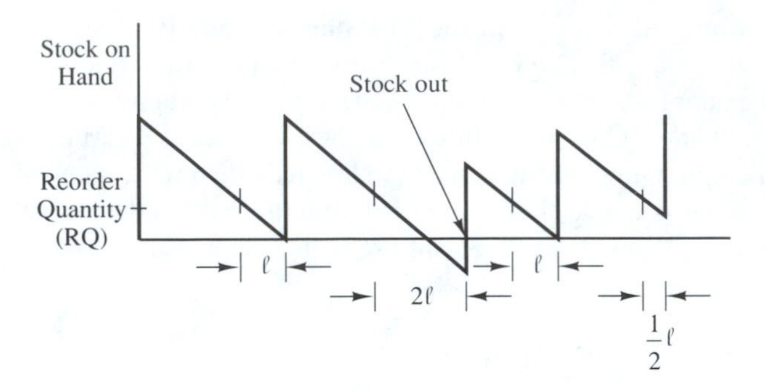

Fig. 14.3. Probabilistic lead time.

Any time random variation is allowed, the model is probabilistic; any time random variation is not allowed, the model is deterministic. The distinction will be used throughout the rest of this book. This chapter discusses some elementary and basic deterministic models. Chapter 15 briefly covers some of the probabilistic models of O.R., while Chapter 16 introduces simulation using probabilistic models.

14.5. Deterministic Operations Research

In this section and the rest of the chapter we shall cover some of the deterministic tools of operations research. We certainly cannot even mention all the areas of deterministic O.R., but we can provide a brief contact with almost all of the important ones and maybe a little insight into a few of the simpler ones. Almost all industrial engineering curricula contain several courses that cover the material presented.

14.6. Mathematical Programming

It is possible to present one general problem formulation that generalizes all deterministic operations research techniques. That problem formulation is called the *mathematical programming problem* and is presented below:

$$\text{Optimize } f(x) \tag{14.8}$$

subject to

$$g_i(x) \leqslant b_i \qquad (i = 1, \ldots, m) \tag{14.9}$$

Here, we have a mathematical function $[f(x)]$ that is to be optimized in some manner (probably minimized or maximized). The function involves some decision variables x, where x is a vector of N components $(x_1,x_2, \ldots ,x_N)$. There are m constraints

$$g_i(x) \leqslant b_i \qquad (i = 1, \ldots ,m)$$

that restrict the choice or combination of variables so that the solution is nontrivial. (Of course, m could be zero and no constraints could be present such that an unconstrained problem like the inventory control order quantity problem results.)

There could also be other constraints. For example, some or all of the x's could be restricted to integer quantities or just to values of 0 or 1 (as in the plant location problem formulated in Chapter 4). The decision variables could be restricted to only positive values (as is usually the case), or other special restrictions could be present. Finally, the constraints are not limited to less than or equal to inequalities. For example,

$$x_1 + x_2 \geqslant 5 \tag{14.10}$$

can be shown as

$$-x_1 - x_2 \leqslant -5 \tag{14.11}$$

and still meet the definition.

14.7. Unconstrained Optimization

The unconstrained optimization problem is one studied in calculus and is simply a mathematical programming problem where $m = 0$ (no constraints are present). The inventory control problem was just that, since the objective was to minimize the total cost equation given by

$$TC = PC\frac{D}{Q} + CC\frac{Q}{2} \tag{14.12}$$

The only constraint present is that positive quantities of Q are required, but we solved the problem without recognizing the constraint and found that there would always be a positive answer.

The unconstrained optimization problem is

$$\text{Optimize } f(x) \tag{14.13}$$

with no constraints. Calculus says that a *necessary condition* for x^* to be an optimum point of $f(x)$ is that the first derivative must be equal to zero.

$$\left.\frac{df(x)}{dx}\right|_{x = x^*} = 0 \tag{14.14}$$

A point x^* satisfying Equation (14.14) is called a *stationary point*. Since we know that it is possible for the derivative to be equal to zero and the point to be a point of inflection, the first derivative being equal to zero is not enough. We then say it is *sufficient* for x^* to be an optimum point if the second derivative at x^* is positive or negative. The point x^* is a minimum point if the second derivative is positive and a maximum point if the second derivative is negative. If the second derivative is zero, then higher-order derivatives must be taken until the first nonzero one is found at, say, the *n*th derivative.

$$\left.\frac{d^n f}{dx^n}\right|_{x = x^*} \neq 0 \tag{14.15}$$

(1) If n is odd, x^* is a point of inflection.
(2) If n is even and

$$\left.\frac{d^n f}{dx^n}\right|_{x = x^*} < 0 \tag{14.16}$$

then x^* is a local maximum. If

$$\left.\frac{d^n f}{dx^n}\right|_{x = x^*} > 0 \tag{14.17}$$

then x^* is a local minimum.

As an example, consider the inventory control problem of Equation (14.12). The *necessary* condition is that

$$\left.\frac{dTC}{dQ}\right|_{Q = Q^*} = \frac{-(PC)(D)}{Q^2} + \frac{CC}{2} = 0 \tag{14.18}$$

and the *sufficient* condition for Q^* to be a minimum point is that

$$\left.\frac{d^2TC}{dQ^2}\right|_{Q = Q^*} = \frac{2(PC)(D)}{Q^3} > 0 \tag{14.19}$$

which is greater than zero for all positive quantities of Q. Therefore, any positive Q^* that satisfies the necessary conditions of Equation (14.18) is a minimum point.

As another example, suppose that the percent of undesirable by-product in a chemical reaction is a function of time as represented by[1]

$$f(t) = 3t^4 - 4t^3 + 2 \tag{14.20}$$

Suppose further that we wish to minimize this percent. The necessary condition is that

$$\frac{df(t)}{dt} = 12t^3 - 12t^2 = 0 \tag{14.21}$$

There are two points that satisfy this relationship:

$$t = 0 \quad \text{and} \quad t = 1$$

[1]Extracted from William W. Claycombe and William G. Sullivan, *Foundations of Mathematical Programming* (Reston, Va.: Reston Publishing Co., Inc., 1975).

The second derivative is given by

$$\frac{d^2f(t)}{dt^2} = 36t^2 - 24t \tag{14.22}$$

At $t = 1$,

$$\frac{d^2f(t)}{dt^2} = 12 \tag{14.23}$$

Therefore, $t = 1$ is a *minimum point*.

At $t = 0$,

$$\frac{d^2f(t)}{dt^2} = 0 \tag{14.24}$$

Therefore we don't know what it is and we should take higher-order derivatives until one is nonzero.

At $n = 3$ and $t = 0$,

$$\frac{d^3f(t)}{dt^3} = 72t - 24 = -24 \tag{14.25}$$

Now, since $n = 3$ is odd, the rule states that $t = 0$ is a point of inflection.

For a geometrical interpretation of the above, consider Figures 14.4, 14.5 and 14.6. In all the figures, x^* satisfies the necessary condition that the first derivative equals zero. In Figure 14.4 the second derivative is positive at x^*; therefore, x^* is a minimum point. Note also that the second derivative would be positive at all points. A function like this is said to be *convex* and any stationary point is a minimum point for a convex function.

In Figure 14.5 the second derivative is negative for all values of x; therefore, any stationary point should be a maximum. There is only one stationary point and that is x^* which is a maximum point. A function whose second derivative is negative for all values of x is said to be *concave*.

Figure 14.6 is slightly different. We can see that for any $x < x^*$, the second derivative is negative; for any $x > x^*$, the second derivative is positive; and at

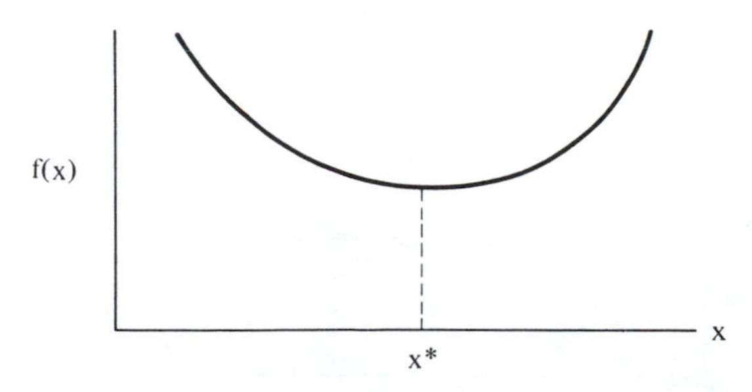

Fig. 14.4. Minimum point-single variable.

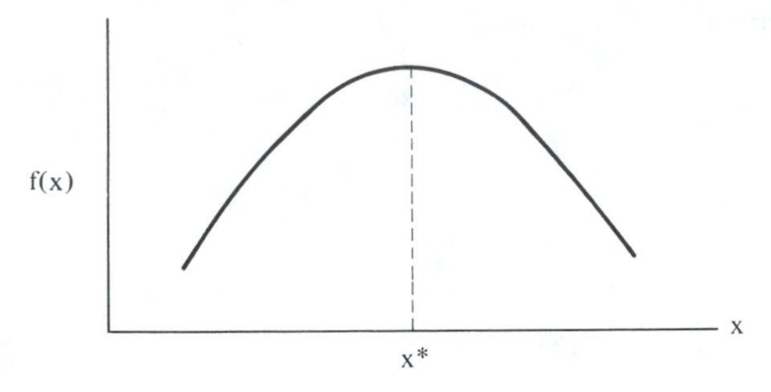

Fig. 14.5. Maximum point-single variable.

$x = x^*$, the second derivative is zero. If continued, we would find that some higher-order nth derivative, where n is odd, evaluated at x^* would be nonzero, signifying that x^* is a point of inflection, which it is.

Finally, it is possible for a function to have several minima, maxima, and points of inflection. Figure 14.7 demonstrates this possibility.

In this figure x_1^*, x_2^*, x_3^*, and x_4^* are all stationary points (first derivative = 0). Points x_1^* and x_3^* are minimum points (second derivative positive), x_2^* is a maximum point (second derivative negative), and x_4^* is a point of inflection (second derivative equal to zero and some higher-odd order derivative not equal to zero). Of course, if the function is convex (concave) for all values of x, there can only be one stationary point and that point is a unique minimum (maximum). When there is only one minimum (maximum), that minimum (maximum) is called the *global* minimum (maximum). When there is more than one minimum (maximum), the smallest of all the minima (largest of all the maxima) is also called the *global* minimum (maximum).

For functions of more than one variable, the same essential relationships hold. For example, the necessary condition for a vector

$$x^* = (x_1, x_2, \ldots, x_n) \tag{14.26}$$

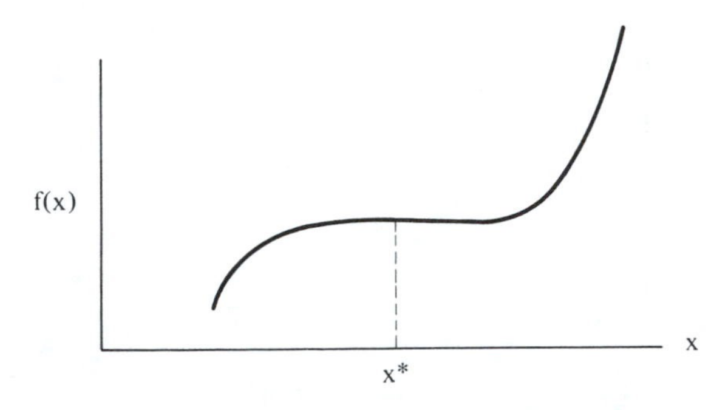

Fig. 14.6. Point of inflection—single variable.

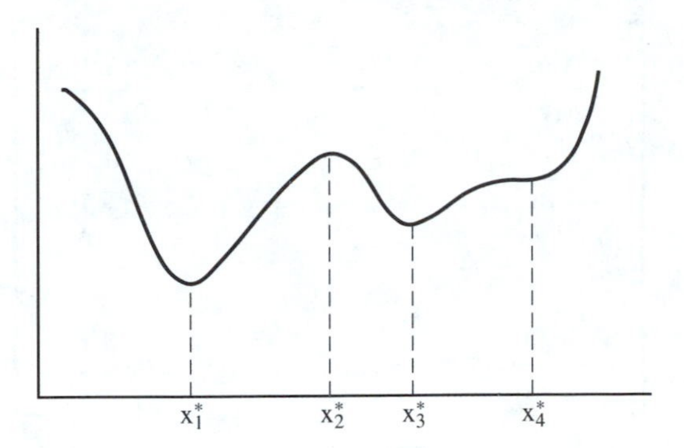

Fig. 14.7. Function with multiple stationary points.

to be a stationary point of a function $F(x)$, all elements of the vector of first partial derivatives (called the *gradient*) evaluated at x^* must be zero.

$$\nabla F(x) = \left(\frac{\partial F(x)}{\partial x_1}, \frac{\partial F(x)}{\partial x_2}, \ldots, \frac{\partial F(x)}{\partial x_n}\right)^T = 0 \qquad (14.27)$$

As an example, suppose that the cost of producing a product in a continuous process is a function of the amount of material added (x_1) and the temperature (x_2) as shown in

$$TC = 2x_1^2 - 2x_1x_2 + x_2^2 - 4x_2 + 5 \qquad (14.28)$$

The necessary condition [Equation (14.27)] states that

$$\frac{\partial TC}{\partial x_1} = 4x_1 - 2x_2 = 0 \qquad (14.29)$$

and

$$\frac{\partial TC}{\partial x_2} = -2x_1 + 2x_2 - 4 = 0 \qquad (14.30)$$

Solving these two equations simultaneously yields

$$x_1^* = 2 \quad \text{and} \quad x_2^* = 4 \qquad (14.31)$$

or

$$x^* = (2, 4) \qquad (14.32)$$

Now, we know that there is only one stationary point, but we do not know whether it is a minimum, maximum, or point of inflection (a point of inflection in more than one dimension is often called a *saddle point*). To find this, we need to evaluate the Hessian matrix, which is the matrix of second partials as shown in Figure 14.8.

The sufficiency test involves testing this Hessian matrix for something called

$$\begin{bmatrix} \frac{\partial^2 f}{\partial x_1^2} & \frac{\partial^2 f}{\partial x_1\, \partial x_2} & \cdots & \frac{\partial^2 f}{\partial x_1\, \partial x_n} \\ \frac{\partial^2 f}{\partial x_2\, \partial x_1} & \frac{\partial^2 f}{\partial x_2^2} & \cdots & \frac{\partial^2 f}{\partial x_2\, \partial x_n} \\ \vdots & \vdots & \cdots & \vdots \\ \frac{\partial^2 f}{\partial x_n\, \partial x_1} & \frac{\partial^2 f}{\partial x_n\, \partial x_2} & \cdots & \frac{\partial^2 f}{\partial x_n^2} \end{bmatrix}$$

Fig. 14.8. Hessian matrix.

positive or negative definiteness (note the similarity to the one-dimensional case). If the matrix is positive definite at a stationary point x^*, then x^* is a minimum point; if the matrix is negative definite, then x^* is a maximum point. Testing for this definiteness is beyond the scope of this book, but it involves a check of the eigenvalues of the matrix or calculation of determinants of the principle minor submatrices. As for the single variable case, we can show that if the Hessian matrix is positive (negative) definite for all values of x, then any stationary point is a minimum (maximum). The function is then said to be convex (concave).

Going back to the example problem of Equation (14.28), we can see that the Hessian matrix is

$$\begin{bmatrix} 4 & -2 \\ -2 & 2 \end{bmatrix}$$

We can show that this matrix is positive definite for all values of x. Therefore, the function is convex and any stationary point is a minimum point. The point

$$x^* = (2, 4) \tag{14.33}$$

is a minimum point.

Sometimes the methods of calculus covered above are not applicable for one of several possible reasons. The most frequently occurring is that the mathematical model of the system is not known. For example, a new chemical process may have several variables, including items such as composition, temperature, and humidity and to run the process may be extremely costly. Therefore, it is virtually impossible to run the system often enough to be able to *fit* a mathematical function to the results. It is necessary, however, to be able to determine *good* if not *optimum* values of the variables. This also occurs when a system is being *simulated* on a computer (to be covered in Chapter 16). Another problem may occur when the mathematical function can be stated, but taking derivatives, etc., would be very difficult. In situations like these the techniques of *search theory* are often used.

Search theory, as the name implies, is a *search* over the feasible domain of the variables in an effort to determine *good* values for the variables. There are two essential search procedures—those that use derivatives and those that don't. Search procedures that use derivatives are limited in scope because derivatives aren't always available and they are more difficult to explain than the procedures that don't require derivatives. Consequently, we shall cover only that class of search procedures that does not require derivatives.

For one-dimensional problems, there are several very efficient search procedures available, for example, Bolzano, Fibonacci, paired block, and golden section. We shall cover here only the paired block procedure. Suppose that we have a production process whose total cost is completely dependent on the temperature of the process and suppose that we wish to minimize this total cost by choosing a good value for the temperature. Suppose also that it is possible to say that the temperature must lie between 50 and 150°F and our controls are capable of distinguishing only full degree differences, not fractional ones. (Minimum spacing for distinguishing between values is called *resolution*.) The paired block procedure places two observations as close together as possible (i.e., one resolution apart) in the middle of the range. In our example the process should be run at 99.5 and 100.5°F for the first two trials. Since we can only run at full degrees, we arbitrarily choose to run at 100 and 101°F, yielding the results shown in Figure 14.9 in which the height of the vertical line represents the total cost of the process at those points. Obviously, 101°F is cheaper and, therefore, most likely (definitely if the function is convex) the best value will lie between 100 and 150°F. We should then try the process at two points close to the middle of the remaining range or

$$100°\text{F} + \left(\frac{1}{2}\right)(150°\text{F} - 100°\text{F}) = 125°\text{F}$$

Arbitrarily, we choose to try 124 and 125°F (125 and 126°F would have worked just as well). The results are shown in Figure 14.10. From these results we can now say that the optimum value should lie between 100 and 125°F. We repeat this process until we can do it no more or we are satisfied with the results.

The simplest of the nonderivative oriented search procedures for multivariate functions is called *sectioning* or *one-at-a-time*. In sectioning, the searcher starts at a point and proceeds in one coordinate direction[2] (holding all other variables constant) in iterative steps until the value observed stops improving. The searcher then chooses another coordinate direction and repeats the process, again holding all other variables constant. This is repeated until successive searches on all N variables do not result in any improvements or the searcher is satisfied with the results. As an example, consider the two variable functions shown in Figure 14.11 in which the starting point is shown and the circular contour lines are various levels of the function. (Imagine a top view of a bowl ''cut'' at various levels.)

---

[2]One step in the positive direction can be tried first; if it does not improve the value, then one step in the negative direction can be attempted.

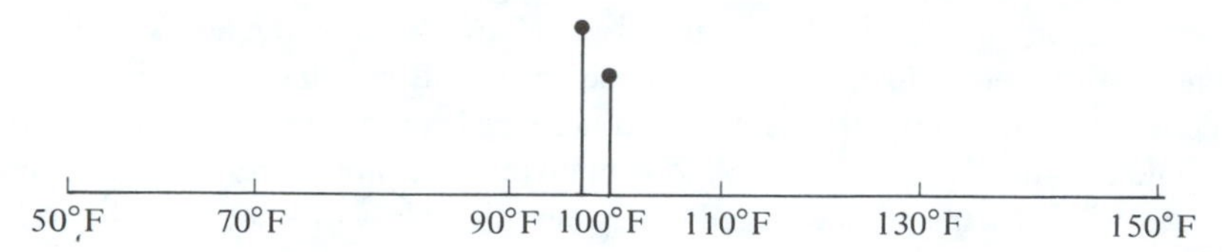

Fig. 14.9. First results of paired block search.

In this case, the searcher started at x^0, held x_2 constant, proceeded in the positive x_1 direction until the function stopped improving at point x^1. Then the searcher proceeded in the positive x_2 direction holding x_1 constant at x^1 and found the optimum point. Sectioning will always yield the optimum answer for perfect functions such as the one shown in Figure 14.11 in N moves, where N is the number of variables (two in this case). Usually, the function is not so well-behaved and many moves are necessary before the searcher is satisfied. Also, each move entails the evaluation of many points as small steps are taken.

As a matter of interest, the function shown in Figure 14.11 is a quadratic sum of squares with no interactive terms and equal multipliers such as

$$x_1^2 + x_2^2 \quad \text{or} \quad 2x_1^2 + 2x_2^2$$

14.8. Linear Programming

One of the first operations research methods was that of linear programming invented by George Dantzig in 1947. Linear programming is still a widely used operations research technique. Many packaged programs are available for solving linear programming problems and the basic theory is very simple, which probably accounts for its popularity.

To define linear programming (L.P.), we must go back to Equations (14.8) and (14.9) that define the general mathematical programming problem. If the objective function $f(x)$ and the constraint set functions $g_i(x)$ are all linear or first-order (the exponent of all variables equals one), then the problem is L.P. Also, since we normally deal with positive values of variables, we can add another constraint set

$$x_i \geqslant 0 \qquad (i = 1, \ldots, N) \tag{14.34}$$

Restating (to show that all functions are linear), we have the following as the L.P. formulation:

$$\text{Optimize } C'X \tag{14.35}$$

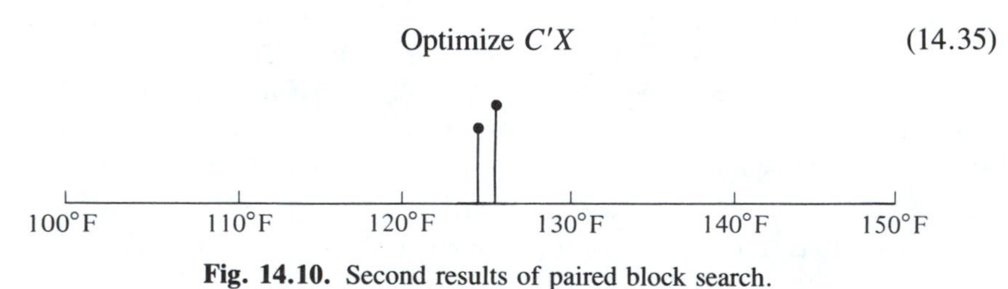

Fig. 14.10. Second results of paired block search.

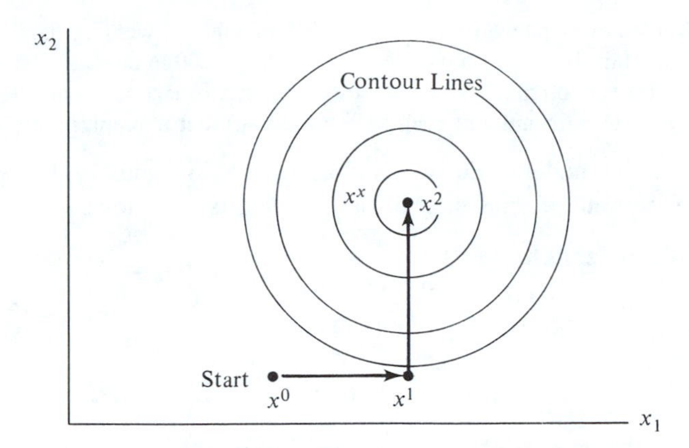

Fig. 14.11. Sectioning procedure.

subject to

$$GX \leqslant B \tag{14.36}$$

$$X \geqslant 0 \tag{14.37}$$

As illustrations, Figure 14.12 presents two linear programming problems and two problems that cannot be classified as L.P. because of the underlined elements that are not linear.

Example 14.1[3]

Gadgets, Inc., is trying to decide how many of each of two products are to be manufactured. A single unit of Gadgets model I requires 2.4 minutes of punch press time and 5.0 minutes of assembly time and yields a profit of \$0.80 per unit. A single unit of model

Linear Programming Problems	Nonlinear Programming Problems
Min. $5x_1 + 3x_2$	Min. $5x_1 + 3x_2$
S.T. $x_1 + x_2 \leqslant 5$	S.T. $\underline{x_1 x_2} \leqslant 5$
$x_1 - x_2 \leqslant 7$	$x_1 - x_2 \leqslant 7$
$x_1 \geqslant 0 \; x_2 \geqslant 0$	$x_1 \geqslant 0 \; x_2 \geqslant 0$
Max. $3x_1 - x_2$	Max. $3x_1 - x_2^2$
S.T. $x_1 + x_2 \leqslant 4$	S.T. $x_1 + x_2 \leqslant 4$
$x_1 - 2x_2 \leqslant 1$	$\underline{\text{Tan } x_1} - 2x_2 \leqslant 1$
$x_1 \geqslant 0 \; x_2 \geqslant 0$	$x_1 \geqslant 0 \; x_2 \geqslant 0$

Fig. 14.12. Sample mathematical programming problems.

[3]Partially extracted from W. J. Fabrycky, P. M. Ghare, and P. E. Torgersen, *Applied Operations Research and Management Science* (Englewood Cliffs, N.J.: Prentice Hall, 1983).

II requires 3.0 minutes of punch press time and 2.5 minutes of welding time and yields a profit of $0.70 per unit. If the punch press department has 1,200 minutes available per week, the welding department 600 minutes, and the assembly department 1,500 minutes per week, what is the product mix (quantity of each to be produced) that maximizes profit?

Following the model of Equations (14.35), (14.36), and (14.37), we see that the linear programming formulation of the problem is as follows:

$$\text{Maximize } 0.8x_1 + 0.7x_2 \tag{14.38}$$

subject to

$$2.4x_1 + 3.0x_2 \leq 1{,}200 \tag{14.39}$$

$$5.0x_1 \leq 1{,}500 \tag{14.40}$$

$$2.5x_2 \leq 600 \tag{14.41}$$

$$x_1 \geq 0 \quad x_2 \geq 0 \tag{14.42}$$

Here x_1 is the number of units of model I to produce and x_2 is the number of units of model II. The objective function [Equation (14.38)] says that the total profit is to be maximized. Constraints 1, 2, and 3 [Equations (14.39), (14.40), and (14.41)] state that the punch press, assembly, and welding departments have only so many hours available per week. The last constraints [Equation (14.42)] say that only positive quantities of products can be produced.

It is a relatively simple matter to solve this problem because it is linear and there are only two variables. A two-variable problem can be drawn as shown in Figure 14.13. Each constraint is first drawn as an equation (straight line); then the appropriate side (''less than'' in this case) is dashed as shown. Finally, the intersection of these areas is the ''feasible region'' for the problem and is shaded in Figure 14.13. For any point to be a possible solution to this problem, that point must lie within the feasible region (shaded area in the figure).

The profit for the problem is given by

$$TP = 0.8x_1 + 0.7x_2 \tag{14.43}$$

This defines a straight line for any value of profit reflecting the different combinations of x_1 and x_2 that yield that value of profit. The straight lines for profits of $100 and $200 are shown in Figure 14.13 by dashed lines.

It should be obvious that the total profit increases as x_1 and x_2 either move away from the origin or increase in value. Also, the *TP* lines are parallel. Since the solution must lie within the shaded region, the optimum point should be the point where the maximum *TP* line just touches the feasible region. With the slope of these *TP* lines, the maximum point should be point *A* where two of the constraints meet. Those two constraints are

$$2.4x_1 + 3.0x_2 = 1{,}200 \tag{14.44}$$

$$5.0x_1 = 1{,}500 \tag{14.45}$$

Solving these simultaneously shows that point *A* is located at (300, 160) and that the total profit at that point is

$$0.8(300) + 0.7(160) = \$352 \tag{14.46}$$

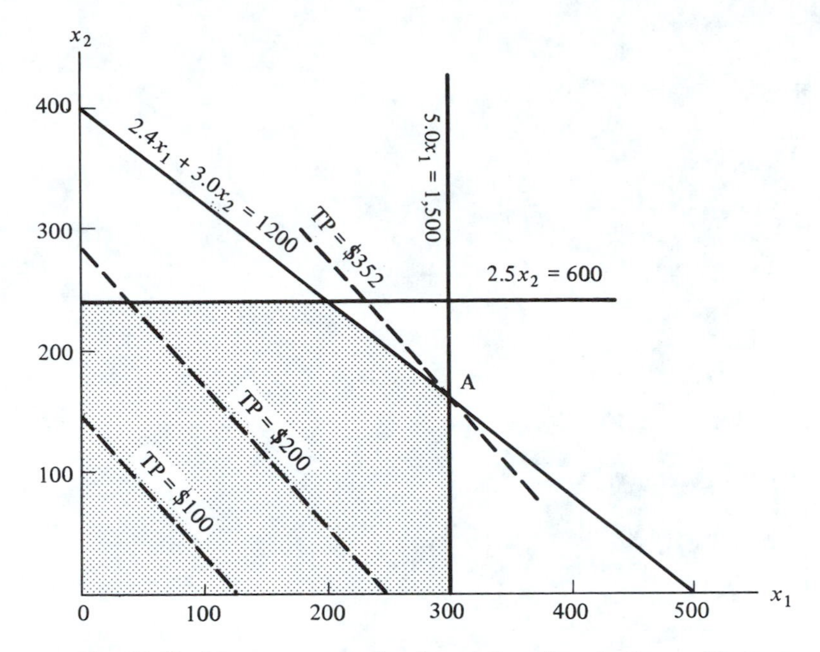

Fig. 14.13. Linear programming formulation of production problem.

That *TP* line is shown by the dashed line segment in Figure 14.13.

A minimization problem is very similar to a maximization problem except that the constraints are usually reversed (greater than or equal to) and it is desirable to bring the total profit line as *close* to the origin as possible. For example, Figure 14.14 represents the following linear programming problem:

$$\text{minimize total cost } TC = 5x_1 + 4x_2 \tag{14.47}$$

subject to

$$5x_1 + x_2 \geqslant 5 \tag{14.48}$$

$$x_1 + 5x_2 \geqslant 5 \tag{14.49}$$

The optimum point is at (5/6, 5/6), which yields a total cost of 7.5.

It is possible, but very difficult, to draw a linear programming problem of three variables (we shall not attempt one here). With three variables, the objective function is a *plane* instead of a straight line and the plane must be moved until it strikes an extreme point (edge or vertice) of the feasible region that yields an optimal answer.

Obviously, since almost all linear programming problems have more than two or three variables, how can we solve them? Since few, if any, L.P. problems are ever solved graphically, the graphical analysis of Figures 14.13 and 14.14 is good only for demonstration. Actually, almost all L.P. problems are solved by using the simplex method.

The simplex method is a mathematical procedure that systematically examines

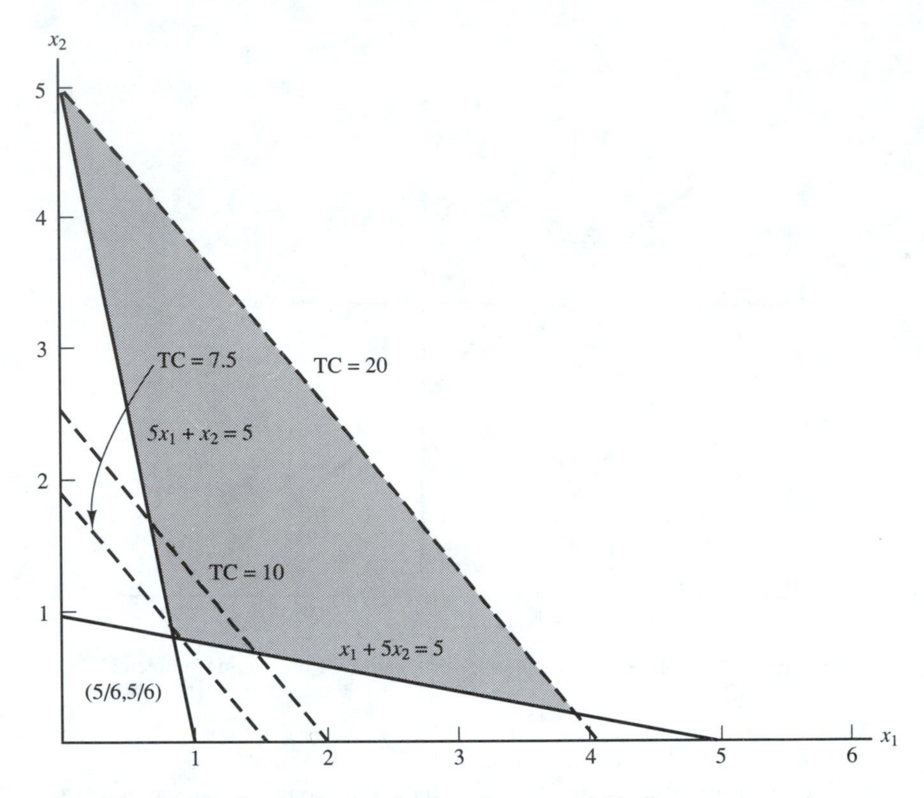

Fig. 14.14. Graphical minimization of a two-variable linear programming problem.

the extreme points of the feasible region and locates an optimum value. We can prove that at least one extreme point is an optimum for any linear programming problem; hence, the simplex method can concentrate on extreme points only. The examples of Figures 14.13 and 14.14 also *show* that the optimum must lie on a boundary, meaning that at least one optimum point is an extreme point.

Although there are many programs available for the simplex method, some people write their own programs because the simplex method is very simple to learn. The simplex method is covered in depth in industrial engineering courses and will not be covered here. Instead, let's look at some specialized linear programming problems that can be solved more easily.

14.8.1 Assignment Problem

The assignment problem is perhaps the simplest of all linear programming problems both to understand and to solve. We encountered the assignment problem in Chapter 5 when we were talking about material handling and transportation, but we did not complete the solution procedure. We shall complete the procedure here, but, first, some review may be helpful.

The assignment problem is encountered any time there are N items or services available at N locations and N other locations (some may be similar) that require one and only one of these N items or services. As an example, suppose that we have four products that can be produced by any one of four machines, and that any machine can produce only one product. Furthermore, because of relative efficiencies on the different products, the cost varies between the assignments as shown in the cost matrix below. The objective is *to assign* the jobs to the machines in order to minimize total cost.

It would be possible to solve the problem above by using a mathematical programming package, but a more efficient method, the *Hungarian method*, has been developed. The first part of the Hungarian method was presented in Chapter 5 and is repeated here for convenience.

(1) Subtract the smallest number in each row from every entry in that row.
(2) In the matrix resulting from step 1, subtract the smallest number in each column from every entry in that column.
(3) In the matrix resulting from step 2, a zero cost assignment (if it can be made) is optimal. If not, complete the Hungarian method.

	Machines			
	A	B	C	D
Jobs 1	16	14	15	18
2	12	13	16	14
3	14	13	11	12
4	16	18	15	17

As step 3 implies, the procedure above frequently is all that is necessary; it is impossible, however, to find a zero cost assignment for the reduced matrix shown in Figure 14.15. Here the assignment matrix has been reduced as far as possible with no zero cost assignment available.

To solve this, it is necessary to enter the next phase of the Hungarian method, which attempts to create more zeros in the matrix so that a zero cost assignment can be made. The steps of the second phase of the Hungarian method are as follows:

	1	2	3	4	5
1	7	3	2	0	0
2	4	0	1	1	1
3	0	0	0	1	1
4	1	1	1	3	0
5	1	1	1	6	0

Fig. 14.15. Reduced assignment matrix with no zero cost assignment.

(4) Draw the minimum number of horizontal and vertical lines (no diagonals) that intersect all the zeros (four in this case as shown below).

	1	2	3	4	5
1	7	3	2	0	0
2	4	0	1	1	1
3	0	0	0	1	1
4	1	1	1	3	0
5	1	1	1	6	0

(5) Subtract the smallest uncrossed number from every other uncrossed number and add it to all elements where the vertical and horizontal lines intersect. (The smallest uncrossed number is 1.)

	1	2	3	4	5
1	7	3	2	0	1
2	4	0	1	1	2
3	0	0	0	1	2
4	0	0	0	2	0
5	0	0	0	5	0

(6) If a zero cost assignment can be made, it is optimal. If not, repeat steps 4 and 5 until one can be made. (In this case, a zero cost assignment can be made as shown by the circled items below. That solution is optimal.)

	1	2	3	4	5
1	7	3	2	(0)	1
2	4	(0)	1	1	2
3	0	0	(0)	1	2
4	(0)	0	0	2	0
5	0	0	0	5	(0)

$\Rightarrow$

	1	2	3	4	5
1				1	
2		1			
3			1		
4	1				
5					1

Solution

The assignment problem is not a difficult problem to solve manually even when the second phase of the Hungarian method is required. It is extremely difficult, however, to program a computer to follow this procedure and do it efficiently.

We have mentioned previously that unbalanced assignment problems (number of sources does not equal number of destinations) can be solved by adding dummy rows or columns. The procedure is very simple, for all we need do is add enough dummy sources or destinations to "balance" the problem. The costs associated with these rows or columns are all zeros to ensure that the solution is not affected.

Figure 14.16 presents two unbalanced assignment problems that have been balanced so that they can be solved by using the standard procedure.

Finally, the objective is sometimes maximizing as opposed to minimizing. In the previous example since max $f(x)$ yields the same *solution point* as min $-f(x)$, the change for maximization is very simple. We replace all the cost elements by their

Unbalanced

	1	2	3	4	5
1	5	3	4	1	6
2	2	5	6	7	3
3	5	2	1	6	4
4	4	5	1	2	1

Add Dummy Row

Balanced

	1	2	3	4	5
1	5	3	4	1	6
2	2	5	6	7	3
3	5	2	1	6	4
4	4	5	1	2	1
5	0	0	0	0	0

	1	2	3
1	5	3	1
2	1	4	2
3	3	1	4
4	6	5	6
5	3	6	2

Add Two Dummy Columns

	1	2	3	4	5
1	5	3	1	0	0
2	1	4	2	0	0
3	3	1	4	0	0
4	6	5	6	0	0
5	3	6	2	0	0

Fig. 14.16. Unbalanced assignment problems.

negative and continue with the minimization procedure. As an example, consider the problem regarding the assignment of salesmen to territories, shown in Figure 14.17, in which each element in the matrix shows profits and the objective is to assign the salesmen to maximize profits.

First, we must replace all elements with their negative. Then we must solve the problem by using the Hungarian method as shown in Figure 14.18. Note that only the first phase of the Hungarian method is necessary and that the solution involves assigning salesman 4 to territory 1, salesman 3 to territory 2, salesman 2 to territory 3, and salesman 1 to territory 4.

14.8.2 Transportation Problem

We have already seen an example of the use of the transportation formulation. It was presented in Chapter 4 and involved the shipment of product from M sources to N destinations.

The transportation problem is actually similar to the assignment problem in that there are M sources with something available and N destinations needing something. The difference is that under the transportation formulation, each source may have more than one unit available and each destination may need more than one unit.

The mathematical formulation is presented below for a typical minimization transportation problem:

$$\text{Minimize } E = \sum_{i=1}^{M} \sum_{j=1}^{N} e_{ij} x_{ij} \tag{14.50}$$

subject to

$$\sum_{i=1}^{M} x_{ij} = b_j \quad (j = 1, \ldots, N) \tag{14.51}$$

$$\sum_{j=1}^{N} x_{ij} = a_i \quad (i = 1, \ldots, M) \tag{14.52}$$

Territory

Salesmen		1	2	3	4
	1	5	3	5	4
	2	2	1	6	3
	3	5	4	6	2
	4	6	3	1	4

Entries Are In 100,000's of Dollars

Fig. 14.17. Maximization assignment problem.

where

$$\sum_{i=1}^{M} a_i = \sum_{j=1}^{N} b_j \tag{14.53}$$

e_{ij} = cost to ship 1 unit from i to j
x_{ij} = number of units shipped from source i to destination j (integers only)
a_i = number units available at i
b_j = number units needed at j

The objective equation (14.50) states that the cost of transporting the units increases linearly with the number of units such that the total cost is obtained by multiplying the cost per unit (e_{ij}) by the number of units (x_{ij}). The first set of constraints [Equation (14.51)] states that each destination needs exactly b_j units so that the sum of units shipped from all sources to the destination must be b_j. Similarly, since each source has exactly a_i units available, the total shipped out must be exactly a_i [Equation (14.52)]. Finally, the algorithm for optimally solving the transportation problem requires that the total number of units available must equal the total number needed [Equation (14.53)]. If this is not true, it can be forced by adding a dummy row or column with zero cost elements and required number of units to balance. For examples of the formulation presented in Equations (14.50) through (14.53), refer to Chapter 4.

The only solution procedure presented thus far is the least cost assignment whereby the analyst chooses the least available cost and assigns as much as possible.

	1	2	3	4	Smallest Number
1	−5	−3	−5	−4	−5
2	−2	−1	−6	−3	−6
3	−5	−4	−6	−2	−6
4	−6	−3	−1	−4	−6

⇒

	1	2	3	4
1	0	2	0	1
2	4	5	0	3
3	1	2	0	4
4	0	3	5	2
Smallest Number	0	2	0	1

⇒

	1	2	3	4
1	0	0	0	(0)
2	4	3	(0)	2
3	1	(0)	0	3
4	(0)	1	5	1

Fig. 14.18. Maximization assignment problem solution.

Next, the analyst corrects the source availabilities and destination requirements and repeats the procedure until all allocations have been made. It should be intuitively obvious that this procedure does not yield an optimum answer but that it does present a very good starting solution that may be improved.

As an example, let us use the transportation problem and solution demonstrated in Figure 4.6 and repeated in Figure 14.19. Although there are other shorter ways, the easiest procedure to understand is to check each combination that is *not* currently in the solution to see if the solution is improved by including it in the solution.

For example, the Tulsa–Washington (1, 1) combination is not in the solution. If one unit were shipped from Tulsa to Washington, then a unit would have to be subtracted from the Tulsa–Lincoln (1, 3) combination or the Tulsa–San Francisco (1, 4) combination to balance the load. If one unit were subtracted from the Tulsa–Lincoln (1, 3) combination, then it would have to be added somewhere in column 3. There is no way of doing this because there are no more entries in column 3. (We cannot *add* a new one since we can prove that the transportation solution can have at *most* $M + N - 1$ combinations in the solution, 7 in this case. Therefore, the only elements that can be added to or subtracted from, *except for the one being examined*, must already be in the solution.) For example, considering the shipment of one additional unit from Tulsa to either Washington, Cleveland, or Phoenix results in cost d_{ij} as follows:

$$d_{11} = c_{11} - c_{14} + c_{34} - c_{31} = 80.00 - 78.00 + 74.50 - 70.80 = \$5.70$$

$$d_{12} = c_{12} - c_{14} + c_{34} - c_{32} = 78.00 - 78.00 + 74.50 - 72.00 = \$2.50$$

$$d_{15} = c_{15} - c_{14} + c_{24} - c_{25} = 77.00 - 78.00 + 71.50 - 70.70 = -\$0.20$$

To / From	1. Washington	2. Cleveland	3. Lincoln	4. San Francisco	5. Phoenix	Plant Capacity
1. Tulsa	80.00	78.00	77.00 4,000	78.00 3,000	77.00	7,000
2. Tempe	76.50	75.00	73.50	71.50 2,500	70.20 3,000	5,500
3. Sparta	70.80 5,000	72.00 6,000	75.00	74.50 1,500	75.00	12,500
	5,000	6,000	4,000	7,000	3,000	

Fig. 14.19. Solution to transportation problem by using least cost assignment.

Shipping a unit from Tulsa to Phoenix (1, 5) would improve the solution by $0.20. Since it is logical to ship as many as possible, we would have to look at the other elements affected. To add to (1, 5), we would have to subtract from (2, 5), add to (2, 4), and subtract from (1, 4). We can see, then, that we can ship 3,000 units from 1 to 5, none from 2 to 5, 5,500 from 2 to 4 (2,500 + 3,000), and none from 1 to 4, yielding the solution shown in Figure 14.20.

To check for optimality, the calculations for all the empty combinations would have to be done again. If all are positive, the solution is optimal. In this case, they are all positive. Therefore, the solution shown in Figure 14.20 is optimal.

We mentioned earlier that the transportation problem cannot have more than $M + N - 1$ positive entries in the solution matrix. It can, however, have fewer. In Figure 14.20 the solution has only $M + N - 2$ positive entries. This is okay, but it does cause some problems in trying to calculate optimality. We shall not discuss that here except to note that we might have difficulty showing that the solution in Figure 14.20 is optimal. It can be done, however.

14.9. Other Techniques

In this chapter we have only begun to skim the surface of deterministic operations research techniques. Only unconstrained optimization and linear programming have been discussed and they haven't been covered in any depth. Some of the other areas are mentioned below in just enough depth to describe them. Curricula in industrial engineering cover almost all of these techniques in detail.

14.9.1 Nonlinear Programming

We spent a large amount of time discussing linear programming because it is the simplest to understand. When the objective function and/or the constraints are

To / From	1. Washington	2. Cleveland	3. Lincoln	4. San Francisco	5. Phoenix	Plant Capacity
1. Tulsa	80.00	78.00	77.00 4,000	78.00	77.00 3,000	7,000
2. Tempe	76.50	75.00	73.50	71.50 5,500	70.70	5,500
3. Sparta	70.80 5,000	72.00 6,000	75.00	74.50 1,500	75.00	12,500
	5,000	6,000	4,000	7,000	3,000	

Fig. 14.20. Revised solution to transportation problem.

nonlinear, the problem is much more difficult to solve and is called nonlinear programming. This terminology covers a large area that actually includes some of the other techniques described below. Basically, nonlinear programming includes any mathematical programming problem that does not have linear constraints and objective function.

14.9.2 Integer Programming

Sometimes a mathematical programming problem has the decision variables limited to integer values. These problems, whether linear or not, are called integer programming problems. The transportation problem is a linear integer programming problem since fractional units cannot be shipped.

14.9.3 Zero–One Programming

When an integer programming problem has the integer variables limited to the values of zero or one, it is called a zero–one programming problem. The assignment problem is an example of zero–one programming.

14.9.4 Quadratic Programming

Sometimes a problem has linear constraints but a quadratic (second-order) objective function. This is called quadratic programming.

14.9.5 Geometric Programming

Often, a mathematical programming problem has a structure such that it can be solved by using a technique called geometric programming. The structure is too complex to explain here.

14.9.6 Other Programming

Some of the other techniques encountered in the study of deterministic operations research methods are *dynamic programming*, *separable programming*, *branch and bound search*, and *mixed integer programming*. Although this list is not exhaustive, it does give an idea of the extensive field covered by deterministic operations research.

14.10. Impact of Computers

It should be obvious that almost all tools of operations research require some mathematical sophistication and a large amount of computation capability and storage space. Without the speed of computers *almost* all operations research techniques would not be applicable to real-world problems. With the storage available on modern computers and the rapid computation time, many techniques can be used today that were only good for textbook discussion just a few years ago.

As an example, linear programming problems have been solved with many hundreds of variables and almost that many constraints. This would be completely impossible without computers. Software programs also exist for all the areas mentioned in this chapter.

Also, the unbelievable growth in power and affordability of personal computers have made larger O.R. models more feasible for the average industrial engineer. It would be rare indeed for an industrial engineer today not to have access to sufficient computing power to enable him or her to use most O.R. models in day-to-day operations. The future is practically unlimited.

DISCUSSION QUESTIONS

1. Define operations research and state how it is similar to and different from industrial engineering.
2. Distinguish between deterministic and probabilistic models of operations research.
3. It has been said that *everything* varies. Since deterministic models do not allow for random variation, why are they still very useful?
4. Why is the large-scale or systems approach so important in industrial engineering and operations research?
5. Define the mathematical programming problem.
6. State the rules for finding the extreme points of a function of one variable and for telling whether the extreme points are minima, maxima, or points of inflection.
7. Graph the general shape of a quadratic function of one variable that has a positive second derivative for all values. Repeat for a quadratic function that has a negative second derivative for all values.
8. Draw a function that has stationary points at which the second derivative is positive, negative, and zero. Clearly label these points.
9. Distinguish between local and global optima.
10. Draw a function for which a search procedure would probably miss finding the global maximum or minimum.
11. Describe *paired block* and *sectioning* search procedures. State when a search procedure might be appropriate.
12. Define a linear programming problem and give a real example. Discuss the objective function and constraints.
13. Define an assignment problem and give a real example.
14. Define a transportation problem and give a real example.
15. Give three examples of tools for decision making that could be called "operations research" or "industrial engineering."

PROBLEMS

1. Assume a situation in which the demand per year for an item is 980 units. The carrying costs are $20 per unit per year and procurement costs are $50 each time an order is issued. If lead time is 15 days, what amount should be ordered each time and what is the reorder point?

2. For the facts given in Problem 1, determine how many short or over the inventory will be when the new order is received if lead time is *twice* as long as usual. Repeat this for *half* as long as usual.

3. Find the extreme points of $2x^3 - 7x + 3$ and state whether they are minima, maxima, or points of inflection.

4. Find the extreme points of $2t^4 + 6t^3 - 5$ and state whether they are minima, maxima, or points of inflection.

5. Graph Problems 3 and 4 and show where the extreme points are. (This would be a good opportunity to use a spreadsheet.)

6. Calculate the Hessian matrix for $x_1^3 + x_2^3 + x_2x_3^2 + 3x_3 - 4$. Evaluate it at the point (0, 1, 1).

7. Use a paired block search to determine the minimum of the equation $\text{TC}(x) = 3x^2 - 40x + 150$ in the range of $0 \leqslant x \leqslant 10$ using increments of 0.5.

Problems 8, 9, 11, and 13 adapted, with permission, from W. J. Fabrycky, P. M. Ghare, and P. E. Torgersen, *Industrial Operations Research* (Englewood Cliffs, N.J.: Prentice Hall, 1972).

8. Solve graphically for the values of x_1 and x_2 that

$$\text{minimize TC} = 7x_1 + 9x_2$$

subject to:

$$18x_1 + 7x_2 \geqslant 250$$
$$3x_1 + 5x_2 \geqslant 84$$
$$9x_1 + 2x_2 \geqslant 60$$
$$x_1 \geqslant 0 \qquad x_2 \geqslant 0$$

9. In the linear programming example given by Equations (14.38) through (14.42) determine the optimum point if the following constraint is added:

$$5x_1 - 2x_2 \leqslant 1{,}000$$

10. A small machine shop has capacity to do turning, milling, drilling, and welding. Two products are under consideration. Each will yield a net profit of $3.50 per unit and will require the following machine time:

	Product		
Operation	*1*	*2*	*Capacity (hours)*
Turning	0.092	0.248	32
Milling	0.224	0.096	16
Drilling	0.080	0.000	32
Welding	0.000	0.240	16

(a) Formulate this as a linear programming problem similar to Equations (14.38) to (14.42).

(b) Solve graphically for the number of units of each product that should be scheduled to maximize profit.

11. Six service technicians are available to assist five customers. The time in hours necessary to travel roundtrip is shown in the assignment matrix. Optimally solve the assignment problem.

Service Technician	*Customer*				
	1	*2*	*3*	*4*	*5*
1	5	4	3	6	7
2	2	1	5	6	6
3	3	8	7	2	5
4	5	1	1	7	8
5	2	4	3	7	8
6	8	7	3	2	3

12. The foreman of a machine shop has five machinists available to assign to jobs for the day. Five jobs are offered with the expected profit in dollars for each machinist on each job being as follows:

Machinist	*Task*				
	A	*B*	*C*	*D*	*E*
1	65	45	100	75	90
2	70	55	105	90	93
3	60	50	100	85	95
4	60	40	120	85	97
5	70	60	110	97	85

Determine the assignment of machinists to jobs that will result in maximum profit.

13. Optimally solve the following transportation problem to allocate parts available to locations needed:

From \ *To*	*1*	*2*	*3*	*4*	*5*	*6*	*Available*
1	2	7	3	5	3	2	11
2	5	7	4	6	3	1	21
3	1	2	5	6	7	4	16
4	3	6	7	5	4	6	31
Needed	9	4	14	9	19	24	

14. Three plants produce parts that are to be shipped to four distribution centers. Plants 1, 2, and 3 produce 24, 34, and 22 shipments worth of parts per month, respectively. Each distribution center needs to receive 20 shipments per month. The distance from each plant to each distribution center is given as follows:

Plant	*Distribution Center*			
	1	*2*	*3*	*4*
1	400	650	200	350
2	550	700	300	500
3	300	600	400	450

The transportation cost for each shipment is \$300 + \$0.75 per mile. How many shipments should be sent from each plant to each of the distribution centers to minimize total transportation costs? Before trying to solve this, be sure to formulate the problem as a transportation problem by constructing the appropriate cost and requirements table.

CHAPTER 15

Probabilistic Models

15.1. Introduction

Chapter 14 provided background into some of the deterministic areas of operations research, those primarily having to do with mathematical programming and search procedures. Very often, however, when we set out to develop a mathematical model of a real-world system, we find that certain important elements are random and we cannot satisfactorily ignore their variation. We then turn to probabilistic operations research models to help us quantify, evaluate, and optimize the system. This chapter includes three areas of probabilistic models including queueing, inventory, and Markov chains. Chapter 16 deals with simulation, which is applicable for use in virtually all areas of probabilistic modeling. The intent, then, of this chapter is to introduce probabilistic O.R. modeling, analysis, and design.

15.2. Queueing Theory

Queueing theory is the study of waiting lines. A waiting line, or queue, is a common occurrence whenever the service facility cannot keep pace with the demand for the service. We are involved in queueing almost every day of our lives. For example, waiting in a grocery checkout line, buying books at the beginning of a school term, or even getting a drink of water from a water fountain constitute queueing situations. In addition, organizations providing services are often called upon to make queueing-related decisions. The number of tool crib clerks to use in a large machine shop, the number of turnpike toll-takers, or the capacity of waiting space for in-process inventory items are examples.

Queueing theory is particularly interesting, for it is almost exclusively probabilistic. That is, no one is capable of predicting exactly when arrivals for service will take place and no one can predict exactly how long a service will take. Scheduling and appointments are often used to smooth the arrival pattern, for example, as in a doctor's office. Arrivals are often either early or late, however, and servers require variable amounts of time, resulting again in a probabilistic situation.

Queueing theory enables us to use mathematics to describe and make decisions about waiting lines under many different conditions. As an example, we could always please customers by providing an excess of servers in order to minimize delay or waiting time. Obviously, this would be costly. We could also provide minimal service and just let the customers wait in line. This, too, would be costly because customers would balk (not enter the line) or decide not to return in the future. We, therefore, seek a balance between these two extremes. Queueing theory provides us information for properly designing our service facility.

15.2.1 Queueing System Structure

A basic queueing system is made up of a *waiting line* and a *service facility*. Arrivals to the queueing system come from the *calling population*. These elements are illustrated in Figure 15.1. Each of these elements has characteristics that define the queueing process.

The *population*, for example, can be infinite (practically speaking, very large) or finite. It is characterized by an arrival rate (λ) and an arrival distribution. It is much easier to model an infinite population instead of one that is finite, because with a finite population the arrival rate is affected by the number of units already in the queueing system. The arrival distribution can vary widely, but it is commonly assumed to be Poisson (i.e., the number of units arriving for service in an interval of time is Poisson distributed).

The *queue* may also be infinite or of limited size. For example, almost all ticket queues can grow as large as may ever be required. A "do-it-yourself" spray car wash, however, has limited waiting room for drivers waiting their turns. Again, the infinite queue is easier to model mathematically. The *service discipline* relates to the way in which units leave the queue for service. The order may be first-come-first-served, last-come-first-served, random, based on priorities, and so on. We usually assume first-come-first-served unless otherwise specified.

The *service facility* is characterized by the number and configuration of servers. There may be only one server or there may be several servers in parallel. If the service requires several steps, there may be servers in series. The service facility is also characterized by a rate of service (μ) and a distribution of service

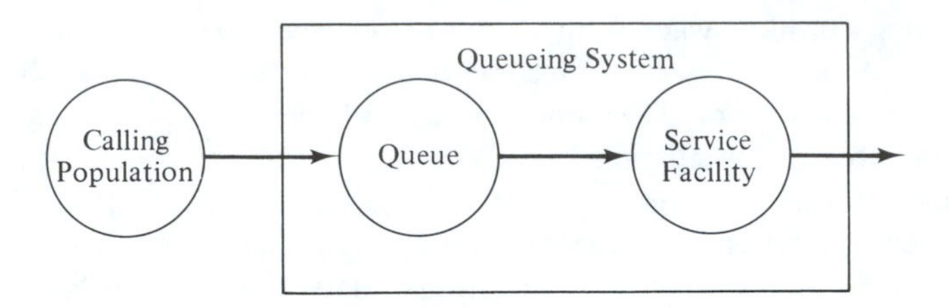

Fig. 15.1. Basic queueing system.

times. Frequently used service time distributions include the gamma (not treated in Appendix A), the exponential, and a constant service time.

15.2.2 Queueing Notation

In the remaining sections on queueing some of the basic, yet most frequently used, models are presented. The notation used in the equations may be summarized as follows:

$P(n)$ = probability that exactly n population units are in the queueing system, including both waiting line and service facility
λ = arrival rate in units per time
μ = service rate in units per time
N = expected number of units in the queueing system, including both waiting line and service facility
N_q = expected number of units in the waiting line
T = expected time in the queueing system, including both waiting line and service facility
T_q = expected time in the waiting line

In each of the following models there are two relationships between expected numbers of units and waiting times. Those relationships are as follows:

$$N = \lambda T \qquad (15.1)$$

and

$$N_q = \lambda T_q \qquad (15.2)$$

Thus, if we calculate the expected number of units, it is easy to get expected time, or vice versa. Also, a relationship between waiting time in the system and waiting time in the queue is as follows:

$$T = T_q + \frac{1}{\mu} \qquad (15.3)$$

This equation states that the expected time in the queueing system equals the expected time in the waiting line plus the mean service time.

15.2.3 Single-Service Channel

Three queueing models will be presented in this section. Each is for a service facility having but one server, an infinite population, and an infinite queue. We shall assume that the number of arrivals occurring over intervals of time are Poisson distributed. The distribution of service times, however, will be exponential, arbitrary, and constant, respectively, in the three sections.

Exponential Service Times. With Poisson arrivals and exponential service times, the probability of no units in the queueing system is

$$P(0) = 1 - \frac{\lambda}{\mu} \tag{15.4}$$

The probability of having a number n units in the system is

$$P(n) = P(0)\left(\frac{\lambda}{\mu}\right)^n \tag{15.5}$$

The expected (average) number of units in the queueing system is

$$N = \frac{\lambda}{\mu - \lambda} \tag{15.6}$$

and the expected time in the system is

$$T = \frac{1}{\mu - \lambda} \tag{15.7}$$

Similarly, the expected number of units in the queue is

$$N_q = \frac{\lambda^2}{\mu(\mu - \lambda)} \tag{15.8}$$

and the expected waiting time there is

$$T_q = \frac{\lambda}{\mu(\mu - \lambda)} \tag{15.9}$$

As an example, suppose that the number of arrivals is Poisson distributed with a mean of 15 per hour. Service times are exponentially distributed with a mean of 3 minutes each (20 services per hour). The probability that no units are in the system is

$$P(0) = 1 - \frac{\lambda}{\mu} = 1 - \frac{15}{20} = 0.25$$

The probability that three units are in the system is

$$P(3) = P_o\left(\frac{\lambda}{\mu}\right)^n = 0.25\left(\frac{15}{20}\right)^3 = 0.1055$$

The other queueing system characteristics of interest include the following:

$$N = \frac{\lambda}{\mu - \lambda} = \frac{15}{20 - 15} = 3 \text{ persons in the system}$$

$$T = \frac{1}{\mu - \lambda} = \frac{1}{20 - 15} = 0.2 \text{ hour in the system}$$

$$N_q = \frac{\lambda^2}{\mu(\mu - \lambda)} = \frac{15^2}{20(20 - 15)} = 2.25 \text{ persons in the queue}$$

$$T_q = \frac{\lambda}{\mu(\mu - \lambda)} = \frac{15}{20(20 - 15)} = 0.15 \text{ hour in the queue}$$

Also, the server is utilized a fraction ρ of the time where

$$\rho = \frac{\lambda}{\mu} = \frac{15}{20} = 0.75$$

or 75% of the time.

Arbitrary Service Times. For arbitrary service times, the mean of the service time distribution, $1/\mu$, and its variance, σ^2, must be known. The actual distribution, however, does not need to be known or stated. The following results apply when $\lambda/\mu < 1$:

$$N = \frac{\lambda^2\sigma^2 + \left(\frac{\lambda}{\mu}\right)^2}{2\left(1 - \frac{\lambda}{\mu}\right)} + \frac{\lambda}{\mu} \tag{15.10}$$

$$T = \frac{N}{\lambda} = \frac{\lambda\sigma^2 + \frac{\lambda}{\mu^2}}{2\left(1 - \frac{\lambda}{\mu}\right)} + \frac{1}{\mu} \tag{15.11}$$

$$N_q = N - \frac{\lambda}{\mu} = \frac{\lambda^2\sigma^2 + \left(\frac{\lambda}{\mu}\right)^2}{2\left(1 - \frac{\lambda}{\mu}\right)} \tag{15.12}$$

$$T_q = \frac{N_q}{\lambda} = \frac{\lambda\sigma^2 + \frac{\lambda}{\mu^2}}{2\left(1 - \frac{\lambda}{\mu}\right)} \tag{15.13}$$

Again consider the previous example where arrivals are Poisson with λ = 15. Let service times be of unknown distribution, but have a mean of 3 minutes each (20 services per hour) and a standard deviation (σ) of 5 minutes (0.083333 hours) and hence $\sigma^2 = 0.006944$. Now

$$N = \frac{\lambda^2 \sigma^2 + \left(\frac{\lambda}{\mu}\right)^2}{2\left(1 - \frac{\lambda}{\mu}\right)} + \frac{\lambda}{\mu} \doteq \frac{15^2\,(0.006944) + \left(\frac{15}{20}\right)^2}{2\left(1 - \frac{15}{20}\right)} + \frac{15}{20} = 5.0 \text{ persons}$$

and values of $T = 0.333$ hour, $N_q = 4.25$ persons, and $T_q = 0.283$ hour follow easily.

We should notice that when the service time distribution is exponential with mean $1/\mu$, and, therefore, with variance $\sigma^2 = 1/\mu^2$, our arbitrary service time distribution results revert to the exponential results in the previous section. Also, we can see that N, T, N_q, and T_q all increase as σ^2 increases. Thus, the variability of the server has an important effect on the performance measures of the queueing system.

Constant Service Times. Many manufacturing systems involve automated processing of certain operations. This automated processing often takes a constant amount of time to perform. In this case, arrivals to the work station may still be Poisson distributed, but service times will be of duration $1/\mu$, with no time dispersion, $\sigma^2 = 0$. In this case, our equations from the preceding section reduce to

$$N = \frac{\left(\frac{\lambda}{\mu}\right)^2}{2\left(1 - \frac{\lambda}{\mu}\right)} + \frac{\lambda}{\mu} \tag{15.14}$$

$$T = \frac{\frac{\lambda}{\mu^2}}{2\left(1 - \frac{\lambda}{\mu}\right)} + \frac{1}{\mu} \tag{15.15}$$

$$N_q = \frac{\left(\frac{\lambda}{\mu}\right)^2}{2\left(1 - \frac{\lambda}{\mu}\right)} \tag{15.16}$$

$$T_q = \frac{\dfrac{\lambda}{\mu^2}}{2\left(1 - \dfrac{\lambda}{\mu}\right)} \tag{15.17}$$

Continuing with our previous example having Poisson arrivals with $\lambda = 15$, but now with constant service at a rate of $\mu = 20$, our performance measures are: $N = 1.875$ persons, $T = 0.125$ hour, $N_q = 1.125$ persons, and $T_q = 0.075$ hour.

15.3. Inventory Control

In Chapter 7 you were introduced to the concept of inventory control, as well as a discussion of the relevant considerations and costs involved. Recall that there is much emphasis and interest today in minimizing inventories and storage by proper sizing of order quantities and maximizing turnover of stock. Inventory models help us determine the costs and consequences of various inventory levels. Some of the models presented in Chapter 7 were deterministic in nature. The demand for an item over a fixed interval of time or the lead time from order to receipt is not always constant, however. That is, demand may vary from day to day in accordance with a probability distribution. Similarly, lead time may be either short or long. If this variability is significant, it may be necessary for us to explicitly take the stochastic nature of demand and/or lead time into account in developing our mathematical model.

Three different inventory models will be discussed in the following sections. The first covers a single period with no setup cost. This is, perhaps, the simplest model that we can use to illustrate some of the considerations required in developing a probabilistic inventory model. The following two sections are devoted to the lot size-reorder point and periodic review models. For these we present only philosophy because of the difficulty of their development.

15.3.1 Single-Period Model—No Setup Cost

A classic problem often used to illustrate the nature of probabilistic inventory models is the single-period model with no setup cost. Units are purchased at a cost of C dollars per item to be sold during a single period for S dollars per item.The salvage value of any leftover items at the end of the period is equal to M dollars per unit. The quantity purchased at the beginning of the period is denoted by Q, and the number of units sold during the period is given by X. Of course, X represents demand for the items and is a random variable described by the distribution $p(X)$.

This model is applicable to the inventory control of an item that becomes obsolete at the end of the period. For example, bakery goods such as bread and donuts have their primary demand shortly after being made. They also have a "salvage" market at a reduced price if some remain unsold after the first day.

Christmas tree inventories are well described by this model, as are Valentine candy and seasonal greeting cards. This has also been called the "newsboy" model, because it applies to the inventory of newspapers that should be maintained by a newsstand on any given day. In each of these cases, the objective is to purchase the optimum quantity of items at the beginning of the period in order to maximize profits over the period.

Let's take a simple example and assume that we belong to a charitable organization which decides to operate a Christmas tree stand to raise money for a worthy cause. We must place an order with a local Christmas tree supplier a month in advance for the number of trees we want to purchase at $10 each. We can then sell the trees individually for $18 each. Any trees left unsold will be removed after Christmas at a cost of $1 each. Note that this represents a negative "salvage value," or $M = -1$. We estimate that from 201 to 300 trees can be sold, with the probability of any number in this range being $p(X) = 0.01$. That is, we believe that a uniform distribution will apply with the probability of selling 201, 202, . . . , 300 trees each being 1/100.[1]

The number we order, Q, should be chosen carefully because for each tree demanded when we are out of stock, a profit of $8 is lost. For each tree left over, however, we have lost not only the $10 originally paid, but also $1.00 for removal. To better define the problem, we should explicitly state the costs and revenues involved as a function of demand, X. This is accomplished in Table 15.1.

It would be a simple matter to decide on a value of the order quantity, Q, and then for a particular demand, X, calculate what the profit would be. Unfortunately, that will not be of much use to us. Instead, since demand is unknown, we must write an expression for the *expected value* of profit for a given order quantity and then find the order quantity that maximizes the expected value.

We recall that the expected value of a random variable X having distribution $p(X)$ is written

$$E(X) = \sum_{X} Xp(X)$$

Similarly, we can write an expression for the expected value of a function of X, say $g(X)$, as

$$E\,[g(X)] = \sum_{X} g(X)p(X)$$

Since profit is a function of demand, we have

$$E(\text{profit}) = E[g(X)] = \sum_{X=0}^{Q} [SX + M(Q - X) - CQ]\, p(X) + \sum_{X=Q+1}^{\infty} (SQ - CQ)\, p(X) \qquad (15.18)$$

[1]This distribution is chosen here for its simplicity rather than for its reality.

Table 15.1. SINGLE-PERIOD INVENTORY COSTS AND REVENUES

Demand X	Revenue	Cost	Profit $g(X)$
$X \leq Q$	$= SX + M(Q - X)$	CQ	$SX + M(Q - X) - CQ$
	$18X - (Q - X)$	$10Q$	$18X - (Q - X) - 10Q$
$X > Q$	$= SQ$	CQ	$SQ - CQ$
	$18Q$	$10Q$	$18Q - 10Q$

In our example problem, since $p(X)$ is defined only over $X = 201, \ldots, 300$, expected profit may be written as follows:

$$E(\text{profit}) = E[g(X)] = \sum_{X=201}^{Q} [19X - 11Q]\, p(X) + \sum_{X=Q+1}^{300} 8Qp(X)$$

One way to optimize the order quantity is to try different values of Q in the expression above. For example, let us see whether we prefer to order $Q = 250$ or $Q = 275$ trees.

At $Q = 250$:

$$E(\text{profit}) = \sum_{X=201}^{250} [19X - 11(250)]p(X) + \sum_{X=251}^{300} 8(250)p(X) = \$1{,}767.25$$

At $Q = 275$:

$$E(\text{profit}) = \sum_{X=201}^{275} [19X - 11(275)]p(X) + \sum_{X=276}^{300} 8(275)p(X) = \$1{,}672.75$$

Obviously, we would rather order 250 trees than 275, but is 250 the true optimum? Continuing as we have above, we would try to find the value of Q which gives a higher expected profit than either $Q - 1$ or $Q + 1$.

Alternatively, we could difference[2] (Equation 15.18) and determine that the optimum order quantity is the largest Q for which

$$1 - F(Q - 1) \geqslant \frac{C - M}{S - M} \tag{15.19}$$

where $F(Q - 1)$ is the cumulative distribution function, or $\sum_{X=0}^{Q-1} p(X)$. For our example, this would be

$$1 - F(Q - 1) \geqslant \frac{10 - (-1)}{18 - (-1)} = \frac{11}{19} = 0.58$$

and the optimum order quantity is $Q_0 = 243$.

---

[2]Differencing a discrete function is akin to differentiating a continuous function.

Sure enough, if we calculate the expected value of profit for $Q = 242$, 243, and 244, we obtain \$1,772.41, \$1,772.43, and \$1,772.26, respectively.

15.3.2 Lot Size—Reorder Point Models

In the lot size–reorder point model an order for Q items is placed when the inventory level drops to a *reorder point*, r. This, of course, requires a data base in which we know the quantity of items remaining on-hand after each transaction. We saw an example of such a model in Figure 14.3 and an extension of that model is considered here.

The example shown in Figure 14.3 had a stockout condition resulting from probabilistic lead time. If we do not wish stockouts, we can reorder earlier than that given by the deterministic model by adding a quantity called safety stock (SS) to the reorder quantity (RQ) and reorder when the stock on hand reaches RQ′ where

$$RQ' = RQ + SS$$

Figure 15.2 shows what might happen in a situation like that in Figure 14.3 when safety stock is used. Notice that the shape of this graph is identical to that of Figure 14.3 except that it has been shifted up by an amount equal to the safety stock, so that stockouts do not occur as often (zero times in this case).

The amount of safety stock necessary can be determined in two ways. First, we can experiment with various levels of safety stock until we get the performance we desire, or we can determine the amount statistically. To determine the amount statistically, we only need to know the distribution of the lead time, its mean and standard deviation, and the service level desired where service level is defined as the percentage of time we will allow stockouts.

For example, suppose that the lead time is normally distributed with a mean of 20 days and a standard deviation of 2 days. Suppose further that management wants us to be sure to have stock available 98% of the time (stated another way, stockouts are allowed only 2% of the time). We can then go to a cumulative standard normal table and find the point that corresponds to 98%.[3] In the normal table of Appendix B we find that this corresponds to approximately $+2.05\sigma$. The reorder point then should be based on demand for a lead time of

$$20 + 2.05(2) = 24.1 \text{ days}$$

If demand is 10 items per day, the reorder point is

$$(24.1 \text{ days})(10 \text{ items/day}) = 241 \text{ units}$$

Another illustration of a lot size–reorder point model is shown in Figure 15.3. Note that the time from placement of the order to receipt of the order is again a random variable. Also, the wavy line indicates that demand is now also a random variable. Under these stochastic conditions, we see that sometimes there is stock

[3]Assuming that the demand rate is constant and lead time is normally distributed.

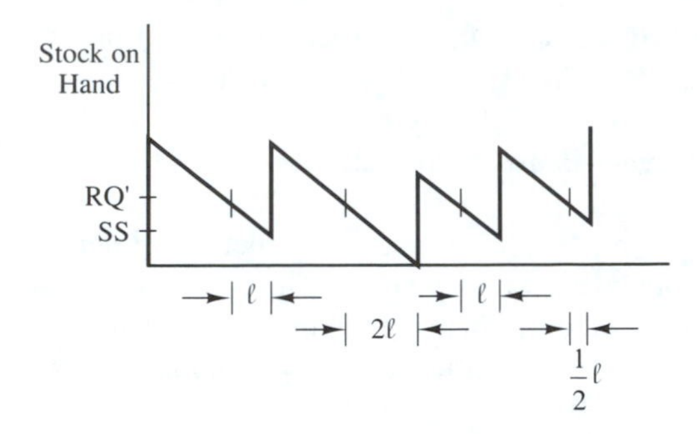

Fig. 15.2. Probabilistic lead time with safety stock.

on hand when a new order is received. At other times, stockouts and back orders or lost sales are incurred.

Expressing this model mathematically is difficult. It is comprised of a setup charge for placing the order (neglected on the single-period model in the previous section), an inventory carrying charge for stock on hand, and a good will or profit loss charge for shortages. The demand and lead time distributions are explicitly included in the model and conventional optimization methods are used to solve for Q and r—the optimum lot size and reorder point.[4]

15.3.3 Periodic Review Models

Rather than have the timing of orders be dictated by on-hand inventory reaching a reorder point, we can employ the *periodic review* strategy. This philosophy requires that we review a particular stock at intervals of time, T. If any units at all have been demanded since the last review, an order is placed. This time, however, the order is sized to raise our "on-hand *plus* on-order" inventory level to an amount R. This philosophy is illustrated in Figure 15.4.

We can see that the order quantity varies in this model, depending on the inventory level at the time our order is placed. Also, although on-hand inventory is not shown, we can have units on hand or shortages, depending on lead times and demands. Once again, our mathematical model reflects setup, inventory carrying, and shortage costs, which we seek to minimize by properly selecting the level R and the periodic review interval T.[5]

---

[4]A good discussion of this model is presented in G. Hadley and T. M. Whitin, *Analysis of Inventory Systems* (Englewood Cliffs, N.J.: Prentice Hall, 1963), pp. 159–175.

[5]This model is also discussed in Hadley and Whitin.

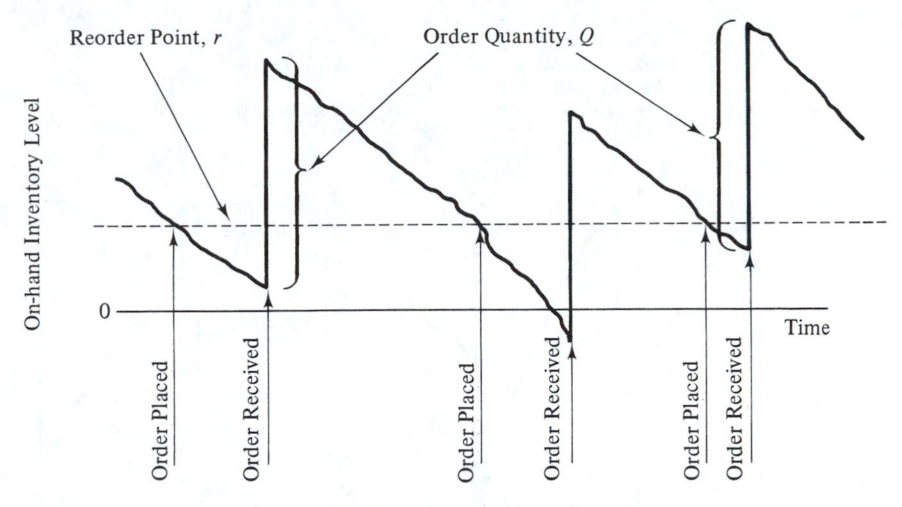

Fig. 15.3. Lot size–reorder point model.

15.4. Markov Chains

A *Markov chain* represents the behavior of a stochastic process[6] that can be in one and only one of M states, 1, 2, . . . , M, at time t, $t = 1, 2, \ldots$. The process is free to switch from one state to another at the next point in time. Such a change, say from state i to state j, between two consecutive points in time is done with *transition probability* p_{ij}.

We can represent a Markov chain as follows:

$$
\text{From state} \quad
\begin{array}{c}
 \\ 1 \\ 2 \\ \vdots \\ M
\end{array}
\overset{\text{To state}}{
\begin{array}{cccc}
1 & 2 & \cdots & M \\
\end{array}}
\begin{bmatrix}
p_{11} & p_{12} & \cdots & P_{1M} \\
p_{21} & p_{22} & \cdots & p_{2M} \\
\vdots & \vdots & & \vdots \\
p_{M1} & p_{M2} & \cdots & p_{MM}
\end{bmatrix}
$$

We should note several important facts about this chain. First, there is only a finite number of states. Second, the probability of making a transition from one state to

[6]A sequence of numbers representing inventory levels or inventory demands over consecutive days or weeks would be considered a stochastic process. Similarly, the numbers of arrivals to or services from a queueing system per unit time represent a stochastic process.

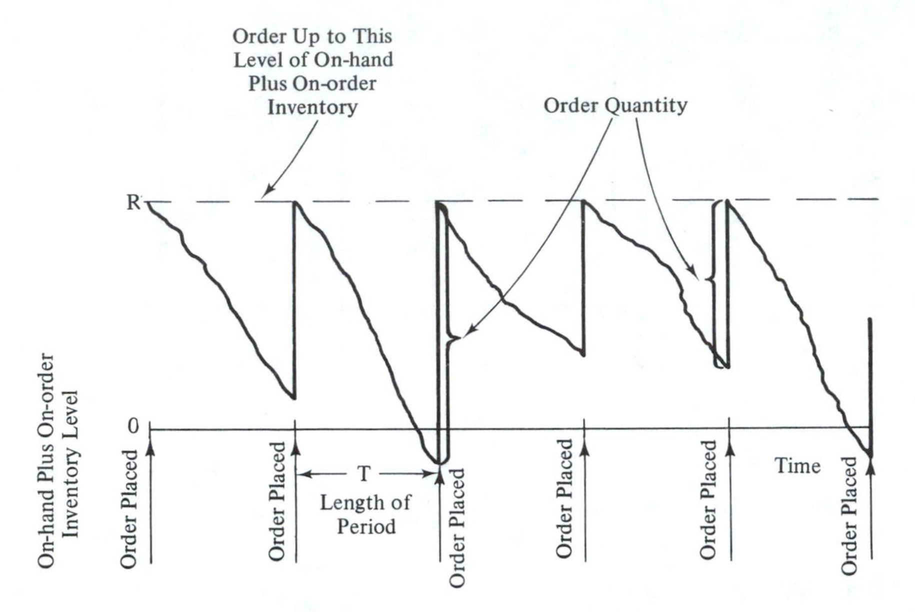

Fig. 15.4. Periodic review model.

another depends only on the present state occupied, not on any earlier history. Third, the transition probabilities p_{ij} are stationary and do not change over time.[7]

15.4.1 Regular Markov Chains

Let us now represent a communications problem as a Markov chain. Suppose that a weather forecast is passed along from person to person. The forecast has three possible states: sleeting (S), rainy (R), and cloudy (C).[8] The transition probabilities for two successive persons are given in the Markov chain transition matrix presented as follows:

$$\text{From state}\quad \begin{matrix} & \text{To state} & \\ S & R & C \end{matrix}$$

$$\begin{matrix} S \\ R \\ C \end{matrix} \begin{bmatrix} 0.6 & 0.3 & 0.1 \\ 0.2 & 0.5 & 0.3 \\ 0.1 & 0.2 & 0.7 \end{bmatrix}$$

We can see that if one person understands the forecast to be, say, rainy, there is a 0.2 chance that the next person told will understand the forecast to be sleeting. Or, there is a 0.3 probability that the forecast will be understood as cloudy. This leaves

---

[7]Markov chains receive an elementary treatment in John G. Kemeny et al., *Finite Mathematical Structures* (Englewood Cliffs, N.J.: Prentice Hall, 1959).

[8]The number of states is held to three to permit ease of presentation. More realistically, we should have several additional weather states or, alternatively, one additional state such as "other."

a 0.5 chance that the forecast will be understood correctly. Of course, there may be many factors that could cause misunderstanding of this sort. Once we understand the meaning and elements of the transition matrix, we can express it more simply as follows:

$$P = \begin{bmatrix} 0.6 & 0.3 & 0.1 \\ 0.2 & 0.5 & 0.3 \\ 0.1 & 0.2 & 0.7 \end{bmatrix}$$

It may be of interest to know the probability that state R is understood as state C after the communication has been through two persons. We can calculate this and all two-step transition probabilities by using a decision tree. Basically, a decision tree enumerates all possible interpretations that can be had at each step. A decision tree for our communications problem is shown in Figure 15.5. It is easy to see that we can go from a forecast of R to C in two steps via three different paths, R–S–C, R–R–C, and R–C–C. The probability of any particular path occurring is given at the end of each path and represents the product of the probabilities of the two transitions along the path. The overall probability that we go from state R to C in two steps is the sum of the probabilities of each path ending in C, or $0.02 + 0.15 + 0.21 = 0.38$.

An easier way to determine the two-step transition probabilities is to calculate the two-step transition matrix. This is done through the matrix multiplication of P times P. For example, P^2 is calculated as follows:

$$\underset{P}{\begin{bmatrix} 0.6 & 0.3 & 0.1 \\ 0.2 & 0.5 & 0.3 \\ 0.1 & 0.2 & 0.7 \end{bmatrix}} \times \underset{P}{\begin{bmatrix} 0.6 & 0.3 & 0.1 \\ 0.2 & 0.5 & 0.3 \\ 0.1 & 0.2 & 0.7 \end{bmatrix}} = \underset{P^2}{\begin{bmatrix} 0.43 & 0.35 & 0.22 \\ 0.25 & 0.37 & 0.38 \\ 0.17 & 0.27 & 0.56 \end{bmatrix}}$$

Now, we can read the two-step transition probability from a forecast of R to one of C directly as .38. In addition, though, we have the two-step transition probabilities from each state to every other state.

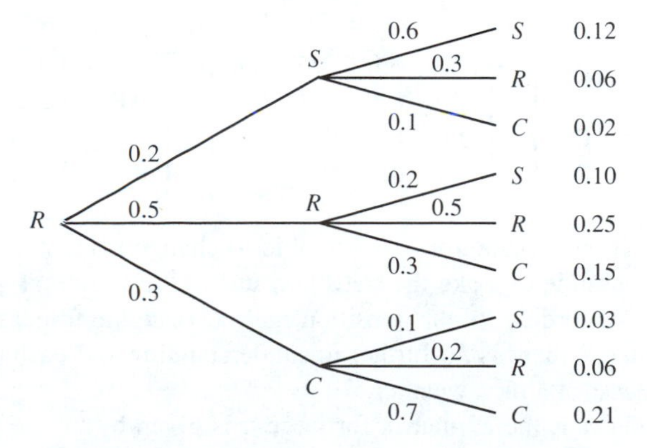

Fig. 15.5. Communications decision tree.

In general, we can calculate n-step transition matrices by using

$$P \cdot P^{n-1} = P^{n-1} \cdot P = P^n \tag{15.20}$$

If we were to calculate P^n for a large value of n in our communications example, the result would be

$$P^n = \begin{bmatrix} 0.265 & 0.324 & 0.412 \\ 0.265 & 0.324 & 0.412 \\ 0.265 & 0.324 & 0.412 \end{bmatrix} \text{ (for large } n\text{)}$$

This result is interesting because it points out the obvious: After our forecast has been passed along many times, the probability that the nth person understands it as sleeting, rainy, or cloudy is independent of the actual original forecast.

15.4.2 Absorbing Markov Chains

Often in modeling with Markov chains we have one or more absorbing states. When the process makes a transition to an absorbing state, it is impossible to leave that state. For example, let us suppose that another state is added to our communications problem. That state can be labeled "nice weather," or N, and it fits into a one-step transition matrix as follows:

$$\text{From state} \quad \begin{array}{c} \\ \end{array} \begin{bmatrix} & \text{N} & \text{S} & \text{R} & \text{C} \\ \text{N} & 1 & 0 & 0 & 0 \\ \text{S} & 0 & 0.6 & 0.3 & 0.1 \\ \text{R} & 0 & 0.2 & 0.5 & 0.3 \\ \text{C} & 0.1 & 0.1 & 0.2 & 0.6 \end{bmatrix}$$

Again, we can calculate the two-, three-, and n-step transition probabilities as before. For example, the two-step transition matrix would be as follows:

$$\underset{P}{\begin{bmatrix} 1.0 & 0.0 & 0.0 & 0.0 \\ 0.0 & 0.6 & 0.3 & 0.1 \\ 0.0 & 0.2 & 0.5 & 0.3 \\ 0.1 & 0.1 & 0.2 & 0.6 \end{bmatrix}} \times \underset{P}{\begin{bmatrix} 1.0 & 0.0 & 0.0 & 0.0 \\ 0.0 & 0.6 & 0.3 & 0.1 \\ 0.0 & 0.2 & 0.5 & 0.3 \\ 0.1 & 0.1 & 0.2 & 0.6 \end{bmatrix}} = \underset{P^2}{\begin{bmatrix} 1.0 & 0.0 & 0.0 & 0.0 \\ 0.01 & 0.43 & 0.35 & 0.21 \\ 0.03 & 0.25 & 0.37 & 0.35 \\ 0.16 & 0.16 & 0.25 & 0.43 \end{bmatrix}}$$

Here we see that even though it is impossible to change from state S or R to N in one step, it is possible to make the transition in two steps by first going to state C and then to N. According to our transition matrix, once the forecast is understood as "nice weather," there is no further misunderstanding and each communication conveys a forecast of "nice weather."

In the long run, the P^n matrix for large n is given by

$$P^n = \begin{bmatrix} 1.0 & 0.0 & 0.0 & 0.0 \\ 1.0 & 0.0 & 0.0 & 0.0 \\ 1.0 & 0.0 & 0.0 & 0.0 \\ 1.0 & 0.0 & 0.0 & 0.0 \end{bmatrix}$$

This verifies our intuition that, after passing through *many* intermediaries, our forecast will be understood as "nice weather" regardless of the original forecast.

15.5. Impact of Statistics and Computers

Nearly all industrial engineering programs require a strong statistics component in which variation is studied. This is important, since virtually everything in engineering (and even life in general) varies. The engineer who has a background in probability and statistics is far out ahead of the pack.

All probabilistic models make an attempt to model a real-world situation more accurately by explicitly addressing the inherent variation. To develop, analyze, or optimize the model, however, requires a background in probability and/or statistics. The added complexity also most often requires use of the computer for any except the most simplified problems. Fortunately, there are personal computer probabilistic O.R. software packages that provide help. Of course, the old standby, spreadsheets, can be used to do a great deal of the work on many small but realistic problems. Finally, when the probabilistic problem becomes so complex that it is not easily modeled analytically, we turn to simulation, the subject of Chapter 16.

DISCUSSION QUESTIONS

1. Describe either a discount store checkout area, post office, or concession stand at a football game in terms of the queueing system, including calling population queue, and service facility. Describe customer reaction to the queueing system (angry, leave queue, etc.). What "costs" are involved from the standpoint of the organization? How might you improve the queueing system considering the costs involved?
2. Suppose that one server works at the rate μ, two servers at the rate 2μ, three at 3μ, and so on. Is there anything wrong with always simply using the smallest number of servers (n) such that $n\mu > \lambda$?
3. Manager A says it is logical that the service rate μ and the arrival rate λ be equal. Manager B says that manager A is wrong—that bad things can happen except under "ideal" conditions.
   (a) Who is right? Why?
   (b) What are the "ideal" conditions referred to by manager B?
4. Why is it important that we know the service time distribution, or at least the variance of that distribution, in addition to knowing the mean service time $1/\mu$?

5. Suppose you operate a business that sells exotic coffees and candies from all over the world.
   (a) List three costs of always having excess inventory.
   (b) List three costs of often having shortages.
6. Select an original example to which the single-period model–no setup cost (Section 15.3.1) applies. Make realistic estimates of the costs involved. Describe, using a formula or a picture, what you think the "demand" distribution should look like (remember, the uniform distribution is not very likely).
7. What are the primary differences between the lot size–reorder point models and the periodic review models in inventory control?
8. For the lot size–reorder point model, what are two ways of being virtually certain that no shortages ever exist?
9. For the periodic review inventory model, what are two ways of being virtually certain that no shortages ever exist?
10. What would happen in any of the inventory models presented if the cost to place an order goes to zero, the lead time is constant, and demand is allowed to vary, but with demand known at least a lead time in advance?
11. Describe why it is possible for a Markov chain with one absorbing state to have an n-step (n large) transition matrix with a column of 1's and all other columns 0's. Is it logical that no matter what state you started in you will eventually be in the absorbing state? Is it possible to have multiple absorbing states?
12. Describe a two-stage gambling game (either a real one such as a lottery or one you make up) as a Markov chain. Describe the possible states and give a realistic transition matrix.
13. Describe some aspect of a sport as a Markov chain. Describe the possible states and give a realistic transition matrix. Is the chain absorbing?

PROBLEMS

1. Arrivals at a single self-service car wash arrive in a Poisson fashion at a rate of 10 per hour. Service time is exponential with a mean of 4 minutes. If we want to be confident that there is enough room for our self-service customers in the system at least 96% of the time, how many cars must we be able to accommodate in the queue?
2. The arrival rate of parts to a machinist is Poisson distributed with a mean of 8 per hour. A part can be machined in an average of 6 minutes, with an exponential distribution. Consideration is being given to replacing this with an automated machine that can machine each part in exactly 7 minutes. What will be the effect on the average number of parts in the queue? The effect on time in queue? What would happen to the average number in the queue if, using the automated machinery, the Poisson arrival rate were increased to 8.50 per hour to almost match the constant service rate?
3. Arrivals at a concession stand are Poisson distributed with a mean of 30 per hour. Service has a mean time of μ minutes. Graph N_q and T_q for values of μ = 1.0, 1.5, and 1.9 minutes, assuming that the service distribution is:
   (a) Exponential.
   (b) Constant.
   (c) Arbitrary with variance one-half that of the exponential variance.

4. Demand for a special health-food bread is uniform (equally likely) from 8 to 12 loaves per day. Our store can buy them for $3.00 and sell them for $7.00 per loaf. If, however, they are not purchased within 24 hours, they can be sold for $1.00 per loaf. How many loaves should our store purchase?

5. Now, suppose that demand for bread in Problem 4 is Poisson distributed with a mean of 10. How many should our store purchase?

6. Suppose that the demand in a lot size–reorder point model is 3 units per day. Lead time is 15 days and you want inventory items to be available (in stock) 97.73% of the time that customers demand them. What is the reorder point if the variance of the lead time is 0 days? If the variance is 4 days? If the variance is 16 days? What is the safety stock in each of the last two questions?

7. Two players each have two silver dollars. They role a die and if a 1 or 2 appears, person 1 takes one of person 2's dollars; if a 3 or 4 appears, vice versa; if a 5 appears, they flip a coin and person 1 takes one of person 2's dollars if a head is showing and person 2 takes one of person 1's dollars if a tail is showing; if a 6 appears, neither takes money from the other. All of the above constitutes one "roll." The game continues until one of the players has no money. Specify the Markov transition matrix for one "roll" of this game; clearly identify the states and probabilities. What is the probability that one player has $1.00 and the other has $3.00 after two rolls? After three rolls? (*Hint:* First prepare a decision tree such as in Figure 15.5.)

8. Two digits, i and j, can each take on the values 0 and 1. They are sent from one receiving station to another via a data link. If either i or j is sent as 0, there is a 0.90 chance that it will be received correctly as 0 instead of 1. If i or j is sent as 1, there is only a 0.70 chance that it will be received correctly. Thus, there are four combinations of ij which can be sent, and for each there are four combinations of ij which can be received. For example, a 10 sent is received as a 00, 01, 10, or 11 with probabilities $(0.3)(0.9) = 0.27$, $(0.3)(0.1) = 0.03$, $(0.7)(0.9) = 0.63$, and $(0.7)(0.1) = 0.07$. Develop the transition matrix. What is the probability that a 01 will be received as a 01 after passing over two data links? If a 01 is received as a 01 over two data links, does that mean it was passed correctly over each link? Explain.

9. A baseball team is composed of players each with a batting average of .250. Set up a Markov chain transition matrix where states represent the number of outs in an inning. What is the probability of no outs after two batters have been to the plate? Is this an absorbing chain? Explain. Assume there are only hits or outs—no walks, errors, etc.

CHAPTER 16

Simulation

16.1. Introduction

The operations research tools introduced in Chapters 14 and 15 are designed for use with problems for which tractable, closed-form solutions may be found. These are typically called *analytical solutions*. In engineering practice, many situations arise involving problems for which no analytical solution can be obtained. Analytical solutions have been derived successfully when dealing with narrowly defined problems residing in relatively small subsystems or components of the total organization. Many problems must be viewed in a larger context, leading to a degree of complexity beyond the ability of analytical solutions to handle.

The limitations of analytical solutions were one of the factors leading to the development of simulation as a means of dealing with complex problem situations. Another motivation for the development of simulation was the desire to be able to examine the details of the dynamics of a complex operating system. In particular, we can often gain valuable insight into system behavior by the detailed tracing of entities moving through the system.

System simulation has become the most widely used tool among industrial and systems engineers.

16.2. Simulation Examples

As an introductory example of simulation, consider the familiar experience of tossing a fair coin. The possible outcomes are "heads and tails," and each is equally likely to occur on a particular toss. Furthermore, the outcome of one toss will in no way affect the outcome of the next toss. Thus successive outcomes are considered *independent*.

Suppose that for some reason we are unable to obtain a fair coin. We do, however, have access to a list of digits (see Table 16.1) 0, 1, 2, 3, 4, 5, 6, 7, 8, 9 that are in completely random order. We can construct a *simulation experiment* that imitates the tossing of a fair coin by assigning exactly half of the digits to correspond to a "head," and the other half to correspond to a "tail."

The probability distribution for the tossing of a fair coin is shown in Figure 16.1a. The cumulative distribution function is shown in Figure 16.1b.

If we assign the digits 0, 1, 2, 3, 4 to correspond to the outcome "head" in our simulated coin-tossing experiment, and the digits 5, 6, 7, 8, 9 to correspond to the outcome "tail," we can generate as many trials as we wish by drawing random digits from our list of random digits in Table 16.1. Digits can be drawn from this table in a variety of ways. For example, we could use the extreme-left single-digit column. The first 10 digits are 3, 8, 0, 3, 0, 9, 9, 1, 2, 2, which would correspond to H, T, H, H, H, T, T, H, H, H. In this sample, we obtain 7 heads and 3 tails: not a very close approximation of the theoretical probabilities for the two outcomes, but an entirely plausible result if we were to toss an actual coin. The problem is one of sample size. The reader should continue down the column and verify the following results:

	Number of Samples				
$p(x)$	*10*	*20*	*30*	*40*	*50*
H	0.7	0.6	0.47	0.45	0.50
T	0.3	0.4	0.53	0.55	0.50

Now, consider the case of tossing an unfair coin. A very large number of experiments has established that this particular coin has a probability of 0.47 of coming up "head," 0.52 of coming up "tail," and 0.01 of landing on its edge. We clearly cannot simulate these probabilities using single-digit random numbers. We must, instead, use two-digit random numbers.

The probability distribution and cumulative distribution functions for this case are shown in Figure 16.2a and b. To perform our simulation experiments, we will use the extreme-left single-digit column and the digit in the adjacent column to form our two-digit random numbers. The first two are 36 and 89. The digits 00–46 will represent a "head," digits 47–98 will represent a "tail," and digit 99 will represent an "edge." The reader should verify the following outcomes:

Table 16.1. HAND-DRAWN RANDOM DIGITS

36550	98056	23736	94727	34879	47261	12538	25757	34258	50167	95802	50562	24401
89796	07620	47787	86649	70937	64585	36855	40407	28721	07273	23007	49725	37646
05983	50001	91826	82914	03235	41536	72104	90021	04880	15607	37422	28986	68156
34415	34393	86438	89537	17041	96369	40309	19173	74179	70516	25286	34218	68319
05648	32013	46902	78910	96560	86010	90506	89154	99909	06901	43675	30794	40072
90740	24858	38917	62201	25396	42331	10262	03001	94234	48033	05899	83159	08628
94697	10437	85219	76188	05292	30189	58429	77799	98158	39084	92137	08623	35271
13199	80107	94071	73054	14619	05779	05967	37654	04510	11663	35196	50849	39540
29622	94207	18958	12650	05160	36521	77625	89042	12010	46724	54309	85235	58677
22504	92303	50156	62504	51028	79390	64878	49920	14898	05145	99219	07882	71913
18354	32762	90334	97817	67887	83203	66395	54369	01498	02864	50063	93443	28268
61978	79644	20051	32057	23904	99112	94519	35138	18440	73147	83445	01340	53947
83773	69718	71520	61986	92535	70426	32567	33471	77371	58059	81488	66028	44411
94864	74318	77919	13728	73108	24038	56958	01136	36537	88984	81581	13442	78662
05174	83531	15425	98442	88428	68645	06533	41574	37378	75647	70959	73571	55543
51602	63904	13242	80612	37318	67352	74609	29663	17106	58806	47142	32473	36843
87636	04993	61505	65126	06915	13441	84627	17411	66290	32692	70359	19974	18478
27932	36660	60331	18861	06090	14290	61498	38418	92613	80466	22598	32021	62896
37237	30304	90098	15925	52730	59982	85657	29982	76637	91637	59043	00661	47037
22683	26320	28745	14753	09337	92193	72088	25069	61236	21574	67589	25827	30987
58178	66217	25095	26812	65662	96589	13613	81059	58664	18280	46826	09001	44014
60628	38526	81246	73004	58480	44765	30804	49710	60768	61120	80775	56051	76291
71783	17403	24911	89961	37526	29216	91847	67142	98210	42045	81985	29970	63734
40720	24312	56871	13170	56642	19816	23803	84668	77796	19555	54025	44325	77635
67671	30334	02293	18773	90515	44496	14903	53865	70311	26413	37990	17265	24121
58878	96540	23409	49122	39104	76370	28587	27108	64678	45937	34707	03351	34034
81323	58722	28368	84258	02462	13631	37409	06978	75540	53513	90336	46498	82952
56005	62356	32617	94365	30817	90307	52348	61557	52833	19295	73552	62830	01247
01045	15189	66867	32947	33578	82258	15844	93901	13277	09747	01625	78190	25780
73551	57253	49467	73742	70703	13393	16199	84230	69048	93226	10983	65077	18264
59850	15350	15526	46134	37527	29436	61470	76626	71550	43323	52173	83869	50438
61217	51080	15270	97588	81424	55470	95942	60926	33425	05774	77328	27584	21160
06249	30048	70208	36398	28538	71455	98580	64586	32889	35665	55124	64594	56457
28627	88560	52007	24839	24453	15843	99620	14738	56309	67421	91906	81140	62461
93447	22741	93380	05956	14780	87760	50253	41125	88069	17036	37914	83832	75007
12669	91704	65678	93611	35240	28095	97447	24843	35980	38654	83066	82453	66846
96252	90334	85665	03746	07454	95413	67678	34295	96265	28131	10493	90775	19611
86346	32782	44391	24731	84592	50880	99935	97335	72085	11823	47547	81435	85442
10147	32169	30187	35742	94051	11313	53564	88306	56333	41244	90012	92130	07836
87898	82498	07194	04596	63575	74322	57189	16562	85534	68843	37290	33940	65369
47158	93692	53645	68529	56366	77735	20208	14514	05314	36795	49869	05092	94981
46915	53514	16696	60012	25648	69902	01822	81820	38032	36275	67910	72884	26626
66512	58580	91015	87784	09244	34806	12695	57722	19418	88528	25450	29532	63215
45444	15844	31068	76729	24516	31870	53130	32184	29121	38728	91528	88755	44495
71801	95118	98230	26211	72970	82872	13659	55413	57741	87062	72094	03092	36030
58688	03020	24492	64330	33429	76947	04468	79148	62968	54914	69992	63234	42738
17282	10399	28449	01839	32812	81561	25847	82339	14984	56133	22374	99746	90023
14817	84395	79463	00923	95064	36755	75044	40233	26347	46475	15462	53778	81041
37253	27111	87904	90105	24805	39220	86661	11434	03596	43846	79784	74490	09823
06625	08048	05250	31351	67023	09525	31724	85009	85535	20758	23600	77615	70627

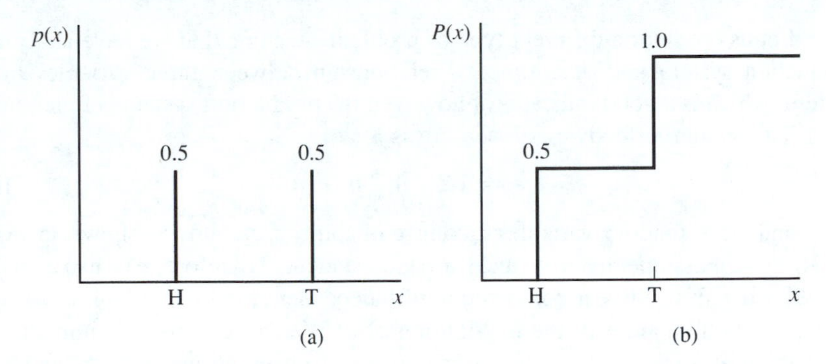

Fig. 16.1. Probability distribution and cumulative distribution functions for tossing of a fair coin.

	Number of Samples				
$p(x)$	*10*	*20*	*30*	*40*	*50*
H	0.7	0.6	0.47	0.45	0.48
T	0.3	0.4	0.53	0.55	0.52
E	0	0	0	0	0

The results are only slightly different from before, because the probabilities are only slightly different. We did not encounter an outcome of "edge" in our 50 sample experiment, but if we continued the experiment for a very long time, we would expect 1% of the outcomes to be "edge."

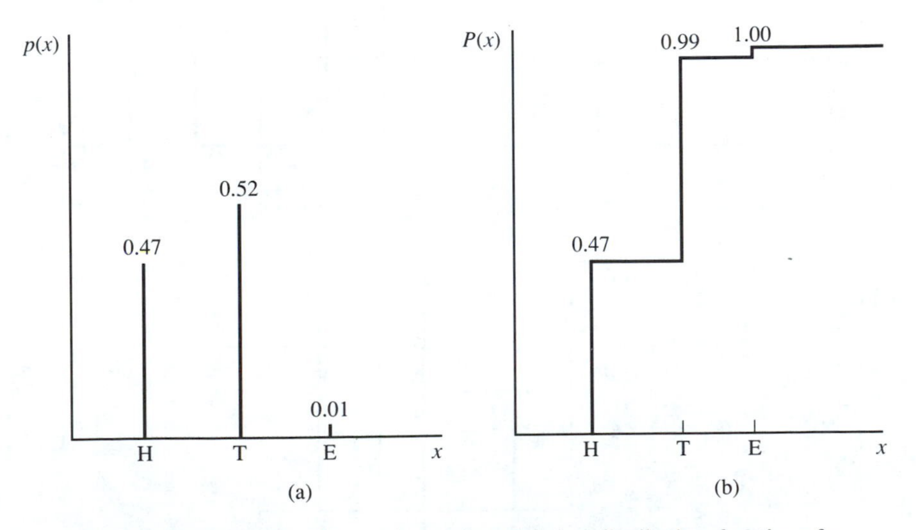

Fig. 16.2. Probability distribution and cumulative distribution functions for tossing an unfair coin that may land on edge.

Let us consider a different type of problem. Assume that we have analyzed a production system and determined a relationship between three variables in the system, which is a good indicator of how well the production system is functioning.

This system effectiveness measure is

$$e = 1.5a + 3b + c \tag{16.1}$$

a, b, and c are random variables having probability functions as shown in Figure 16.3. We will assume that a, b, and c are independent. Therefore, e is also a random variable since it is a linear combination of independent random variables. We wish to determine the nature of the distribution of e. We can do this via simulation by conducting an experiment in which a number of observations of a, b, and c are generated using the general procedure employed in the previous example. These sets of values will be substituted into Equation (16.1), giving a number of values of e. We can then analyze these values of e, just as we would any set of statistical data collected from the "real world." We can plot the frequency distribution, calculate the mean and standard deviation, and determine whether the distribution of e is of some standard form.

Using the same scheme employed in the previous example, we will assign digit groupings to correspond to the probabilities of the possible outcomes of the random variables. For variable a, digit 0 will correspond to "8," digits 1–2 to

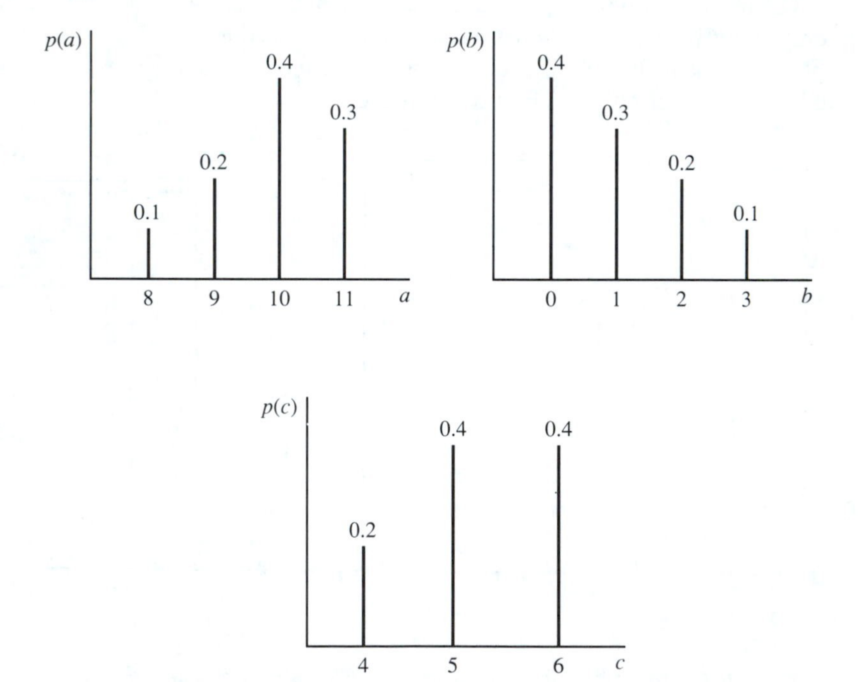

Fig. 16.3. Probability distributions for variables a, b, and c.

"9," digits 3–6 to "10," and digits 7–9 to "11." Similar assignments are made for variables b and c.

Since we are now sampling from three probability distributions, we should use a different "stream" of random digits from Table 16.1. This can be done in a number of ways. For this case, we will select three columns of five-digit numbers and use the rightmost digit from each group. Let us select the first three columns. Column 1 is used to sample from variable a, column 2 from variable b, and column 3 from variable c. The first set of digits are [0,6,6], the second set are [6,0,7], the third set are [3,1,6], and so on.

The results of a 20-sample experiment are shown in Table 16.2. The resulting probability distribution of e is shown in Figure 16.4. Obviously, a much larger sample size would be required to obtain a realistic probability distribution.

As a more realistic example, suppose that we have a set of machines and, with equal likelihood, either none or one breaks down per hour. Half of the breakdowns require 1½ hours to repair, and the other half require 2¼ hours to repair. For each minute a machine is out of service the company loses $1. Repairpersons for these machines are each paid $6 per hour, whether or not the repairpeople are working. Only one repairperson may work on a given machine at a time. The objective of our experiment is to determine the number of repairpeople we should employ in order to minimize total costs.

We shall simulate the performance of this machine–repairperson system for a period of 10 hours. First, we shall perform the simulation with one repairperson

Table 16.2. SIMULATION EXPERIMENT FOR EQUATION (16.1)

Sample Number	a			b			c		
	R.N.	a	$1.5a$	R.N.	b	$3b$	R.N.	c	$e = 1.5a + 3b + c$
1	0	8	12.00	6	1	3	6	6	21.00
2	6	10	15.00	0	0	0	7	6	21.00
3	3	10	15.00	1	0	0	6	6	21.00
4	5	10	15.00	3	0	0	8	6	21.00
5	8	11	16.50	3	0	0	2	5	21.50
6	0	8	12.00	8	2	6	7	6	24.00
7	7	11	16.50	7	2	6	9	6	28.50
8	9	11	16.50	7	2	6	1	4	26.50
9	2	9	13.50	7	2	6	8	6	25.50
10	4	10	15.00	3	0	0	6	6	21.00
11	4	10	15.00	2	0	0	4	5	20.00
12	8	11	16.50	4	1	3	1	4	23.50
13	3	10	15.00	8	2	6	0	4	25.00
14	4	10	15.00	8	2	6	9	6	27.00
15	4	10	15.00	1	0	0	5	5	20.00
16	2	9	13.50	4	1	3	2	5	21.50
17	6	10	15.00	3	0	0	5	5	20.00
18	2	9	13.50	0	0	0	1	4	17.50
19	7	11	16.50	4	1	3	8	6	25.50
20	3	10	15.00	0	0	0	5	5	20.00

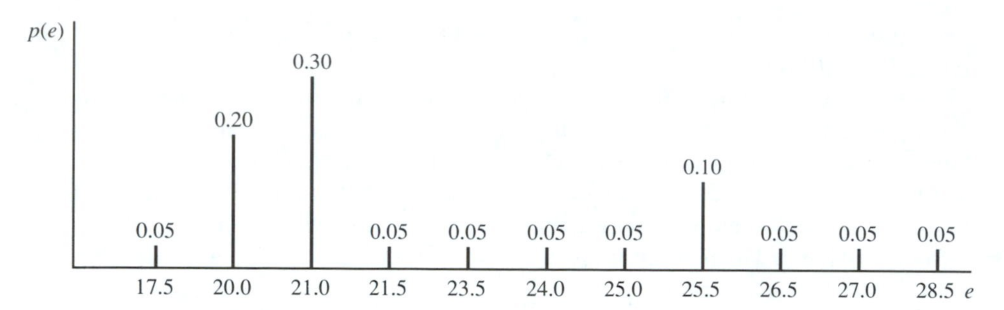

Fig. 16.4. Observed probability distributions of e from Equation (16.1).

and measure the resulting total cost; then we shall use two repairpeople and measure the resulting total cost; and so on. The equation for calculating total cost for various numbers of repairpeople, based on a 10-hour simulation, is

$$TC_k = (\$6)(10)(k) + (\$1)(DT) \tag{16.2}$$

where TC = total cost
k = number of repairpeople employed
DT = number of minutes of machine downtime

We shall obtain a value of TC for several different numbers of repairpeople and select the policy that results in lowest TC.

The simulation procedure is simple. We shall flip coin 1 and if it comes up tails, we shall assume that no machines break down. If it comes up heads, we shall assume that one machine breaks down at the beginning of the hour. If a machine breaks down, we shall flip coin 2 to determine whether the repair time is 1.5 hours (tail) or 2.25 hours (heads). Notice that one repairperson should be able to handle the job since, on the average, only one machine is breaking down each 2 hours and the average repair time is only 1.875 hours. With only one repairperson, however, we are taking a chance that several machines will break down in a short period of time or that several will require a long repair time, causing avoidable machine downtime and increasing the cost.

Table 16.3 is a tabular model of the system with which we are experimenting. We see that with one repairperson there are 11.25 hours of machine downtime (9.75 hours repair time plus 1.5 hours avoidable downtime), bringing the total cost to $60 for the employee plus $675 for machine downtime, for a total cost of $735. With two repairpeople there is no avoidable downtime in our simulated example; therefore, the cost is $120 plus $585 or $705. Although three repairpeople were not considered, their cost would have been $765. The results indicated that we should hire two repairpeople, even though they will not be kept busy all the time (actually, not even half the time). Of course, these conclusions are based on a short time period, and it is customary to simulate a much longer period in order to gain confidence in the results.

Table 16.3. SIMULATED MACHINE REPAIR SYSTEM

Hour of Operation	Coin 1	Number of Breakdowns	Coin 2	Machine Repair Time (Hours)	Start Repair		End Repair		Avoidable Machine Downtime (Hours)	
					1 Man	2 Men	1 Man	2 Men	1 Man	2 Men
1	T	0								
2	H	1	T	1.5	2	2	3.5	3.5	0	0
3	H	1	H	2.25	3.5	3	5.75	5.25	0.5	0
4	T	0								
5	H	1	T	1.5	5.75	5	7.25	6.5	0.75	0
6	T	0								
7	T	0								
8	H	1	H	2.25	8	8	10.25	10.25	0	0
9	T	0								
10	H	1	H	2.25	10.25	10	12.5	12.25	0.25	0
				9.75					1.5	0

16.3. Random Number Generation

Several assumptions were made in the previous example to vastly simplify the simulation so that it could be done by hand. For example, there were only two equally likely choices (0 and 1) of breakdowns per hour. Also, breakdowns occurred only at the beginning of the hour and repair times included only two equally likely choices (1.5 and 2.25 hours). In both cases, the value of the random variable (0 or 1; 1.5 or 2.25) was determined by a coin flip. Simulations normally involve much more complex models, and the process requires a computer.

During a simulation we frequently require the computer program to select a value of a random variable from its underlying distribution. For example, suppose that during a simulated experiment we require several "typical" values of a Poisson distributed random variable X with mean 1.5. From Appendix B we can see that the random variable will take on values of 0 with probability 0.223, 1 with probability 0.335, 2 with probability 0.251, and so on.

If we use the computer's ability to generate many three-digit *random numbers* ranging from 001, 002, up to 999 and 000, each having the same probability of occurrence, these can be used to randomly select values for our Poisson random variable X.

Consider the cumulative Poisson distribution shown in Figure 16.5. Suppose that the first three-digit number we generate is between 001 and 223, inclusive. This will "map" onto a value of $X = 0$. If the number generated ranges from 810 to 934, it will "map" onto a value of $X = 3$. For example, the number 876 is shown in Figure 16.5 to call for the value $X = 3$.

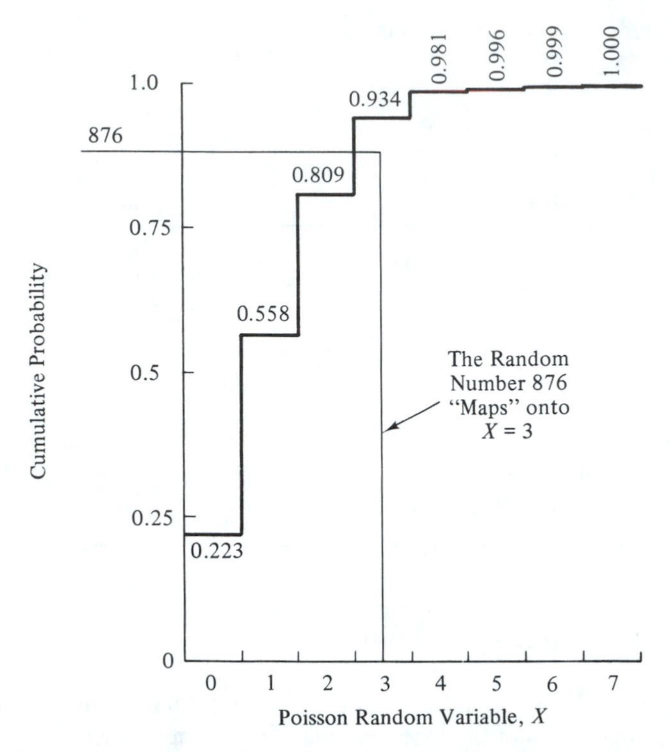

Fig. 16.5. Cumulative distribution function of Poisson distribution (mean = 1.5).

Thus far, we have considered only discrete probability distributions. Many random variables are inherently continuous. Sampling from continuous distributions can also be simulated.

Consider the exponential random variable. Its density function, $f(x)$, and cumulative distribution function, $F(x)$, are shown in Figure 16.6. We can obtain simulated sample values from the exponential by generating a random value from a uniform distribution having range [0,1]. This value can be transposed to the vertical axis of the cumulative distribution function, projected horizontally until it intersects the $F(x)$ curve, and then projected vertically downward to the x axis. The point of intersection is a randomly selected value of the random variable (see Figure 16.7). Note that this process is essentially identical to the one we employed for sampling from the discrete Poisson random variable in Figure 16.5.

In order to sample from continuous distributions in a computer simulation model, we need to be able to perform the operations just illustrated computationally. In other words, we need to develop an equation for generating random values that can be included in our computer program. This turns out to be very easy to do for the exponential random variable.

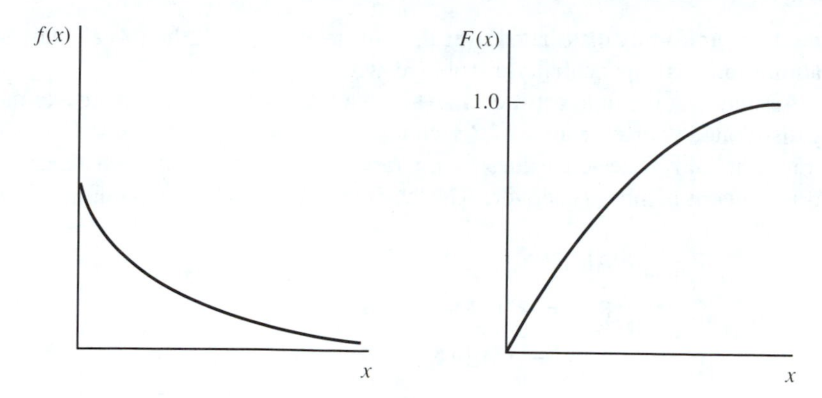

Fig. 16.6. Density function and cumulative distribution function for an exponential random variable.

The exponential random variable has a single parameter, its mean, μ. The cumulative distribution function is

$$F(x) = 1 - e^{-x/\mu} \tag{16.3}$$

An examination of Figure 16.7 reveals that $r = F(r)$ and that $F(r) = F(x)$ (both being defined on the unit interval). This allows us to form the relation

$$r = 1 - e^{-x/\mu} \tag{16.4}$$

Solving this equation for x gives

$$x = -\mu \ln(1 - r) \tag{16.5}$$

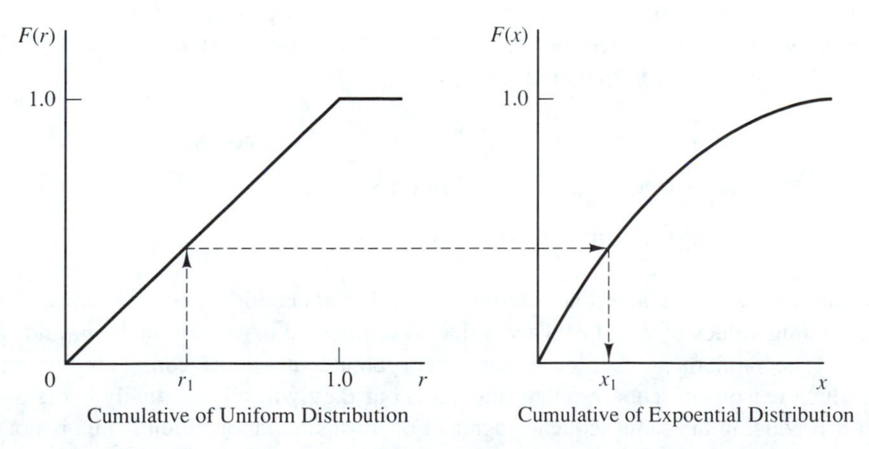

Fig. 16.7. Illustration of simulated sampling from a continuous distribution.

Hence if r is uniformly distributed over the unit interval [0,1], then x calculated by Equation (16.5) is exponentially distributed with a mean of μ.

Assume that the time between arrivals to a toll booth is found to be exponentially distributed with a mean of 37.3 seconds. The following FORTRAN code will generate randomly selected values of interarrival times given the existence of a uniform random number generator, RNUM (FORTRAN function name):

```
AMEAN = 37.3
RN = RNUM( )
X = -AMEAN*ALOG (1.0-RN)
```

The exponential function is a very easy function to work with. Others, such as the density function for the normal random variable, are more difficult, and cannot be derived in closed form. Approximations have been developed for the normal and other intractable density functions. Essentially all simulation languages include routines for automatically generating random values from a wide variety of statistical distributions, both discrete and continuous.

All such process generators depend on a random number generator that generates random values on the unit interval [0,1]. Many approaches have been attempted to develop good random number generators. It would be inconvenient to be restricted to using tables of previously determined random numbers, such as those in Table 16.1. We need a way to calculate a random number whenever we need one.

The most widely used random number generators are those that employ the multiplicative congruential method

$$r_{n+1} = k\, r_n \pmod{m} \tag{16.6}$$

where k and m are positive integers with $k < m$. This generator must be given a "starting value," r_0, called the "seed value." For example, if we specify $k = 5$, $m = 8$, and $r_0 = 7$, Equation (16.6) gives

$$r_1 = kr_0 \pmod{m} = (5)(7) \pmod{8} = 35 \pmod{8} = 3$$

$$r_2 = (5)(3) \pmod{8} = 15 \pmod{8} = 7$$

$$r_3 = (5)(7) \pmod{8} = 35 \pmod{8} = 3$$

Notice that we have a serious problem with this generator. It is only capable of generating values of 7 and 3. This is due to our choice of values for k, m, and r_0.

The random number generators incorporated into most computer libraries produce reasonably good random numbers, but they will all eventually cycle and start producing the same sequence again. For most simulation studies, this is not a serious problem.

16.4. Time-Flow Mechanism

Since almost all simulation studies are concerned with systems whose events are time-ordered, the computer program must be written in such a way that it moves the model through simulated time. That is, events are caused to occur in the proper order and with a proper time interval between successive event occurrences. Our machine repair example used a *time-step incrementation*, stepping through time in equal (1-hour) increments, determining what events have occurred during the time increment, and recording this information as summarized in Table 16.3.

A more popular mechanism available to represent the flow of time in a simulation model is *event-step incrementation*. Event-step incrementation calls for a simulation to proceed from one event to the next. In this case, the incremental time steps are generally uneven. The simulation begins at time zero, and the occurrence times of the events resulting from the simulated performance of all system components are determined. The *master clock* is updated to the time of the earliest event occurrence. Simulated time advances to this point, the new system state is determined, and subsequent event occurrence times are determined for the processes for which events have just occurred. The simulation then advances to the time of the next earliest event occurrence, and the process repeats itself throughout the simulation.

An example of event-step incrementation is the flow chart of a queueing system having one service facility and infinite queue capacity as illustrated in Figure 16.8. When a customer arrives for service, before being assigned to a server, we first generate the time of our next arrival and enter that arrival chronologically in the *event list*. Then, if the server is free, we assign our customer to the server, show the server as busy, and generate the length of the service. We then enter the end-of-service time in the event list. If the server is busy, we increase the queue size by one and our customer must wait. When a service ends, the server is removed from the busy status, the queue is checked, and if another customer is waiting, the line is reduced by one and service begins. If there is no one in the queue, we simply advance time to the next event in the chronological list.

16.5. Simulation Languages

Although we now have a feel for simulation, we must realize that simulation is not merely an exercise in operating the model of a system. Instead, we try to collect important system performance measures as the simulation proceeds. For example, in the queueing system we may want to know the maximum length of the queue, the utilization of the server, or the average waiting time in the system for given arrival and service distributions. These things can be determined by building into the computer program appropriate counters or registers to gather these statistical data. Better yet, we can make use of available simulation languages to do this work

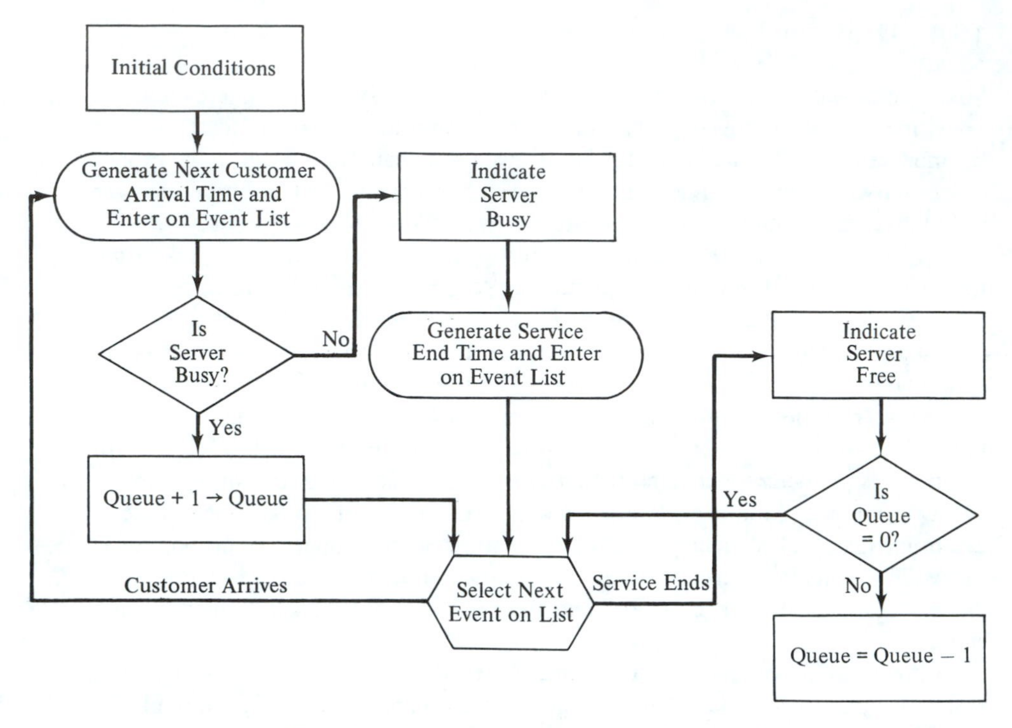

Fig. 16.8. Variable increment simulation flowchart.

for us. Simulation languages provide specific concepts and procedures for describing a system *and* moving it from state to state over time. Three relatively well-known simulation languages are GPSS, SLAM, and SIMSCRIPT.[1]

GPSS (General-Purpose Simulation System) has broad appeal due to its modeling simplicity. A GPSS model is developed by forming a block diagram from over 40 standard operational blocks, such as "generate," "terminate," and "advance." GPSS programming consists primarily of learning the functions of the GPSS blocks and how to combine them logically to represent the desired system.

SLAM (Simulation Language for Alternative Modeling) is an advanced FORTRAN-based language that represents the state of the art in simulation languages of interest to the I.E. at the time. It provides network symbols for building graphical models, and it contains subprograms for discrete event (as presented in this chapter) and continuous modeling developments. Extremely complex problems of interest to practicing industrial engineers can be well handled using SLAM.

SIMSCRIPT II is a simulation language divided into five levels:

---

[1] A review of these and other simulation languages is presented in A. Alan, B. Pritsker, *Introduction to Simulation and SLAM*, 3rd ed. (New York: Halsted Press, 1986).

(1) A simple teaching language to introduce programming concepts.
(2) Statement types of power comparable to FORTRAN.
(3) Statement types of power comparable to ALGOL or PL/I.
(4) Statement types that provide a structure for modeling.
(5) Statement types for advancing time, processing events, generating samples, and accumulating data statistics.

SIMSCRIPT programs are easy to read and tend to be self-documenting, due to their English-like form.

PROBLEMS

1. Repeat the experiment in Section 16.2 dealing with the tossing of an unbalanced coin using the following probabilities: $p(x)$ = head = 0.45, $p(x)$ = tail = 0.50, $p(x)$ = edge = 0.05. Use the same sequence of random digits from Table 16.1 as was used in the earlier experiment.

2. Write a simple computer program to replicate the experiment in Section 16.2 dealing with the system effectiveness measure e, Equation (16.1). Employ the random number generator available on the computer you use. First, run an experiment with 20 samples and compare your results with those in Table 16.2 and Figure 16.4. Explain why they are different. Now run an experiment with 100 samples and compare with your first experiment. What general conclusions can you draw?

3. Draw a flowchart of simulation for the lot size–reorder point model of Section 15.3.2 and Figure 15.2. Insert a step that calculates the shortage (if any) during each cycle.

4. Set up the process generator that you would use to simulate the data transmission of Problem 6 in Chapter 15 (assume that you will use two-digit random numbers).

5. Consider the following systems effectiveness measure:

$$e = 5a + 3b + 2ab$$

The probability distributions for a and b are shown in Figure P16.5.

(a) Determine the probability distribution of e using the leftmost two digits of the rightmost two columns of five-digit numbers in Table 16.1. The first three pairs of random numbers are [50,24], [49,37], and [28,68]. Generate 30 sample values of e.

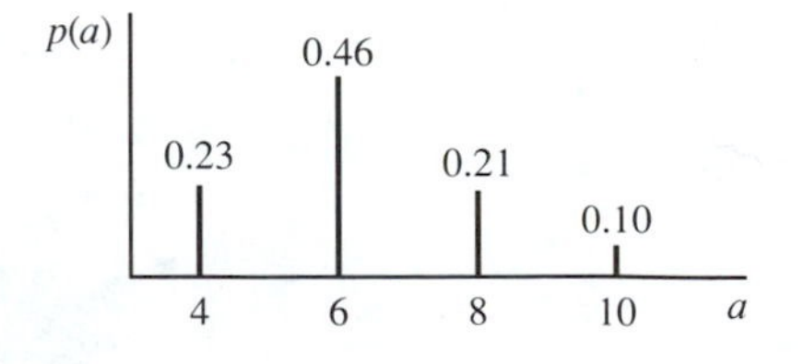

Fig. P16.5.

(b) Conduct the same experiment using a computer program, and generate 100 sample values of e. Then generate 1000 sample values of e and compare the results.

6. Consider the machine breakdown and repair example in Section 16.2. The cost factors are modified as follows:

Out-of-service cost: $6.50/min.
Repairperson salary: $13.20/hr.

The distribution of repair time is modified as follows:

Time to Repair	*Probability*
2.5	3.0
0.5	0.5

Note: It may be necessary to consider more than 2 repairpersons. If so, additional columns need to be inserted into Table 16.3.

7. An electronics manufacturer produces transistors that are packaged in lots of 500. The quality control inspector draws a sample of 5 transistors from each lot, with each transistor in the sample classified as good or bad. Historical records show that the production equipment produces defective transistors at an average rate of six per lot under normal conditions. Simulate the inspection of 30 lots, and estimate the average number of defective transistors per lot detected by the inspector. In addition, estimate the probability distribution of the number of defective transistors per lot.

8. Assuming that uniformly distributed random digits are available in any quantity, describe a method for the random sampling of library books. It is desired that the books in the sample be readily available from the sampling process. Use your general knowledge of library stacks and be imaginative.

9. Generate random numbers by the multiplicative congruential method [Equation (16.6)] using $m = 2^5$, $r_0 = 5$, and $k = 11$ until the complete cycle of numbers is obtained.

10. Develop a general process generator for uniformly distributed random variables with range [A,B], comparable to Equation (16.5) for the exponential random variable. Develop a computer program to validate your generator.

CHAPTER 17

Project Management

17.1. Introduction

Management of a project should be treated as if it were a one-time task. By "project" we mean a major undertaking that is unlikely to be repeated in exactly the same way at a future time. Any mistakes made are likely to be costly to rectify, and we are very concerned about performing the job tasks correctly. This is in contrast to managing the production of goods or services in which the process is ongoing. That is, production management is repetitive in nature, whereas project management is more of a single, major job.

Examples of project management might include building a football stadium, administering a large research contract, performing major transplant surgery, establishing a production line, earning a college diploma, or raising a child. All of these projects have some similarities. Each is comprised of many tasks or activities, the number of activities depending on the detail in which we wish to describe the project. The activities of each project have precedence relationships. That is, certain activities cannot be started until others have been completed. Each activity has a duration. Sometimes we can predict with great confidence how long an activity will take; at other times we are very uncertain about our estimates. Finally, the activities of each project must be scheduled in such a way that the project is brought to a successful completion.

Example 17.1

As president of Gadgets, Inc., you have been very successful. Through industrial engineering techniques, you have planned properly, anticipated problems, corrected problems, have a high level of employee morale, and now are faced with excess profits. In order to take advantage of tax write-offs, you have decided to build a new plant. It will be bigger and more modern than your current facility, and all of the arrangements from start to finish of your project seem to be more than you can coordinate in your head.

You have already decided to oversee and coordinate the project. You have decided to separate the project into two phases. The first phase will include obtaining marketing and production factors such as sales forecasts, production levels, storage needs, and so on. This phase will also include plant location and layout. This type of project you have been involved in before, and you need a project management tool in which you can estimate the activity times very closely. The second phase is the construction phase of the project about which you feel much less certain. Here you need a project management tool that allows some flexibility or uncertainty in making time estimates.

In both phases of the project you plan to trade direct costs off against indirect costs to achieve a duration for each phase of the project which provides a minimum total cost. You also want to keep a good cost record of each activity for tax and control purposes. Finally, you desire to balance or level your manpower resource needs over time in both phases of the project.

You recognize that CPM and PERT network activity models offer you the needed techniques for properly planning your expansion. Before launching into your project, however, you take time to review CPM and PERT.

Numerous methods have been devised for project management. One that you worked with in Chapter 7 is the Gantt chart. The Gantt chart shows the time needed for activities, as well as the earliest times when activities can begin. Unfortunately, the Gantt chart is not ideally suited to the management of large projects. The interrelationships of the activities are not shown and it is difficult to determine from a Gantt chart how long the start of a project can be delayed without adversely affecting the project. Also, as the project proceeds and we have deviations from the original plan, adjustments and rescheduling are difficult using a Gantt chart if there are many activities.

In an attempt to overcome some of the shortcomings of the Gantt chart, critical path systems were developed in the late 1950s. We shall look at two of the most popular—critical path method (CPM) and program evaluation and review technique (PERT). CPM and PERT have many similarities. They both show activity duration and indicate the earliest activity start times, as does the Gantt chart. In addition, they clearly show precedence interrelationships between activities, allowing us to determine the most critical activities in the chain of events leading to project completion. Activity modifications and progress updating are relatively easy to do with either CPM or PERT, especially now that computer software is available for use on mainframes and personal computers, taking the drudgery out of this work.

Of course, CPM and PERT are likely to take more time to use than a Gantt chart, and therefore, they are more costly to use. On small projects the Gantt chart may be preferred, but on large projects critical path systems will almost certainly be required. The choice between CPM, PERT, Gantt charts, and other project management tools must be made by the user and depend on the size and complexity of the project, as well as the software available.

17.2. Project Planning Networks

For analysis by CPM or PERT each project must have three properties:

(1) The project consists of a set of well-defined activities or tasks which, upon completion, signify the end of the project.
(2) The activities may be started and stopped independently of each other within a given sequence.
(3) The activities have precedence relationships and must be performed in the proper order.

Both CPM and PERT utilize a network representation to describe and structure the activities of a project. A critical path network is a graph or picture that shows each task to be performed, its predecessor, and its successors. We can best explain the development of a critical path network by developing one using a simple example.

Suppose that we are in charge of coordinating the installation of a precision rain gauge/wind sensor station. Telemetering equipment will be housed inside a small building to continually record all readings. Not all of the activities needed to install the station can be done at once. Neither need they be done one at a time. In order to determine which activities may be done concurrently, we must specify all precedence relationships. These relationships and our estimates of individual activity durations are given in Table 17.1.

Table 17.1. RAIN GAUGE/WIND SENSOR INSTALLATION DATA

Activity	*Description*	*Predecessor*	*Duration (Days)*
A	Layout sensing station	None	5
B	Procure rain gauge/wind sensor	None	25
C	Procure telemetry equipment	None	18
D	Construct first 50% of building	A	14
E	Construct remainder of building	D	14
F	Install rain gauge/wind sensor	B,D	4
G	Install telemetry equipment	C,D	5
H	Connect telemetry equipment to rain gauge/wind sensor	F,G	3
I	Final test	E,H	1

We can now draw a picture of this information in the form of a network as shown in Figure 17.1. This figure tells us a great deal once we know how to read it. First, let us define the essential characteristics of the graph as follows:

○ A circle represents a *node*. Arrows, or activities leaving a node, cannot be started until all activities incoming to the node have been completed. The completion of all activities incoming to a node is considered an *event*, as is the start of the project.

⟶ An arrow represents one of the *activities* of the project. Its length is of no significance.

-- → A dashed arrow represents a *dummy activity*. Dummy activities are used to represent precedence relationships. They are not activities in the real sense and have a duration of zero.

It is common to number the nodes, but the number given to a node is irrelevant and does not indicate precedence or duration. It is also common to indicate each activity with a code letter followed by its duration. The code letter is used to maintain a neat graph instead of trying to write each activity description over its respective arrow.

Having drawn the network and defined its elements, we can see, for example, that installation of the rain gauge/wind sensor (F) cannot begin until they have been procured (B) and the building is 50% completed (D). Clearly, the last half of the building cannot be constructed (E) until the first half is completed (D). In this case, a dummy activity is required because activity F requires both B and D to be completed, while E requires only that D be finished. It is also easy for us to observe the total time required for various paths through the network. For example, the path consisting of activities C, G, H, and I (nodes 1, 5, 6, 7, 8) is estimated to consume 27 days. We can also see that the longest path through the network includes activities A, D, E, and I (nodes, 1, 2, 3, 7, 8) requiring 34 days. This would be called the *critical path*.

We can see that critical path techniques offer a number of advantages for project planning, scheduling, and control, for example:

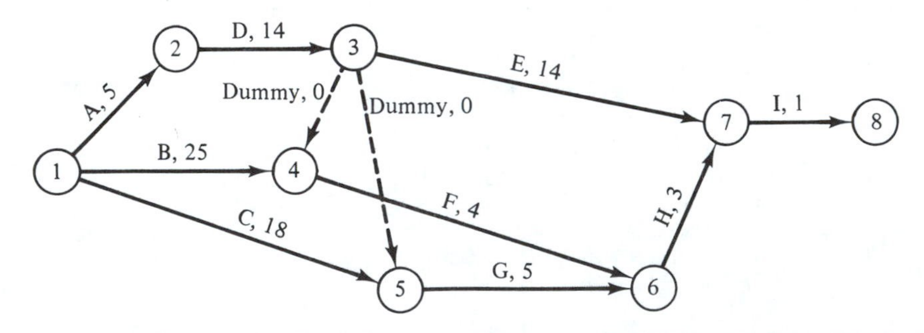

Fig. 17.1. Critical path network of rain gauge/wind sensor installation.

(1) They encourage a logical approach to project management.
(2) They encourage long-range and detailed planning.
(3) They provide a standard method of communicating project plans, schedules, and time performance.
(4) They focus management attention on the usually small percentage of critical activities.
(5) They can illustrate the effects of technical and procedural changes on the project schedule.

We shall now look at CPM and PERT in more detail.

17.3. Critical Path Method

Once we have reduced a project to a network of activities and nodes, we are ready to determine the critical path. In the above example we were easily able to identify the critical path as that sequence of activities requiring the longest time to accomplish. As the size of a project grows, however, it becomes almost impossible to identify the critical path through the network. CPM offers a systematic procedure for selecting the critical path. In addition, the amount of slack or free time on noncritical paths may be determined. Knowing the slack on noncritical paths may permit us to trade off manpower and equipment resources from noncritical activities in order to concentrate on and shorten the critical path.

CPM requires that we begin by identifying all project activities, precedence relationships, and times. As an example, we may have a project described as follows:[1]

(1) A, B, and C are concurrent and begin at the start of the project.
(2) D and E are concurrent and can begin upon the completion of A.
(3) F may begin upon the completion of B.
(4) G starts after C is finished.
(5) H follows D.
(6) I can begin when E and F are complete.
(7) Completion of G, H, and I marks the end of the project.

Once we have activities and precedence relationships, we can draw the basic CPM network. This allows *planning*, but in order to do *scheduling* we must have activity time estimates. Some rules for estimating activity times are as follows:

(1) Obtain time estimates from the best possible source. Estimating must be done by experienced personnel, often the person, group, or subcontractor responsible for accomplishing the activity.

---

[1]Taken from an example used by R. J. Craig and W. C. Turner in teaching extension courses through Virginia Polytechnic Institute and State University, and Oklahoma State University.

(2) Estimates for activities will be on an independent basis. That is, the estimate for each arrow is made irrespective of estimates made for every other arrow in the diagram.
(3) The time estimate is based on the usual practical resources available. This means using a regular working crew, a normal working day, and so on.
(4) Consistent time units must be used throughout.

Suppose that we estimate the activity durations for our example as follows:

A:	8 days	F:	9 days
B:	20 days	G:	10 days
C:	33 days	H:	8 days
D:	18 days	I :	4 days
E:	20 days		

The CPM network for our project can now be illustrated as in Figure 17.2.

We are now ready to analyze the network in detail. To do this, we will use the following nomenclature:

t = single estimate of mean activity duration
T_E = earliest event occurrence time
T_L = latest allowable event occurrence time
ES = earliest activity start time
EF = earliest activity finish time
LS = latest allowable activity start time
LF = latest allowable activity finish time
S = total activity slack

Using this notation, we shall now discuss forward and backward passes through the network, total activity slack, and the critical path.

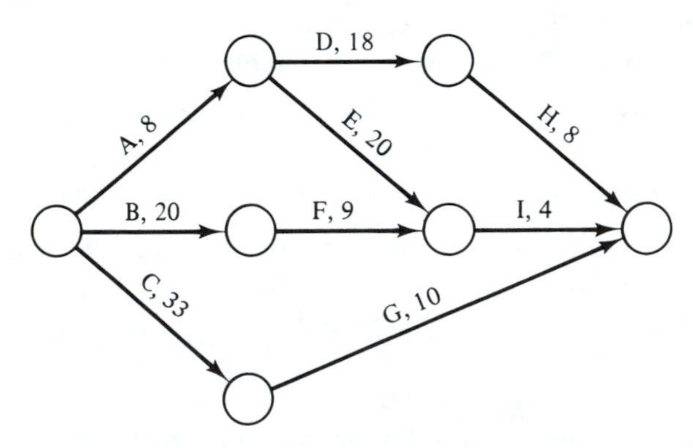

Fig. 17.2. CPM network.

17.3.1 Forward Pass

The computations from a forward pass through the network provide information on the earliest time that an event can be completed. In the forward pass calculations we begin with the first activity (or activities) and proceed along each path of the network observing the following rules:

(1) The earliest occurrence time of the (single) initial event (node) is taken as zero. That is, $T_E = 0$ for the initial event.
(2) Each activity begins as soon as its predecessor event (node) occurs. That is, $ES = T_E$ (for its predecessor event). Since the duration of the activity is t, its earliest finish time is $EF = ES + t = T_E + t$.
(3) The earliest event (node) occurrence time is the largest of the earliest finish times of the activities merging to the event in question. Mathematically, $T_E = \max(EF_1, EF_2, \ldots, EF_n)$ for an event with n merging activities.

We can illustrate the forward pass notation pictorially as shown in Figure 17.3. In this figure activity K has a duration of t. It connects events (nodes) i and j. The earliest event occurrence time for node i is T_{Ei}, this also being the earliest start time, ES, for activity K. The earliest finish time for activity K is EF which is t time units after ES. The earliest event time for node j is T_{Ej}, which is the maximum of the earliest finish times of the incoming activities. If only one activity is incoming to node j, as shown, $T_{Ej} = EF$.

Let us now redraw our network by using the forward pass. This is done in Figure 17.4. From this drawing, we can see the earliest start and finish times for each activity, as well as the earliest time at which our events (nodes) may be accomplished.

17.3.2 Backward Pass

The backward pass calculations tell us the latest times at which events can be completed and still meet the final schedule date. The backward pass is identical in nature to the forward pass, except that the backward pass starts with the last node and works toward the starting event. The backward pass rules are as follows:

(1) The latest allowable occurrence time of the (single) terminal event is set to the same value as the earliest occurrence time that we calculated in the forward pass. That is, $T_L = T_E$ for the last event.

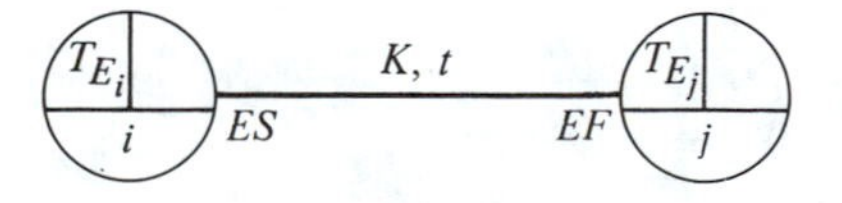

Fig. 17.3. Forward pass notation.

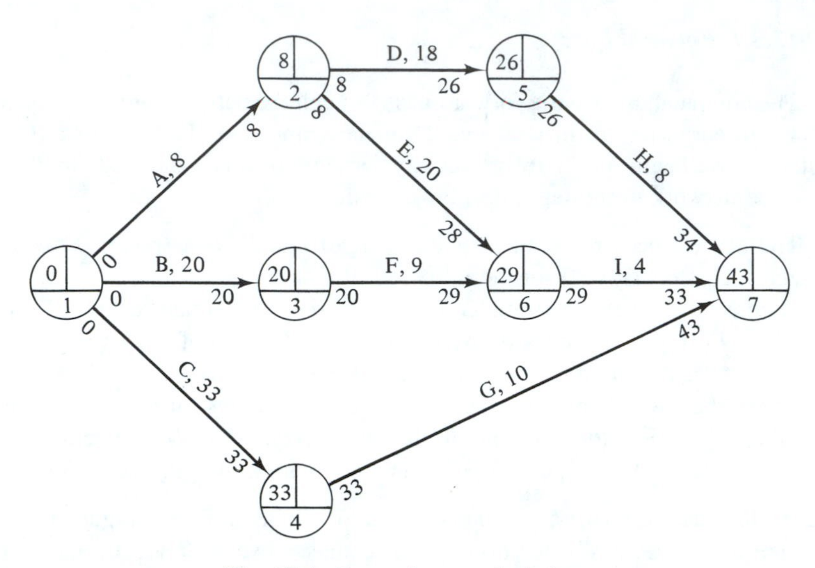

Fig. 17.4. Forward pass calculations.

(2) The latest allowable finish time for an activity is its successor event's latest allowable occurrence time. For this we have $LF = T_L$ for the successor event. The latest possible activity start time is its latest allowable finish time less the duration of the activity in question. Mathematically, $LS = LF - t = T_L - t$.

(3) The latest allowable time for an event is the smallest of the latest allowable start times of the activities emanating from the event of interest. That is, $T_L = \min (LS_1, LS_2, \ldots, LS_n)$ for an event with n emanating activities.

Let us now add the backward pass notation to that for the forward pass. This is done in Figure 17.5. The latest event time for node j is T_{Lj}, this also being the latest finish time, LF, for activity K. The latest start time for activity K is LS which is t time units before LF. The latest event time for node i is the minimum of the latest start times for the emanating activities. If only one activity leaves node i as shown, $T_{Li} = LS$.

We can now modify Figure 17.4 to give both the forward and backward pass data. This modification is performed in Figure 17.6.

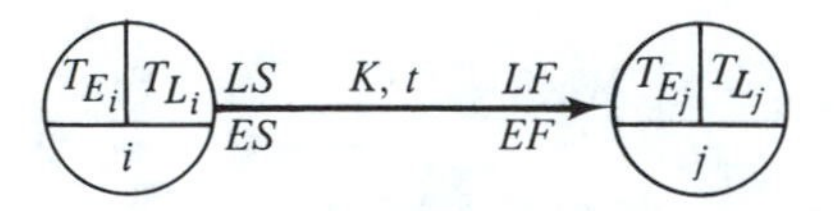

Fig. 17.5. Forward and backward pass notation.

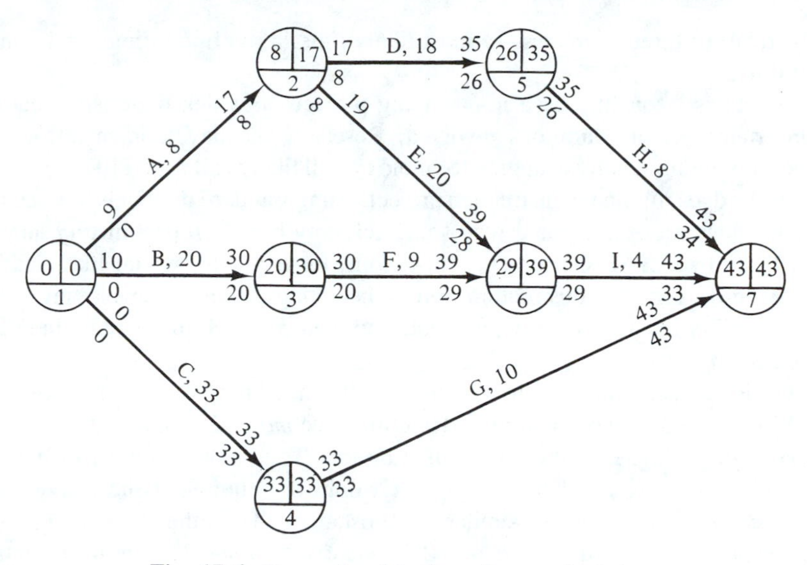

Fig. 17.6. Forward and backward pass calculations.

17.3.3 Total Activity Slack

Total activity slack equals the latest allowable occurrence time of an activity's successor event, less the earliest finish time of the activity of interest. That is, $S = T_L - EF$. The total activity slack tells us the amount of delay we can have for a task without affecting the earliest start of any activity on the critical path. For example, activity E has 11 days of slack. It can be started as early as day 8 or as late as day 19 without extending the project duration beyond 43 days.

17.3.4 Critical Path

The critical path may be identified as the sequence of activities having no slack. We can see that $ES = LS$ and $EF = LF$ for each activity on the critical path, and $T_E = T_L$ for each node. The critical path is one in which a delay in any activity on the path will cause a delay in the project. Our critical path is easily identified as activities C, G, or nodes 1, 4, 7.

17.4. Program Evaluation and Review Technique

PERT was developed in an independent effort at about the same time CPM was established. It utilizes a project network, has a critical path, and maintains the notion of total slack. In what way do they differ? The basic difference is in the way activity durations are estimated. Whereas CPM requires a single estimate for each activity,

PERT requires three estimates—a most likely time, a minimum time, and a maximum time.

PERT is frequently used in managing projects in which there is a great deal of uncertainty in the durations involved. Research and development efforts and projects involving rapidly changing technology will likely call for a PERT approach. Houses, bridges, or other construction projects using standard materials have activity elements that can be estimated with some certainty based on past history and will normally utilize CPM. Construction of a dam, for example, would likely call for CPM; if, however, the construction were to be taking place in a developing nation, the availability of manpower, equipment, and materials may necessitate the PERT approach.

PERT calls for the user to assume that the uncertainty of each activity's time may be described by a probability distribution. The *most likely time* estimate would be the mode or highest value of the distribution. This is our estimate of the most frequently occurring number of days that would be required if the activity were performed many times under similar circumstances. The other two estimates are sometimes called the *optimistic time* and the *pessimistic time*. The optimistic activity time would be the shortest reasonable time for activity completion, that is, if nearly everything goes right. The pessimistic time would be the time required if Murphy's law[2] were encountered, that is, if most everything goes poorly. For highly uncertain activities, the time between estimates is likely to be large; for virtually certain activity durations, all three time estimates may be equal.

Before continuing with our discussion of PERT, we shall define some additional notation as follows:

$$\begin{aligned} t_o &= \text{optimistic time} \\ t_m &= \text{most likely time} \\ t_p &= \text{pessimistic time} \\ t_e &= \text{expected (average) time} \end{aligned}$$

The first three times can be combined in a weighted average to determine an expected time. This weighted average is

$$t_e = \frac{t_o + 4t_m + t_p}{6} \tag{17.1}$$

Four times the weight is given to the most likely activity time than to the other two estimates. The expected time, t_e, is the average time the activity would take if it were repeated many times. This differs from t_m, which is the most frequently occurring duration in many repetitions of the activity.

Not only are we interested in an average activity duration, but we would also like a measure of the reliability of the estimate. Stated differently in a statistical

---

[2] A version of Murphy's law is as follows: "If there is even the most remote chance that something may go wrong, it will."

sense, we would like to know the variability or spread of the probability density function underlying the different times that may be taken to complete an activity. We recall that two standard measures of distribution spread are the variance and its square root, the standard deviation. The standard deviation may be estimated by the formula

$$S_t = \frac{t_p - t_o}{6} \tag{17.2}$$

This indicates that the standard deviation is $\frac{1}{6}$ of the spread between the maximum time and the minimum time. This is not a bad assumption in that we frequently consider virtually all of a unimodal (one peak) distribution to be contained within a spread of six standard deviations. Since the standard deviation is the square root of the variance of a distribution, the variance of t is given as follows:

$$V_t = \left(\frac{t_p - t_o}{6}\right)^2 \tag{17.3}$$

As an example of the PERT estimating procedure, suppose we determine that our three times for an activity are

$$t_o = 7, \qquad t_m = 9, \qquad t_p = 13$$

The expected or average time by the PERT formula is

$$t_e = \frac{t_o + 4t_m + t_p}{6} = \frac{7 + 4(9) + 13}{6} = 9.33 \tag{17.4}$$

The variance of the activity time distribution is found to be

$$V_t = \left(\frac{t_p - t_o}{6}\right)^2 = \left(\frac{13 - 7}{6}\right)^2 = 1$$

The standard deviation is therefore also equal to one. The distribution of time for this example is illustrated in Figure 17.7.

Now let us return to our example having activities A through I. But now let us assign PERT estimates instead of the single CPM activity duration. Table 17.2 indicates PERT estimates for each activity in our network. The three PERT activity

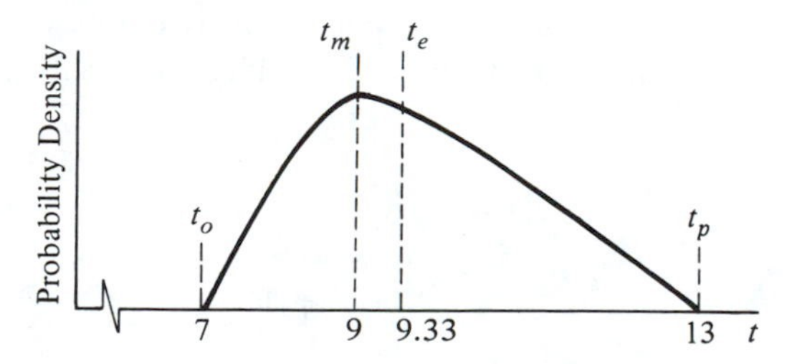

Fig. 17.7. Example activity time distribution.

Table 17.2. EXAMPLE PERT ACTIVITY TIME ESTIMATES

Activity	*Optimistic Time,* t_o	*Most Likely Time,* t_m	*Pessimistic Time,* t_p	*Expected Time,* t_e	*Variance,* V_t
A	5	7	15	8	2.778
B	14	21	22	20	1.778
C	21	33	45	33	16.000
D	14	16	30	18	5.444
E	18	19	26	20	1.778
F	7	9	11	9	.444
G	7	9	17	10	2.778
H	8	8	8	8	.000
I	3	4	5	4	.111

times must be selected based on the estimator's best judgment. In Table 17.2, however, we "handpicked" the times in such a way that the expected time for an activity, t_e, equals the CPM estimate for the same activity. In this way, we shall be able to better compare CPM and PERT through the same example.

In PERT the expected length of a sequence of activities, T_E, is equal to the sum of the respective activity expected times. That is,

$$T_E = \sum_{\substack{\text{all activities}\\ \text{along a path}\\ \text{of interest}}} t_e \tag{17.5}$$

This gives us an easy way of determining the length of any path through the network. In fact, it is identical to the method used for CPM! Therefore, the network representation, forward and backward pass calculations, total slacks, and critical path found for our CPM example are directly applicable to our PERT example.

With PERT, though, we can glean even more information from the network. If, in addition to knowing the mean or expected time along the critical path, we knew the variance along the critical path and the overall probability distribution of time, we could calculate the probability of project completion by a certain time. Fortunately, the variance of a sum of independent activity times is equal to the sum of their individual variances. This sum of variances we shall denote as V_T and write as

$$V_T = \sum_{\substack{\text{all activities}\\ \text{along a path}\\ \text{of interest}}} V_t \tag{17.6}$$

Naturally, the path standard deviation, S_T, is written as

$$S_T = \sqrt{V_T} \tag{17.7}$$

Further, from the central limit theorem we can deduce that the distribution of the sum of n independent activities approaches the normal distribution as n grows large. We already saw that the normal distribution is used in quality control work with subgroups as small as size 4 or 5. The typical project requiring a critical path approach will normally have at least this number of activities along its critical path.

Let us now calculate the mean and variance of the critical path in our example. Only two activities, C and G, comprise the critical path.[3] The total expected time is

$$T_E = \sum_{C,G} t_e = 33 + 10 = 43$$

The variance along the critical path is

$$V_T = \sum_{C,G} V_t = 16.000 + 2.778 = 18.778$$

The standard deviation equals

$$S_T = \sqrt{V_T} = \sqrt{18.778} = 4.333$$

We are now ready to calculate the probability of project completion by, say, 47 days. Graphically, our problem is illustrated in Figure 17.8 in which a normal distribution having mean 43 and standard deviation 4.333 is shown. We want to know the area under the crosshatched portion of the normal curve. This represents the probability of project completion by time 47. First, we calculate the standard normal deviate, Z:

$$Z = \frac{X - \mu}{\sigma} = \frac{X - T_e}{S_T} = \frac{47 - 43}{4.333} = 0.923$$

Now, from the cumulative normal table in Appendix B, we find that the probability of a value less than $Z = 0.923$ is 0.822. In other words, there is an 82.2% chance of completing the critical path projects in 47 days. A similar calculation may be made for any number of days. Of course, the probability of completing the critical path by 43 days is 50%.

---

[3] With only two activities comprising the critical path, it is probably not a good idea to put much stock in the assumption of a normal distribution from the central limit theorem. However, the procedure will be continued for the sake of presenting an example of the method used.

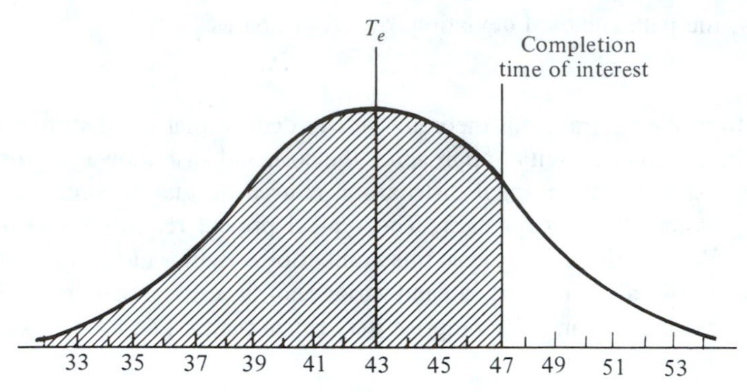

Fig. 17.8. Critical path probability distribution.

17.5. Time–Cost Trade-offs

In both CPM and PERT we have been using activity time estimates based on normal working conditions such as a regular work day or standard crew size. Almost all jobs, however, can be reduced in duration if additional resources are applied. Additional manpower, overtime, heavy equipment, and other resources increase the direct cost of an activity, however, and it only pays to use them if there is some incentive to do so. Often that incentive exists, and it may be present in many different forms. Indirect costs such as management, equipment, rentals, fixed expense allocation, and other "overhead" items normally increase as a function of project duration. Another form of indirect cost may be a contract performance clause providing for cost penalties for time overruns as well as monetary rewards for completion ahead of schedule. For example, it is not uncommon in the aerospace industry to have contract penalty or reward clauses of from \$50,000 to \$500,000 per day for late or early delivery, respectively.

We can see that project schedules influence both direct and indirect costs. Our objective as planners and schedulers is to trade off direct cost increases against indirect cost decreases, in search of an overall minimum total cost. We begin by estimating for each activity the absolute minimum activity completion time if all the necessary resources are available. This minimum time is called the *crash time*. A typical direct cost curve is illustrated in Figure 17.9. In this example the solid curved line may be our best estimate of direct cost vs. activity time; however, it is often sufficiently accurate to use a straight line between the crash and normal time-cost points. Here we have a slope of

$$\frac{\$9{,}000 - \$4{,}000}{10 \text{ days} - 5 \text{ days}} = \$1{,}000 \text{ per day}$$

which represents the direct cost of decreasing the activity time by 1 day. We assume that any integer day from 5 days through 10 days may be achieved. Returning to

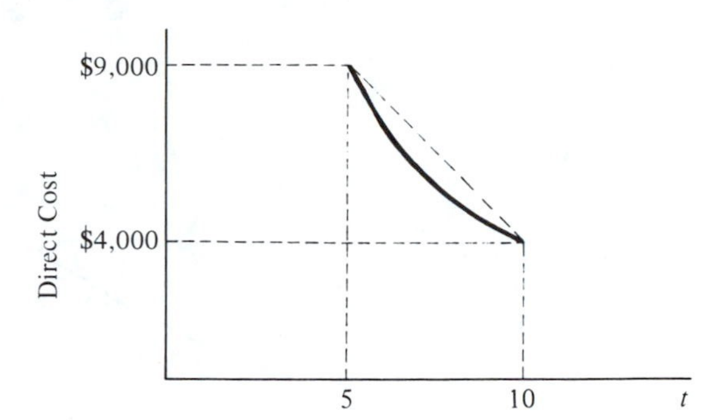

Fig. 17.9. Direct cost vs. activity duration.

our CPM example, we see that the normal and crash times and costs and the activity direct cost slopes are summarized in Table 17.3.

Once we have established normal and crash times, normal and crash costs, and direct cost slopes for each activity, we must establish the indirect costs as a function of project time. The total indirect cost over time for our project is shown in Figure 17.10. The indirect costs are also tabulated in Table 17.4. The indirect costs given include penalty and performance clauses, rentals, wages, and so on, as estimated from various contracts and financial records.

Having specified our direct and indirect costs as a function of time, we can begin systematically reducing activity times to see if a lower total project cost may be achieved. For example, in our problem the project duration using normal activity times is 43 days, as illustrated in Figure 17.6. If we can shorten the project duration from 43 days to 42 days, our indirect costs will drop from $152,500 to $142,500, according to Figure 17.10.

The critical path through the network consists of activities C and G. Either activity can be reduced by 1 day, dropping the project duration to 42 days. Of course, we do not want to randomly reduce the activity duration of either C or G. We want to choose the one that will increase direct costs by the least amount. From Table 17.3, we decide to decrease C by 1 day at a cost of $1,000 instead of decreasing G by 1 day for $3,000. Reducing C from 33 days to 32 days in duration increases total direct costs from $110,000 to $111,000, while reducing indirect costs from $152,500 to $142,500, providing a net savings of $9,000.

Table 17.3. ACTIVITY NORMAL AND CRASH TIMES AND COSTS (Costs are in thousands of dollars)

Activity	*A*	*B*	*C*	*D*	*E*	*F*	*G*	*H*	*I*
Normal time	8	20	33	18	20	9	10	8	4
Crash time	6	16	21	14	17	7	7	8	3
Normal cost	10	22	30	20	6	4	9	3	6
Crash cost	15	38	42	26	7.5	8	18	3	9.5
Slope (cost/day)	2.5	4	1	1.5	.5	2	3	–	3.5

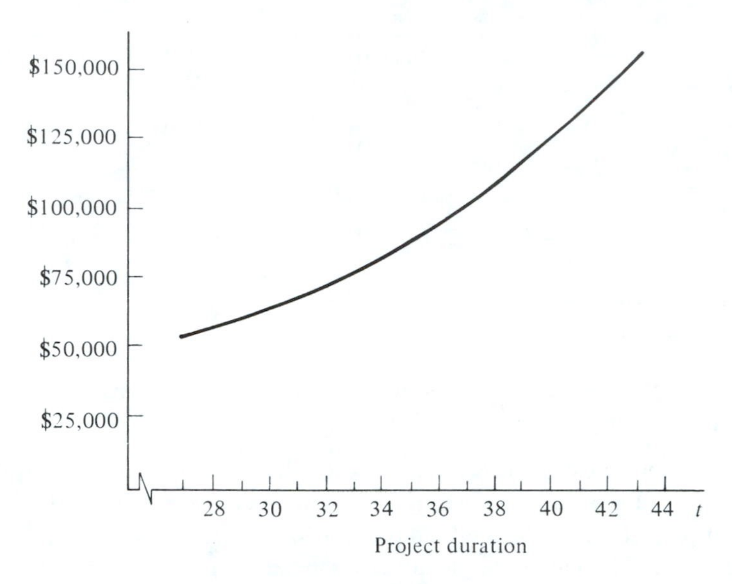

Fig. 17.10. Example indirect costs.

We now consider reducing the project duration further, and we analyze the network in the same way as before. Again, we decide to reduce activity C by another day. In fact, we continue reducing C day by day (from 32 days to 24 days), providing a project duration of 34 days with a direct cost of $119,000 and an indirect cost of $80,500. At a project duration of 34 days, illustrated in Figure 17.11, we have multiple critical paths. That is, both the C, G and A, D, H activity paths have durations of 34 days. Reducing the project length from this point in an optimal (least total project cost) manner is no longer a simple task.

There are several options open to us in which we can reduce project time from 34 days to 33 days. We can reduce each of the activities in any of the following pairs by 1 day each to achieve a project duration of 33 days: A, C; D, C; H, C; A, G; D, G; and H, G. Those pairs containing activity H can be ruled out immediately since we see from Table 17.3 that H cannot be reduced. Further, since C can be reduced even more at a direct cost of $1,000 per day, we shall choose it over reducing activity G in the lower path. Finally, we decide to reduce activity D in the upper path at a cost of $1,500 per day instead of shortening activity A at a cost of $2,500 per day. The result of our deliberations is to reduce activity D from 18 days to 17 days and activity C from 24 days to 23 days, increasing direct cost from $119,000 to $121,500 and decreasing indirect cost from $80,500 to $75,000, for a net gain of $3,000.

We have now achieved a project duration of 33 days and there are three critical paths: C, G; A, D, H; and B, F, I. In reducing the network duration further, we find that the direct cost of reducing the network tends to increase at a faster rate. This is because activities such as C with a low cost slope of $1,000 per day

Table 17.4. TIME–COST TRADE-OFFS FOR EXAMPLE NETWORK
(Costs are in thousands of dollars)

Project Duration	*Activity Changes*	*Incremental Cost $1,000's*	*Direct Cost $1,000's*	*Indirect Cost $1,000's*	*Total Cost $1,000's*
43	None	None	110	152.5	262.5
42	C: 33 to 32	1	111	142.5	253.5
41	C: 32 to 31	1	112	133	245
40	C: 31 to 30	1	113	124	237
39	C: 30 to 29	1	114	115.5	229.5
38	C: 29 to 28	1	115	107.5	222.5
37	C: 28 to 27	1	116	100	216
36	C: 27 to 26	1	117	93	210
35	C: 26 to 25	1	118	86.5	204.5
34	C: 25 to 24	1	119	80.5	199.5
33	C: 24 to 23	1			
	D: 18 to 17	1.5	121.5	75	196.5
32	C: 23 to 22	1			
	D: 17 to 16	1.5			
	F: 9 to 8	2	126	70	196
31	C: 22 to 21	1			
	D: 16 to 15	1.5			
	E: 20 to 19	.5			
	F: 8 to 7	2	131	65.5	196.5
30	D: 15 to 14	1.5			
	G: 10 to 9	3			
	I: 4 to 3	3.5	139	61.5	200.5
29	A: 8 to 7	2.5			
	B: 20 to 19	4			
	G: 9 to 8	3	148.5	58	206.5
28	A: 7 to 6	2.5			
	B: 19 to 18	4			
	G: 8 to 7	3	158	55	213

eventually reach their crash time and cannot be reduced further and we must resort to shortening more costly activities such as G with a cost slope of $3,000 per day. A complete summary of all project duration reductions and costs is given in Table 17.4.

The shortest project duration possible is estimated to be 28 days. That is, the critical path(s) cannot be reduced below 28 days since all the activities along at least one of the paths have been decreased to their crash times. For example, activity C is at 21 days and activity G is at 7 days. Neither activity may be reduced further. We should also note that the minimum total cost of $96,000 occurs at a project length of 32 days. We have performed a "trade-off of costs," initially swapping increases in direct cost for larger decreases in indirect costs. This continued until at 31 days the increase in direct cost exceeded the decrease in indirect cost. Thus the

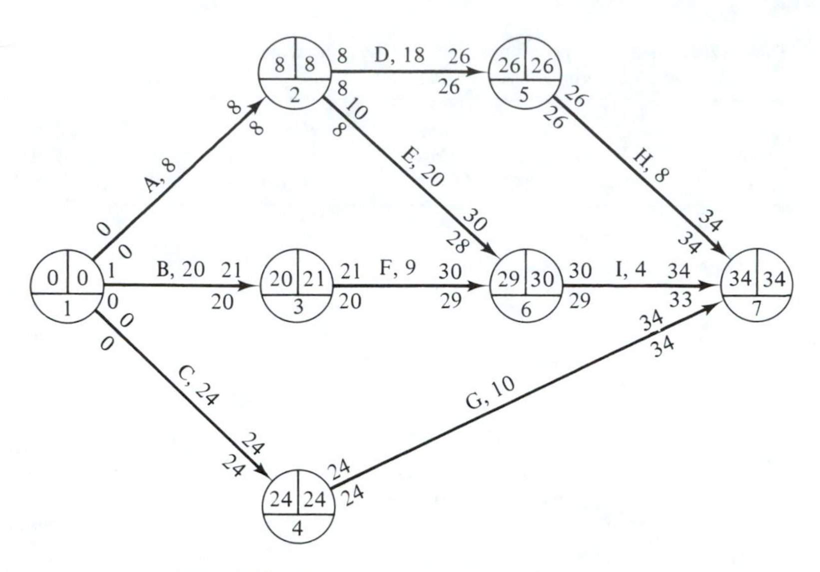

Fig. 17.11. Network compressed to 34 days.

optimum project length is 32 days. The concept of "trade-off of costs" is illustrated in Figure 17.12.

The procedure used for reducing the network in our example was largely one of complete enumeration and common sense. We can see, however, that for a network that has hundreds of activities, this approach would be infeasible. Numerous computer packages suitable for computer use on large projects have been developed.

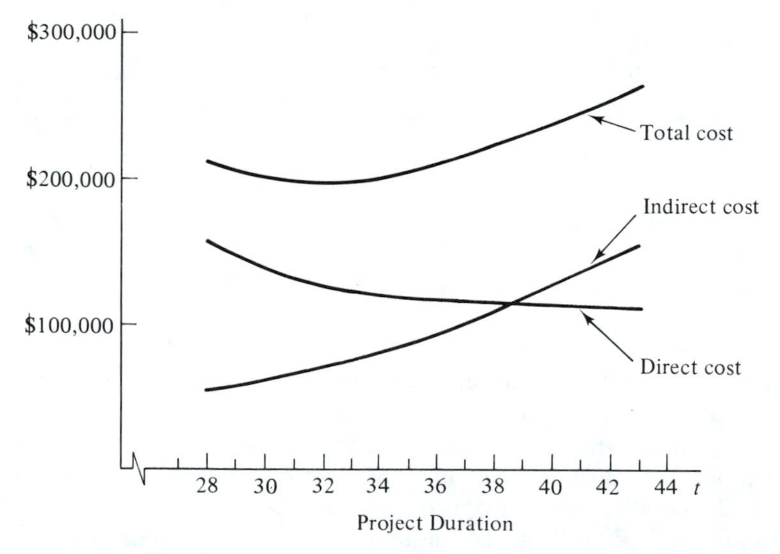

Fig. 17.12. Trade-off of costs in example network.

17.6. Resource Leveling

Often when we schedule the manpower resources for the activities of a project, some days require a large number of workers and other days require very few. The reason for this is that we have probably scheduled our manpower based on each activity beginning at its "earliest start" time. We should remember, however, that each activity may also begin as late as its "latest start" time. This flexibility, which we previously called slack, may be used to delay noncritical projects in such a way that fluctuations in manpower needed may be greatly reduced. That is, our resource needs over time may be leveled.

The obvious advantage of resource leveling is that we need not maintain a work force based on the highest manpower requirements originally scheduled. Further, through resource leveling we can minimize worker idle time. The procedures for resource leveling are beyond the scope of this book.

DISCUSSION QUESTIONS

1. Describe the differences between: project management and production management; activity and node; CPM and PERT; earliest start time and latest start time; path and critical path.
2. Give the equations relating: earliest event occurrence time, earliest activity start time, and earliest activity finish time; latest allowable event occurrence time, latest allowable activity start time, and latest allowable activity finish time; earliest activity start time, latest activity start time, and total activity slack.
3. Suppose that you need to know an activity duration. The person most knowledgeable about that activity claims to be unable to estimate its duration (this is not uncommon). How would you help this person provide an estimate?
4. Give three examples not in the text where you would prefer CPM to PERT; PERT to CPM.
5. If a "dummy" activity is not really an activity and has zero time, of what use is it in CPM and PERT?
6. What distribution of time is assumed when determining the probability of completing all of the activities along the critical path? What role is played by the central limit theorem? (See Appendix A.)
7. What can you say about the earliest start time and latest start time along a critical path? Along a noncritical path?
8. Assuming that all time estimates are correct, will the longest path actually achieved ever be other than the critical path identified using CPM? Using PERT? Explain.
9. Suppose that you are partially into a project and find that some of your time estimates were in error, and perhaps it now appears that even a different path may be critical. What would you do? Do you need to throw out your planning network?
10. In Figure 17.9 we used a straight line to approximate the direct cost increase in going from normal time to crash time. How would you use the curved line if you felt that additional accuracy were needed?

PROBLEMS

1. Redraw Figure 17.1 using the node and activity representation shown in Figure 17.6. Modify the times as: A, 3; B, 30; C, 22; D, 12; E, 12; F, 1; G, 6; H, 8; I, 2. Make both the forward and backward passes through the network and identify the total activity slacks.
2. Determine the critical path of the network shown in Figure P17.2.
3. Write the precedence matrix as in Table 17.1 for the network in Problem 2.
4. A small plant layout job consists of 10 steps. Their precedence relationships and activity times are identified as follows:

Step	*Predecessor*	*Time (Hours)*
A	None	9
B	None	13
C	None	16
D	A	18
E	C	19
F	C	8
G	B,F	11
H	D,G	9
I	E,H	26
J	B,F	35

Draw the network and complete the forward and backward passes. What activities make up the critical path? Which activity has the most slack?

5. Suppose that the network tasks in Problem 4 have the following PERT time estimates:

	PERT Time		
Step	t_o	t_m	t_p
A	6	9	12
B	7	13	19
C	12	15	24
D	6	20	22
E	15	19	23
F	8	8	8
G	2	12	16
H	5	8	17
I	20	26	32
J	30	35	40

What is the expected time duration along the project's critical path? What is the variance of project time along the critical path? Within how many hours does the critical path have an 84.13% chance of completion?

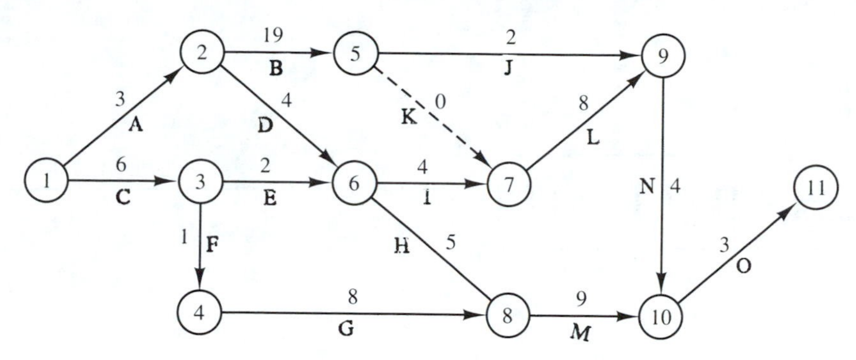

Fig. P17.2.

6. If a project has a PERT expected duration of 40 days and a standard deviation of 7 days, what is the probability that the project will be completed in the interval from 30 to 50 days?
7. Determine the critical path for the network shown in Figure P17.7, where the numbers shown are t_0–t_m–t_p. What is the time of completion for the critical path that has an 80% chance of being achieved?
8. If you know the variance of a PERT activity time is 4.00, and the most likely time is one less than the expected time of 25, what are t_o, t_m, and t_p?
9. For the time–cost trade-off problem of Section 17.5, determine the optimum project duration letting indirect cost (in thousands of dollars) decrease as follows:

Project Duration	*Indirect Cost*	*Project Duration*	*Indirect Cost*
43	152.5	37	113.5
42	143.5	36	110.5
41	135.5	35	108.5
40	128.5	34–28	107.5
39	122.5		
38	117.5		

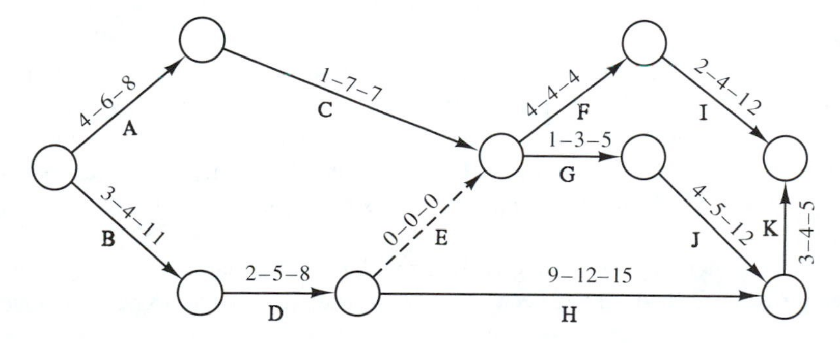

Fig. P17.7.

PART 3 INTEGRATED SYSTEMS DESIGN

CHAPTER 18

Systems Concepts

18.1. Introduction

In Chapter 2[1] we proposed the notion that the industrial engineering profession is entering a new era, one that could be called "industrial and *systems* engineering." It was further suggested that one of the important characteristics of this new era will be the *scope* of the systems with which practicing industrial and systems engineers must deal. In particular, I.&S.E.s must be more concerned about the design of *total systems*. This implies a shift of focus from a micro view of systems to a macro view.

In Chapters 3 through 17 a large number of tools and techniques were introduced which are essential for performing specific industrial engineering functions. I.&S.E.s of the future will still need to have full knowledge of these tools and techniques. What now must be added are knowledge and methodologies that will facilitate the I.&S.E.'s role in analyzing, designing, controlling, and enhancing total systems.

We see then that the traditional I.E. role has been expanded to I.&S.E. None of the previous functions have been replaced or deleted. Relative to new knowledge

[1]It is suggested that Chapters 1 and 2 be reviewed before proceeding with this chapter.

and new tools and techniques, we now need a *unifying framework* for dealing with total systems. Such a framework is still evolving in the I.&S.E. profession. Our treatment of it here will necessarily be elementary and introductory.

18.2. Introduction to Systems Thinking

18.2.1 Origin of Systems Thinking

The origin of systems thinking may have begun as human beings developed the rather distinctive characteristic of categorizing objects according to common features. Early in human experience, for example, primitive thought processes began to perceive that it was useful to distinguish living objects from nonliving ones.[2] Later refinements of this concept led to further distinctions within each of these categories. Living objects, for example, can be further classified into animals and plants, because all objects within each of these categories share certain common features. This process of finer and finer classification continues to this day. Human beings seem to have an innate ability (need?) to perceive reality in such definitional terms.

Another aspect of the origin of systems thinking was the realization that particular objects are composed of components and that these components are interrelated and interdependent. This kind of thinking is fairly recent in human history, perhaps as recent as approximately 4,000 years ago. Only in the last 200 years or so have we been able to understand the *nature* of the interrelationships and interdependencies among the components making up an object. Our ability to exercise *control* over these interrelationships and interdependencies has existed for less than a century. Notice that the discussion in this paragraph has an *inward* focus relative to a particular object.

Yet another conceptual development relative to systems thinking was the realization that groups of individual (perhaps different) objects also exhibit interrelationships and interdependencies. Many aspects of human experience led to this realization, one of which was observations of the interrelationship between the moon's phases and the ocean's tides. Notice that the discussion in this paragraph has an *outward* focus relative to a particular object.

One of the critical developments relative to the origin of systems thinking is that of *cause and effect*. This represents a further refinement of interdependencies among the objects that are related in some way. The concept of cause and effect comes from the recognition that when a particular object behaves in a certain manner, a different object in the related group will behave in some predictable manner. The behavior of the affected object may be immediate or delayed. It may be similar or opposite to the behavior of the initiating object. It may be deterministic

---

[2]It is interesting to note that not only was this the genesis of taxonomy (the science of classification), it was also the genesis of set theory.

(the response behavior always occurs) or probabilistic (the response behavior may occur with a certain probability). Thus there are many possible ways in which effect is related to cause. The important thing is that the nature of the relationship is measurable and predictable.

A final aspect that we will discuss regarding the origin of systems thinking is that of *synergism*, a concept that has been recognized intuitively for centuries but has yielded to some degree of formalization and quantification only in recent years. Synergy is the phenomenon that occurs when separate components of a system, working together, have greater total effect than the sum of their individual effects working separately (independently).

Systems thinking represents a more comprehensive and cohesive way of conceiving our experiences. By identifying and characterizing the interactions of influences among components of an object, we begin to recognize that we cannot understand the object's behavior accurately simply by adding up the behaviors of its components. The object's behavior can only be understood by identifying and characterizing the impact of the components on each other and the net influence of the components on the overall object. "Systems thinking enriches our awareness of nearly all dimensions of life. It explains complexity in a more comprehensive way."[3] We could add that systems thinking explains complexity in a more *comprehensible* way.

Systems thinking has had, and will continue to have, a dramatic impact on society. Some of this impact is extremely beneficial and some of it is detrimental. Consider, for example, large-scale exploration for oil and gas. Systems thinking at a high level has resulted in an extremely efficient process of locating these hydrocarbon deposits, extracting them, transporting them throughout the world, and refining them into products for human consumption. Concomitant with the beneficial results of this process are the detrimental results that have occurred: massive oil spills, destruction of animal habitat, pollution of the air, and so on.

Some people argue that our obsession with large-scale technology (made possible by our ability to engage in large-scale systems thinking) has led to the many major problems facing humankind today. Yet the only reasonable approach to solving these problems appears to be the application of systems thinking. Perhaps our earlier applications of the systems concept employed system boundaries that were too restrictive.

18.2.2 Hierarchical Nature of Systems

Consider the many ways in which we use the term "system": production *system*, wage incentive *system*, material handling *system*, inventory control *system*, inspection *system*, information *system*, computer *system*, programming *system*,

---

[3]Robert Wright, *Systems Thinking: A Guide to Managing in a Changing Environment* (Dearborn, Mich.: Society of Manufacturing Engineers, 1989), p. 5.

scheduling *system*, work measurement *system*, and so on. Careful examination of these items reveals that some are clearly subsets of others. For example, a material handling system is clearly a subset of a production system. Is the term ''system'' being misused in some or all of the examples cited above? Not at all! As we shall see in the next section, each of the cited examples fully qualifies as a system.

Whether a particular item is properly considered a system depends on the specific context in which interest in it is being expressed at that time. A particular item may be a system in one context and a component of a system in another context. Careful analysis of any component of a system will reveal that the component itself is composed of several elements which collectively constitute a system. Conversely, a particular system defined for one purpose may be a component of a higher-level grouping of entities (i.e., a system). We can begin to see emerging the concept of a *hierarchy* relative to system identification.

Let us consider an example. Shown in Figure 18.1 is a slice of an abstract system called the ''world economy.'' Before starting down the hierarchy, recognize that the world economy becomes one of several components of a higher-level system that is concerned not only with economic matters, but also with the natural environment, equitable distribution of political power, and so on. Also recognize that this structure illustrates the manner in which any individual worker, say John, is related to every higher-level system in the hierarchy. John's individual performance *does* make a difference! Finally, recognize that John is an element of many other systems (bowling team, family, church, civic club, etc.), and that John himself is composed of many systems (circulatory system, respiratory system, digestive system, nervous system, etc.).

Consider the system ACME Transmission Company in Figure 18.2. Line B in shop 2 manufactures gears. Suppose that you are a member of an I.&S.E. team that is responsible for a complete analysis of the entire line B. In such a case, the *system with which you are concerned* would be the one enclosed within the dotted line.

Shop 2 in Figure 18.2 contains all the heavy-duty metal cutting equipment. A second I.&S.E. team is concerned with analyzing the total capacity of shop 2. The system with which the second team is concerned is the one enclosed within the dashed line.

As a final example of a system definition, depending on the particular purpose at some point in time, consider the situation in which ACME is considering building a special-purpose transmission for Ford Motor Company. This will involve a major effort in engineering design and a significant investment in advanced machining equipment that would be located in line C. Quality and reliability considerations are of paramount importance. It is also critical that a particular raw material be available for purchase at reasonable prices. The ''system'' pertinent to this situation might consist of the elements enclosed within the dashed line in Figure 18.3. This is an example of a system that will function only until the decision is made whether to proceed with the new transmission. If the decision is positive, the ''normal'' systems take over.

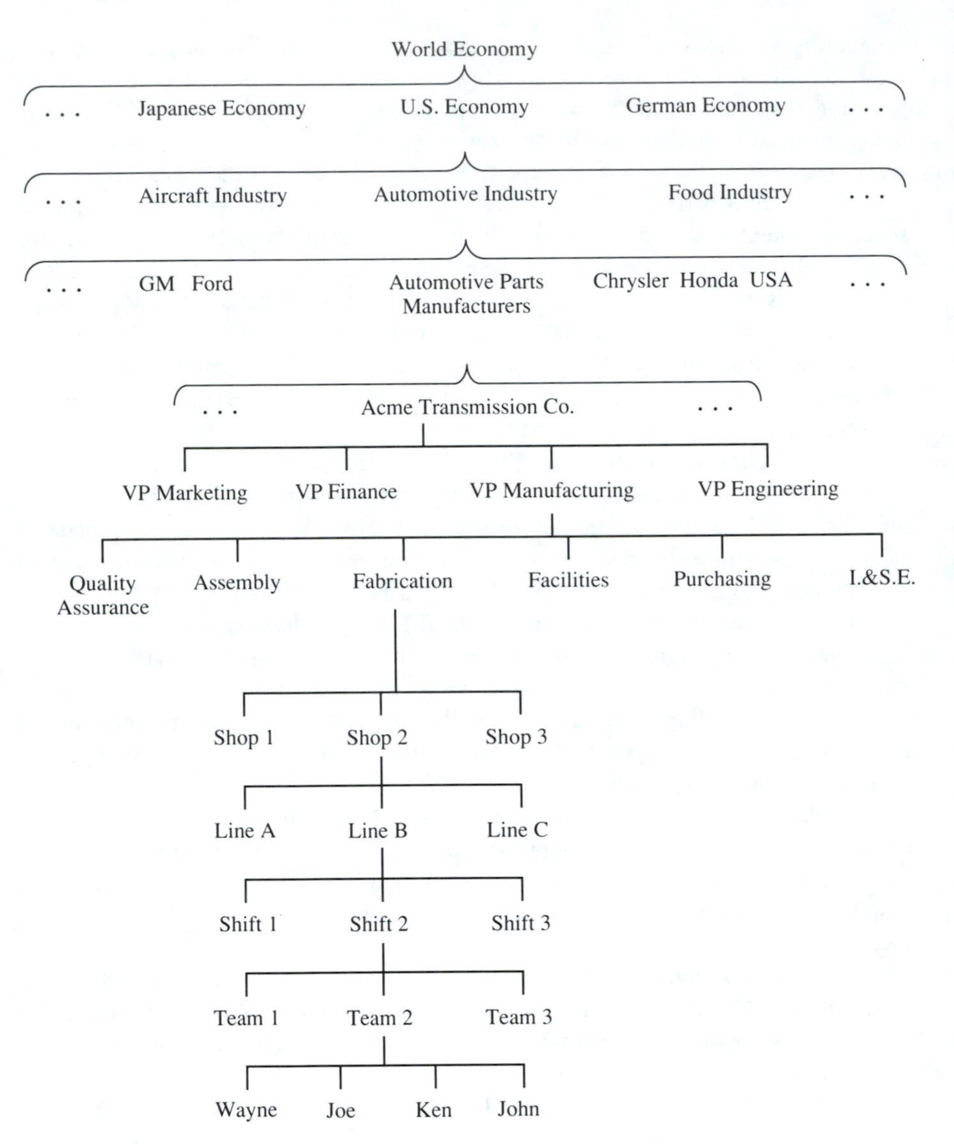

Fig. 18.1. Hierarchical structure of systems.

18.3. Definitions and Terminology

We have been using many terms in our discussion (such as "system boundary") without defining or explaining their meaning. We will now become more rigorous in our use of terms by providing definitions relative to our study of "systems concepts."

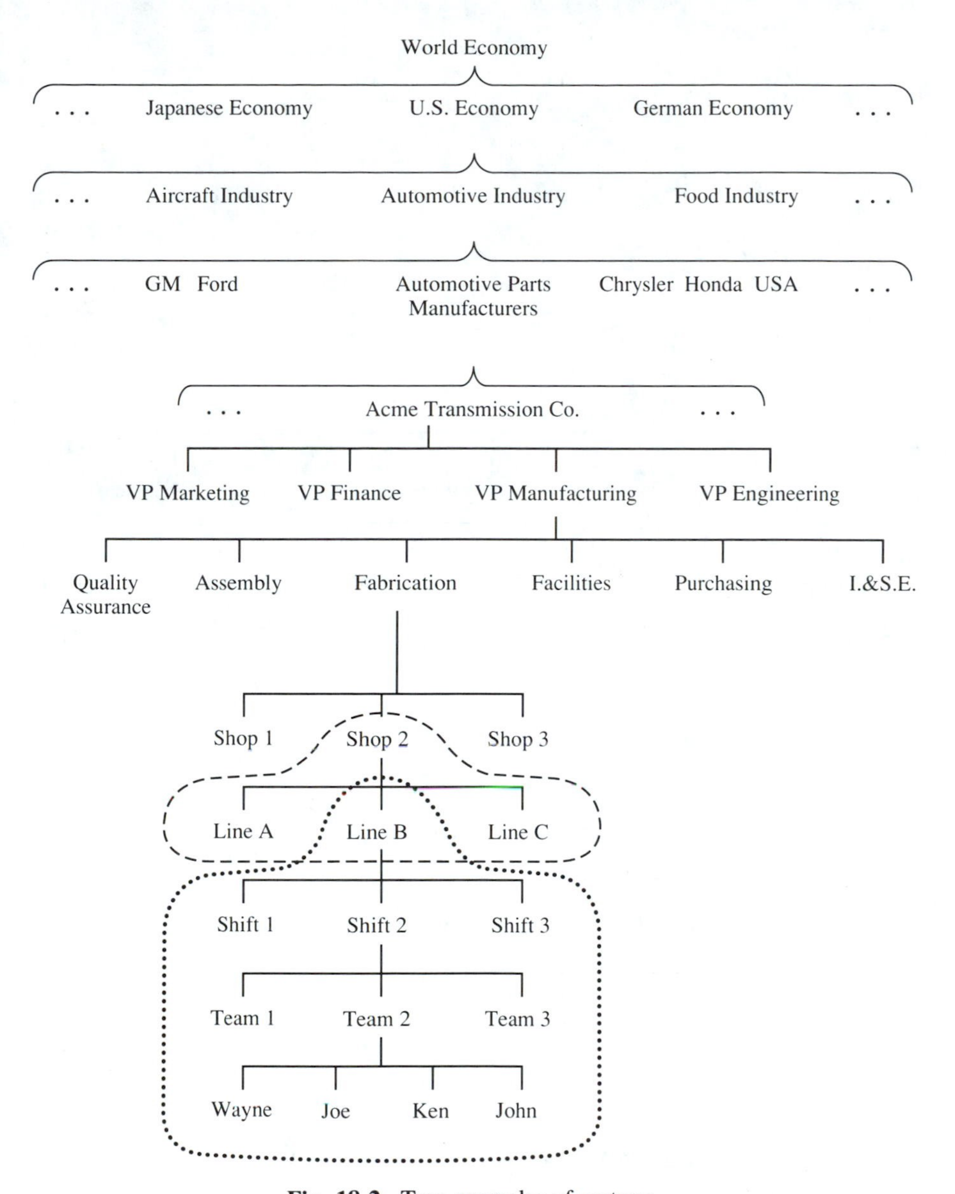

Fig. 18.2. Two examples of systems.

- *System*—A set of components or subsystems related by some form of interaction, acting together to achieve some common purpose.
- *Components*—The individual parts, or elements, that collectively constitute a system.
- *Subsystem*—If elements from two hierarchical levels are involved in a particular system, the elements at the lower level are called *subsystems*. Parts of the

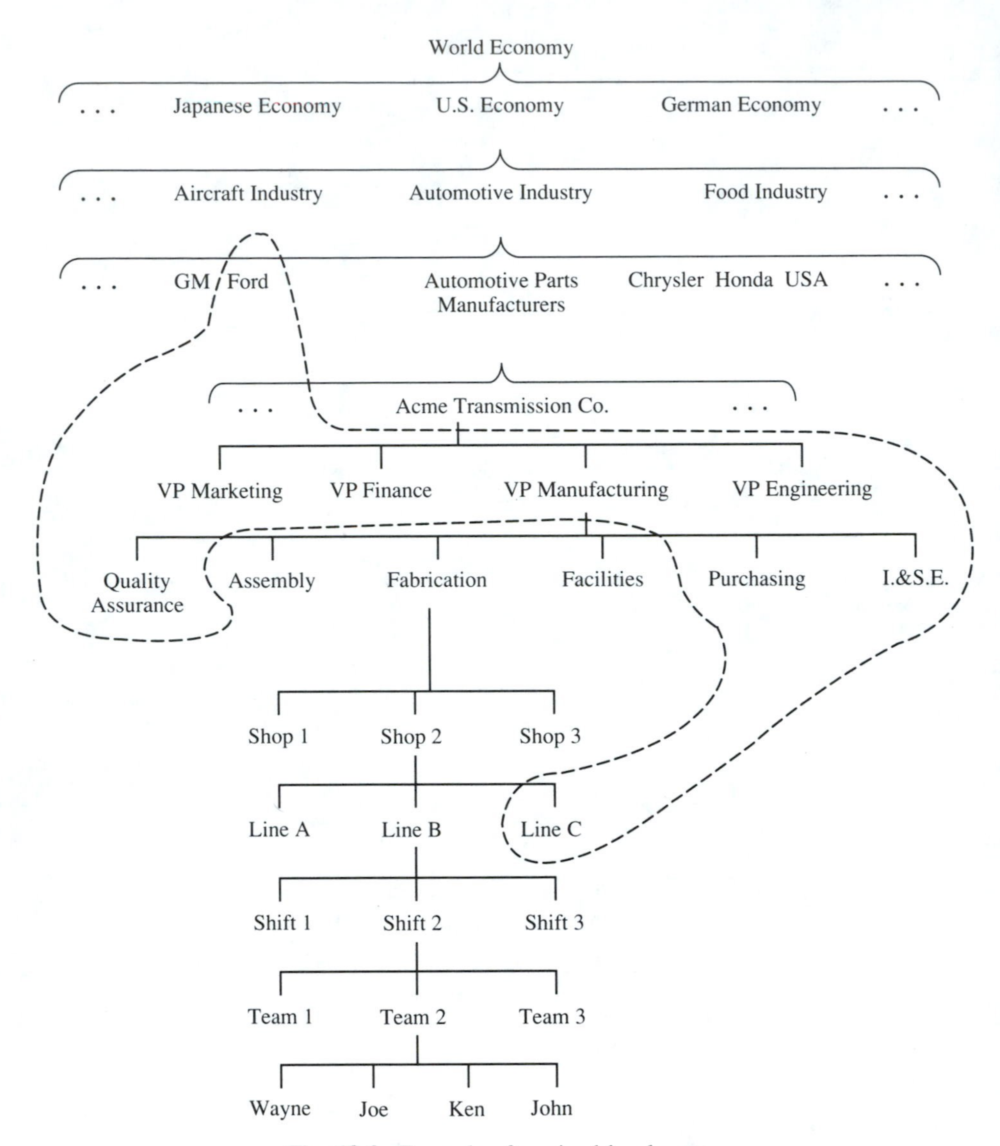

Fig. 18.3. Example of a mixed-level system.

subsystems are called components.[4] Subsystems are those processes necessary to the operation of a total system (defined below).

- *Relationship*—A functional or logical dependency between components of a system.

---

[4]The terms "system," "subsystem," and "component" are relative. A system at one level could be a subsystem or component at another level.

- *Interaction*—Mutual interplay between two or more system elements operating concurrently (e.g., human–machine interaction).
- *Purpose*—The fundamental reason for the system's existence (its *raison d'etre*).
- *Function*—A purposeful action performed by a system (e.g., information processing over a networked computer system).
- *Attribute*—A characteristic or property of a system component that is pertinent to the system of interest (e.g., the dimensional tolerance that a particular machine is capable of holding).
- *Environment*—A set of all objects, within specified limits, that may have influence on, or be influenced by, the operation of the system of interest.
- *Boundary*—A specified demarcation that prescribes a limit within which the components, attributes, and their relationships are sufficiently explained.
- *Open System*—A system that regularly exchanges across its boundary materials, energy, or information with its environment.
- *Closed System*—A system that operates with (theoretically) no interchange of materials, energy, or information with its environment.
- *Input*—That which passes from the environment, across the boundary, and into the system during the time frame of interest.
- *Output*—That which passes from the system proper, through the boundary, and into the environment.
- *Throughput*—That which passes into a system, is usually modified in some way, and then passes out of the system.
- *Process*—The set of system components, their attributes, and interrelationships required to produce a given result. Processes may be physical (cutting metal, moving parts between work centers) or abstract (accounting procedures, planning).
- *Transformation*—The changing of inputs (e.g., material, labor, capital, energy) into outputs (usually products or services). Within manufacturing firms, the transformation results in *added economic value*. Within service organizations, the transformation yields *added social value*.[5]
- *Constraint*—A limit imposed upon the system (e.g., walls of a plant that may not be changed).
- *Feedback*—Information relative to the output of a process that is returned to the source of the process controller, such that the actual output may be compared with the intended output, and any appropriate corrective action determined prior to subsequent operation of the process. More generally, feedback exists in a system when a closed sequence of cause–effect relationships exist between system elements.
- *Control*—That which guides, directs, regulates, or constrains, with the particular control action determined from feedback.

---

[5]Wright, p. 28.

- *System State*—The condition of the system at any instant, indicated by the values of the system variables.
- *Dynamic System*—A system whose behavior varies with time.
- *Total System*—The set of all subsystems, components, attributes, and relationships necessary to achieve the basic purpose of a system given its constraints. The importance of the "total system concept" is to recognize that to optimize the performance of one component or subsystem without regard to its effect on the larger system will very often lead to suboptimal performance of the total system.

18.4. Systems Engineering

18.4.1 Systems Analysis and Design

We distinguished between analysis and design in Chapter 1. There we learned that *analysis* is the investigation of the properties of a given (existing) system, while *design* involves the choice and arrangement of system components to perform a specific function. Furthermore, we described *synthesis* as the creation and structuring of components into a whole, so as to obtain optimal performance from the total system.

This suggests that there are two basic approaches to system design:

- Design via analysis
- Design via synthesis

Design by analysis is accomplished by modifying the characteristics of an existing or standard system configuration. Design by synthesis is performed by defining the form of the system directly from its specifications.

Although we would like to always perform systems design generically (by synthesis), in reality, most system design work is related to ongoing systems that we "inherit." For these cases, considerable analysis is required before we are able to begin our design work. Most design work, therefore, involves both analysis and synthesis.

18.4.2 The Systems Design Process

The systems design process can be considered a series of identifiable stages. These stages are not sharply demarcated and linearly sequenced, but they do represent a definite progression of activity. Our development here will generally follow that of Blanchard and Fabrycky.[6]

We will outline a 10-stage system design process. By combining some of the

---

[6]For further details, see B. S. Blanchard and W. J. Fabrycky, *Systems Engineering and Analysis*, 2nd ed. (Englewood Cliffs, N.J.: Prentice Hall, 1989), Chap. 2.

stages, other authors outline design processes having fewer or more stages than this one. It is all a matter of perspective.

(1) Recognition of need and definition of system purpose.
- Identify specific *needs*.
- Determine system *operational requirements*.
- Focus on *decisions* required for successful system operation.

(2) System functional analysis.
- Determine *functional requirements*.
- Specify *generic operational functions*.
- For each generic function, identify *alternative specific functions*.

(3) Preliminary system optimization.
- Identify *relative trade-offs* among alternatives.
- Evaluate *alternative configurations* in terms of functional requirements.

(4) Preliminary system design.
- Specify preliminary *system configuration* and arrangement of chosen system components.
- Develop detailed *system specifications*.

(5) Detailed system design.
- *Detailed design* of functional system.
- Formally *document* all design data.
- Perform a detailed *design review*.

(6) System prototype.
- Develop/construct system *prototype*.
- Perform *testing* and *evaluation* of prototype.
- Perform systems analysis and *evaluation* on test data.
- Specify *modifications* as required.

(7) System acquisition (for purchased items).
- Develop *Requests for Proposals* (RFPs).
- Evaluate submitted RFPs on appropriate criteria.
- *Select vendor(s)*.
- Negotiate delivery *schedule* of system components.

(8) System implementation.
- Develop detailed *master plan* for system implementation.
- Prepare the organization for *change*.
- Perform *education* and *training*.
- Arrange for *system conversion*.

(9) System operation and maintenance.

(10) System upgrades, improvements.

At some point, the system will need to be replaced. All systems have a finite life cycle.

As was mentioned earlier, the 10 stages outlined above are not usually performed in a strictly linear progression. After certain stages have been performed, it may be necessary to return to an earlier stage. Iterative cycling of this type may be required throughout the entire design process.[7]

18.5. System Representation

Humans have always sought ways of picturing various elements of their world, through graphic illustration. We will now show how the rather abstract concepts of systems engineering may be represented graphically.

18.5.1 Block Diagrams[8]

The *block diagram* is a commonly used visualization aid of engineers. Figures 2.4 through 2.7 are examples of block diagrams. We will explore an expanded version of this concept. In particular, we will add a controller block, as shown in Figure 18.4. This system is said to be governed by *open-loop control*, since there is no feedback of the actual output.

A *closed-loop control* system utilizes a measure of the actual output, compares it with the desired output response, and then conveys appropriate corrections to the controller. Adjustments are made to the control signal, which affects the process. Such a system is portrayed in Figure 18.5.

Most of the systems with which I.&S.E.s must deal involve multiple inputs and outputs. A block diagram depicting a *multivariable control system* is shown in Figure 18.6. A review of Figure 7.3 will show that it is such a system.

To crystallize some of these ideas, consider one of the video games that simulates antiaircraft warfare.[9] The player controls two handlelike devices that have several rotational degrees of freedom. As the game progresses, various enemy aircraft move swiftly across the screen. The game objective is to ''kill'' as many enemy aircraft as possible. This game is very realistic in its capturing of the essential features of antiaircraft warfare.

Considering antiaircraft warfare (the real thing) as a system, one of the critical ''components'' is a human being. The person performs two primary functions in this system:

(1) Observes or senses the input and the output to determine the error; then
(2) Generates a control signal that drives the process, hopefully in the direction to reduce the error toward zero.

---

[7]The authors recognize that this rather abstract process is difficult to comprehend by those readers who have not yet been involved with ''real-world systems.'' Nevertheless, it is an important framework for later professional activities.

[8]Some of the concepts in this section are based on Richard Dorf, *Modern Control Systems* (Reading, Mass.: Addison-Wesley Publishing Co., 1967), Chap. 1.

[9]See John Trauxal, *Introductory Systems Engineering* (New York: McGraw-Hill Book Company, 1972), pp. 86–87.

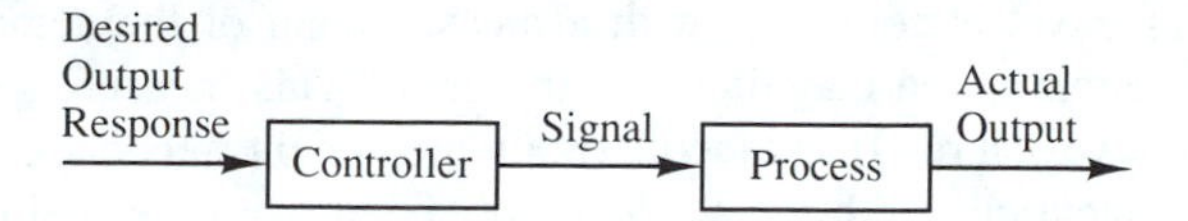

Fig. 18.4. Open-loop control system.

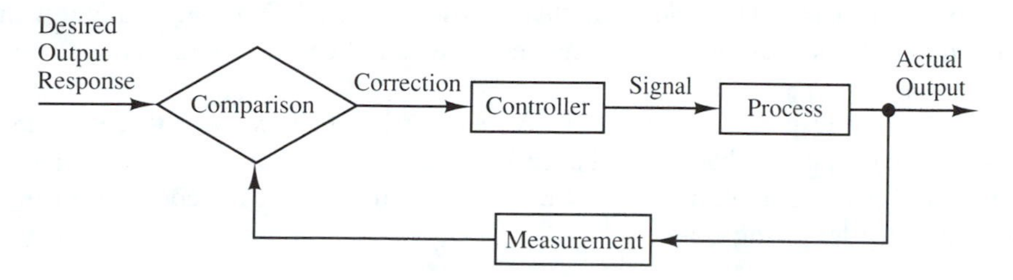

Fig. 18.5. Closed-loop feedback control system.

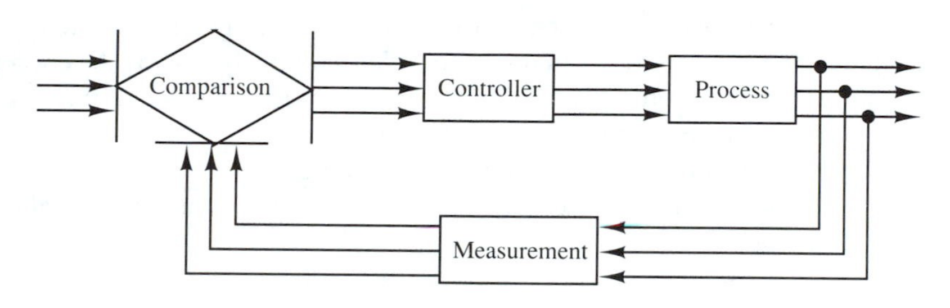

Fig. 18.6. Multivariable control system.

Figure 18.7 shows the role of a human in a control system, such as the antiaircraft example. The human is observing the output, comparing it to the target, and initiating corrective action. In this case, the *process* is the gun mount with its inertia, friction, chatter, and other dynamic characteristics. The *output* is the gun direction. (Once the projectile leaves the gun barrel, it is in open-loop mode; there is no control over its direction.) The *measurement* is the perceived deviation of the projectile and the target. The *input* is the target position predicted ahead in time so as to provide the time required to compensate for the time required for the projectile to reach the moving target.

As another example of a human directly in the feedback control loop of a complex system, consider the ordinary experience of driving a car.[10] A block diagram of an automotive steering control system is shown in Figure 18.8. The

[10]This example is adapted from Dorf, p. 5.

desired path of travel is compared with a measurement of the actual path, which yields an indication of the magnitude of the error. This measurement is acquired through visual (eyes on road) and tactile (body movement) feedback. This measurement must be integrated with the feedback obtained from the feel of the steering wheel by the hands.

The role of a manager attempting to guide his or her organization toward the achievement of its goals and objectives is very analogous to that of an automobile driver. Some managers would claim that their job is more like driving at night with only dim lights in a driving hailstorm over a road filled with washouts, fallen trees, and so on.

It is the responsibility of the industrial and systems engineer to design the management systems that will enhance the chances of an organization having a successful and profitable journey. We explore some elementary concepts in this regard in the following chapter.

18.5.2 Transfer Functions

Engineers who design feedback control systems such as guidance systems for missiles, oil refineries, and chemical processing plants routinely employ the concept of the *transfer function*. Although a full treatment is beyond the scope of this book, the basic concept can be introduced at an elementary level.

Consider the elementary system illustrated in Figure 18.9. Consider the case for which a component is described in terms of its input and output by an algebraic equation of the form

$$O(t) = Ki(t) \tag{18.1}$$

where K is a constant. The product of $i(t)$ with K gives $O(t)$. Therefore,

$$K = \frac{O(t)}{i(t)} \tag{18.2}$$

The constant K can be regarded as a *transfer function* for the component, in the sense that K times the input gives the output. For example, if the "component" is a production line manufacturing printed circuit boards, and K is known to be 0.79, we then know that of all the boards started into production as input, 79% will be successfully completed.

Unfortunately, most components in industrial systems are described by differential equations rather than algebraic equations. Thus it is not sufficient to define the transfer function as $O(t)/i(t)$. The Laplace transform is typically employed for this purpose since the Laplace transformation converts a differential equation into an algebraic equation in s, where s is a complex variable having the real and imaginary components: $s = \sigma + j\omega$, $j = \sqrt{-1}$. The transfer function, frequently denoted $T(s)$, can then be defined for a linear component as the ratio of $\mathcal{L}[O(t)]$ to $\mathcal{L}[i(t)]$. Initial conditions are assumed to be zero for defining the transfer function.

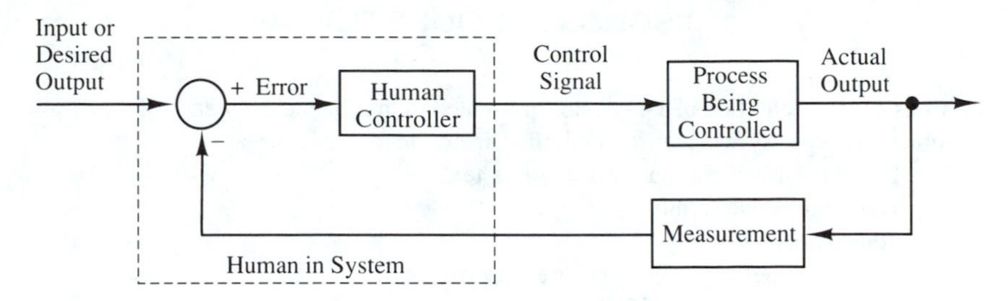

Fig. 18.7. Human being in the loop of a control system.

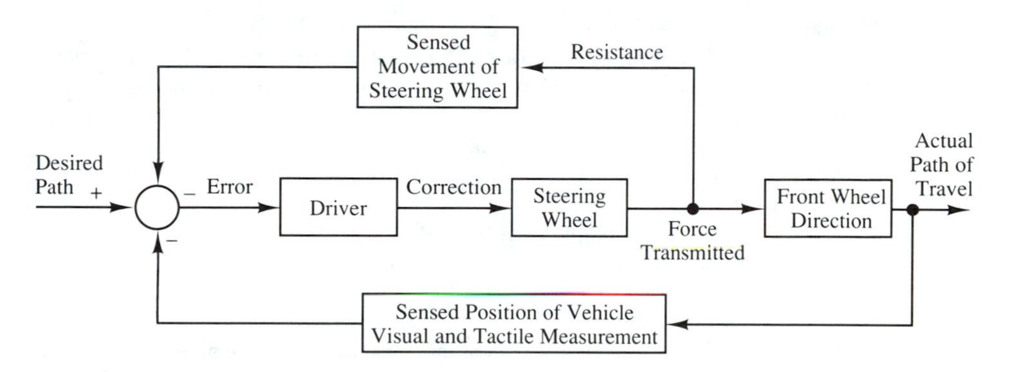

Fig. 18.8. Steering control system of an automobile.

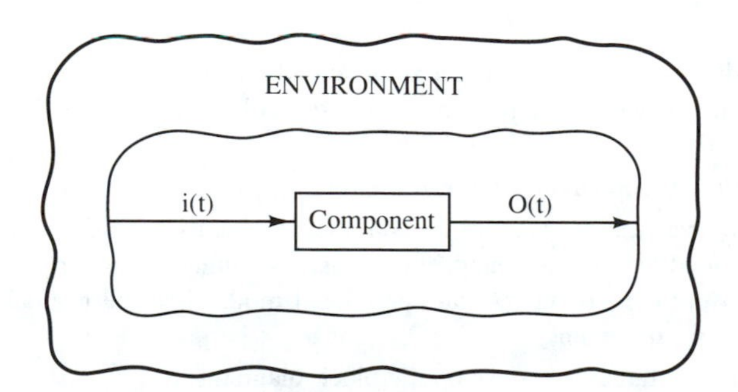

Fig. 18.9. A system within its environment.

DISCUSSION QUESTIONS

1. Consider the definition of a "system" presented in this chapter. For each of the following situations, specify whether the definition is applicable and why:
   (a) The class in which you are using this text.
   (b) Your immediate family.
   (c) Your extended family.
   (d) The department in which you are majoring.
   (e) Your student chapter of IIE.
   (f) Your barber, your dentist, and the mechanic who keeps your car running.
   (g) This textbook and the last bottle of soft drink you enjoyed.
   (h) You, the instructor in this class, and the authors of this text.

2. For each of the situations described in Question 1 that you believe is a system, describe how two components within that system could be working at cross purposes.

3. For each of the situations described in Question 1 that you believe is a system, specify three or more performance criteria.

4. For each of the following systems, use the definitions from this chapter to specify components, subsystems (if any), relationships, interactions, purpose, functions, inputs, outputs, throughput (if any), and processes:
   (a) An automatic clothes washing machine.
   (b) A laundromat.
   (c) A fast-food restaurant.
   (d) A drive-in bank with multiple lanes.
   (e) A university bookstore.
   (f) A microwave oven.
   (g) A hairstyling shop.
   (h) Traffic control system at an intersection that must accommodate pedestrians as well as vehicles.
   (i) A commode in a toilet.

5. (a) Describe the components and input/output variables of the biological control apparatus involved when a person walks in a prescribed direction.
   (b) Why is walking a closed-loop operation?
   (c) Under what conditions would the human walking apparatus become an open loop?
   (d) Would the answers to parts (a), (b), and (c) be different if a cat had been specified rather than a person?

6. Devise a simple control system that automatically turns on a light at dusk and turns it off at dawn. It must operate properly at any time of the year. Show a schematic of the system you devise.

7. Devise a closed-loop automatic bread toaster.

8. Explain the operation and identify the components, inputs, and outputs of an automatic, radar-controlled antiaircraft gun. Assume that no human operator is required except initially, to put the system into an operational mode. Use the example given in this chapter as a starting point.

9. For Questions 5 through 8, develop the block diagrams.

10. When a human being is an active component of a control system, he or she must receive various inputs. What are the generic forms of input that people are capable of receiving?

11. What are the consequences if the boundary defining a system changes over time such that the components constituting the system change?

12. Develop a hierarchical diagram of the human body analogous to that in Figure 18.1. Be as complete as possible.

13. Explain how the economic concept called the "law of supply and demand" can be interpreted as a feedback control system. Select the market price (selling price) of a particular item (e.g., a barrel of oil, a gallon of gasoline, etc.) as the output of the system, and assume that the system objective is to maintain a stable price.

CHAPTER 19

Management Systems Design

19.1 Introduction and Perspective

We presented the notion in Chapter 2 that I.&S.E.s design systems at two levels. The first level was called *human activity systems* and is concerned with various aspects of the physical workplace at which human activity occurs.[1] The second level was called *management control systems* and is concerned with processes and procedures related to planning, measuring, and controlling all activities within the organization.

Much of the material in previous chapters was concerned with defining the I.&S.E. role in the design of human activity systems. This chapter is concerned with the management systems of an organization.

To establish a perspective, consider these thoughts from Blair and Whitson:[2]

---

[1]With the rapid advances in automation, one could seriously question the term "human activity system." In a totally automated environment, no hands-on human effort is involved, yet many activities are performed. The I.&S.E. is just as concerned with the design of such systems as with the design of manual systems. Perhaps a better term is "physical activity system."

[2]Raymond N. Blair and C. Wilson Whitson, *Elements of Industrial Systems Engineering* (Englewood Cliffs, N.J.: Prentice Hall, 1971), pp. 272 and 276.

> Management is the brain and nervous system of an organization. It is the overall control system. What is called "organizational structure" is the framework of the system.
>
> Management, viewed as a feedback control system, must have goals and plans for attaining the goals, must issue authoritarian orders to effect the plans, must compare the actual organizational performance with planned performance, and must take compensatory actions to counteract unfavorable deviations. Such activity, in varying proportions, must occur at every level of management.

Clearly, this view of management and management control systems is totally consistent with our development of more general feedback control concepts in Chapter 18.

19.2. A Systems View of an Organization

Picking up on the quote from Blair and Whitson in the preceding section, we now explore the question: How does an engineer actually go about designing systems for the management of an organization? Using the systems concepts developed in Chapter 18, we shall develop a comprehensive framework for representing the functioning of an organization within its environment.

19.2.1 Gaining a Perspective

In Chapter 18 we defined a system as a set of components or subsystems related by some form of interaction to achieve some common purpose. A management system is, therefore, a specific set of components or subsystems, each of which has some relationship to at least one other component or subsystem in the set, working together for the common goals of the organization.

A typical manufacturing firm has among its major components product design, marketing and sales, production, finance, procurement, personnel, distribution, and management control. A manager has primary reponsibility for each of these functions of the company. Particular policies and practices that would make one manager look good may make another look bad. The important consideration is the effect of particular policies on the total company, not on an individual manager.

Each of the above-mentioned functions is a component in the total management system, but each component can be viewed as a subsystem at one level lower in the company organization. Each subsystem has its own set of components working together toward the common goals of the subsystem. Again, particular policies or practices that would cause one component to look good might cause another to look bad. As before, the proper practice or policy is the one that is best for the total subsystem.

This process of defining a component at one level as a "system" at a lower level can be continued as far down in the organization as desired. In terms of personnel, the "lowest" level of this subdivision process would be that in which the "components" are individual workers. Starting back up, the individual workers are in groups, each having specific objectives. Several groups may form another grouping at a higher level. This continues until the entire company is considered a "group" at the highest level.

We have just described the general process of organization design, or management system design. At the highest level, organization objectives are crisply defined in measurable terms. At each successive lower level, subobjectives are defined which are consistent with and supportive of the objectives at the next higher level.

> *The industrial and systems engineer of the future will have the responsibility for designing total management systems which assure that all system elements at all levels are performing for the maximum benefit of the entire organization.*

The previous discussion concentrated on the importance of having common objectives and on the entire organization's working together to achieve those objectives. Simply having the common objectives is not enough, however. In addition, we must continuously monitor the actual performance of each organizational unit to assure that objectives and performance goals are being met.

It is instructive to visualize a management system as a feedback control system. Such a system involves a set of objectives, a specification of performance criteria, a means of evaluating actual performance, and a specification of corrective action when actual performance deviates significantly from planned performance.

If the system also includes a means for modifying its performance criteria, it can be regarded as an adaptive control system. Figure 19.1 portrays such a system for any management organization.

Notice that the control system structure shown in Figure 19.1 is applicable to any system or subsystem at any level within the organization. Consequently, a total management control system for an organization would consist of several interconnected diagrams similar to that shown in Figure 19.1.

This concept of management control is applicable to staff functions as well as to operating departments. For example, it is applicable to the management of an industrial and systems engineering department, as shown in Figure 2.2. Performance criteria for the industrial and systems engineering department could be similar to those listed in Figure 2.3.

19.2.2 Finding a Starting Point

Practically every reflective person has at least once in his or her life asked the question, "What is the purpose of my existence?" Organizations have also found that this profound question is one that needs to be addressed explicitly. "Who are we?" "What are we about?" "What is our fundamental reason for being?"

These questions may at first seem trivial. Everyone "knows" that the funda-

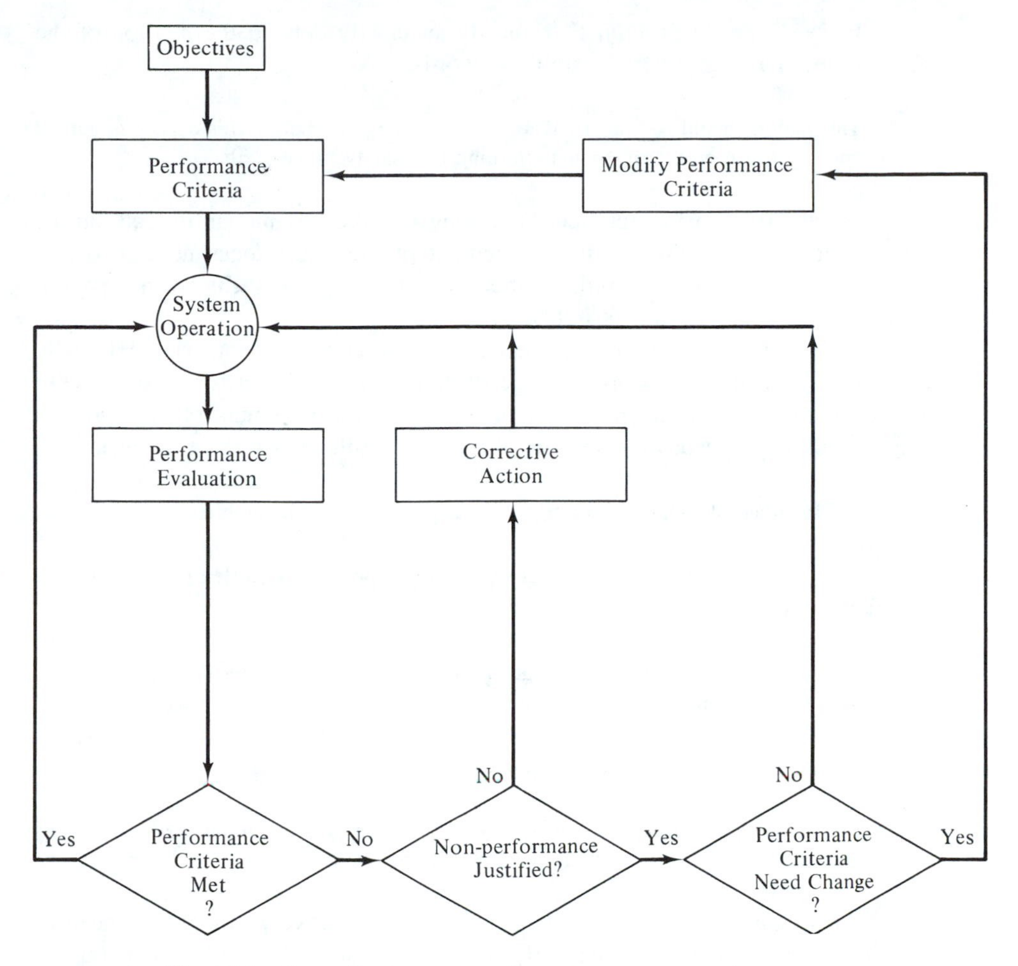

Fig. 19.1. Generalized structure of a management control system.

mental reason for an organization's existence is to make all the money it can. If this were true, the airlines that make scheduled flights between U.S. cities and certain countries in Central and South America should contrive to capitalize on the lucrative trade in illicit drugs. If profit maximization was the only reason for existence, why would any company spend money maintaining an attractive appearance or a clean and pleasing work environment for employees?

It turns out that profit (or loss) is an *outcome* that results from an organization's pursuit of its fundamental reason for existing. We shall use the term *mission* as an organization's *raison d'etre* (reason for being). It is extremely important that an organization craft a *mission statement* which describes in one complete sentence the organization's justification for existence.

In 1961, President John F. Kennedy succinctly defined the mission of the *Apollo* lunar landing program with these words:

> This nation should commit itself to achieving the goal, before this decade is out, of landing a man on the moon and returning him safely to the earth.

In July 1969, astronaut Neil Armstrong was the first human to walk on the surface of the moon. The mission statement provided the focus needed to pull together the efforts of hundreds of thousands of people to achieve the greatest technological accomplishment in history.

Writing a meaningful mission statement in one complete sentence is extremely difficult. Companies frequently engage in animated discussion for several weeks before finalizing their statements. Here is the mission statement of an industrial engineering department in a company that manufactures gas turbine engines:

> We facilitate the manufacture of turbine engine parts and assemblies.

The following mission statement was developed for a College of Veterinary Medicine within a large land-grant university:

> We generate and transmit knowledge in veterinary medicine and the biological sciences and provide clinical and diagnostic services to selected publics.

The mission statement for Apple Computer Company is:

> To help people transform the way they work, learn, and communicate by providing exceptional personal computing products and innovative customer services.

An organization's mission statement should be relatively "timeless," needing updating or revision very infrequently. It should be so basic and fundamental that every person within the organization, as well as its clients, will understand and relate to it. It should only state *what* the organization's basic role is, not *how* the role will be accomplished.

19.2.3 Universal Outcome Goals

Peter F. Drucker has been a leading scholar in management theory for over 40 years. Among his many contributions, Drucker proposed in 1954[3] that every for-profit organization had a set of common, overriding "objectives of performance and results." These have become known as "key results areas" (KRAs). There are eight in number:

---

[3]Peter F. Drucker, *The Practice of Management* (New York: Harper & Row, 1954), Chap. 7.

- Market standing
- Innovation
- Productivity
- Physical and financial resources
- Profitability
- Manager performance and development
- Worker performance and attitude
- Public responsibility

Drucker argued that an organization should regularly and rigorously measure its actual performance against these universal yardsticks of organizational performance. Notice that Drucker included three intangible KRAs, the last three.

It should be noted that some companies have modified these basic KRAs to better express their priority concerns. This has worked well as long as the companies defined explicit KRAs, set targets for each, measured company performance against the targets, and implemented corrective action as appropriate.

There, we see it again! What we have just described is a feedback control system. If you are beginning to feel that you cannot get away from feedback systems, you are absolutely correct.

As another example of KRAs, let us consider a not-for-profit organization, the Institute of Industrial Engineers. A meaningful set of KRAs for IIE would be those which measure the real factors of successful operation over a long period of time. IIE's KRAs could be the following:[4]

- Information dissemination and technology transfer
- Formal education
- Research and development of methodology
- Image/perception of IE profession
- Public responsibility
- Membership marketing
- Financial performance
- Crafting strategic initiatives

Clearly, if IIE set aggressive targets in each of these areas and consistently demonstrated outstanding performance in each, its overall performance would be judged superior.

19.2.4 Determining Goals and Objectives

There are usually multiple goals associated with each KRA. For each goal, there can be multiple objectives. It is important to distinguish between goals and objectives.

---

[4]The reader should be aware that these are the authors' recommended KRAs for IIE; they have not been adopted by the organization.

- *Goal*—A fundamental, nonspecific statement that is future oriented and time independent. It stresses desirable ends without specifying the means to achieve the ends.
- *Objective*—Defines the desired outcome in specific, quantifiable terms, with specific dates and assigned responsibility.

Consider, for example, the KRA of *market standing*. There can be several general goals concerning market standing. An example goal would be: *remain among the top three companies in the luxury tier of the North American market*. For this goal, several objectives can be specified. These objectives are statements of explicit actions that can be initiated within a specific time frame which, if successfully accomplished, will contribute to the attainment of the stated goal. An example objective would be: *During the next 18 months, establish a direct customer hot line for communicating concerns and suggestions; responsibility—Bob Burk*. A general guideline for formulating objective statements is suggested in Figure 19.2.

One final step remains to be done: Set up an action plan to accomplish each objective. The action plan consists of a detailed listing of all activities that must be accomplished to complete the objective. A target completion date is set for each activity. Figure 19.3 shows a form that could be used for laying out an action plan for each objective.

One of the important activities of an I.&S.E. is the design of incentive systems. All employees in an organization are given regular performance appraisals, and each person's rewards (raises, promotions, awards, etc.) are based on the results of the appraisal. It is important that each person's performance be measured against that person's individual contributions to the goals and objectives of the organization. The general manner in which all these things relate is illustrated in Figure 19.4. We call this *linking planning to action*.

19.2.5 A Unified Framework

We are now in a position to pull together all the components of an organization into a unified framework that is true to the concepts of feedback control systems. Such a structure is shown in Figure 19.5. Notice that the corrective action feedback

Infinitive	*Noun*	*Goal**	*Date*
To install	Position evaluation	All exempt jobs	February 1, 1993
To increase	Undergraduate enrollment	20–25%	September 1, 1993
To reduce	Duplicating costs	15–20%	June 30, 1993
To develop	Five-year development plans	All faculty	December 31, 1994
To meet	Affirmative action targets	90–100%	June 30, 1995

*Quantity, quality, cost, time.

Fig. 19.2. Format for stating objectives. Objective statements consist of an infinitive, followed by a noun, with a goal or target, and a completion date.

GOAL/OBJECTIVE/PROGRAM SHEET

No.

Key Results Area:			
Goal:			
Objective: Priority			
Measurement/Indicators:			
Program of Activities:	Responsibility	Target Date	Completion Date

Fig. 19.3. Form for specifying each objective. [Format motivated by W. J. Reddin, *Effective Management by Objectives* (New York: McGraw-Hill Book Company, 1970).]

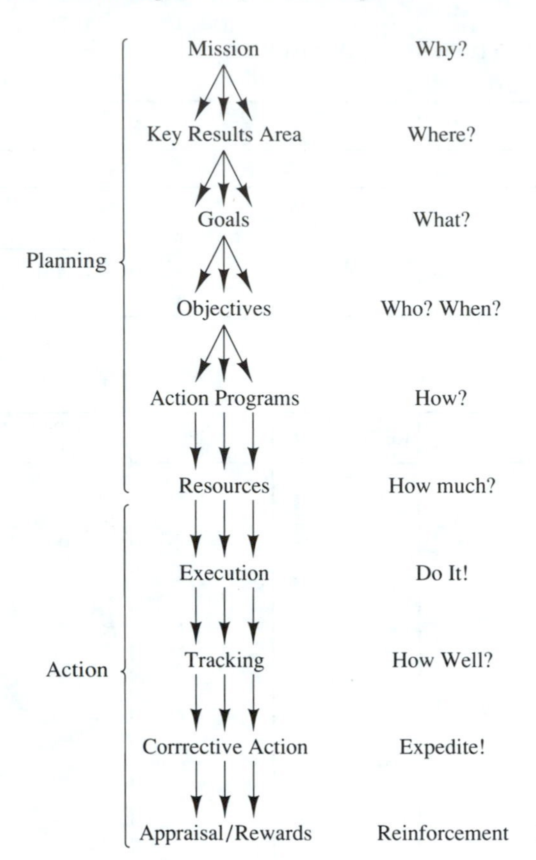

Fig. 19.4. Linking planning to action.

loop is geared to serve the three different time frames of operational planning (monthly), tactical planning (quarterly), and long-range planning (annually).

19.3. Organization Design[5]

One of the functions that many experienced industrial and systems engineers perform is that of *organization design*. This function is concerned with analyzing the total programs and activities of a company or enterprise and determining the most effective arrangement of these activities.

As president of Gadgets, Inc., for example, you cannot directly supervise each individual employee. Similarly, you cannot make every decision pertaining to purchasing, scheduling, marketing, inventory management, facilities layout, and so forth. The total task of the business must be divided into a number of smaller,

[5]Portions of this section are based on Amrine/Ritchey/Moodie, *Manufacturing Organization and Management*, 5th ed. © 1987, pp. 38–61. Adapted by permission of Prentice Hall, Englewood Cliffs, New Jersey.

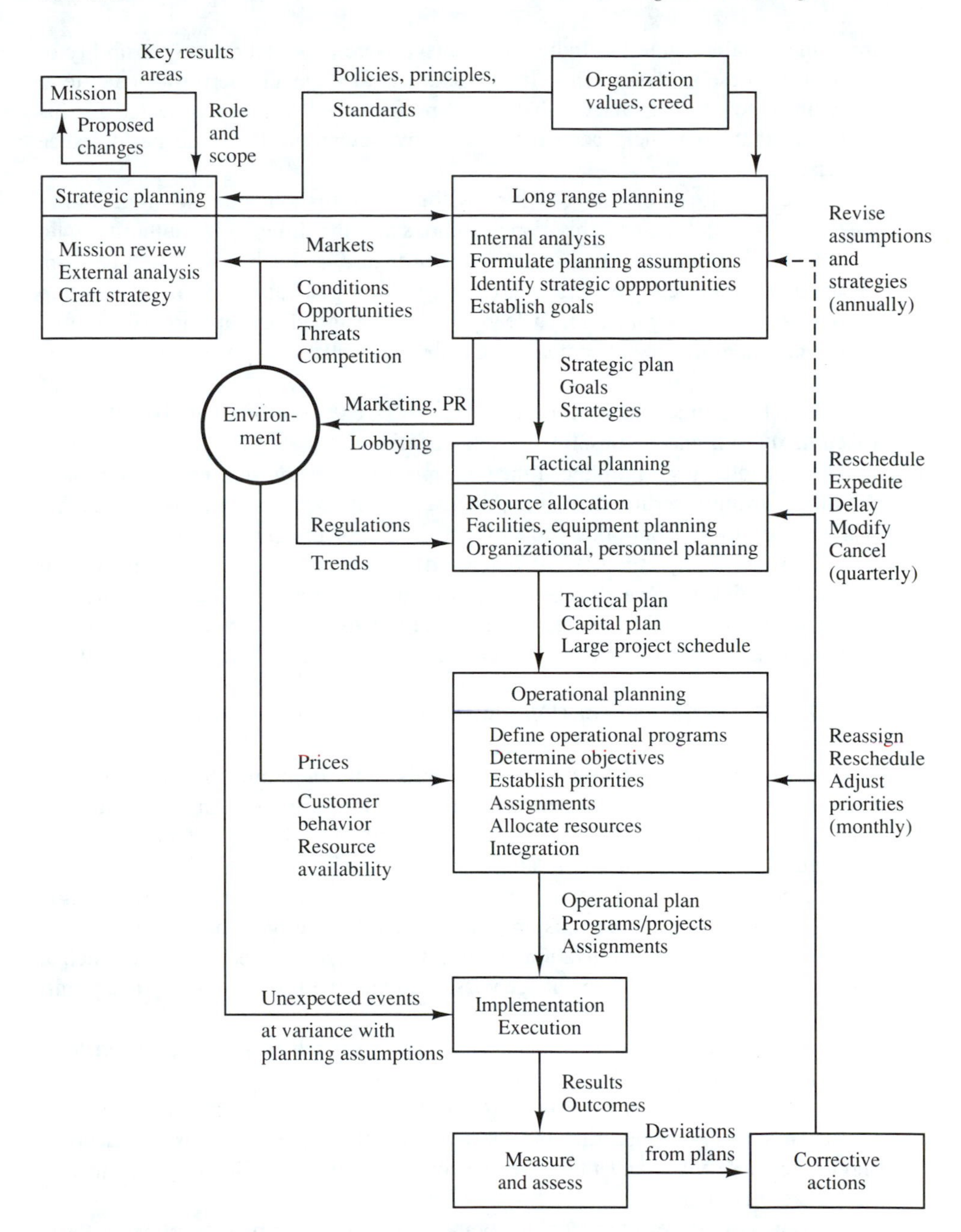

Fig. 19.5. Unified framework for the functioning of an organization within its environment.

more manageable subtasks. Individual managers are assigned the responsibility of performing these subtasks. This, in turn, implies that specific performance criteria are defined for each manager, that actual performance data are collected for his unit, and that he is held accountable for any deviations from the performance criteria.

Each employee of Gadgets, Inc., will perform his job better when he knows clearly what his job is, what his responsibilities are, the limits of his authority, who his boss is, and what the total organization structure is. Each person will perform his job better if he clearly understands his role in the organization and his relationship to all others in the organization. Clearly, a well-designed organization is essential for effective communication among its members as well as for the efficient execution of its mission.

Good organizations do not just "happen." They are brought about through much hard work and purposeful conceptualization.

In Chapter 1 we introduced the concept of *systems design*; indeed, we made the rather strong assertion that "the principal activity engaged in by engineers that distinguishes them from other professionals is that of the *design of systems*."

Through many centuries of organized human activities some principles of organization design have emerged. Although not as "scientific" as the principles associated with, say, the design of a chemical process, they, nevertheless, provide a sound basis for approaching the difficult task of designing a "good" organization.

19.3.1 Specification of Objectives

An organization is formed to accomplish one or more objectives that are common among two or more persons. This common purpose is the essential integrating element of the organization, and it establishes a framework within which individuals cooperate and interrelate.

Specifying explicit objectives is the initial step in the process of organization design. Almost all organizations engaged in private enterprise have as their most obvious objective the maximization of profit. Other organizations, such as municipal governments, state governmental agencies, churches, and so on, do not have profit, per se, as an objective.

Even private enterprises should not place so much emphasis on short term profit maximization that they jeopardize their futures.

An objective that is shared by essentially all organizations is that of survival. It can be strongly argued that the most important objective of *any* organization should be to survive, and that other objectives, such as profitability and market share, are secondary.

Some companies only want to compete in a particular market; others want to compete in a particular geographical region. Some companies insist on producing only high-quality merchandise; others wish to compete only in the high-volume, low-quality market.

Organizations must sometimes specify objectives pertaining to external factors, such as attracting investor capital or maintaining maximum flexibility to capitalize on shifting governmental programs.

An organization should be designed to assure the maximum attainment of each objective, insofar as possible. Often, two or more objectives will conflict, thus necessitating a compromise.

19.3.2 Determination of Functions

Once the objectives of an organization have been specified and agreed upon, the next logical question is, "How can we achieve these objectives?" Suppose, for example, that we are designing an organization to produce very high-quality, long-lasting automobiles. What are some of the things we must do if we are to survive and prosper?

One of the first things we must do is find adequate capital to build plants, buy materials, purchase machines, and so forth. Another vital thing we must do is to maintain a competent staff of engineers who are capable of designing the many components and subsystems that will eventually comprise the high-quality product we wish to produce.

Another thing we must do is determine particular features that potential customers would like to have in their car. Other tasks include hiring and training workers, establishing dealerships, designing maintenance and repair procedures, providing for spare parts, and setting up production facilities.

The process we have just been through is that of determining the *functions* that the organization must perform in order to accomplish its objectives. Typical functions for a manufacturing organization are production, sales, marketing, distribution, finance, procurement, personnel, and engineering. Typical functions for a municipal government are police protection, water and sewer, garbage collection, street maintenance, parks and recreation, and welfare.

In the final organizational design the above described functions may very well be broken down further into subdepartments. In designing a particular organization the industrial engineer must determine which functions for this particular enterprise are significant enough to be separately identified.

19.3.3 Grouping the Functions

The next step in designing an organization is to determine the relationships among the several functions and to group the functions accordingly. For example, suppose that a particular city provides fire protection, police services, and ambulance services. Since these functions have many similar characteristics, they might be grouped together and report to one person (perhaps the director of emergency services), who in turn reports to the city manager.

Logical questions at this point might be, "Why group the functions at all? Why doesn't each function simply report to the top executive?" There may be 20 or 30 separate departments in a city government. The city manager is limited in how many subordinates he is capable of supervising directly.

This introduces the principle of *span of control*. Few managers are capable of supervising directly a large number of subordinates. Some management theorists claim that the ideal number is about seven. In practice, however, the number depends on such factors as the level within the organization, the type of problems encountered by the subordinates, the nature of the particular organizational unit, and the particular abilities and "style" of the individual manager.

At the conclusion of this step, the basic structure of the organization chart is essentially complete.

19.3.4 Functional Objectives

Once the above step has been completed, particular organizational units having identifiable labels will be in place. The next step is to specify explicit objectives for each organizational unit to assure that the overall objectives of the total organization will be accomplished. This involves the final specification of responsibilities of each function.

The outputs of each organizational unit should be measurable in terms of cost, schedule, and quality so that the supervisor of that unit can be held accountable for the performance of the unit. This step is necessary for "overhead" functions, such as personnel, as well as "line" functions, such as production.

19.3.5 Job Descriptions

Notice that our design process becomes progressively more detailed. We are now to the point where we must analyze the total work load assigned to each organizational unit and determine the particular individual *jobs* that collectively are required to perform the work. A job is related to an individual, even though several individuals may be required to perform essentially identical jobs. (The makeup of job descriptions was covered in Chapter 9.)

19.3.6 Management Controls

The design of an organization cannot be considered complete until the designer has specified the manner in which the operation of the organization will be monitored and corrected as required.

In the preceding section a framework was presented for controlling the internal operations of an organization. This same basic approach is applicable to the total organization.

For each organizational unit, the following control framework should be established:

(1) *Objectives of the unit* (described earlier).
(2) *Specification of performance effectiveness measures* (what specific things do we measure so that we can know whether or not the objectives are being achieved).
(3) *Corrective action procedures* (accountability for deviations of actual performance from goals).

19.3.7. Organization Design Is Continuous and Dynamic

It would be erroneous to believe that once an organization design has been decided upon it will continue unchanged forever. No organization has ever remained static. The Catholic Church is one of the oldest formal organizations in existence, but it has experienced many changes and will continue to do so.

The industrial engineer working in organization design faces an ever-changing and dynamic environment. Markets change, products change, executives change, the company grows. In addition, new ideas are continually being developed regarding the proper design of organizational structures.

Consequently, the designer of an organization should attempt to build into the design as much flexibility as possible.

19.3.8. Organization Structures

A convenient way to represent and portray the relationships among the various organizational units is through the use of an *organization chart*. The organization chart shows lines of authority and accountability.

In almost all organizations the functions are grouped into two broad classifications: *line functions* and *staff functions*. Line functions are those directly involved with the operations of the organization. Staff functions are those concerned with policy development, coordination, support services, and other activities not associated directly with producing the primary output of the organization.

Figure 19.6 portrays a typical organization chart. It contains typical line and staff functions. It is not intended to portray a complete organization. Staff relationships are shown with dashed lines.

Organizations are subdivided according to a number of rational bases. The previous discussion in this chapter used *function* (manufacturing, sales, etc.) as a basis for subdividing.

Some companies are organized by *product line*. For example, a meatpacking firm may choose to organize as shown in Figure 19.7. Each type of meat has a separate line organization. In this case, soap products are a major product line, using the by-products of the other three lines as raw material.

Other companies are organized by *location*, with a major line organization at each geographical location. Still other companies organize by *class of customer*, with a separate line organization for each class (such as commercial, government, and military).

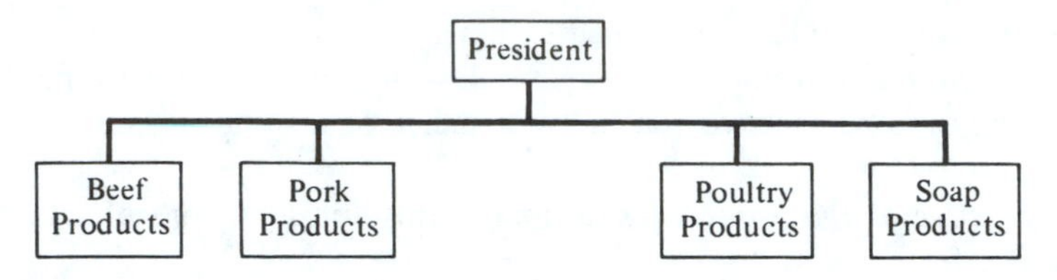

Fig. 19.6. Typical organization chart.

19.3.9. Coordination within the Organization

The efforts and activities of all the units within an organization must be properly coordinated in order to achieve efficient and effective performance. The organization chart is one means of achieving coordination.

The organization chart is only an outline of the enterprise. It usually shows the names of departments (with job titles) and the relationships among the units. Many companies supplement the organization chart with an *organizational manual*, which provides detail on the duties and responsibilities of each job shown on the chart.

The organization manual is the final documentation of the organization design process. It clearly delineates the objectives of each organizational unit, the scope of authority of the manager of each unit, and the relationship to other units. If relationships to organizations outside the company are part of a particular job, these relationships are also spelled out. The organization manual is also very useful for orienting new personnel to their jobs.

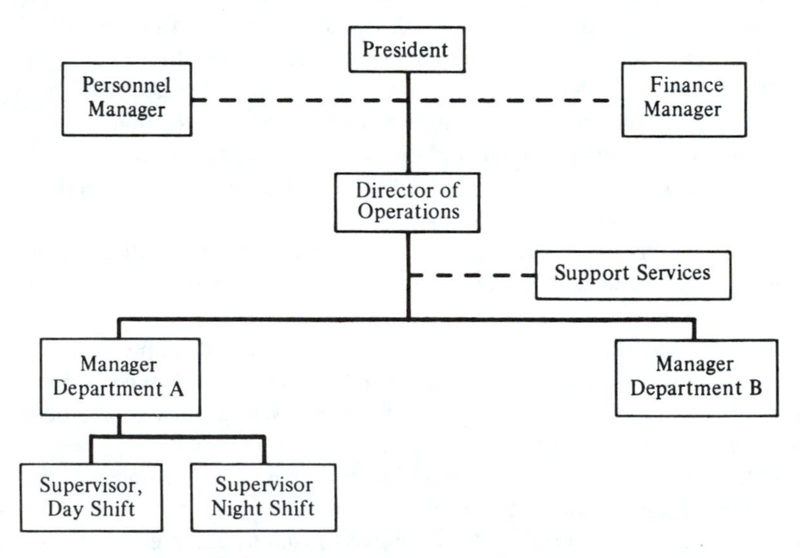

Fig. 19.7. Possible organization of a meat packer.

Finally, *standard operating procedures* are employed to assure consistency and efficiency within an organization. These procedures give explicit detail on how a particular job is to be performed.

19.3.10. Keeping the Design Current

Needless to say, the organization chart, the organization manual, and the standard operating procedures should be kept current. This is no easy task, and all too often is not achieved. Consequently, all the effort that went into their preparation is essentially wasted.

Organizations of all types should continuously review their organizational structure and maintain current, accurate documentation. If they do not, gross inefficiencies and lack of synchronization between organizational units will creep into the system and can become very damaging.

Organization design should be considered a vital, continuing activity, at least as important as the design of new products. It should be performed by professional industrial and system engineers who have a sound background in the design process and who have adequate experience in dealing with complex systems.

19.4. Providing Management Controls

The management system design activities described in Sections 19.2 and 19.3 result in a management system structure. This structure must now be energized and activated. There remains one critically important task for the I.&S.E. design team to perform. Management controls must be specified which provide to the system manager the tools and processes that will be needed to guide the organization day after day, month after month, and year after year.

The system design developed by the design team should result in successful system operation. However, the system structure should not be regarded as permanently fixed. As the system moves into full operation, actual performance measures become available to the system manager. This feedback mechanism can be used to continuously fine-tune both the performance of the system and the system structure. A review of Figure 19.5 shows that a self-correcting mechanism has been included explicitly in the system structure.

The manager[6] of the system serves the role of *system calibrator*,[7] continuously monitoring system performance, looking for possible imbalances among the components, sensitive to any threatening changes in the environment, and in general attempting to maintain stability within the system. This role is directly analogous to that of managing a petroleum refinery, a large chemical processing plant, or an

---

[6]We use the word "manager" in the singular, but in a large system there will be dozens, even hundreds, of managers.

[7]Robert Wright, *Systems Thinking: A Guide to Managing in a Changing Environment* (Dearborn, Mich.: Society of Manufacturing Engineers, 1989), pp. 127–133.

electric power generating plant. Each of these systems is managed from a *control room* that is equipped with many dials, gauges, digital readouts, and recording instruments. Conceptually, then, this section is concerned with providing a *management control room* that will facilitate keeping management's finger on the important pulses of the organization. Ideally, the manager would be guiding the system using *on-line, real-time data access*.

During the operation of the system, changes will occur in the relative demands for different products. These, in turn, will cause an imbalance of work flow. The federal government may issue a safety regulation that results in an additional 15% work load in the quality control department. Initially, this will cause a bottleneck to occur until the system has been restructured to accommodate the additional work.

New equipment becomes available that would greatly improve product quality, but will cause a major redeployment of the work force. The I.&S.E. department determines an improved method that will increase productivity but will require a major rearrangement of equipment. A vendor suddenly goes bankrupt, and the company buys the vendor's equipment, hires some of the vendor's employees, and begins producing the component in-house. Each of these actions will cause major disruptions to the system. A system that has been operating roughly in equilibrium now must go through the shock of each of these major perturbations. As *system calibrator and stabilizer*, the manager must find ways to fine-tune the new structure and bring the system back to equilibrium as quickly as possible.

We need to add an additional capability to our control room. The manager needs to be able to ask a wide variety of "what if" questions. The I.&S.E. team needs to design a *system simulator* for this purpose. Such a simulator would permit the manager to assess the probable impact of implementing alternative decisions, policies, structural changes, and so on. It would be equivalent to a *management wind tunnel*.

19.5. The Organization Life Cycle

The concept of a product life cycle is well known. Essentially, every product has a beginning and an ending. The phases of a product's life distinguish themselves in a pattern common to most products. This pattern, shown in Figure 19.8, has five phases. The actual duration of each phase varies considerably among different products, but the general shape of the life-cycle curve is the same for almost every product.

19.5.1 Life-Cycle Stages of an Organization

Applying the life-cycle concept to an organization is a fairly recent innovation in management thinking. It is somewhat difficult to create an illustration of an organizational life cycle, since we have no measure for the vertical axis. Neverthe-

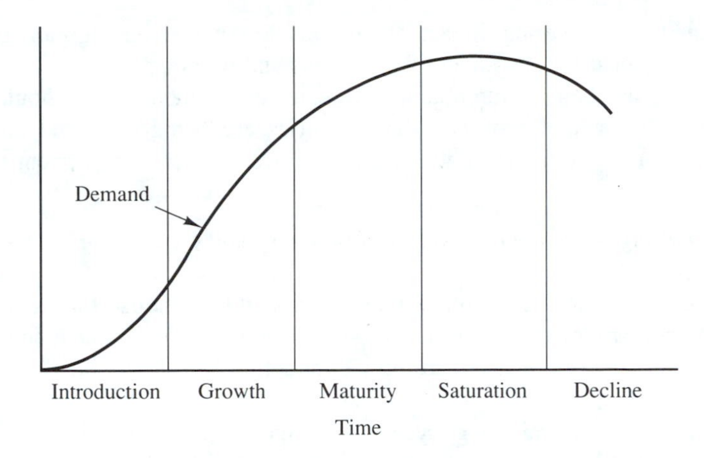

Fig. 19.8. Life cycle of a product.

less, we can illustrate the concept by identifying the stages and then characterizing each stage.

Since we frequently regard an organization as being similar to a living organism, it may be helpful to describe its life cycle in related terms. We will consider the stages to be *conception*, *adolescence*, *young adulthood*, *mature adulthood*, *aging adulthood*, and *demise*. Each of these stages is characterized in Figure 19.9 on three factors: *management style*, *primary driving force*, and *nature of systems*.

19.5.2 Organizational Renewal and Redesign

Is ''demise'' necessarily the fate of every organization? Many industrial organizations in Europe are more than 250 years old. The Catholic Church has existed for almost 2,000 years. There have been numerous accounts of large organizations

Stage	*Management Style*	*Primary Driving Force*	*Nature of Systems*
Conception	Dictatorial	Technological innovation	Minimal
Adolescence	Authoritative	Quality	Informal
Young adulthood	Delegation	Efficiency	Flexible
Mature adulthood	Consultative	Effectiveness	Decentralized
Aging adulthood	Detached	Risk aversion	Bureaucratic
Demise	Abdication	Resolved to fate	Rigid

Fig. 19.9. Life cycle of an organization.

faltering and then recovering. In fact, all organizations have their "ups and downs." Survival is the most basic instinct of organisms and organizations.

Volumes are being written about strategies that organizations should follow in order to survive and prosper. Many theories are being proposed and many organizations are experimenting with these theories. Some of the current concepts are:

- *Downsizing*—Greatly reducing staff functions and other "non-value-adding" functions.
- *Outsourcing*—Rather than maintaining capability in-house for all functions, contract many of them to vendors who specialize in particular functions. For example, some companies are now contracting for all their data processing services.
- *Streamlining*—Eliminating layers of management.
- *Decentralizing*—Redeployment of staff functions such as I.&S.E. from a central organization to the various operating units.
- *Integrating*—Bringing together critical functions such as engineering and manufacturing, so as to "tear down the walls" between functions.
- *Empowering*—Greatly reducing the number and role of supervisors, resulting in "self-managed teams."

There are many other "movements" under way, some of which are contradictory to others. We must view such initiatives with caution and respectful skepticism. Any concept advertised as the ultimate solution most assuredly is not. Some of the very concepts that work well during the early 1990's may need to be reversed in the late 1990's due to technological innovations or other reasons.

In many ways, an organization should always be in a state of renewal and redesign. This is the meaning of "system calibration" described in Section 19.4.

19.5.3 The Learning Organization

Much research is currently being devoted to the development of "intelligent machines," machines that can "learn" from their mistakes and their past performance, and then change themselves so as to deliver better performance in the future. Practical devices of this nature will probably not be available for many years.

Organizations, however, should certainly be capable of learning. After all, they contain humans as components. As individuals within the organization experience particular outcomes as a consequence of particular actions, they quickly "learn" which action combinations produce superior results. The superior practices become institutionalized through their incorporation into method descriptions, standard practice manuals, operating procedures, policies, and so on.

While the phenomenon just described can legitimately be called "organizational learning," it is at best a crude form of learning and not as effective as it might be. One of the major deficiencies has been the absence of a feasible *corporate memory*. Some would claim that the manuals, procedures, and so on, described above constitute the corporate memory. Perhaps they do, but the current systems have major weaknesses. In particular, there is a lack of a common format for recording the information. It currently resides in file cabinets, notebooks, desk drawers, closets, and so on. It is very difficult to associate a particular item in one document with a related item in a different document.

In computer terminology, current manual systems lack an *associative ability*. Work is being done on this problem, and at the current time, *relational data bases* seem to hold the most promise. This concept is explored further in Chapter 20.

DISCUSSION QUESTIONS

1. Determine if your sorority, fraternity, or any other club of which you are a member has a mission statement. If so, critique it in terms of the criteria for writing mission statements presented in this chapter. If no such statement exists, develop one.
2. Consider the generic key results areas proposed by Drucker and those suggested for the Institute of Industrial Engineers. Develop a set of KRAs that would be appropriate for each of the following organizations:
   (a) Your university.
   (b) A bank.
   (c) A hospital.
   (d) A civic organization.
   (e) A grocery store.
   (f) A fast-food restaurant.
   (g) Your student government association.
3. (a) For each of the organizations listed in Question 2, select one of the KRAs you developed, and determine two goals that pertain to that goal.
   (b) For one of the goals of each case, develop two objectives, using the format shown in Figure 19.2.
   (c) For one of the objectives of each case, complete the goal/objective/program sheet shown in Figure 19.3.
4. In Figure 19.5, consider the circled component labeled "environment." For each of the organizations listed in Question 2, specify the major environmental factors.
5. What is meant by the following terms?
   (a) Span of control.
   (b) Line function.
   (c) Staff function.
6. Try to locate a copy of the organization chart of your university. Analyze it. Does it adhere to the basic principles of organization design?
7. Can "organization design" legitimately be called an engineering activity? Why?
8. Develop an "ideal" organization design for your university chapter of IIE (or other appropriate organization). If you make any modifications to the existing design, defend your modifications.
9. Apply the concepts in Section 19.3 to the organization design of a bookstore.

10. Apply the concepts in Section 19.3 to design an "ideal" organization for the city in which you live.

11. For the systems in Questions 8, 9, and 10, design the management controls that are necessary (objectives, performance criteria, corrective action procedures).

12. Why would organization design be a good career path to top management?

13. Draw an organization chart for a company that is organized by location.

14. Draw an organization chart for a company that is organized by type of customer.

15. Apply the principles of organization design described in Section 19.3 to a nonprofit organization in your community. Potential organizations, would be churches, civic clubs, youth organizations, or charitable organizations.

16. The concept of life cycle can be applied to an entire industry (e.g., the automotive industry). For each of the following industries, indicate the stage it is now in in each of the countries listed. Use the stages from Figure 19.8. Leave an entry blank if the country has no activity in a particular industry.

Industry	*U.S.*	*Japan*	*Mexico*	*Kenya*
Steel				
Textile				
Petroleum				
Aircraft				
Pharmaceutical				
Agriculture				
Space				
Computer				
Consumer electronics				
Information				

17. Using the stage names in Figure 19.9, specify the current stage of each of the following organizations:

(a) Your university.
(b) American Motors Company.
(c) General Motors Corporation.
(d) Apple Computer Company.
(e) MCI Communications.
(f) Fox TV Network.

CHAPTER 20

Computers and Information Systems

20.1. Perspective

Industrial engineers were among the first people to make extensive use of computers when they became widely available in the late 1950's and early 1960's. Several of the previous chapters have been heavily computer oriented.

Industrial and systems engineers of the future will utilize computers in the following ways:

- *Problem Solving*—Developing and using statistical and mathematical routines to analyze data and otherwise assist in solving specific problems.
- *Optimization*—Developing and using mathematical programming and search techniques to obtain the optimal values of decision variables for specifically formulated decision problems.
- *Process Control*—The programming of numerical control machine tools to achieve specific geometrical dimensions of machined parts, and the use of programmable controllers to direct material flow and other aspects of factory work.
- *Factory Management*—Developing and using production and inventory control packages to assist in the planning and control of the manufacturing operations of a firm.

- *Simulation*—Developing and using computer simulation models to provide managers at all levels with the capability to ask "what if" questions regarding policy alternatives and other decision problems.
- *Management Information Systems*—Developing and using management information systems to support the complete realm of the decision structure throughout an organization.
- *Integrated Communication Systems*—Developing and using communication networks that enhance the effective flow of all types of communication throughout an organization (see Section 10.6).
- *Artificial Intelligence*—Developing and using knowledge bases which are the foundation for artificial intelligence procedures, aimed at enhancing human decision processes within complex decision environments.

Some of these categories are under development and will remain so for many years to come. Many of the categories have been covered in previous chapters, either explicitly or implicitly.

For the purpose of this text, the category of computer utilization that needs further elaboration is that of management information systems.

20.2. Basic Concepts of Information Systems

Every organization, large or small, depends heavily upon information. It needs information about its customers, its market, its suppliers, and its competitors. It also needs information about trends in government regulation, tax laws, environmental constraints, and the general economic climate. All the things mentioned thus far are external to the organization. An even larger set of internal information is required in the organization. It must have current, accurate information on its employees, their job skills, productivity levels, and attendance patterns. The organization needs information on its facilities, its equipment capabilities, and the location of perhaps thousands of tools. A large factory may have thousands of work orders at various stages of completion at any one time. Detailed information regarding the location and completion status of each work order is needed. Information is maintained on quality records and breakdown patterns for each machine or process. Extremely detailed information is generated, processed, and maintained to serve the needs of the organization's accounting system.

The information items just described are not intended to represent an exhaustive set of such items for any organization. Rather, the intent is to illustrate the very broad nature of the different types of information that are vital to the successful operation of a firm.

The total cost of a typical firm's information system is quite large, representing 10 to 15% of the retail price of a product or service. This proportion will increase as companies move more toward automation. Direct labor costs will decrease substantially, while the information content of a product's cost will increase. It is clearly to a firm's advantage to view information as a valuable resource and to manage it as such.

The terms "data" and "information" are often used interchangeably, but they are inherently different. *Data* can be thought of as raw facts. After data have been processed and interpreted, they become information. *Information* consists of data that have been translated into a meaningful context for the recipient. An *information system* is the set of organized procedures and associated equipment which translates data into information for decision making and control of the organization.

There are several logical processing steps, or operations, required to translate data into information. These *fundamental data operations* are:[1]

- *Capturing*—The recording of generic data at its source, such as sales orders, labor reporting at a machine center, inspection results, machine breakdowns, generating personnel records, reporting students' grades, and so on.
- *Verifying*—The checking of data to ensure that they were recorded correctly.
- *Classifying*—The act of grouping data into specific categories to aid the user in understanding its meaning, such as student grade reports by major.
- *Sorting*—Arranging data elements in a specified sequence, such as alphabetically sequenced class lists.
- *Summarizing*—Combining or aggregating data elements, such as identifying all students whose grade-point average is greater than 3.0.
- *Calculating*—Arithmetically combining or manipulating data for specific purposes, such as determining a student's cumulative grade-point average.
- *Storing*—Placing data onto some storage medium for later retrieval, such as entering the last semester's course grades into a student's transcript file.
- *Retrieving*—Searching for, identifying, and gaining access to specific data elements from the medium on which they are stored.
- *Reproducing*—Duplicating data, such as for file backup.
- *Disseminating/Communicating*—The transfer of data from one place to another, such as the mailing of semester grade reports to students and/or their parents.

The goal of any information system is to provide (disseminate) accurate information to the end users in a timely manner. Any information system may be thought of as being composed of some combination of the 10 fundamental data operations just discussed.

It is worth noting that not all information systems involve the use of computers. Information systems existed long before computers existed. The 10 fundamental data operations described above are equally applicable to manual information systems.

---

[1]Adapted from J. G. Burch and G. Grudnitski, *Information Systems: Theory and Practice*, 5th ed. (New York: John Wiley & Sons, Inc., 1989), pp. 7–8.

Today, essentially all information systems in organizations are based on computer usage. Even in very small organizations (a three-person dentist office, for example), personal computers are used to perform most of the data operations discussed above. Even homeowners are now putting their records on computer files. Computer-based information systems are, indeed, becoming "household words."

20.3. The Process of Designing Information Systems

Industrial engineers frequently have the responsibility for designing the information system for a firm. A fairly well-defined process for performing this design has emerged over the last several years. This process is illustrated in Figure 20.1 and explained in further detail in the following subsections.

20.3.1 Feasibility Study

The idea for a new or improved information system is usually stimulated by a *need* to improve some or all aspects of the organization's information-processing system. This leads to a *preliminary survey* to determine the feasibility of developing a system capable of addressing this need. *Existing procedures* are analyzed, and *potential cost/benefits* are calculated for the new system. The management responsible for providing information services to the organization then makes the *final decision on feasibility* of the proposed project. The following subsections presume a positive outcome to this feasibility assessment.

20.3.2 Systems Analysis

Information is useful in an organization primarily to support decision making at all levels. The initial step in analyzing the system, therefore, is to *define the decision structure* throughout the organization. This step is closely tied, but prerequisite to, the *definition of user's problems and needs*. What information is needed? By whom? When? To support what decisions? In what form? Answers to these questions for the total organization help *define the scope of the system*, that is, what the new or improved system is to include.

Study facts for this phase come from a very wide variety of sources. Accounting records, management reports at all levels, performance reports, transaction files, historical reports—all of these are excellent sources of study facts. The most useful technique for gathering study facts pertaining to the decision structure of a firm is the personal interview. Although this technique is very time consuming, there is no substitute for gathering this critical information. Some of the questions that are asked are: "What is your responsibility?", "What specific measures of performance are you evaluated on?", "What information do you need, from whom, in order to perform your job?", "What information do you provide to others?". These ques-

FEASIBILITY STUDY

Clarification of Need
Preliminary Survey
Existing Procedures
Potential Costs/Benefits
Final Decision on Feasibility

SYSTEMS ANALYSIS

Defining the Decision Structure
Definition of User's Problems/Needs
Defining System Scope
Gathering of Study Facts
Collection of Data on Volumes, Input/Output, Files
Analysis of Study Facts

GENERAL SYSTEMS DESIGN

Defining System Goals
Developing a Conceptual Model
Applying Organizational Constraints
Defining Basic Data Processing Operations
Developing Design Alternatives

SYSTEMS EVALUATION AND JUSTIFICATION

Request for Systems Proposals
Evaluation of Proposals
Cost/Benefits Analysis
Selection of General Systems Design

DETAIL SYSTEMS DESIGN

Organizational and Procedural Controls
Processing Logic
File Design, Data-Base Considerations
Data Backup and Security Issues
Input/Output Specifications
Programming Requirements
Personnel, Facility, and Capital Requirements

SYSTEMS IMPLEMENTATION

Training and Educating Personnel
Detailed Programming/Testing
Facility Preparation and Installation
Testing the System
System Conversion
System Validation and Acceptance

SYSTEMS OPERATION AND MAINTENANCE

Post Implementation Audit
Managing System Updates
Continuing Follow-Up and Evaluation
Recognizing Need to Repeat Complete System Design Process Again

Fig. 20.1. Information system design methodology.

tions lead to critical data on the *volume of information flow* between all points in the organization, as well as *input/output* and *data file* considerations.

20.3.3 General Systems Design

The design of any system must be preceded by a clear specification of the *goals of that system*. The goals of an information system are, in general, to provide accurate, timely, useful, usable information to those who need it throughout the organization. We express these goals by specifying the types of reports to be generated, to whom they are distributed, with what frequency, and so on.

Developing a conceptual model is the next step in designing an information system. This is best done by breaking the total system down into subsystems, such as accounts payable, payroll, and so on. For each subsystem, specify the input requirements, the processing required, and the outputs. Gross flowcharts can aid in performing this conceptual task.

Organizational constraints must be dealt with in a realistic manner. Capital availability is the most common constraint. Other resources that are in limited supply are people (different skill levels), space, facilities, and management attention.

The *basic data processing operations* discussed in Section 20.2 are now defined for the general system design as it currently stands. This allows the designer to specify a set of reasonable *design alternatives* for management's consideration.

20.3.4 Systems Evaluation and Justification

At this point in the procedure, we are ready to solicit proposals from hardware and software vendors. (Frequently, one company will submit a proposal covering both hardware and software.) The *request for proposal* includes performance requirements for both hardware and software. It is the designer's responsibility to state the performance requirements in such a manner as to assure that the overall output requirements of the system are achieved. Some of the important performance requirements are processor speed, response time for on-line inquiries, number of terminals supported simultaneously, file storage, operating system, and special languages.

Each vendor who chooses to submit a proposal will put together a system configuration that addresses the needs of the information system being designed. Every component of the proposed system will be described in great detail, including costs. These proposals must now be *evaluated* to determine which system best meets the needs of the organization. Benchmark tests are run on each proposed system to provide a basis for comparing their performances. An attempt is made to calculate the *expected total costs* and the *expected benefits* of the alternative systems being considered. The final *selection* also includes intangible factors, such as the reputation of each of the vendors for reliability, customer service, and system support.

20.3.5 Detail Systems Design

The general systems design covered in Section 20.3.3 was expressed at a conceptual level. This conceptual design must now be translated into an integrated system of personnel, equipment, and procedures that collects and processes *data* to produce *information* needed to support decision making throughout the organization. We will discuss some of the items that must be included in the detail systems design.

One of the first considerations in detail systems design is that of *organizational and procedural controls*. Organizational controls are simply the management components that must be carefully put in place for the information system function within the organization. Information system departments must be well managed, just as we manage the production operations or any other function. Procedural controls are the detailed checks and balances that we put in place to assure that all functions of the information systems group are executed properly. These include such things as forms design, data verification procedures, programming standards, documentation, and so on.

The *processing logic* must be developed in detail at this point. This consists of the actual computer programs, written in a specific computer language, to carry out the detailed processing steps. Closely associated with this step is that of detailed *file design*. What will our overall data base consist of? What are the major files? How are they related?

Any special *programming requirements* must be recognized and provided for. For example, will we need to purchase a particular software package to perform statistical analysis of data?

We must also recognize the skills needed for all the *personnel* who will be involved in operating this system. We will need a certain number of systems analysts, utility programmers, systems programmers, computer operators, maintenance personnel, and supervisors.

We may have to make *facility* modifications for the new system. We need adequate space, proper environmental controls, sufficient electrical power, and a power backup supply. Finally, we must recognize and provide for the *capital* we will need to acquire, operate, and maintain this system.

20.3.6 Systems Implementation

It is highly likely that the new system will require extensive *training and education* of many personnel within the organization. This will include many people outside the information systems department, such as operating people who receive instructions over the system and provide input data to the system, and management/administrative personnel at all levels who will receive enormous amounts of information from the system in the way of reports or on-line inquiry.

Each program module must be individually implemented and tested. Sometimes this can be done at another site which has comparable equipment, before the hardware is actually delivered. The *physical facility must be prepared*, and the *equipment must be installed* when it arrives.

After installation, a period of thorough *testing of the system* must occur. Here, for the first time, the individual program modules must be tested together, to be sure that all interfaces have been properly designed.

Once the testing has been satisfactorily completed, we can begin the *system conversion* phase. In a carefully phased manner, we will implement modules of the new system and discontinue appropriate portions of the old system. In some cases, we run the new and old systems in parallel for a while to be certain the new system is producing the outputs we expected.

Once all modules of the new system have been implemented, we run *system validation tests* to satisfy ourselves that the system is meeting the performance requirements that we specified earlier. The system is *formally accepted*, once these requirements have been met.

20.3.7 Systems Operation and Maintenance

It is wise to conduct a *post-implementation audit* to assess what actually occurred as opposed to what was projected to occur during the development phase. We need to be sure that all personnel have been trained properly, that programs are documented, and that the information outputs are acceptable and appropriate to the users.

Another important consideration is that of *maintaining* the information system as it changes. Information systems do not stay static for very long. They are in a state of continuous modification. As systems are modified, it is important to update all documentation, inform all affected users, and to assure that all other programs and files related to the one being changed are updated appropriately.

An information system will never be completely satisfactory for all present and future needs. We must *continually follow up and evaluate* the system's performance and be responsive to changing user needs.

It is the nature of information systems that at some point in the future the ''new'' system will become the ''old'' system. Better hardware and software is continuously being developed by vendors. There will come a time when it is more effective to go to a new system than to try to modify the present one. At this point, we are faced with going back to the first phase and *repeating the complete cycle again*.

20.4. Data-Base Management Systems

Until the late 1970s, the traditional approach to designing files was for each function within the organization to develop its own filing system. Separate, independent data files existed for accounting, personnel, production, purchasing, marketing, engineering, quality control, and so on. The *data-base approach* to data file design takes a different approach. It attempts to interrelate the several files within an organization to increase the associative ability of the data base.

The data-base approach has a number of important advantages over the traditional approach: (1) more standardization of record formats, (2) common data elements can be associated in a logical manner, (3) redundant data elements among files can be minimized, (4) file updates can be synchronized such that all files are assured of being included in the update transaction, and (5) users may interrogate the data base directly and make inquiries that are essentially unanticipated.

A term that has become widely used is *data-base management system* (DBMS). A DBMS is a set of software packages that provides the user with a relatively easy method for creating, maintaining, and accessing a complex data base.

The purpose of a data-base management system is to facilitate the creation of data structures, thereby relieving the user of the problems of dealing with complicated files. A DBMS separates the definition of data from the programs that access that data.

A data-base language consists of a set of commands for entering, updating, processing, and accessing data. To the user, the internal operations (program logic) and data structures are transparent.

A typical data-base language consists of the seven functions shown in Figure 20.2. In the right column are typical commands available to the user for executing each of the associated functions. Each one of the commands shown (only a small proportion of all commands are shown here) calls into action a ''canned'' computer program that is resident in the host computer on which the data-base language is implemented. These rather complex programs and all the detailed file linkages are transparent to the user. This approach greatly enhances the productivity of the person responsible for designing and maintaining the data base.

The concept of *relational data base* has emerged in recent years, and is an extension to the basic concepts just discussed. In a relational data base, data is organized as a set of ''relations'' or two-dimensional tables. The logical associations among data elements are shown by organizing the data into columns and rows. The process of converting a data file organized in the traditional manner to one organized in the relational mode is known as *normalization*.

To illustrate how the normalization process works and to give us a better feel for the concept of a relational data base, we will consider the data file shown in Figure 20.3, organized in the traditional way. The file in Figure 20.3 contains a repeating group—the job change information. Each time any employee changes jobs, there will be a new entry, showing date of change, new job title, and new job salary. The other data in the file remains unchanged. We can normalize this file by separating out the repeating group.

The normalization process results in two relations, shown in Figure 20.4. The first relation, named EMPLOYEE, contains each employee's number, name, and address. These are items that rarely, if ever, change. The second relation, named JOB/SALARY HISTORY, contains the information concerning each employee's

Function	*Typical Command*
Design Data Base Define Data Elements, Master Files, Detail Files, Pointers, and Passwords	SALES, DETAIL (1, 4/4, 8)
Create Schema Define Structural Relations	ACCOUNT (CUSTOMER(PURCH-DATE))
Create Data Base Initialize Each Data Set in Data Base	CREATE STORE/PASSWORD
Insert Data Into Data Sets Enter Data	UPDATE ADD
Data Set Access Retrieve Selected Data Delete Selected Data Modify Selected Data	FIND UPDATE DELETE UPDATE REPLACE
Report Generation Individual Data Element Summarize Information	LIST BADGE # F-NAME, L-NAME (Several commands required)
Data-Base Maintenance Backup, Recovery, Restructuring, Data-Base Security	DBLOCK, DBUNLOCK, DBEXPLAIN, DBCONTROL

Fig. 20.2. General components of a data-base language.

job–salary changes. Notice that the employee number is included in each relation, in order to identify each record uniquely.

Relational data files are, in general, easier to use than traditional files, especially in updating operations. For example, compare the two file structures in terms of modifying the file to reflect a new address for an employee. The biggest advantage of relational data structures is that they allow a user to query the data base in an

EMPLOYEE MASTER FILE

EMP. #	EMP. NAME	ADDRESS	DATE OF CHANGE	JOB-TITLE	SALARY
6271	Brown	14 Oak St.	03/01/86	QC Supv.	$29,800
6271	Brown	14 Oak St.	09/01/81	QC Technician	21,500
8019	Green	10 Pecan Dr.	04/15/84	Prod. Mgr.	42,800
8019	Green	10 Pecan Dr.	06/01/79	Sr. Prod. Engr.	38,000
8019	Green	10 Pecan Dr.	12/01/76	Prod. Engr.	32,100

Fig. 20.3. Data organized in traditional manner.

EMPLOYEE

EMP. #	EMP. NAME	ADDRESS
6271	Brown	14 Oak St.
8019	Green	10 Pecan Dr.

JOB/SALARY HISTORY

EMP. #	DATE OF CHANGE	JOB-TITLE	SALARY
6271	03/01/86	QC Supv.	$29,800
6271	09/01/81	QC Technician	21,500
8019	04/15/84	Prod. Mgr.	42,800
8019	06/01/79	Sr. Prod. Engr.	38,000
8019	12/01/76	Prod. Engr.	32,100

Fig. 20.4. Data organized as relational files.

unstructured way. Through "relational operators," pieces of data are pulled together from various relations to provide the information requested.

20.5. Data Communications Networks

When computers were first utilized in organizations, computing was highly centralized. Most organizations had one large computer (today we would call it a "mainframe") that operated in batch mode (sequential processing of individual jobs). The information system had to be designed to conform with this mode of processing. Thousands of individual computer programs were written to perform certain data-processing tasks. Each program required data as input. Often, the output of one program became the input to another. Programs submitted to the computer had to be sequenced carefully. Data input was primarily in the form of punched cards. Intermediate data exchange (from one processing stage to another) was usually in the form of magnetic tape. Output was almost entirely on paper. Large backlogs of jobs waiting to be processed were all too common. Turnaround time was often 2 or 3 days.

Through time, the clumsiness and inefficiencies of centralized batch processing led to the development of smaller computers, called *minicomputers*, that were employed within a particular department. All the local processing could be done within the department, and only the processing related to the total organization had to be submitted to the centralized computer.

A major breakthrough occurred when *timesharing* was introduced. Several users could be connected to the mainframe simultaneously via *remote terminals*.

There was still no way for the central mainframe computer to communicate with, or share data with, the small minicomputers within the organization.

Computer networks were another major step forward in computing technology. These allowed computers to communicate among themselves and users to access the data bases scattered around the organization.

Distributed computing was made possible by the development and refinement of computer networks. Actually, a more appropriate term is *data communications networks*.

There are two general categories of data communications networks: *local area networks* (LANs) and *wide area networks* (WANs). In general, LANs are intended to serve users who are physically located in close proximity, such as the people on one floor of a building. By placing a restriction on the maximum distance between users, special devices can be used that transmit data at very high speeds.

WANs are somewhat similar to a telephone network. They are used to allow communications between people in different buildings, different cities, and even in different countries. A message traveling between two points in a WAN may be routed through several interim points before reaching its final destination. Obviously, the data transmission rate is much slower in a WAN than in a LAN. We will not discuss WANs any further in this book, concentrating instead on LANs.

Each individual point within a network that can communicate through that network is called a *node*. Each node is assigned a unique *address*. In this manner, a destination address can be incorporated into each message and the message can then be forwarded to the correct recipient.

Many different types of devices can be attached to the network: computers, personal computers, dumb terminals, printers, plotters, programmable controllers, robots, data collection devices, bar code readers, CAD terminals, FAX machines, time/attendance stations, production monitoring equipment, vision systems, and so on.

Local area networks consist of three primary components: the transmission medium, network interface units, and protocol software. Notice that these do not include the devices attached to the network. The *transmission medium* is the wire or cable (metal or fiber optic) that physically carries messages between devices attached to the network. *Network interface units* are microprocessors that provide the physical interface between the various user devices and the transmission medium. *Protocol software* are computer programs that execute in the network interface units to provide the various data transfer functions, such as message forwarding, destination address decoding, file transfer, electronic mail services, and so on.

The physical configuration of the LAN is called the *network topology* and may be one of several arrangements: *point-to-point networks*, *star networks*, *ring networks*, *bus networks*, and so on. These are illustrated in Figure 20.5.

Point-to-point networks require a wire or cable between each pair of nodes. This arrangement is more or less permanent and allows faster response. Although it is expensive, it has the advantage of independent linkages. That is, even if one

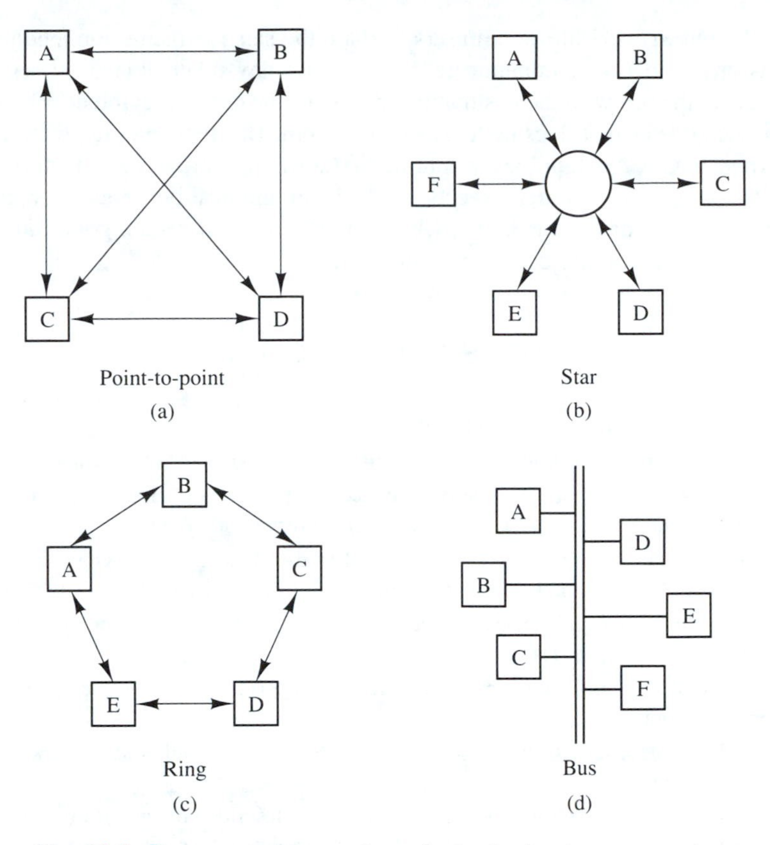

Fig. 20.5. Four common network topologies for local area networks.

link is broken, or if one processor on the network goes down, the rest of the system is unaffected.

The star topology employs a controlling, central supervisor (host) and transmits messages along single conductors to and from various receivers and transmitters (user devices). Each device communicates directly with the host. The star network is limited by the speed of the host. One can see that messages will frequently have to wait while the host is occupied with other nodes. The entire system becomes inoperable if the host device is disabled. Telephone systems in office buildings are structured as star networks when all incoming calls come through a central point.

In the ring configuration, transmitted messages move from one node to the next around the ring. When a node in the ring recognizes a message addressed to it (remember, messages include a destination address), it accepts the message. If a particular message is not intended for the node, it forwards the message to the next station in the ring. This continues until the message reaches its destination. Notice that if one link is broken, the entire network becomes inoperable.

The *bus network* allows all users to share the same transmission medium and permits any device to communicate directly with any other device by sending a message via the bus with the destination address of the desired recipient. The failure of a single node or link does not disable the system. Bus networks facilitate growth and expansion. A bus topology would be the most appropriate for a factory LAN.

There are many other aspects of data communications networks that are beyond the scope of this book. I.&S.E.s of the future will need a good foundation in digital electronics and data communications.

DISCUSSION QUESTIONS

1. Distinguish between "data" and "information."

2. Do some library research and trace the development of management information systems.

3. From current literature, learn the basic concepts of "decision support systems." How are these concepts different from management information systems?

4. Would the same general procedures outlined in the chapter for designing information systems apply to a completely manual (no computer) operation? Why or why not?

5. Discuss the major differences between the traditional approach and the data-base approach to data-base design.

6. Conceive of a situation in which a particular message is data to one person and information to another.

7. Identify and describe the basic data processing operations. Could these be applicable to a manual system? Discuss.

8. Consider how we have used the word "design" to describe how an information system is created. Is this usage consistent with the way we defined design in Chapters 1 and 2?

9. In Chapter 2 we made the statement that industrial and systems engineers design systems at two levels: human activity systems and management control systems. Into which category does information system design fall? Discuss thoroughly.

10. Define the desirable characteristics of "good information."

11. Define the desirable characteristics of "a good information system."

12. Use the basic principles of this chapter to design an information system for your university chapter of the Institute of Industrial Engineers.

13. Determine who at your university is responsible for data communications networks. Interview this person or his or her assistant and determine the network topologies employed. Ask why these were selected. Inquire about reliability and adequacy of the networks. Are there significant delays in response time? Do the current networks provide for expansion and growth? How many categories of devices are attached to the network?

CHAPTER 21

Personnel Management

21.1. Introduction

The most important asset of any organization is its people. Therefore, it is impossible to overstate the importance of sound personnel management. The functions of a personnel department vary tremendously between organizations, but there is a select set of functions normally called *personnel management* regardless of who does the actual work. Usually, industrial engineering is heavily involved in some or all of these functions either by administrating, doing the actual work, and/or acting in a consulting role. These functions are as follows:[1]

Selection.
Testing.
Placement.
Performance appraisal.
Promotions, layoffs, etc.
Training.
Job analysis.

---

[1]This list and much of the discussion to follow was adapted from Dale S. Beach, *Personnel* (New York: Macmillan Publishing Company, 1975).

Job evaluation.
Labor relations.
Safety programs.
Benefits and services.
Management—motivation, supervision, and communications.

Almost all of these are discussed in this chapter, but first a revisit to Corvallis, Oregon, is in order.

Example 21.1

Being a modern manager, you are astutely aware of the importance of employee-manager relations. Therefore, you want to be very careful in the personnel area. You want to start off with good employees who have been screened, selected, and placed in the proper positions. Using the material presented in this chapter, you accomplish this with sound *personnel selection, testing, and placement programs*.

Since you also recognize the importance of promoting from within, you want to be able to pick out employees who have potential for advancement and prepare them for higher positions. You do this with sound *performance appraisal, training, and educational programs*.

Since you have already performed the job analysis and job evaluation, you really don't need to spend any more time on this and you move on to the critical task of *labor relations*. With what you learn in this chapter, you are at least prepared to understand why unions and your employees react the way they do and are that much closer to good labor relations.

You then decide that the safety of your employees is very important to you and their families. You design a sound *safety program* by using the material in this chapter. Also, the Occupational Safety and Health Act (OSHA) requires that you have a good safety program and keep accurate safety records.

Next, you want to provide good *benefits and services* for your employees because you recognize the need for them, your corporate social responsibility makes you want to do it, and because you recognize that good benefits and services help attract and keep good employees.

Finally, you realize that you and your entire management staff must become experts in the art and science of *management*. This includes *motivation*, good *supervision*, and good *communications*. The last section of this chapter discusses these.

After all this, you feel that your personnel management is in good shape and you are ready to move into the next area.

21.2. Selection, Testing, and Placement

This area of personnel management is very important since it is the beginning of a manager–employee or company–employee relationship that may last many years. Personnel management should be performed correctly from the beginning because mistakes could be critical and become problems that a manager would have to live with for years.

It is important for an organization to have both short-range and long-range plans for employee needs. Short-range plans are needed to fill immediate vacancies, and long-range plans are needed to insure a constant labor force of well-qualified personnel. A forecast should be made of future manpower needs, and the people to fill these needs should be hired as qualified people become available. Job descriptions and job specifications are very important and should be used extensively by the personnel department in its recruiting and selection efforts.

There are many sources of personnel to fill the needs demonstrated by these long- and short-range plans. For the nonprofessional needs, the sources include the following:

(1) Existing personnel:
   (a) Management review of personnel.
   (b) Job vacancy announcements within the plant.
(2) Employment agencies (both public and private).
(3) Union hiring halls (where unions keep lists of qualified people).
(4) Mass-media recruiting (newspaper ads, television ads, and so on).

For the professional needs, each of the above is a possible source, as well as the following:

(1) Colleges.
(2) Management consulting firms.
(3) Professional meetings.
(4) Executive recruiters.

Personnel managers go to these various sources to find the needed employees. The actual hiring process begins when the employee shows up for an interview. Following are the usual steps in the employment process:

(1) *Person applies for a job.* He or she may answer an ad, or the company may call him or her. Some way the prospective employee knows about the company and comes to seek a job.
(2) *A short interview with the personnel director or secretary occurs.* This is a quick interview to screen out any obviously unqualified employees.
(3) *Applicant completes an application form.* The person fills out a standard application form that asks for data on health, education, employment history, and so on.
(4) *Employee takes one or more selection tests.* The purpose here is to check qualifications and look for interest areas and skills that may help in placement.
(5) *Personnel manager interviews applicant.* A professional (usually the personnel manager) conducts the interview, and preliminary attempts at placement are made.
(6) *Personnel department conducts reference checks.* A check is made of references.
(7) *Final interview and selection are made by immediate supervisor.* The personnel department has decided where to place the employee, but the supervisor almost always makes the final yes or no decision.

(8) *Employee is given a medical exam.* This is the last possible rejection in the selection phase. A medical exam also provides data for later medical claims and it sometimes discloses unknown problems.

(9) *Employee is inducted.* Benefits and services are answered, and all papers are completed.

(10) *Employee either enters the training program or goes to work.* Usually, there is a trial period where the company carefully observes the employee, and vice versa. This is a very critical period for all concerned and should be done carefully.

In the process above, the relationship may be terminated in any of the steps 2, 3, 4, 5, 6, 7, 8, or 10. By the time the individual is hired (step 9), he or she has successfully completed the screening process and should be a good, capable employee.

The industrial engineer is very active in the selection process. He or she is the one best qualified to determine forecasted labor needs, and this forecast is critical to a sound personnel program. The I.E. usually writes or helps write the job description and job specification so vital to the personnel manager. Some industrial engineers are very active in the design of forms, questionnaires, and tests of which application blank and selection test are two examples.

21.3. Performance Appraisal, Training, Education, and Promotions

This next area of personnel management might be called the *daily operations* of personnel management. In the first phase, well-qualified employees were hired; now the employees should be evaluated, good performance should be rewarded, and the employees should be trained and educated for future growth.

Perhaps the most important part of this phase is performance appraisal, for it determines training and education needs as well as potential for development. The emphasis *should always be on development of the person* who is being appraised to make him or her a more valuable person both to the company and personally. Therefore, a *definition of performance appraisal* could be *the systematic evaluation of employees with respect to job performance and development potentials.*

It is important that this definition not mention *personality.* More than likely, a supervisor has little if any effect upon an employee's personality, and therefore, the supervisor should not appraise personalities. He should appraise results that can be gauged by goals that the employee and supervisor have established in prior meetings.

There are various methods of appraising performance, but following are some of the more popular:

(1) *Rating Scale*—The employee is evaluated on certain characteristics by checking a continuous or discontinuous scale. Figure 21.1 is an example of a scale

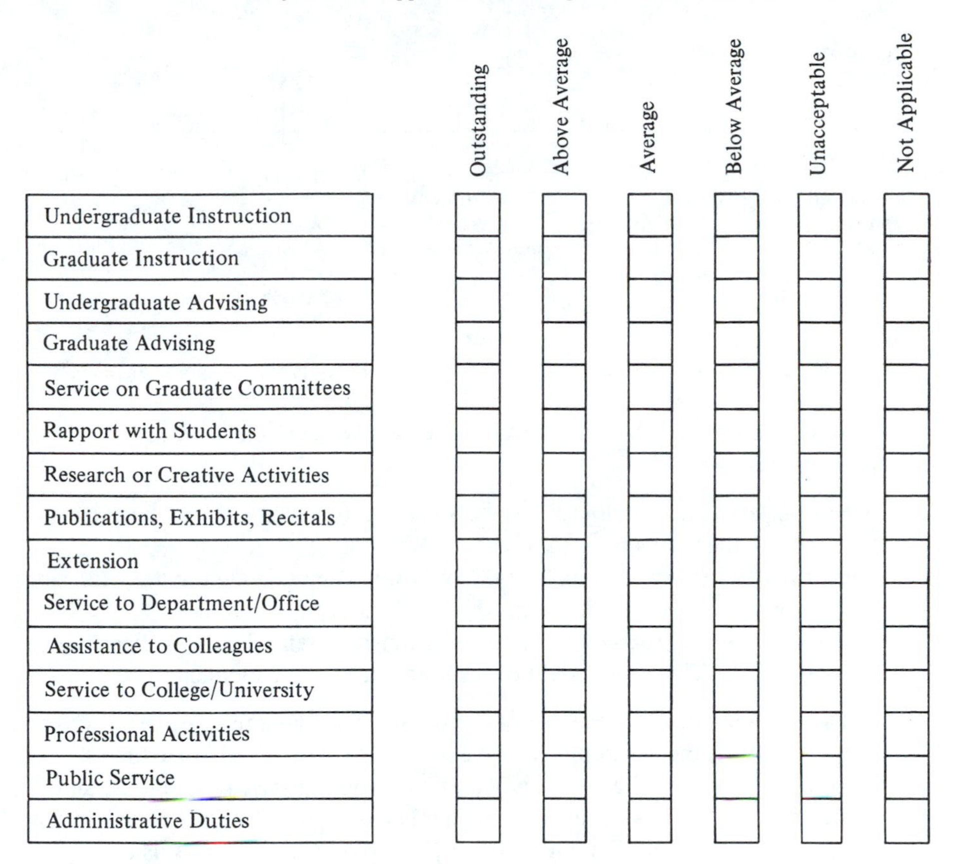

	Outstanding	Above Average	Average	Below Average	Unacceptable	Not Applicable
Undergraduate Instruction						
Graduate Instruction						
Undergraduate Advising						
Graduate Advising						
Service on Graduate Committees						
Rapport with Students						
Research or Creative Activities						
Publications, Exhibits, Recitals						
Extension						
Service to Department/Office						
Assistance to Colleagues						
Service to College/University						
Professional Activities						
Public Service						
Administrative Duties						

Fig. 21.1. Example of a discontinuous scale used in performance appraisal.

used for college professors. The figure shown is for a discontinuous scale, but a continuous scale could also be developed, as in Figure 21.2.

(2) *Employee Comparison*—In this method the rater is forced to compare his employees to each other. In ranking plans the rater ranks the employees in order of value to the organization. In other methods he or she may be forced to fit a distribution (e.g., upper quarter, average, lower quarter).

(3) *Checklist*—A series of statements are written describing certain types of behavior and job performances. The rater reads through the list and checks those items that he or she believes apply to the employee. Each of the statements carries a certain weighted value so that a composite score can be derived.

(4) *Other Methods*—Other methods include (a) the *critical incident*, in which the rater keeps a log of all effective and noneffective practices; (b) the *field review*, in which the rater and the employee do the rating together; (c) the *essay*, in which the rater describes the employee's performance in essay form; and (d) the *group*, in which several people familiar with the employee do the rating.

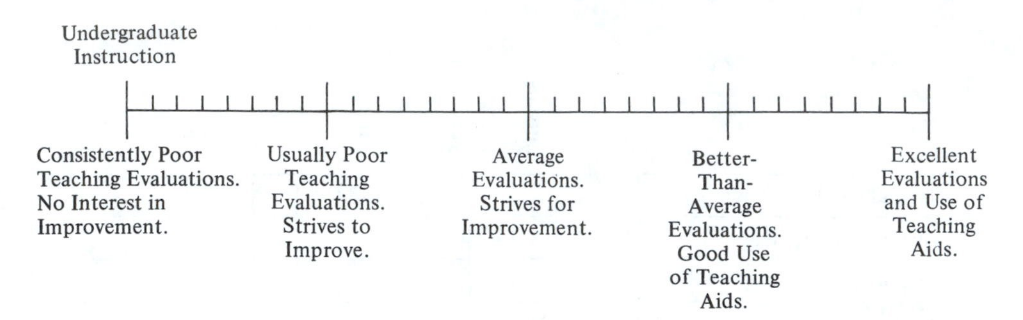

Fig. 21.2. Example of a continuous scale used in performance appraisal.

Some common problems that raters frequently encounter and should try to avoid are as follows:

(1) *Halo Effect*—An employee may be so strong (or weak) in one trait that the rating of other traits may be affected by it.
(2) *Lack of Sensitivity*—Raters tend to not evaluate high enough on the high side and low enough on the low side. Figure 21.3 depicts this.
(3) *Aversion to Criticism*—No one likes criticism, even when it is meant to be constructive. The rater should concentrate on positive elements.

The interpretation and the use of the performance appraisal are critical. The primary objective is the development of the individual. Therefore, some time should be devoted to discussing the appraisal with no mention of pay, promotions, and so forth. Another objective is to encourage better performance. Therefore, it is critical to be sure that the appraisal results and the promotion and pay rewards are highly correlated.

One of the results of good appraisals is a promotion. Since promotions are exceptionally important to almost all people, promotions should be handled properly

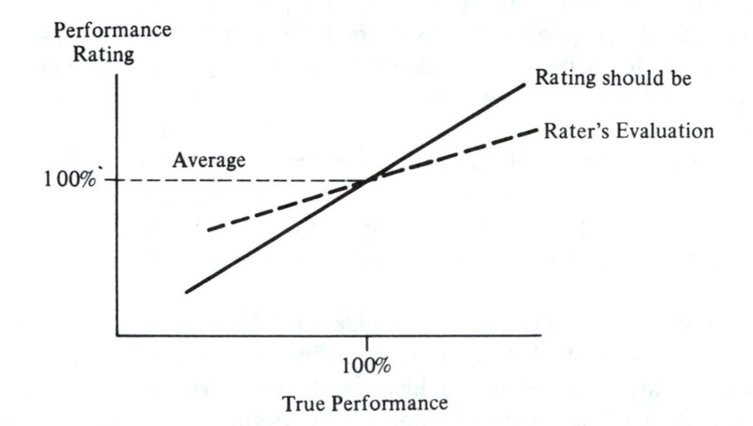

Fig. 21.3. Lack of sensitivity illustration.

and carefully and should be the result of good performance appraisals. There are, however, other considerations. Almost all unions strive for seniority as a basis for promotion as well as compulsory posting of vacancies for potential job applications. Whether or not a union is present, these two factors should be considered by personnel management. Also, with almost all promotions there are disappointed candidates whom personnel must consider and handle carefully.

Other personnel changes that occur with some regularity are as follows:

(1) *Transfer*—Reassignment to another job of similar pay and responsibility.
(2) *Layoff*—Indefinite separation from payroll as a result of factors beyond employee's control.
(3) *Demotion*—Reassignment to job of lower status.
(4) *Discharge*—Permanent separation from payroll.

These changes must be handled judiciously by personnel management because each situation can be very sensitive. It is important that promotions and appraisals correlate.

Since the basic objective of performance appraisals is the development of the individual, all *training and education programs* should be designed to mesh with the performance appraisals. When appraisals disclose specific weak areas, programs can be implemented to strengthen these areas.

Distinctions should be made between training, education, and retraining.

(1) *Education* is learning for the development of the individual with no immediate use plans. The study of English or sociology in college is an example.
(2) *Training* is learning for immediate use toward a specific goal. The study of welding by an employee who is or will be welding is an example.
(3) *Retraining* is learning to keep up to date and is aimed toward specific goals. Its counterpart in education is called *continuing education*.

Education, training, and retraining are usually administered by the personnel management, but the design, development, and, sometimes, teaching are often handled by industrial engineers who are particularly active in designing training programs.

In these training programs the objective is to teach either a motor skill or concepts and attitudes. I.E.s have found that workshops with "hands-on" experience (participants actually perform or do what is being taught) where the material is taught in phases are good for motor skills and that lecture and discussion groups are good for concepts and attitudes. "On-the-job" training programs have been particularly successful for motor skills, but there are other methods of training. For example:

(1) *Vestibule*—Lecture followed by on the job experience.
(2) *Conference*—Smaller group operating in a democratic fashion, discussion groups, and free interchange of ideas.
(3) *Case Study*—Students examine and together solve a typical problem.

(4) *Role Playing*—Similar to case study except that students must act out the individual roles.
(5) *Programmed Instruction*—Students learn in small units and pace themselves through the course.
(6) *Demonstration*—Instructor solves a particular problem; often used in conjunction with on-the-job training.
(7) *Simulation*—Students solve a problem that is symbolic of the real problem.

Industrial engineers involved in the design and control of training programs must become familiar with these various methods of training and their advantages and disadvantages.

I.E.s also become involved in *continuing education* and *management development* courses—often as students. Management development courses are systematic *training* and growth programs in which students learn how to become more effective and efficient managers. Almost all management development programs include a large element of classroom (lecture), conference, and case study methods of instruction, but the programs also may include *work experience related* methods. Following are examples of work experience related methods:

(1) *Understudy*—Employee works as an assistant to a person whose job he is learning.
(2) *Coaching*—Supervisor *encourages* development by allowing the employee to perform on the job and then working with the employee to improve performance.
(3) *Job Rotation*—Employee is rapidly shuttled to and from different jobs so that he can learn them all.
(4) *Advisory Boards*—Groups of employees are allowed to study major high-level problems and make recommendations.

21.4. Job Analysis and Description

This is an important area of personnel management, as was demonstrated in Chapter 9. The functions are sometimes administered by personnel management, but even then industrial engineers often do the actual work or act as consultants. Since this material was adequately covered in Chapter 9, no more will be said here.

21.5. Labor Relations

Labor relations deals with the labor force collectively as a group. Over the years the term has come to imply management–union relations in which the labor force is represented by a union. Unions caused great changes in the lifestyles of both employees and management. The employees' standard of living and status improved as a result of union bargaining (at least at first). Management could no longer deal directly with employees and could no longer freely make the decisions they had been used to making.

There are many reasons why an employee might join a union; for example, the person may want more pay and better benefits, may wish to be heard, may want less favoritism and discrimination, friends and family may exert social pressure on the person, he or she may be frustrated in his or her present situation, the person may have to join if compulsory membership in the union exists, or the person may wish to join hoping to be able to work less and earn more. If management understands these reasons, management will be closer to having good labor relations.

Unions first appeared in this country around 1800. Their growth rate was fairly slow at first, accelerating to astounding speed between 1935 and 1950, but then the rate slowed down dramatically. (In some areas union memberships are declining.) Some of this variation in growth and decline has been a direct result of federal actions. The Wagner Act of 1935 established the National Labor Relations Board and guaranteed the right of employees to form and join unions. In 1947 the Taft-Hartley Act imposed restraints on unions by guaranteeing the right of the employee to *not* join a union and by defining unfair union practices.

Collective bargaining is the relationship between a company and its employees through the union. The process is shown in Figure 21.4. The union representatives and management meet periodically to draft a new contract or agreement. Once an agreement is reached, management becomes the administrator of the contract. The employees live by the contract, but they can file *grievances* and eventually go to arbitration (usually).

A *grievance* is a complaint, the handling of which is described in the contract. In *arbitration* a third (impartial) party analyzes the grievances and gives an opinion. The opinion may or may not be binding, depending on the contract. Why do grievances occur and why is arbitration sometimes necessary? One reason is that

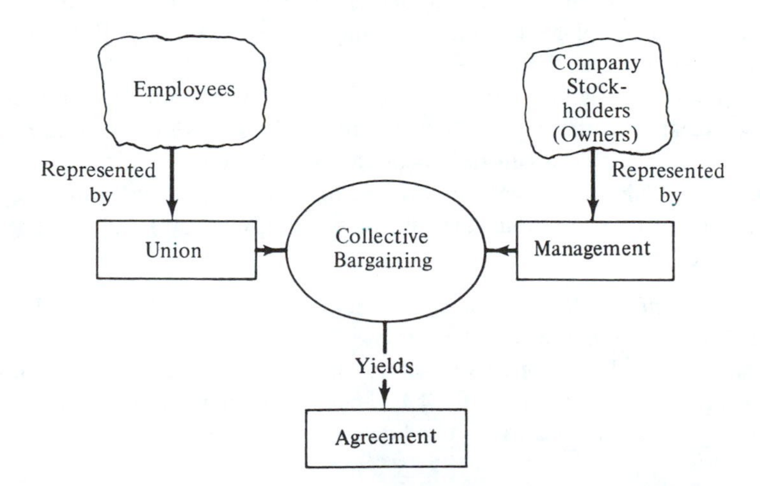

Fig. 21.4. Collective bargaining.

since union officials are elected by and represent the employees, the officials must strive for what the employees desire. Sometimes the employees' demands may appear illogical to management and cause much hardship at collective bargaining time.

The industrial engineer plays a very active role in the collective bargaining process. First, since almost all grievances are over time studies, job evaluations, or pay scales, the I.E. automatically becomes involved. Second, many companies have industrial engineers on their bargaining teams (so do many unions). Third, work goes on between bargaining sessions compiling statistics and preparing reports for the next session much of which is done by industrial engineers. Finally, many arbitrators are industrial engineers who are true labor experts.

21.6. Safety Programs

We have already stated that the most important asset of any organization is its people. It is only logical that the organization should do everything feasible to insure the health and safety of its people. That is exactly the objective of a well-run safety program—to establish and maintain a good safety record.

In 1970 there were over 2,000,000 disabling injuries and over 14,000 accidental deaths in employment related accidents in the United States. The cost of these was approximately $9,000,000,000.[2] This in itself is reason why management should support safety programs. Humanitarianism interests and governmental regulations are two other reasons. The strongest governmental regulation is the Occupational Safety and Health Act (OSHA) of 1970.

OSHA, which became effective in April 1971, designed safety and health standards for industry and established a regulatory mechanism consisting of federal inspectors and a civil and criminal judiciary system for violations. Although the basic plan is a federal plan, state plans are allowed, but if and only if the state plan is at least as effective as the federal plan.

The industrial engineer can and should become deeply involved in the safety program. The smaller the organization, the more likely it is that he or she will be involved because smaller organizations often do not have safety directors. Even in large companies, industrial engineers become involved—some on a full-time basis. The functions of a safety group that are often performed by industrial engineers include the following:

(1) *Administration*—As mentioned above, in smaller companies the I.E. may often direct the entire safety program.
(2) *Committees*—In a successful safety program committees are established throughout the plant. Since the I.E. is deeply involved in the manufacturing facility, he or she is a likely candidate for these committees.

---

[2]Beach, p. 721.

(3) *Engineering*—The two basic causes of accidents are *physical* (unsafe conditions) and *mental* (unsafe acts). Physical causes can usually be eliminated through engineering of safety devices. The I.E. is a likely person to design these devices.
(4) *Statistics and Analysis*—A good safety program usually keeps good statistics on accidents and accident causes. Because of his or her statistical background, the I.E. can design and administer this phase.
(5) *Education and Training*—To minimize the mental causes of accidents, education and training programs are helpful. Since an I.E. often becomes involved in other training programs, he or she is also a likely candidate for the safety training program.
(6) *Inspections*—Because of the I.E.'s knowledge of machines and manufacturing processes, he or she makes a good safety inspector.

There are personnel functions that may not involve industrial engineers. These functions include enforcement, medical treatment, rehabilitation, and mental health counseling (e.g., alcohol- and drug-related problems).

21.7. Benefits and Services

Although industrial engineers are not usually actively involved in these programs, benefits and services are an important part of personnel management.

Usually, the term *benefits* implies a monetary return or potential. Insurance is a good example of a benefit in which substantial amounts of money may be involved. *Services* implies that no direct flow of money is involved. Ball fields, counseling services, and Christmas parties are all examples of services.

There is some disagreement on how much benefits and services cost a firm (mainly because of the difficulty in defining benefits and services; e.g., some people would consider overtime a benefit). Published studies show an upward trend, with a probable average today of somewhere around 30% of direct payroll.[3] Obviously, this is a very costly program for any company.

If these programs are so costly, why do so many companies offer them? There are many reasons, but the main ones are listed below:

(1) *Recruiting*—Good benefits and services packages aid in obtaining good personnel.
(2) *Employee Relations*—Good programs also help maintain good employees and may help motivate, or at least keep from negatively motivating, employees.
(3) *Collective Bargaining*—Almost all unions strive to improve benefit packages at collective bargaining time.
(4) *Social Responsibility*—Many companies have good benefits and services programs because they feel responsible for their employees.

[3]Beach, p. 757.

(5) *Economics*—Often, benefits packages can be purchased cheaper in bulk than by individuals. Insurance is a good example.
(6) *Taxes*—Benefits are usually tax free.

The industrial engineer does not normally become actively involved in a benefits and services program (except as a participant), but there are minor exceptions. For example, bonuses and profit-sharing plans are sometimes considered benefits. The industrial engineer is trained to design and implement these plans. Other than this, almost all benefits and services programs are designed and administered by personnel management.

21.8. Motivation, Supervision, and Communications

Motivation—the willingness to work toward a goal—is one of the most important problems facing every manager, yet it is probably the least understood function of management. Communication (or the lack of) is also a daily problem facing management. A little is known about the science of communication, but much more is needed. Supervision is another problem. We have all seen natural leaders who have made good supervisors, and natural leaders who have not. What makes a person a natural leader or a good supervisor?

21.8.1 Motivation

There are many basic theories on motivation, but one of the most popular theories is that of Maslow. His theory contends that a person has a hierarchy of needs that must be satisfied in a certain order. The needs are presented in Figure 21.5.

Human beings start at the bottom of this hierarchy and first strive to meet their *physiological* needs (food, shelter, clothing). Until a person is satisfied at this level, he does not proceed higher. But once these needs are satisfied, he strives for the next level—*safety and security*. Here, he may be concerned with such things as job security, seniority, adequate retirement plans, and so on. Once he feels that these needs are relatively satisfied, he moves to the next level—*love and*

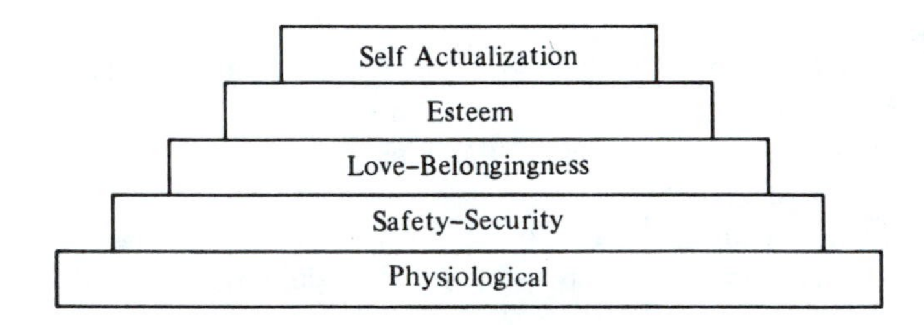

Fig. 21.5. Maslow's hierarchy of needs.

belongingness. At this level he is looking for acceptance by his fellow workers. Does he fit in? Do they like him? Not until he moves to the next level—*esteem*—is he concerned about his "status" in the group. Do the people respect him? Is his opinion valued? The final level is entitled *self-actualization*. This is the highest level in the chain and one obtained by few people. Here, the person is concerned with his opinion of himself. Has he been successful? Is his work worthwhile to humankind and of a lasting nature? Artists and writers frequently achieve these levels—sometimes by having fewer needs at the lower levels.

In Maslow's theory management should be aware of a person's progress in this ladder of ascent and should concentrate on the active goals. For example, if the employee is seeking esteem, then management should strive to help him reach that goal.

Another theory (similar to Maslow's) is that expounded by Herzberg, who contends that there are two factors that affect motivation. These are the *hygenic* and *motivating* factors. Hygenic factors really do not motivate, but the lack, or absence, of hygenic factors has a negative influence on motivation. In other words, the hygenic factors are those factors that *must* be present for motivation to occur, but the presence of them does not motivate. Examples are salary or wages paid, working conditions, and benefits packages.

Motivating factors definitely do motivate or have a positive influence on motivation. The motivators, however, cannot motivate unless the hygenic needs are first satisfied. Job satisfaction, achievement, recognition, responsibility, and advancement are all motivating factors.

There are other theories of motivation. Some are similar, and some are different, but, essentially, they all try to *model* a human being and attempt to predict his or her actions based on the supervisor's actions. Gellerman, McClelland, McGregor, Fein, and Argyris are just a few of the many who have written on motivation theory.

Although there does not appear to be any theory yet that adequately models man, there are some actions that have proven to affect motivation, for example:

(1) *Integration of Goals*—When individual and organizational goals are aligned, motivation tends to be higher.
(2) *Positive Rewards*—Punishment seems to inhibit certain behavior, not encourage it. Positive rewards, however, tend to motivate.
(3) *Job Enlargement*, *Job Rotation*—Enlarging the responsibilities associated with a job and periodic rotation to other jobs seem to motivate almost all people.
(4) *Participation*—Having employees participate in decision making and soliciting their opinions also seem to motivate almost all people.
(5) *Money*—Incentives and bonuses seem to help motivate some, even though this idea is contrary to many motivation theories.
(6) *Competition*—Competition also seems to help in many situations, but competition can also be very dysfunctional.

Nearly every industrial engineer takes at least one course in motivation theory, because an industrial engineer works with and through people every day and should be able to understand their actions. Also, many industrial engineers quickly move into management where they need motivational skills.

21.8.2 Supervision

One definition of supervision is "accomplishing work through others." This implies that the supervisor must be able to *understand*, *lead*, and *guide* his people. Not only must he understand motivation, he must also be a good leader and understand the field of *group behavior*.

A leader is a person who shows the way or guides and uses his influence to get others to help him obtain certain goals. A supervisor may or *may not* be a leader, but a group of employees almost always has a leader who probably is not the supervisor. A leader is usually a person who is intelligent, socially sensitive and active, and a good communicator.

A *good* supervisor is able to discern who in the group is the *informal leader* and is able to convince the informal leader to work with him or her. The others then follow more readily.

As with motivation, the theories of good supervision have yet to be proven adequately valid to help the average supervisor. Below are listed traits of good supervisors that are worth exploring:

(1) *General Supervision*—A good supervisor usually appraises more on results and less on how the job is done. General supervision instead of close supervision is used.
(2) *Employee Sensitive*—A good supervisor is usually sensitive to the needs and desires of employees and can usually identify with his people.
(3) *Considerable Influence*—A good supervisor usually exerts considerable influence throughout the organization. Her people sense this and are proud to work with him or her.
(4) *Good Communicator*—For obvious reasons, a good supervisor is usually a good communicator.
(5) *Group Cognizant*—A good supervisor usually realizes that his employees are not only individuals, but also a group or groups of individuals. He or she, therefore, attempts to work with the employees as individuals and as groups.
(6) *Management Oriented*—Usually, a good supervisor is management oriented. That is, she concentrates on the problems of management and lets the employees do the work. This does not imply that she is afraid to get his or her "hands dirty," however.

Now, one of the most perplexing problems for those who study supervision is that some supervisors have violated from one to six of the traits above and yet have proven to be very successful. This perhaps demonstrates that good supervision is an individual trait and that supervisors should concentrate on their *strong points* and develop them.

Industrial engineers deal with supervision every day, particularly at the first-line level (foremen). It is important then that industrial engineers understand the basic theories and philosophies of supervision. Usually, the same course in the I.E. curriculum that discusses motivation also discusses leadership and supervision.

21.8.3 Communications

Obviously, it is important for every member of an organization to be a good communicator. This is compounded when the member must deal through and work with others as both managers and industrial engineers do. Industrial engineers must periodically issue reports and make presentations. Therefore, it is important that they understand the art and science of communication. Usually, this is done through one course that spends some time on communication. This course can be the same course that examines motivation and supervision, as described above.

Although we cannot develop valid mathematical models of communication, it is possible to show a graphic model of the process of communicating. One such model involving only one sender and one receiver is shown in Figure 21.6.

Figure 21.6 shows that mistakes can be made at any one of the four phases of communication. Also, any time we insert other people or destinations in the chain, the probability of error grows. Given all this, it is a wonder more communicating problems are not encountered.

To improve communications, the following observations or suggestions are appropriate:

(1) *Words are only symbols*. Since words have no meanings in themselves, the communicator should be very careful in his choice of words to select those that leave concrete, universal meanings.
(2) *Emotions affect sending and translating*. If the receiver becomes angry or emotionally upset, the intended message may not be received. Also, an emotional sender cannot be sure that he is transmitting properly.
(3) *Messages are receiver dependent*. The receiver often interprets a message as he wishes it to be or hears what he wants to hear. Also, he hears as he is used to hearing. A sender who is usually highly critical would have a difficult time getting a compliment across to the receiver.
(4) *Feedback is helpful*. To be sure that the destination received the appropriate message, it is useful to have the destination repeat or send back the message.

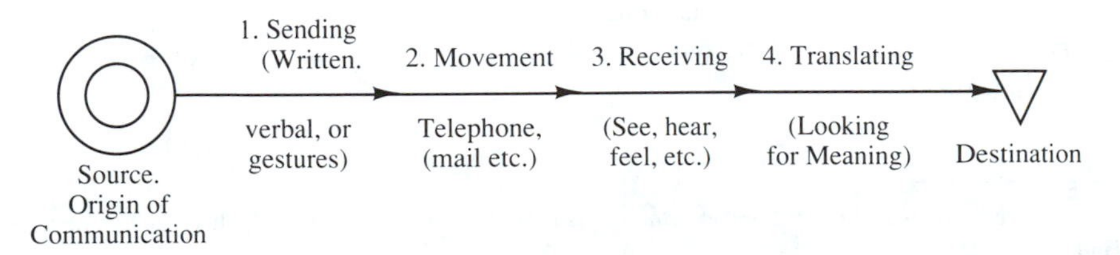

Fig. 21.6. Model of communication.

(5) *Actions speak louder than words.* The sender should always be sure to back up whatever she communicates. Over time, this builds the confidence of the receiver in the sender, which, in turn, improves communication.
(6) *Vagueness is received as more vagueness.* The sender must be absolutely sure that he understands what he is communicating.
(7) *Redundant systems are good.* A written message accompanied by a verbal summary improve the probability that the message will be received correctly. Other methods of redundancy are also good.

21.9. Engineering Management

In recent years there has been a movement that generally carries the name ''engineering management.'' Most of the educational programs in engineering management are at the master's degree level and are options within industrial engineering departments. There is a small number of departments carrying the engineering management label, and a few of these offer undergraduate degrees.

The motivation for the emergence of engineering management programs stems primarily from the increasing need for people having technical backgrounds in engineering, plus a knowledge of management principles. It is very common for engineers to move into management roles. The engineering management programs are intended to provide formal education in management topics.

It is expected that an increasing number of engineers will be moved into management positions in the future, especially as products and services become increasingly technical and sophisticated. We are beginning to hear the phrase ''high-tech, high-touch'' management.

It is the view of the authors that engineering management is a specialty area within industrial engineering (analogous to operations research and quality control), rather than a separate, unique engineering discipline. The reader may wish to pursue this subject in more depth by looking at a book on the subject.[4,5]

DISCUSSION QUESTIONS

1. List the functions of personnel management discussed in this chapter and discuss briefly the role of the industrial engineer in each.

2. Distinguish between training and education.

3. Distinguish between benefits and services. Discuss the importance of good benefits and services programs.

---

[4] J. M. Amos and B. R. Sarchet, *Management for Engineers* (Englewood Cliffs, N.J.: Prentice-Hall, Inc., 1981).

[5] R. E. Shannon, *Engineering Management* (New York: John Wiley & Sons, Inc., 1980).

4. Discuss the differences in approach for recruiting professional and managerial people as opposed to recruiting nonprofessional people.
5. Do you agree or disagree that the personnel manager should do the preliminary screening of employees? Defend your viewpoints.
6. Distinguish between checklist and rating scale methods of performance appraisal.
7. Can there be a negative halo effect? Discuss.
8. Can you think of any reason why a union steward might push or vote for something he doesn't personally believe in? Discuss.
9. Distinguish between supervision and leadership and discuss the characteristics of a good supervisor.
10. Compare Maslow's theory to that of Herzberg.
11. Explain the following statement: "With the employment process of successive hurdles, most employees should be good employees."
12. Is continuing education important to industrial engineers? Why or why not?
13. In Section 21.3, a significant difference is drawn between training and education. Can you think of any subject matter that you have had to date that is (to a large extent) training?

PROBLEMS

1. Design a training program for a forklift truck driver. State the methods of instruction that should be used (be broad; do not be too specific).
2. Contact a local company (maybe the university) and find out what safety program they have. Describe the program in some detail.
3. Design a management training program. Briefly state the subject area that would be covered and the method(s) of instruction that would be used.
4. In Figure 21.1, a discontinuous scale method for rating college professors is given. Develop a similar method or chart for a first-line supervisor in a fast-food restaurant.
5. There is a "game" used in many management training programs whereby the first person whispers a message to person 2, who whispers a message to person 3, and so on. Then person N tells the message out loud. Using Figure 21.6, explain why the message is so very different at the end. How might the result be improved? (You might want to try this as a class exercise first.)

APPENDIX A

Probability and Statistics[1]

A.1. Introduction

Many of the problems we face daily as industrial engineers have elements of risk, uncertainty, or variability associated with them. For example, we cannot always predict what the demand will be for a particular inventory item. We cannot always be sure just how many people will shop at a grocery store and desire to check out during a particular hour. We cannot always be sure that the quality of raw materials from one of our suppliers will be consistent. We are not prophets who can accurately predict results in advance. But we as industrial engineers are educated in the use of applied probability and statistics to make intelligent engineering decisions despite our lack of complete knowledge about future events.

The material presented in this appendix is meant to introduce you to some of the basic laws of chance. Although the treatment is elementary, it is sufficient to allow you to grasp the material on quality control, probabilistic models, and project management (PERT).

Further grounding in probability and statistics will come as more specialized courses on these and other topics are studied.

---

[1]Much of the material for this appendix is taken from Joe H. Mize and J. Grady Cox, *Essentials of Simulation* (Englewood Cliffs, N.J.: Prentice Hall, 1968), Chaps. 2 and 3.

A.2. Basic Probability Theory

An *experiment* is any process that results in an observation. Almost all probabilistic experiments have more than one possible *outcome*. The number of outcomes of an experiment may be finite or infinite. For example, when we visually inspect a resistor, the outcomes are typically finite (either good or bad), but the exact theoretical resistance measurement could take on an infinite number of values.

A.2.1 Sample Space

The *sample space* of an experiment is the set of *all* possible outcomes pertinent to the experiment. As an example, suppose that we select two electric light bulbs and test them in order to determine whether or not they light properly. If we are trying to distinguish completely all outcomes, we might list the following set:

Bulb 1	*Bulb 2*
Light	Light
Light	Not light
Not light	Light
Not light	Not light

We note that there are four possible distinct outcomes to this experiment.

A.2.2 Events

An *event* is a subset of the sample space of an experiment. It may consist of none of the outcomes (void), some of the outcomes, or all of the outcomes. Returning to our light bulb example, let us consider the event that exactly one bulb lights. The event that one bulb lights represents a subset of two of the outcomes of the sample space.

Two events are said to be *mutually exclusive* if the occurrence of one excludes the occurrence of the other; that is, they do not possess any points in common from the sample space. For example, it is clear that the event of neither bulb lighting and the event of both bulbs lighting are mutually exclusive.

As a further example linking sample space and events, a group of 100 items is taken for inspection. We are interested in the number of nonconforming items in this group. If 3 or fewer are found, we shall conclude that the manufacturing process from which the items were taken is operating satisfactorily. The *sample space* for our inspection would include 101 different possible outcomes—that 0, 1, 2, . . . , 99, or all 100 items are nonconforming. The *event* that we conclude that our manufacturing process is operating satisfactorily might be represented by the set of 2 outcomes—that 0 or 1 items are found to be nonconforming.

A.2.3 Probability of an Event

An event may consist of any combination of possible outcomes of an experiment. It is up to the experimenter to define events that are meaningful in the experiment. Let us consider the experiment of drawing one ball from a box containing ten balls, numbered 1, 2, . . . , 10. The balls numbered 1 through 5 are green; those numbered 6 through 10 are white. The probability of drawing any particular one of the ten balls on any performance of the experiment is 0.1. We can define many events relating to the experiment of drawing one ball from the box as follows:

E_1 is the event of drawing an even-numbered ball.
E_2 is the event of drawing a green ball.
E_3 is the event of drawing an even-numbered green ball.
E_4 is the event of drawing a ball larger than 6.
E_5 is the event of drawing a ball less than or equal to 4.

Many other events could be defined for this experiment, but we shall discuss only these five events and the way in which each event could occur. There are five outcomes by which event E_1 can be realized. We shall say that five of the ten possible outcomes of the experiment are *favorable* to E_1. Since each outcome is equally likely to occur in this example, the probability of event E_1 is 0.5. Notationally, we have

$$P(E_1) = 0.5$$

By similar reasoning, we see that the probabilities of the remaining events are

$$P(E_2) = 0.5$$
$$P(E_3) = 0.2$$
$$P(E_4) = 0.4$$
$$P(E_5) = 0.4$$

In general, the probability of an event E is the sum of the probabilities of the outcomes of which E is comprised.

A.2.4 Rules of Operation

In order to begin to calculate probabilities, we need a few more definitions. Two events are said to be *equally likely* if each event has the same probability of occurrence. Two events are said to be *independent* if the occurrence of one event in no way affects, or is affected by, the occurrence of the other event. The *intersection* of the events A and B (denoted by AB) consists of all the sample space points corresponding to *both* A and B. The *union* of the events A and B (denoted by $A + B$) consists of all the sample-space points corresponding to *either* A or B (*or both*).

We express the *multiplication theorem* for events A and B as follows:

$$P(AB) = P(A \mid B)P(B) \tag{A.1}$$

where $P(A \mid B)$ is read "the probability that A occurs *given that B has occurred.*" If events A and B are independent, then $P(A \mid B)$ equals $P(A)$ and Equation (A.1) can be written in the following form:

$$P(AB) = P(A)P(B) \tag{A.2}$$

The *conditional probability* of the event A, given that event B has occurred, also comes from Equation (A.1) and is given as follows:

$$P(A \mid B) = \frac{P(AB)}{P(B)} \tag{A.3}$$

The *addition theorem* for probability is indicated by the following:

$$P(A + B) = P(A) + P(B) - P(AB) \tag{A.4}$$

If A and B are mutually exclusive, then we obtain the simplified form of Equation (A.4) as follows:

$$P(A + B) = P(A) + P(B) \tag{A.5}$$

These rules must be extended whenever more than two events are involved.

As an example, imagine a subsystem comprised of two main components, A and B. They are produced independently, and the nonconforming proportions of A and B are, respectively, 0.06 and 0.10. The probability of the event that both A and B are nonconforming in a subsystem is given by the multiplication theorem in Equation (A.2) and equals $(0.06)(0.10) = 0.006$. The probability that either component A or B (or both) is bad, resulting in a nonconforming subsystem, is given by Equation (A.4) and equals $0.06 + 0.10 - 0.006 = 0.154$.

A.2.5 Combinations

In considering problems involving finite sample spaces of equally likely outcomes, we are often tempted to count the frequencies of interest. This procedure is perfectly valid, but counting may be time consuming and tedious. Therefore, we most often use a formula to determine numbers of combinations. The number of *combinations* of n things taken k at a time is expressed by the following notation and formula:

$$C_{n,k} = \binom{n}{k} = \frac{n!}{k!\,(n - k)!} \tag{A.6}$$

For example, we may want to know the number of ways in which two nonconforming items can be observed in a sample of four parts. If (1, 2) indicates that the nonconforming items were the first and second items observed, we can extend this notation to show the other possible ways the inspector might have encountered nonconforming items as (1, 3), (1, 4), (2, 3), (2, 4), and (3, 4). In all, we count six ways. Using Equation (A.6), we can solve this problem by using $n = 4$ (sample size) and $k = 2$ (number of nonconformances) as $4!/2!2! = 4\cdot3\cdot2\cdot1/2\cdot1\cdot2\cdot1 = 6$.

A.3. Random Variables

A *random variable* is a numerically valued variable defined on a sample space. For each point of the sample space, the random variable would be assigned a value. For our light bulb example with four outcomes in the sample space, we might define the random variable x as follows:

$x = 0$ for the sample point (not light, not light)
$x = 1$ for the sample point (not light, light)
$x = 2$ for the sample point (light, not light)
$x = 3$ for the sample point (light, light)

Or we might decide that the random variable x should be the number of bulbs that lit of the two tested; thus, for the sample space, x would be defined as follows:

$x = 0$ for (not light, not light)
$x = 1$ for (not light, light, or vice versa)
$x = 2$ for (light, light)

In considering our magnetic tape example we might define the random variable to be the number of "bad" spots on a reel of tape. The set of values that such a random variable would have consists of zero and the *positive integers*.

The examples above illustrate *discrete* random variables since the values that x can take on can be individually identified and are *countable* (e.g., in the three examples of the previous paragraph, there are 4, 3, and ∞ individual discrete values, respectively, that x can assume. Many examples we will see, however, involve a *continuous* random variable in which the values that x takes on are *not* individually identifiable and not countable. Instead, the continuous random variable can assume an infinite number of values. For example, the number of tires on an approaching vehicle is a discrete random variable, while the air pressure in any given tire is a continuous random variable.

The word *random* in "random variable" means that the variable will assume its values in a chance manner. Consider a problem concerned with machine breakdowns. Our random variable might be the number of breakdowns per day, the time between breakdowns, the number of simultaneous breakdowns, the time required to repair breakdowns, and a host of other possibilities. Every problem must be individually analyzed and the pertinent random variable must be chosen and defined.

Some examples of experiments whose outcomes are random variables are as follows:

(1) Number of nonconforming items observed in a sample (discrete).
(2) Average dimension of items sampled and measured (continuous).
(3) Rain or no rain on any given day (discrete).

(4) Amount of rainfall on any given day (continuous).
(5) Launch or abort on any attempted missile shot (discrete).
(6) Success or failure on a missile launch (discrete).
(7) Accuracy of a missile launch (closeness to intended target)(continuous).
(8) The temperature in a petroleum refining process (continuous).
(9) The number of whole barrels of gasoline produced by a petroleum refining process in a given time period (discrete).
(10) The voltmeter reading of a residential electrical circuit (continuous).

The examples above are of random variables because the value of any particular performance of each of the experiments cannot be predicted with certainty.

A.4. Estimating Probabilities

Let us consider an experiment to be the inspection of a reel of magnetic tape produced by a certain process. Each *repetition* of this experiment will consist of the examination of a different reel of tape, the possible decisions being accept, rework, and reject. The number of repetitions corresponds to the number of reels of tape, and the possible decisions number only three; thus, both the number of repetitions and the number of decisions are finite.

We let

n_i = number of repetitions of the experiment that will result in decision i
k = number of different possible decisions (three in this example)
N = total number of repetitions

It follows that the ratio n_i/N is the fraction of repetitions favorable to decision i (frequency ratio of decision i). If the random variable x can take on the values x_1, x_2, and x_3 corresponding to accept, rework, and reject, respectively, then we can *estimate* probability $p(x_i)$ as the frequency ratio n_i/N. That is,

$$p(x_i) = \frac{n_i}{N} \qquad (i = 1,2, \ldots, k) \tag{A.7}$$

Our probability function has the following characteristics:

$$p(x_i) \geqslant 0 \qquad (i = 1,2, \ldots, k) \tag{A.8a}$$

and

$$\sum_{i=1}^{k} p(x_i) = 1 \tag{A.8b}$$

Returning to our magnetic tape example, suppose that 100 reels are inspected and that 50 are accepted, 30 are to be reworked, and 20 are rejected. Figure A.1 portrays this probability function graphically. The properties expressed by Equations (A.8a) and (A.8b) are seen to be true.

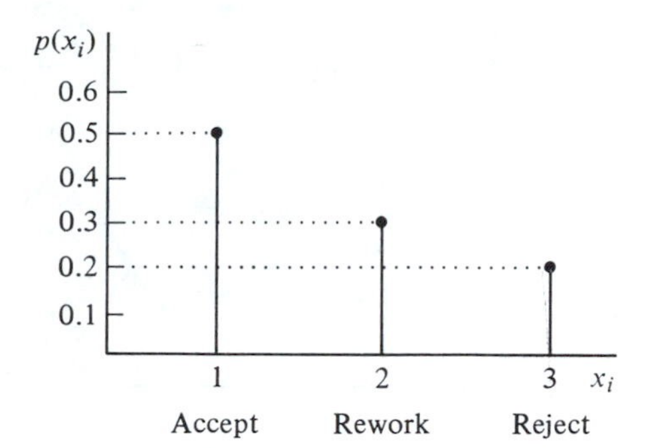

Fig. A.1. Probability function.

In the example above, we thought of the number of outcomes k as being relatively small. Actually, k can tend toward infinity. For example, we might be concerned with the *number* of "bad" spots on a reel of tape. If these spots can be infinitesimally small, yet distinct and countable, it is theoretically possible to have infinite "bad" spots. In this case, k is infinite, but Equations (A.7), (A.8a), and (A.8b) are still correct.

We normally think of the number of repetitions N as being finite. As the number of repetitions of an experiment increases, we feel more comfortable or confident about the results observed; this feeling does have a basis in mathematical fact. We often like to think of the probability $p(x_i)$ as

$$p\,(x_i) = \lim_{N \to \infty} \frac{n_i}{N} \qquad (i = 1, 2, \ldots, k) \tag{A.9}$$

That is, $p(x_i)$ equals the fraction of outcomes favorable to x_i in an infinite number of experiments, if that were possible. For practical reasons, we usually have to strike a balance between the number of experiments we can afford and the number of repetitions N that gives us a good probability estimate.

The most difficult case for us to handle conceptually occurs when the random variable x can take on an infinite number of possible values that we cannot count. Even when we talked about the number of "bad" spots on a reel of tape (perhaps infinite), at least we could "count" or identify them with the real numbers (e.g., 0, 1, 2, 3, . . .). That is, the random variable was *discrete*. *Continuous* random variables do not take on values that are countable. Continuous random variables have a probability density function, $f(x)$, as illustrated in Figure A.2, instead of having distinct probabilities or weights, $p(x)$, at points, as in Figure A.1.

Since we cannot count the possible values that a continuous random variable can take on, we normally talk about subintervals. The amount of rain in May is an example of a continuous random variable. We might speak of 0–2, 2–4, 4–6, and so on, centimeters of rain. These *subintervals* are countable, and we can estimate the probabilities associated with each. For example, the probability that the random variable x takes on a value between a and b may be estimated from data as

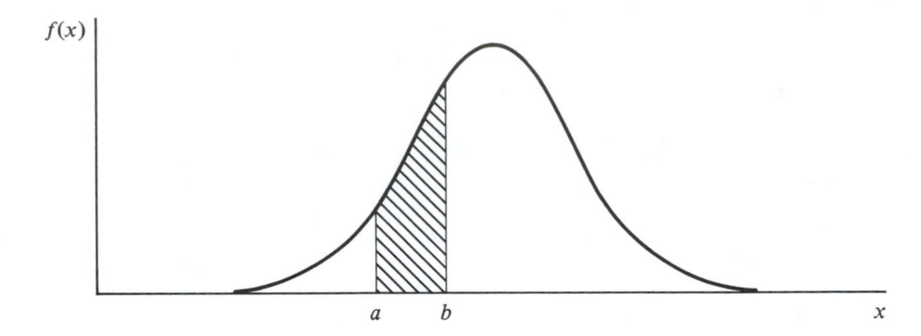

Fig. A.2. Probability density function.

$$P(a < x \leqslant b) = \frac{n}{N} \tag{A.10}$$

where n is the number of repetitions of the experiment in which the observed value of the random variable is between a and b.

Our new probability density function $f(x)$ has the following properties:

$$P(a < x \leqslant b) = \int_a^b f(x)dx \qquad \text{for } a < b \tag{A.11}$$

$$f(x) \geqslant 0 \tag{A.12}$$

$$\int_{-\infty}^{\infty} f(x)\, dx = 1 \tag{A.13}$$

Equations (A.12) and (A.13) present the same basic properties for continuous random variables as those given in Equations (A.7) and (A.8) did for discrete random variables.

A.5. Some Important Probability Distributions

Associated with every random variable is a *probability distribution*. As a matter of fact, we have already encountered two probability distributions in Figures A.1 and A.2. A probability distribution of a random variable is just that—a distribution, or portrayal, of the probabilities of occurrence of each value (or interval of values) of the random variable. The probability distribution that we have already encountered in Figure A.1 is "discrete" because the particular random variable is discrete (accept, rework, reject). The distribution shown in Figure A.2 is obviously continuous. The presentation in this section will include both the discrete and continuous cases. We should always bear in mind that the *nature of the random variable* determines whether a distribution is discrete or continuous.

In addition to the mathematical representation of a distribution, we are often interested in measures that further describe the properties of a population under study. Two very popular characteristics of a distribution are the *mean* and *variance* of the random variable. The mean is a measure of central tendency and represents somewhat of an average value. The variance is a measure of dispersion or distribution spread. The mean is often represented as μ and the variance as σ^2. We often speak of the *standard deviation*, which is the positive square root of the variance. The standard deviation is represented as σ.

A.5.1 Discrete Distribution Properties

A *discrete probability distribution* is a function $p(x)$ of the discrete random variable x yielding the probability $p(x_i)$ that x will take on the value x_i. Such a function is often called the *frequency function* of a discrete random variable. The defining characteristics of such a function are given by the following mathematical statements:

$$p(x_i) \geqslant 0 \qquad (i = 1, 2, \ldots, k) \tag{A.14}$$

and

$$\sum_{i=1}^{k} p(x_i) = 1 \tag{A.15}$$

Of course, Equations (A.14) and (A.15) are repeats of Equations (A.8a) and (A.8b).

Much of our work with distributions involves cumulative probabilities, that is, the probability that the random variable will assume one of a set of possible values. The *discrete cumulative distribution* is defined as follows:

$$F(a) = P(x \leqslant a) = \sum_{x_i \leqslant a} p(x_i) \tag{A.16}$$

$F(a)$ is the probability that x will have any value less than or equal to a. A useful result of Equation (A.16) is that

$$p(x_i) = F(x_i) - F(x_{i-1}) \tag{A.17}$$

Thus, the discrete probability distribution and the discrete cumulative distribution are equivalent, since either can be obtained from the other. Figure A.3 is the discrete cumulative distribution corresponding to Figure A.1. Note that according to our definition it is a step function.

In practice, it has been found that a few discrete probability distributions closely approximate many naturally occurring distributions. The next three sections will briefly present some of these distributions.

A.5.2 Binomial Distribution

Suppose that we consider n independent experiments, each of which has only two possible outcomes. For example, inspecting and classifying n items as good or bad meets our description. If our random variable is the number of occurrences of

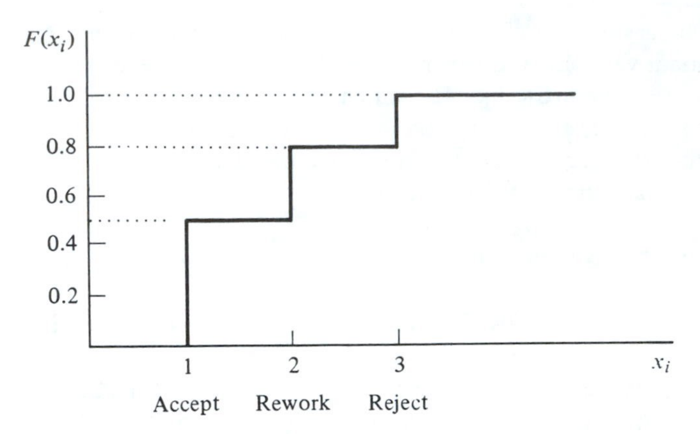

Fig. A.3. Discrete cumulative distribution.

a particular outcome, such as bad items, we can see that the possible values of the random variable are 0, 1, . . . , n.

For convenience, let us agree to refer to the occurrence of a particular outcome (e.g., bad items) as a "success" and the alternative outcome as a "failure." Now, by considering n independent and identical "two outcome" experiments, with a "success" having the probability p' and a "failure" having the probability $q' = 1 - p'$, we can explore the probability of "x successes," where $x \leqslant n$. First, the number of ways we can obtain x successes is the number of combinations of n things taken x at a time. That is,

$$C_{n,x} = \binom{n}{x} = \frac{n!}{x!\,(n - x)!} \tag{A.18}$$

By recalling our multiplication rule for independent events, we note that each such combination has a probability of $p'^x q'^{n-x}$ of occurring. Thus, the probability of x successes in n independent and identical experiments may be written as follows:

$$p(x\,;\,n,p') = \binom{n}{x} p'^x q'^{n-x} \tag{A.19}$$

where $x = 0, 1, \ldots, n$.

The probability distribution expressed by Equation (A.19) is called the *binomial probability distribution*. Its cumulative binomial probability distribution is as follows:

$$F(a\,;\,n,p') = \sum_{x=0}^{a} p(x\,;\,n,p') \tag{A.20}$$

for $a = 0, 1, \ldots, n$.

Also, the mean and variance of the binomial are given by the following:

$$\mu = np' \tag{A.21}$$
$$\sigma^2 = np'q' \tag{A.22}$$

As an example of the binomial distribution, assume that we randomly select six transistors from a production line. Past data have indicated that 10% of the transistors inspected are bad. We can determine the probability of finding exactly two bad transistors by using Equation (A.19) as

$$p(2\,;6,\,0.1) = \binom{6}{2}\left(0.1^2\right)\left(0.9^4\right) = 0.0098$$

Similarly, using Equation (A.20), we can find the probability of detecting two or fewer bad transistors as:

$$F(2;6,\,0.1) = \sum_{x=0}^{2} p(x;6,\,0.1) = 0.984$$

The binomial distribution for this example is graphed in Figure A.4.

A.5.3 Poisson Distribution

The Poisson distribution is applicable in many situations in which some kind of event (such as a "flaw" or a "change") occurs randomly in time or over distances, areas, or volumes. To be consistent with our earlier terminology, we shall continue to call the occurrence of an event a "success." The *average rate of occurrence* of the event is considered constant in a Poisson process and is denoted by λ. The probability of x successes in a system having a constant average occurrence rate λ is

$$p(x;\,\lambda) = e^{-\lambda}\,\frac{\lambda^x}{x!} \tag{A.23}$$

where $x = 0, 1, 2, \ldots$. In addition to being a distribution in its own right, the Poisson distribution is a good approximation to the binomial when p is small. Some writers recommend this approximation if $np \leqslant 5$ and $p \leqslant 0.1$.

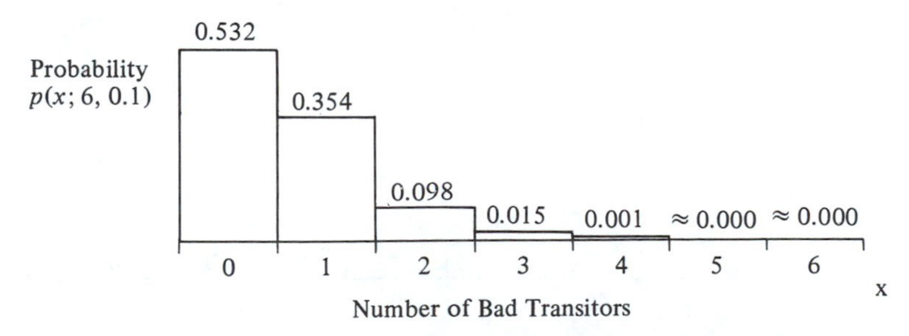

Fig. A.4. Binomial distribution.

We can express the cumulative Poisson as follows:

$$F(a; \lambda) = \sum_{x=0}^{a} p(x\ ;\ \lambda) \qquad \text{(A.24)}$$

Since the cumulative Poisson is used extensively in practice, we have included a table of values as Table B.1 in Appendix B. The table considers values of λ from 0.02 to 10.

For the Poisson distribution, we obtain values of the mean and variance as follows:

$$\mu = \lambda \qquad \text{(A.25)}$$

and

$$\sigma^2 = \lambda \qquad \text{(A.26)}$$

In other words, the variance and mean are identical.

Some examples of use of the Poisson distribution are provided by the number of imperfections on a sheet of metal, the number of diseased spots on a tree, the number of weeds on a plot of land and so forth. When we use the Poisson, we should remember that the random variable can assume the set of numbers 0, 1, 2, . . . , which is a countably infinite discrete set.

As a numerical example, suppose that we have many rolls of magnetic tape and that the average rate of occurrence of "bad spots" is six per roll. What is the probability of having no more than three bad spots on a particular roll? Using Table B.1 in Appendix B, we see that the cumulative Poisson probability for $F(3; 6)$ equals 0.151. This example demonstrates the Poisson as a distribution in its own right.

Now, let's use the Poisson as an approximation to the binomial by solving the earlier example concerning six transistors. Using the binomial distribution, we found the cumulative probability of two or fewer bad transistors to be 0.984. Using the Poisson and letting $\lambda = np = 6(0.1) = 0.6$, we can read the probability of 2 or fewer defectives from Table B.1 as $F(2; 0.6)$, which equals 0.977. The Poisson distribution for this example is illustrated in Figure A.5 for comparison with the

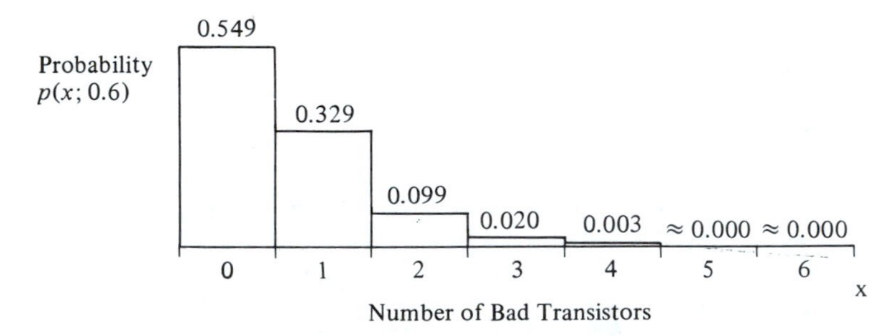

Fig. A.5. Poisson distribution.

binomial distribution. The probabilities shown are determined from the Poisson cumulative distribution in Table B.1 and using Equation (A.17).

A.5.4 Uniform Distribution

Uniform distribution is a very easy distribution to analyze and one of the most common. We shall define our random variable so that it can assume a finite and discrete set of values. We define the uniform distribution as follows:

$$p(x;\, n) = \frac{1}{n} \tag{A.27}$$

for $x = 1, 2, \ldots, n$. We can see that the cumulative distribution is

$$F(a) = \sum_{x \leq a} p(x;n) = \frac{k}{n} \tag{A.28}$$

where k is the number of values of x less than or equal to a.

The mean and variance of the uniform distribution are given as follows:

$$\mu = \frac{n + 1}{2} \tag{A.29}$$

and

$$\sigma^2 = \frac{n^2 - 1}{12} \tag{A.30}$$

We can see that the real distinguishing characteristic of the uniform distribution is that each value of the random variable has the same probability of occurring. Despite the obvious character of the uniform distribution, we should recognize it as a distinct probability distribution.

As an example, suppose that four finalists have been selected for a drawing. The finalists are numbered 1, 2, 3, and 4. Only one top prize is to be given. The probability that finalist number 3, say, is chosen is given by Equation (A.27) as 1/4. The probability that finalists 1, 2, or 3 are selected is given by Equation (A.28) as $F(3) = 3/4$. The probability of any of the four finalists being chosen is illustrated in the uniform density function graphed in Figure A.6.

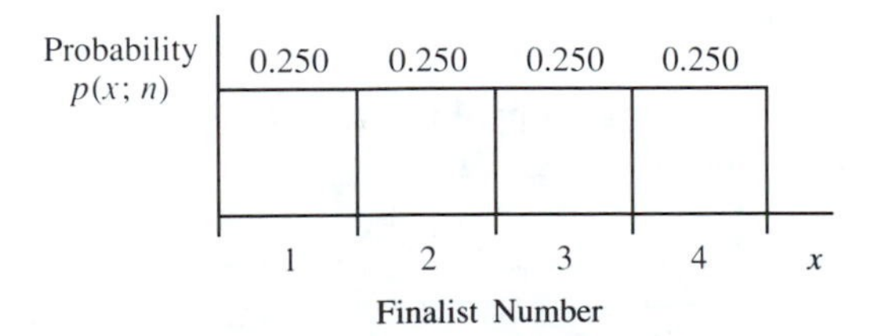

Fig. A.6. Uniform distribution.

A.5.5 Continuous Distribution Properties

A *continuous probability distribution* is a function $f(x)$ of the continuous random variable x that possesses the following properties:

$$P(a < x \leqslant b) = \int_{-\infty}^{b} f(x)\, dx \qquad \text{for } a < b \tag{A.31}$$

$$f(x) \geqslant 0 \tag{A.32}$$

$$\int_{-\infty}^{\infty} f(x)\, dx = 1 \tag{A.33}$$

These properties are, of course, just a repeat of Equations (A.11), (A.12), and (A.13). Such a function is often called either a *density function* or a *probability density function*.

The *continuous cumulative distribution* is defined as follows:

$$F(a) = \int_{-\infty}^{a} f(x)\, dx \tag{A.34}$$

$F(a)$ is the probability that the random variable x will have any value less than or equal to a. If the derivative of F exists, we have the following:

$$F'(x) = f(x) \tag{A.35}$$

Thus, with nice mathematical properties the existence of either $f(x)$ or $F(x)$ determines the other.

We should recall that $f(x)$ is a density type of function instead of a pure probability function. By integration we actually obtain probability. That is, we can specify the probability that a random variable will assume a value *between two points*; however, *the probability for any single point is zero.*

In considering our integral definition of $F(x)$ we may find in practice that our continuous probability distribution $f(x)$ is defined only over a part of the real-number axis. In such cases, we simply extend the definition over the total axis by assigning $f(x)$ the value of zero elsewhere.

In the same spirit as for the discrete case, we shall present three continuous probability distributions that closely approximate many naturally occurring phenomena.

A.5.6 Normal Distribution

We shall define the general form of the *normal distribution* for the continuous random variable x as follows:

$$f(x) = \frac{1}{\sqrt{2\pi}\,\sigma} e^{-(x-\mu)^2/2\sigma^2} \tag{A.36}$$

for $-\infty < x < \infty$.

This distribution is one of the most interesting and useful that we can study. It has a single peak at the mean and is symmetrical about that point. If we plot an example of a normal distribution, it will be readily apparent that it is *bell-shaped* with mean μ and variance σ^2.

In practice, many distributions are well approximated by the normal distribution. Some examples include bolt diameter, construction errors, resistance of a specified type of wire, weight of a packaged material, and so on.

We can show mathematically that if a random variable is distributed normally with mean μ and variance σ^2, then the *standardized normal random variable* $z = (x - \mu)/\sigma$ is distributed with zero mean and unit variance. By using the transformation of the standardized normal random variable, we can derive the standard normal distribution, which is

$$f(z) = \frac{1}{\sqrt{2\pi}} e^{-z^2/2} \tag{A.37}$$

for $-\infty < z < \infty$. Since we can always make the standard transformation in practice, we shall use this function in Table B.2. Since the cumulative distribution function $F(x)$ of Equation (A.37) cannot be derived in closed form, the tables represent the result of numerical integrations.

For example, the resistance measures of many thousands of carbon composition resistors are normally distributed with mean $\mu = 1{,}000$ and variance $\sigma^2 = 900$. What proportion of the resistors have measurements in excess of 1,060 ohms? Unfortunately, we do not have a table of areas under the normal curve for $\mu = 1{,}000$, $\sigma^2 = 900$. But we do have a standard normal table for $\mu = 0$, $\sigma^2 = 1$. Converting our data to the standardized normal random variable z, we have

$$z = \frac{x - \mu}{\sigma} = \frac{1{,}060 - 1{,}000}{30} = 2$$

Considering Table B.2, we find that the area under the normal curve for $z \leq 2$ is 0.9773. This corresponds to the fraction of resistors with values measuring 1,060 ohms or fewer. Therefore, the proportion of resistors reading in excess of 1,060 ohms is 0.0227. The normal density function depicting this example is illustrated in Figure A.7.

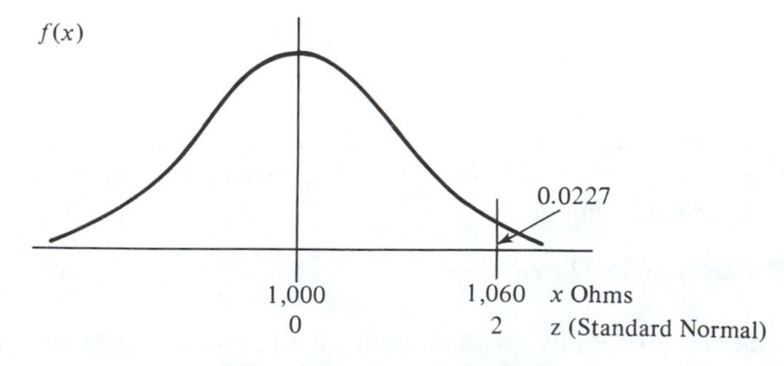

Fig. A.7. Normal distribution.

A.5.7 Exponential Distribution

One very useful continuous probability distribution is called the *exponential distribution*. It is most commonly used when we are interested in the distribution of the interval (measured in minutes, for example) between successive occurrences of an event. The probability function associated with the exponential random variable is as follows:

$$f(x) = \lambda e^{-\lambda x} \qquad \text{for } x > 0 \qquad \text{and} \qquad \lambda > 0 \tag{A.38}$$

and $f(x) = 0$ for $x \leqslant 0$. The cumulative distribution of the exponential random variable is as follows:

$$F(a) = \int_0^a \lambda e^{-\lambda x}\, dx = 1 - e^{-\lambda a} \tag{A.39}$$

The mean and variance of the exponential distribution are as follows:

$$\mu = \frac{1}{\lambda} \tag{A.40}$$

and

$$\sigma^2 = \frac{1}{\lambda^2} \tag{A.41}$$

There is an interesting relationship between the exponential distribution and the discrete Poisson distribution in problems involving the occurrence of events ordered in time. In the Poisson case, there are *changes* occurring intermittently in the process in which we are interested. The Poisson distribution describes the number of such changes in a unit time interval. The exponential distribution describes the time spacing between such occurrences.

Suppose that the life in hours of a certain type of tube is a random variable having an exponential distribution with a mean of 1,000 hours. What is the probability that such a tube will last at least 1,250 hours? To solve this, we use the cumulative distribution expression in Equation (A.39) as follows:

$$\begin{aligned} F(a) &= 1 - e^{-\lambda a} \\ &= 1 - e^{-\frac{1}{1{,}000}(1{,}250)} \\ &= 0.713 \end{aligned}$$

This is the probability that the tube will last 1,250 or fewer hours; hence, the answer to our questions is $1 - F(1{,}250) = 0.287$. The exponential distribution for this example is illustrated in Figure A.8.

A.5.8 Rectangular Distribution

A continuous probability distribution that has constant density over the range of values for which the density of the random variable is nonzero is called a

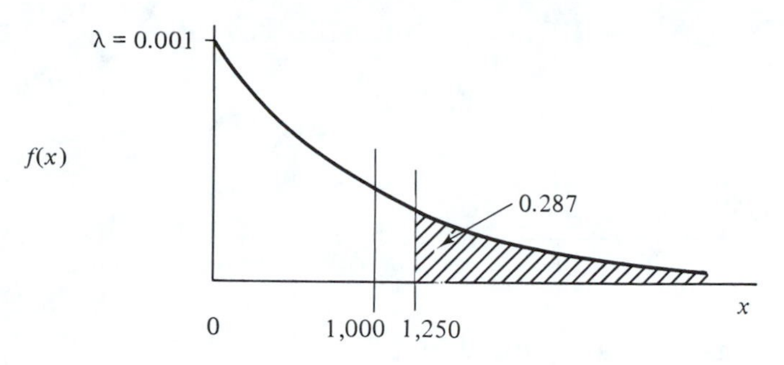

Fig. A.8. Exponential distribution.

rectangular distribution. The probability distribution associated with the rectangular random variable is as follows:

$$f(x) = \frac{1}{d - c} \qquad \text{for } c \leq x \leq d \tag{A.42}$$

and $f(x) = 0$, otherwise. The cumulative distribution of the rectangular random variable is as follows:

$$F(a) = \int_c^a \frac{1}{d - c}\, dx = \frac{a - c}{d - c} \tag{A.43}$$

The mean and variance of the rectangular distribution are given by the following equations:

$$\mu = \frac{c + d}{2} \tag{A.44}$$

and

$$\sigma^2 = \frac{(d - c)^2}{12} \tag{A.45}$$

Suppose that the demand for a particular bulk fluid inventory item is rectangular between 100 and 1,100 gallons per day. Using Equation (A.43), we see that the probability that no more than 700 gallons will be required is

$$F(700) = \frac{700 - 100}{1{,}100 - 100} = 0.6$$

The rectangular distribution for this example is graphed in Figure A.9.

A.5.9 Distribution Summary

Table A.1 summarizes the important points of the common distributions treated in this section. The formula parameters, range, mean, and variance are presented in the table for easy reference.

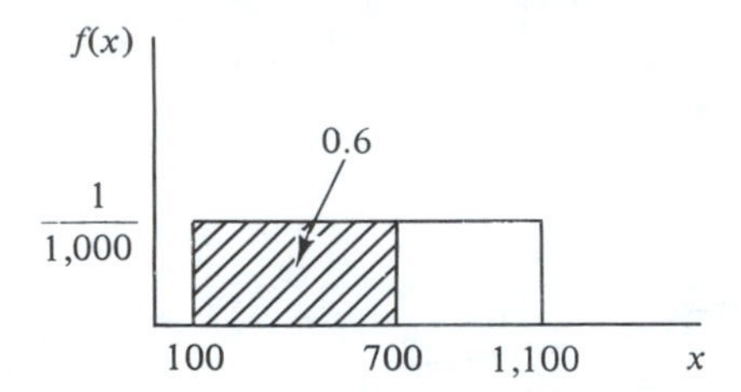

Fig. A.9. Rectangular distribution.

A.6. Expected Values and Variability

We earlier introduced the distributional characteristics known as mean and variance. In this section we shall elaborate upon those measures from the viewpoint of *expected values*.

A.6.1 Mean

The *mean* or expected value of a discrete random variable x is denoted by the letter μ and is defined as follows:

$$\mu = E(x) = \sum_i x_i\, p(x_i) \tag{A.46}$$

where $p(x_i)$ is the probability that x takes on the value x_i. In an analogous manner, we define the mean or expected value of a continuous random variable as follows:

$$\mu = E(x) = \int_{-\infty}^{\infty} xf(x)\, dx \tag{A.47}$$

where $f(x)$ is the continuous probability distribution of x. It is interesting to note that the mean need not equal a value that the random variable may assume. For example, the expected value or mean of many dice rolling experiments is

$$\mu = \sum_i x_i\, p(x_i)$$

$$= 1\left(\frac{1}{6}\right) + 2\left(\frac{1}{6}\right) + \cdots + 6\left(\frac{1}{6}\right)$$

$$= 3.5$$

The value 3.5 is clearly not a value that can be assumed on a roll of the die.

A.6.2 Variance

The measure of variability that we have considered is called the *variance* (σ^2), and it is defined as follows:

Table A.1. COMMON DISTRIBUTION SUMMARY

Distribution of Random Variable x	*Formula*	*Parameters*	*Range of x*	*Mean*	*Variance*
Discrete					
1. Binomial	$f(x) = \binom{n}{x} p^x(1 - p)^{n-x}$	n, p	$x = 0, 1, 2, \ldots, n$	np	$np(1 - p)$
2. Poisson	$f(x) = \frac{e^{-\lambda}\lambda^x}{x!}$	λ	$x = 0, 1, 2, \ldots$	λ	λ
3. Uniform	$f(x) = \frac{1}{n}$	n	$x = 1, 2, \ldots, n$	$\frac{n + 1}{2}$	$\frac{n^2 - 1}{12}$
Continuous					
1. Normal	$f(x) = \frac{1}{\sqrt{2\pi}\,\sigma} e^{-(x-\mu)^2/2\sigma^2}$	μ, σ	$-\infty < x < \infty$	μ	σ^2
2. Exponential	$f(x) = \lambda e^{-\lambda x}$	λ	$0 < x < \infty$	$\frac{1}{\lambda}$	$\frac{1}{\lambda^2}$
3. Rectangular	$f(x) = \frac{1}{d - c}$	c, d	$c \leq x \leq d$	$\frac{c + d}{2}$	$\frac{(d - c)^2}{12}$

$$\sigma^2 = E(x - \mu)^2 \tag{A.48}$$

This definition holds for both discrete and continuous random variables. The reader should note that the following computational convenience holds:

$$E(x - \mu)^2 = E(x^2) - \mu^2 \tag{A.49}$$

Using Equation A.49, we can determine the variance of the distribution for rolling a die in a manner similar to finding the mean. That is,

$$\begin{aligned}\sigma^2 &= E(x^2) - \mu^2 \\ &= \sum_i x_i^2\, p(x_i) - \mu^2 \\ & 1^2\left(\frac{1}{6}\right) + 2^2\left(\frac{1}{6}\right) + \cdots + 6^2\left(\frac{1}{6}\right) - (3.5)^2 \\ &= 2.917\end{aligned}$$

For the reader familiar with mechanics we can point out that the mean is analogous to the first moment about the origin, and the variance is equivalent to the second moment about the mean.

A.7. Populations and Samples

Much of the work of the applied professions involves the study of only a subset of the total items of interest, in the hope of making statistical inferences about the total. An engineer might collect data on machine utilization for 1 month, hoping to infer from it machine utilization information for many months or years. An automobile manufacturer might test a small number of automobiles and then make generalized statements about all the automobiles produced during that model year. An inspection team might use destructive inspection on a small percentage of items in order to infer characteristics of the total number being produced. In order to describe this process accurately, we must clearly understand the meaning of population and sample.

A.7.1 Population

A *population*, in the broadest sense, is the total set of elements about which knowledge is desired. Some populations are relatively small, for example, the number of space shuttles; other populations are large, for example, all the electric light bulbs now in existence and to be produced in the future. All elements of a population do not have to be in existence, as the last example indicates. The important thing to remember is that the population must be definable.

The definition of population clearly indicates that it contains the elements in which we have an interest. Why, then, do we not study the complete population? The answer is simple: The population is usually too large, or too complex, or not available, or the expense of considering all of it is too high. Any investigator would measure all the elements of his defined population if it were not prohibitive in some manner. As a result of the impossibility or impracticality of always considering all elements of a population, we are forced into a consideration of a sample (or samples) from that population.

A.7.2 Sample

A *sample* is a subset of a population. In extreme situations the sample may be the complete population or it may consist of no elements at all. Of course, this latter sample would yield no information and we shall not consider it further. Remember that the purpose of a sample is to yield inferences about the population from which it was taken.

The two most important features of a sample are its size and the manner in which it was selected. Much of the study of sampling statistics concerns the determination of these two characteristics. As expected, this determination is based on the specific conditions prescribing the purpose of the sample.

A.7.3 Sample Statistics

A *sample statistic* is a value calculated from a sample that may be used to estimate a *population parameter* such as a mean or variance. Since samples from a population are not identical, it is immediately apparent that sample statistics are not always the same, that is, they vary from sample to sample. Thus, a sample statistic is a random variable with its own frequency function.

Two important sample statistics are the sample mean and the sample variance. The sample mean is defined as follows:

$$\bar{x} = \sum_{i=1}^{n} \frac{x_i}{n} \tag{A.50}$$

where n is the number of measurements in the sample and the x_i's are the values of the random variable x in the sample. The sample variance is defined as follows:

$$s^2 = \sum_{i=1}^{n} \frac{(x_i - \bar{x})^2}{n - 1} \tag{A.51}$$

For computational purposes, the sample variance can also be written as follows:

$$s^2 = \frac{\sum_{i=1}^{n} x_i^2 - \dfrac{\left(\sum_{i=1}^{n} x_i\right)^2}{n}}{n - 1} \tag{A.52}$$

In this form, s^2 is much easier to calculate. As we might expect, $\bar{x}$ is an estimate of the population mean μ and s^2 is an estimate of the population variance σ^2.

Other measures of central tendency include the *median*, which is the middle value in an ordered set of data, and the *mode*, which is the value that occurs most frequently. The *range*, denoted by R, is a particularly useful measure of dispersion in quality control work. It is simply the largest value in a sample minus the smallest value:

$$R = x_{\text{largest}} - x_{\text{smallest}} \tag{A.53}$$

A.7.4 Distribution of Sample Means

We often make inferences about a population from the average value of a sample. This usually requires that we know the parameters of the distribution of means. Naturally, the expected value of the sample average is μ, the same mean value as held by the population. The variance of the sample means, $\sigma_{\bar{x}}^2$, differs from the population variance and is given by the following:

$$\sigma_{\bar{x}}^2 = \frac{\sigma^2}{n} \tag{A.54}$$

It is reasonable to expect that the distribution of sample means to have a smaller variance, since the larger the sample, the closer one would expect the average to fall to the population mean, giving rise to a smaller distribution spread.

We can return to our example in which resistors are normally distributed with mean $\mu = 1{,}000$ and variance $\sigma^2 = 900$. If we take samples of size $n = 9$ and average the ohmmeter readings, it is virtually impossible that the sample average will be as high as 1,060 which had a 0.0227 probability for a single resistor. In fact, the likelihood that the sample average will exceed 1,030 is very small. To calculate the exact probability, we must first determine the mean and variance of the sample average distribution. The mean μ will remain at 1,000, but the variance is given by

$$\sigma_{\bar{x}}^2 = \frac{\sigma^2}{n} = \frac{900}{9} = 100$$

Finding the standard normal variable, we have

$$z = \frac{\bar{x} - \mu}{\sigma_{\bar{x}}} = \frac{1{,}030 - 1{,}000}{10} = 3$$

Consulting our Table B.2 values in Appendix B, we can see that the probability that z exceeds 3 is $1 - 0.99865 = 0.00135$. Correspondingly, this is also the probability that $\bar{x}$ will exceed 1,030. The distributions of x and $\bar{x}$ for this example are graphed in Figure A.10.

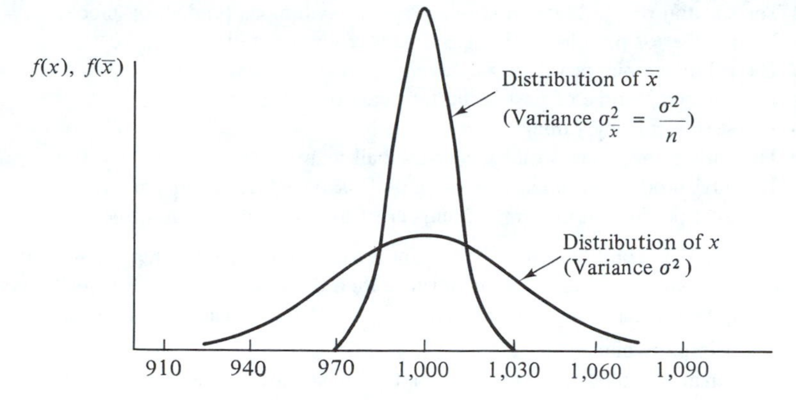

Fig. A.10. Population and sample mean distribution.

A.8. Central Limit Theorem

In the preceding example we implicity assumed the distribution of sample means to be normal. In fact, this is true if the population is normally distributed. Although the sample mean distribution is not truly normal if the population is other than normally distributed, we frequently treat the sample means as if they were. The reason we can do this is stated in the central limit theorem, which, in essence, says: *If* x *has a distribution with a finite variance* σ^2, *then the random variable* $\bar{x}$ *has a distribution that approaches normality as the sample size tends to infinity.* Fortunately, for many population distributions often encountered, sample sizes as low as $n = 4$ produce sample average distributions which are workably close to normal. We use the central limit theorem extensively in quality control, probabilistic models, or project management.

DISCUSSION QUESTIONS

1. (a) List the outcomes possible in terms of score when two baseball teams (A and B) play a complete game and the sum of their scores is 8 or less (e.g., 1,0 and 0,1 would be possible outcomes).
(b) Which outcomes correspond to the event that team A wins?

2. How many outcomes are possible when examining a production run of 100 items, classifying them as good or bad, and listing as the outcome the number bad?

3. How many outcomes are possible when examining a large roll of denim (used to make overalls) and counting the number of nonconformities such as runs, spots, and so on?

4. Identify each of the following as either discrete or continuous random variables:
(a) Amount of energy used by a refinery in 1 hour.
(b) The fraction of a finite number of brackets that are bad.

(c) The fraction of bad brackets coming from a continuous production process.
(d) The number of persons buying gasoline at a service station.
(e) The gallonage pumped during a day at a service station.
(f) The percentage time of possession for your football team.
(g) The season rushing yardage for your football team.
(h) The number of games won by your football team.
(i) The total production in terms of weight of usable apples from a tree.
(j) The total production in terms of number of usable apples from a tree.

5. You are an I.E. consultant to a bank. The bank has been receiving complaints from customers about the waiting times involved at their drive-up windows. Suggest some data that might be collected and studied to learn more about the cause of the waiting.

6. Consider six parts numbered 1, 2, 3, 4, 5 and 6.
(a) How many combinations of these parts can be made in groups of four? Use the formula, and also list the combinations.
(b) How many combinations of these letters can be made in groups of two? Use the formula.
(c) Explain why your answers to parts (a) and (b) came out as they did.

7. Match each description of a random variable to the best probability distribution describing that random variable.

(a) Length of parts cut to a specified dimension.	(1) Poisson.
(b) Number of on-the-job accidents during a month.	(2) Normal.
(c) Time between arrivals of persons to a post office.	(3) Uniform.
(d) Number of nonconforming parts produced on a continuous production line.	(4) Exponential.
(e) Person to be selected in a random draw of tickets out of a bowl.	(5) Binomial.

8. Tell whether each description (a) to (e) in Question 7 is for a discrete or a continuous random variable.

9. In your electrical science, thermodynamics, statics, strength of materials, fluids, and other engineering science courses, random variation is seldom, if ever, mentioned. Rather, formulas are given and we simply plug and chug through them, often to several places after the decimal. Are these areas immune from variation? Explain why they are taught as deterministic (nonvarying).

PROBLEMS

1. Plot the following discrete probability distribution and the corresponding cumulative distribution:

x	$p(x)$
2	0.10
3	0.15
8	0.40
12	0.35

2. Suppose that you have two fair four-sided dice. Their sides are marked 1, 2, 3, and 4.
(a) Determine the probability that the sum of their "down" side is 1, 2, . . . , 8.
(b) Plot the corresponding discrete probability distribution.
(c) Plot the corresponding cumulative distribution.

3. Consider the continuous probability distribution defined as follows:

$$f(x) = \begin{cases} \dfrac{3x^2}{8} & \text{for } (0 \leq x \leq 2) \\ 0 & \text{elsewhere} \end{cases}$$

(a) Plot this distribution and the corresponding cumulative distribution.
(b) What is the probability that x will have a value ≤ 1?
(c) What is the probability that $1 < x \leq \frac{3}{2}$?

4. A survey of workers' ages is taken in a department store. The summary results are as follows:

Age	*Number*
Teens	3
20's	18
30's	15
40's	7
50's	3
60's	4

(a) Are workers' ages discrete or continuous?
(b) If you were to randomly select the name of a worker, what is the probability that the person is in his or her 20's? 30's? 20's or 30's?

5. Ten balls are identical except for their numbers of 0, 1, 2, 3, 4, 5, 6, 7, 8, and 9. Four balls are independently and randomly selected, each ball being replaced after its number is noted. Determine the probabilities of obtaining 0, 1, 2, 3, and 4 of these balls with a number greater than or equal to 6. Use the binomial distribution to solve this.

6. Eight lab instruments are delivered to a customer. Experience shows that 10% break down within the first 90 days. What is the probability that none will break down? Two or fewer?

7. Experience has shown that an average of 6.6 students get sick each day at a small university.
(a) What is the probability that 8 or fewer get sick on a given day?
(b) What is the probability that from 4 to 10, inclusive, get sick?
(c) What is the probability that exactly 7 get sick?

Use the Poisson distribution tables to solve this.

8. Cloth is inspected before cutting and stitching in a clothing factory. The average number of nonconformities (or defects) per 100 yards of cloth is 4.2.
(a) What is the probability of having no nonconformities on the next 100 yards of cloth?
(b) What is the probability of from 2 to 7 nonconformities, inclusive?
(c) What is the probability of *more* than 2 nonconformities?

9. A foundry produces valve castings having weights that are normally distributed with mean 100 and standard deviation 2.0. Determine:
(a) $P\ (x \leq 97)$
(b) $P\ (98 < x \leq 104)$
(c) $P\ (x > 103)$

10. If a sample of size $n = 4$ is taken from the normal distribution of Problem 9, determine:
(a) $P(\bar{x} \leq 98.5)$
(b) $P(99 < \bar{x} \leq 102)$
(c) $P(\bar{x} > 101.5)$

11. Viscosity is normally distributed with $\mu = 70.0$ and $\sigma = 1.7$. A sample of product is checked on a viscometer. Determine:
(a) $P(x < 65.0)$.
(b) $P(68.0 < x \leq 72.0)$.
(c) $P(x > 68.3)$.
(d) $P(x = 70.0)$.

12. Let x be the number of pounds of cake sold in 1 day by a bakery. Assume that x is rectangularly distributed between 200 and 250 pounds.
(a) Write out the density function for this random variable.
(b) Develop the cumulative distribution function of the random variable.
(c) Plot the density function and the cumulative distribution functions.
(d) Determine the probabilities of the following events:
(1) $x \leq 234$
(2) $220 < x \leq 245$
(3) $150 < x \leq 210$
(e) Determine the mean and variance of the random variable.

13. Suppose that the life in hours of a certain integrated circuit is a random variable having an exponential distribution with a mean of 5,000 hours.
(a) Determine the expression for the cumulative distribution function.
(b) What is the probability that the IC will last more than 5,000 hours?
(c) What is the probability that the IC will fail between 1,000 to 2,000 hours?
(d) If it is desired to replace the IC when its probability of failure exceeds 0.10, after how many hours should it be replaced?

APPENDIX B

Tables

Table B.1. POISSON DISTRIBUTION—CUMULATIVE

SUMMATION OF TERMS OF THE POISSON DISTRIBUTION

Entries in body of table give the probability (decimal point omitted) of X or less defects (or defectives), when the expected number is that given in the left margin of the table.

μ or np \ X	0	1	2	3	4	5	6	7	8	9
0.02	980	1,000								
0.04	961	999	1,000							
0.06	942	998	1,000							
0.08	923	997	1,000							
0.10	905	995	1,000							
0.15	861	990	999	1,000						
0.20	819	982	999	1,000						
0.25	779	974	998	1,000						
0.30	741	963	996	1,000						
0.35	705	951	994	1,000						
0.40	670	938	992	999	1,000					
0.45	638	925	989	999	1,000					
0.50	607	910	986	998	1,000					
0.55	577	894	982	998	1,000					
0.60	549	878	977	997	1,000					
0.65	522	861	972	996	999	1,000				
0.70	497	844	966	994	999	1,000				
0.75	472	827	959	993	999	1,000				
0.80	449	809	953	991	999	1,000				
0.85	427	791	945	989	998	1,000				
0.90	407	772	937	987	998	1,000				
0.95	387	754	929	984	997	1,000				
1.00	368	736	920	981	996	999	1,000			
1.1	333	699	900	974	995	999	1,000			
1.2	301	663	879	966	992	998	1,000			
1.3	273	627	857	957	989	998	1,000			
1.4	247	592	833	946	986	997	999	1,000		
1.5	223	558	809	934	981	996	999	1,000		
1.6	202	525	783	921	976	994	999	1,000		
1.7	183	493	757	907	970	992	998	1,000		
1.8	165	463	731	891	964	990	997	999	1,000	
1.9	150	434	704	875	956	987	997	999	1,000	
2.0	135	406	677	857	947	983	995	999	1,000	

Reprinted by kind permission from E. C. Molina, *Poisson's Exponential Binomial Limit* (New York: D. Van Nostrand Co. Inc., 1947).

SUMMATION OF TERMS OF THE POISSON DISTRIBUTION (CONT.)

μ or np \ X	0	1	2	3	4	5	6	7	8	9
2.2	111	355	623	819	928	975	993	998	1,000	
2.4	091	308	570	779	904	964	988	997	999	1,000
2.6	074	267	518	736	877	951	983	995	999	1,000
2.8	061	231	469	692	848	935	976	992	998	999
3.0	050	199	423	647	815	916	966	988	996	999
3.2	041	171	380	603	781	895	955	983	994	998
3.4	033	147	340	558	744	871	942	977	992	997
3.6	027	126	303	515	706	844	927	969	988	996
3.8	022	107	269	473	668	816	909	960	984	994
4.0	018	092	238	433	629	785	889	949	979	992
4.2	015	078	210	395	590	753	867	936	972	989
4.4	012	066	185	359	551	720	844	921	964	985
4.6	010	056	163	326	513	686	818	905	955	980
4.8	008	048	143	294	476	651	791	887	944	975
5.0	007	040	125	265	440	616	762	867	932	968
5.2	006	034	109	238	406	581	732	845	918	960
5.4	005	029	095	213	373	546	702	822	903	951
5.6	004	024	082	191	342	512	670	797	886	941
5.8	003	021	072	170	313	478	638	771	867	929
6.0	002	017	062	151	285	446	606	744	847	916

	10	11	12	13	14	15	16
2.8	1,000						
3.0	1,000						
3.2	1,000						
3.4	999	1,000					
3.6	999	1,000					
3.8	998	999	1,000				
4.0	997	999	1,000				
4.2	996	999	1,000				
4.4	994	998	999	1,000			
4.6	992	997	999	1,000			
4.8	990	996	999	1,000			
5.0	986	995	998	999	1,000		
5.2	982	993	997	999	1,000		
5.4	977	990	996	999	1,000		
5.6	972	988	995	998	999	1,000	
5.8	965	984	993	997	999	1,000	
6.0	957	980	991	996	999	999	1,000

SUMMATION OF TERMS OF THE POISSON DISTRIBUTION (CONT.)

μ or np \ X	0	1	2	3	4	5	6	7	8	9
6.2	002	015	054	134	259	414	574	716	826	902
6.4	002	012	046	119	235	384	542	687	803	886
6.6	001	010	040	105	213	355	511	658	780	869
6.8	001	009	034	093	192	327	480	628	755	850
7.0	001	007	030	082	173	301	450	599	729	830
7.2	001	006	025	072	156	276	420	569	703	810
7.4	001	005	022	063	140	253	392	539	676	788
7.6	001	004	019	055	125	231	365	510	648	765
7.8	000	004	016	048	112	210	338	481	620	741
8.0	000	003	014	042	100	191	313	453	593	717
8.5	000	002	009	030	074	150	256	386	523	653
9.0	000	001	006	021	055	116	207	324	456	587
9.5	000	001	004	015	040	089	165	269	392	522
10.0	000	000	003	010	029	067	130	220	333	458
	10	11	12	13	14	15	16	17	18	19
6.2	949	975	989	995	998	999	1,000			
6.4	939	969	986	994	997	999	1,000			
6.6	927	963	982	992	997	999	999	1,000		
6.8	915	955	978	990	996	998	999	1,000		
7.0	901	947	973	987	994	998	999	1,000		
7.2	887	937	967	984	993	997	999	999	1,000	
7.4	871	926	961	980	991	996	998	999	1,000	
7.6	854	915	954	976	989	995	998	999	1,000	
7.8	835	902	945	971	986	993	997	999	1,000	
8.0	816	888	936	966	983	992	996	998	999	1,000
8.5	763	849	909	949	973	986	993	997	999	999
9.0	706	803	876	926	959	978	989	995	998	999
9.5	645	752	836	898	940	967	982	991	996	998
10.0	583	697	792	864	917	951	973	986	993	997
	20	21	22							
8.5	1,000									
9.0	1,000									
9.5	999	1,000								
10.0	998	999	1,000							

Table B.2. NORMAL DISTRIBUTION—CUMULATIVE

Proportion of total area under the normal curve from $-\infty$ to $Z = \frac{x - \mu}{\sigma}$

Z	0.09	0.08	0.07	0.06	0.05	0.04	0.03	0.02	0.01	0.00
−3.5	0.00017	0.00017	0.00018	0.00019	0.00019	0.00020	0.00021	0.00022	0.00022	0.00023
−3.4	0.00024	0.00025	0.00026	0.00027	0.00028	0.00029	0.00030	0.00031	0.00033	0.00034
−3.3	0.00035	0.00036	0.00038	0.00039	0.00040	0.00040	0.00043	0.00045	0.00047	0.00048
−3.2	0.00050	0.00052	0.00054	0.00056	0.00058	0.00060	0.00062	0.00064	0.00066	0.00069
−3.1	0.00071	0.00074	0.00076	0.00079	0.00082	0.00085	0.00087	0.00090	0.00094	0.00097
−3.0	0.00100	0.00104	0.00107	0.00111	0.00114	0.00118	0.00122	0.00126	0.00131	0.00135
−2.9	0.0014	0.0014	0.0015	0.0015	0.0016	0.0016	0.0017	0.0017	0.0018	0.0019
−2.8	0.0019	0.0020	0.0021	0.0021	0.0022	0.0023	0.0023	0.0024	0.0025	0.0026
−2.7	0.0026	0.0027	0.0028	0.0029	0.0030	0.0031	0.0032	0.0033	0.0034	0.0035
−2.6	0.0036	0.0037	0.0038	0.0039	0.0040	0.0041	0.0043	0.0044	0.0045	0.0047
−2.5	0.0048	0.0049	0.0051	0.0052	0.0054	0.0055	0.0057	0.0059	0.0060	0.0062
−2.4	0.0064	0.0066	0.0068	0.0069	0.0071	0.0073	0.0075	0.0078	0.0080	0.0082
−2.3	0.0084	0.0087	0.0089	0.0091	0.0094	0.0096	0.0099	0.0102	0.0104	0.0107
−2.2	0.0110	0.0113	0.0116	0.0119	0.0122	0.0125	0.0129	0.0132	0.0136	0.0139
−2.1	0.0143	0.0146	0.0150	0.0154	0.0158	0.0162	0.0166	0.0170	0.0174	0.0179
−2.0	0.0183	0.0188	0.0192	0.0197	0.0202	0.0207	0.0212	0.0217	0.0222	0.0228
−1.9	0.0233	0.0239	0.0244	0.0250	0.0256	0.0262	0.0268	0.0274	0.0281	0.0287
−1.8	0.0294	0.0301	0.0307	0.0314	0.0322	0.0329	0.0336	0.0344	0.0351	0.0359
−1.7	0.0367	0.0375	0.0384	0.0392	0.0401	0.0409	0.0418	0.0427	0.0436	0.0446
−1.6	0.0455	0.0465	0.0475	0.0485	0.0495	0.0505	0.0516	0.0526	0.0537	0.0548
−1.5	0.0559	0.0571	0.0582	0.0594	0.0606	0.0618	0.0630	0.0643	0.0655	0.0668
−1.4	0.0681	0.0694	0.0708	0.0721	0.0735	0.0749	0.0764	0.0778	0.0793	0.0808
−1.3	0.0823	0.0838	0.0853	0.0869	0.0885	0.0901	0.0918	0.0934	0.0951	0.0968
−1.2	0.0985	0.1003	0.1020	0.1038	0.1057	0.1075	0.1093	0.1112	0.1131	0.1151
−1.1	0.1170	0.1190	0.1210	0.1230	0.1251	0.1271	0.1292	0.1314	0.1335	0.1357
−1.0	0.1379	0.1401	0.1423	0.1446	0.1469	0.1492	0.1515	0.1539	0.1562	0.1587
−0.9	0.1611	0.1635	0.1660	0.1685	0.1711	0.1736	0.1762	0.1788	0.1814	0.1841
−0.8	0.1867	0.1894	0.1922	0.1949	0.1977	0.2005	0.2033	0.2061	0.2090	0.2119
−0.7	0.2148	0.2177	0.2207	0.2236	0.2266	0.2297	0.2327	0.2358	0.2389	0.2420
−0.6	0.2451	0.2483	0.2514	0.2546	0.2578	0.2611	0.2643	0.2676	0.2709	0.2743
−0.5	0.2776	0.2810	0.2843	0.2877	0.2912	0.2946	0.2981	0.3015	0.3050	0.3085
−0.4	0.3121	0.3156	0.3192	0.3228	0.3264	0.3300	0.3336	0.3372	0.3409	0.3446
−0.3	0.3483	0.3520	0.3557	0.3594	0.3632	0.3669	0.3707	0.3745	0.3783	0.3821
−0.2	0.3859	0.3897	0.3936	0.3974	0.4013	0.4052	0.4090	0.4129	0.4168	0.4207
−0.1	0.4247	0.4286	0.4325	0.4364	0.4404	0.4443	0.4483	0.4522	0.4562	0.4602
−0.0	0.4641	0.4681	0.4721	0.4761	0.4801	0.4840	0.4880	0.4920	0.4960	0.5000

Table B.2. (CONT.)

Z	0.00	0.01	0.02	0.03	0.04	0.05	0.06	0.07	0.08	0.09
+0.0	0.5000	0.5040	0.5080	0.5120	0.5160	0.5199	0.5239	0.5279	0.5319	0.5359
+0.1	0.5398	0.5438	0.5478	0.5517	0.5557	0.5596	0.5636	0.5675	0.5714	0.5753
+0.2	0.5793	0.5832	0.5871	0.5910	0.5948	0.5987	0.6026	0.6064	0.6103	0.6141
+0.3	0.6179	0.6217	0.6255	0.6293	0.6331	0.6368	0.6406	0.6443	0.6480	0.6517
+0.4	0.6554	0.6591	0.6628	0.6664	0.6700	0.6736	0.6772	0.6808	0.6844	0.6879
+0.5	0.6915	0.6950	0.6985	0.7019	0.7054	0.7088	0.7123	0.7157	0.7190	0.7224
+0.6	0.7257	0.7291	0.7324	0.7357	0.7389	0.7422	0.7454	0.7486	0.7517	0.7549
+0.7	0.7580	0.7611	0.7642	0.7673	0.7704	0.7734	0.7764	0.7794	0.7823	0.7852
+0.8	0.7881	0.7910	0.7939	0.7967	0.7995	0.8023	0.8051	0.8079	0.8106	0.8133
+0.9	0.8159	0.8186	0.8212	0.8238	0.8264	0.8289	0.8315	0.8340	0.8365	0.8389
+1.0	0.8413	0.8438	0.8461	0.8485	0.8508	0.8531	0.8554	0.8577	0.8599	0.8621
+1.1	0.8643	0.8665	0.8686	0.8708	0.8729	0.8749	0.8770	0.8790	0.8810	0.8830
+1.2	0.8849	0.8869	0.8888	0.8907	0.8925	0.8944	0.8962	0.8980	0.8997	0.9015
+1.3	0.9032	0.9049	0.9066	0.9082	0.9099	0.9115	0.9131	0.9147	0.9162	0.9177
+1.4	0.9192	0.9207	0.9222	0.9236	0.9251	0.9265	0.9279	0.9292	0.9306	0.9319
+1.5	0.9332	0.9345	0.9357	0.9370	0.9382	0.9394	0.9406	0.9418	0.9429	0.9441
+1.6	0.9452	0.9463	0.9474	0.9484	0.9495	0.9505	0.9515	0.9525	0.9535	0.9545
+1.7	0.9554	0.9564	0.9573	0.9582	0.9591	0.9599	0.9608	0.9616	0.9625	0.9633
+1.8	0.9641	0.9649	0.9656	0.9664	0.9671	0.9678	0.9686	0.9693	0.9699	0.9706
+1.9	0.9713	0.9719	0.9726	0.9732	0.9738	0.9744	0.9750	0.9756	0.9761	0.9767
+2.0	0.9773	0.9778	0.9783	0.9788	0.9793	0.9798	0.9803	0.9808	0.9812	0.9817
+2.1	0.9821	0.9826	0.9830	0.9834	0.9838	0.9842	0.9846	0.9850	0.9854	0.9857
+2.2	0.9861	0.9864	0.9868	0.9871	0.9875	0.9878	0.9881	0.9884	0.9887	0.9890
+2.3	0.9893	0.9896	0.9898	0.9901	0.9904	0.9906	0.9909	0.9911	0.9913	0.9916
+2.4	0.9918	0.9920	0.9922	0.9925	0.9927	0.9929	0.9931	0.9932	0.9934	0.9936
+2.5	0.9938	0.9940	0.9941	0.9943	0.9945	0.9946	0.9948	0.9949	0.9951	0.9952
+2.6	0.9953	0.9955	0.9956	0.9957	0.9959	0.9960	0.9961	0.9962	0.9963	0.9964
+2.7	0.9965	0.9966	0.9967	0.9968	0.9969	0.9970	0.9971	0.9972	0.9973	0.9974
+2.8	0.9974	0.9975	0.9976	0.9977	0.9977	0.9978	0.9979	0.9979	0.9980	0.9981
+2.9	0.9981	0.9982	0.9983	0.9983	0.9984	0.9984	0.9985	0.9985	0.9986	0.9986
+3.0	0.99865	0.99869	0.99874	0.99878	0.99882	0.99886	.099889	0.99893	0.99896	0.99900
+3.1	0.99903	0.99906	0.99910	0.99913	0.99915	0.99918	0.99921	0.99924	0.99926	0.99929
+3.2	0.99931	0.99934	0.99936	0.99938	0.99940	0.99942	0.99944	0.99946	0.99948	0.99950
+3.3	0.99952	0.99953	0.99955	0.99957	0.99958	0.99960	0.99961	0.99962	0.99964	0.99965
+3.4	0.99966	0.99967	0.99969	0.99970	0.99971	0.99972	0.99973	0.99974	0.99975	0.99976
+3.5	0.99977	0.99978	0.99978	0.99979	0.99980	0.99981	0.99981	0.99982	0.99983	0.99983

Index

Absorbing Markov chain, 392
Absorbing state, 392
Accounting, 345, 392
Accounting equations, 330–32
Accreditation Board for Engineering and Technology, 10–12, 19
Activities, 413–14
Activity relationship diagram, 107
Addition theorem, 504
AIIE, 17
Aldep, 113
Allowances, 167
Alloys, 58
American Assoc. of Cost Engineers, 17
American Production and Inventory Control Society, 17
American Society for Quality Control, 17
Amos, J.M., 498
Amrine, H.T., 456
Analysis, 5–7
Apple, J.M., 127, 131
Arbitrary service times, 382
Armstrong, Neil, 452
Arrival distribution, 379–84
Arrival rate, 379–84
Artificial intelligence (AI), 48
Assembly, batch, 67
Assembly, continuous, 67
Assets, 330
Assignment problem, 367
Assignment programming, 134
Associates in Process Improvement, 212
Association for Computing Machinery, 17
AT&T, 241
Attributes data, 243
Automatic guided vehicle (AGV), 130, 288
Automatic storage & retrieval systems, 131, 286
Automation, 296–97

Babbage, Charles, 13
Back orders, 388
Backward pass, 417–18
Balance sheets, 329–31, 333–34
Baldrige (*See* Malcolm Baldrige National Quality Award)
Balk, 379
Beach, D., 483, 492–93
Beam welding, 65
Bedrock quality concepts, 214
Belcher, D.W., 261
Benchmarking, 221

Bending, 60
Benefits (personnel), 493
Berry, W.L., 203
Best of the best, 221
Between subgroups, 236
Bill of materials, 47, 204
Binomial distribution, 509–11
Blair, R.N., 26, 448
Blanchard, B.S., 36, 440
Block diagram, 442
Brainstorming, 226
Brazing, 67
Broaching, 62
Brooks, G.H., 183
Burch, J.G., 471
Bus network topology, 480

c chart, 235, 244
CAD/CAM, 277
Calling population, 379
Capital stock, 330
Carrying charge, 388
Carrying cost, 198
Case, Kenneth E., 213
Casting, 58
Category, 213
Cause-and-effect diagram, 226
Center line, 235
Central composite designs, 233
Central limit theorem, 236, 523
Checksheet, 228
Cheek, 58
Chemical engineering, 4, 22
Christmas tree model, 385
Civil engineering, 4, 22
Clark, G., 141
Claycombe, William W., 356
Closed-loop control, 442
Closing, 333
Collective bargaining, 491
Combinations, 504, 510
Common cause, 235
Communications, 497
Competition, 213
Competitiveness, 215
Computers, 393
Computer-aided design (CAD), 20, 279–82
Computer-aided factory management, 290
Computer-aided manufacturing (CAM), 20, 282–91
Computer-aided process planning (CAPP), 282
Computer-aided testing & inspection, 288
Computer applications, 73
Computer integrated manufacturing (CIM), 297–98
Concave, 357
Conditional probability, 504
Constancy of purpose, 217
Constant service time, 381
Continuous distribution, 514–18
Continuous improvement, 214–15
Continuous random variable, 505–8, 514–20
Control charts, 234
Control limits, 235–37, 243–44, 247
Convex, 357
Conveyors, 127
Cope, 58
Corrective action, 188, 203, 451, 457
Corrosive resistance, 58
Cost accounting, 330, 333
Cost accounting system, 334
Cost estimating, 50–55, 70–71
Cost of goods made, 335
Cost of goods sold, 335
Cotton, Frank, 17
Council of Industrial Engineering Academic Departments, 19
Countable, 505, 512
Cox, J. Grady, 501
CPM, 412–13, 415, 419–20
Craig, R.J., 415
Cranes, 127
Crash cost, 425
Crash time, 424
Critical path, 414, 419, 422–23, 425, 427
Critical path method (*See* CPM)
Crosby, Philip, 220
Cumulative distribution, 509–14, 516–17
Cumulative distribution function, 386
Curtailable rate schedules, 319
Curtin, Frank, 45
Customer driven, 215
Customer focus and satisfaction, 214
Customer needs, 218
Customer requirements, 218
Customer satisfaction, 212
Customer-supplier relationship, 216
Customers:
  external, 212
  internal, 212
Cycle time, 215

Dantzig, George, 362
Data, 471
Data-base language, 477
Data-base management systems, 476–79
Data collection form, 227
Data communications networks, 479
Daywork, 267
Deatherage, B.H., 308
de Camp, L.S., 2
Decision tree, 391
Defects, 215
Demand, 384
Demand distribution, 388
Deming, W. Edwards, 216

Design, 5–7, 25
Designed experimentation, 233
Designing information systems, 472–76
Determination of organizational functions, 459
Deterministic, 352–54
Direct cost, 424
Direct labor, 70, 335
Direct material, 70, 335
Discrete distribution, 509–13
Discrete event modeling, 408
Discrete random variable, 505–13
Dispatching & progress control, 188, 203
Distributed computing, 480
Distribution, 137
Dorf, R.C., 296, 442
Doyle, Lawrence E., 59–66, 68
Drag, 58
Drawing and stretching, 60
Drilling, 61
Drucker, Peter, 14, 452
Dummy activity, 414

Earliest activity finish time, 416–17
Earliest activity start time, 416
Earliest event occurrence time, 416–17
Economic order quantity (EOQ), 200, 351
Eddy-Rucker-Nickles Co., 271
Effectiveness measures for I. E. function, 34–37
Efroymson, M.A., 91
Electric arc welding, 64
Electrical engineering, 4, 22, 25
Elevators, 127
Emerson, H.P., 12
End of the year, 337
Energy management, 312, 314
Energy rate schedules, 316
Engelberger, J.F., 293
Engineer in training (E.I.T.), 9
Engineering, early developments, 2–3
Engineering economy, 329–30, 336, 345
Engineering ethics, 8
Engineering management, 498
Engineering process, 5–7
Engineering profession, 6–8
Ergonomics, 164
Errors, 215
Estimated probability, 506
Euclidean distances, 83, 96
Evans, W.O., 113
Event, 414, 502–3
Event-step incrementation, 407
Expected (average) time, 420
Expected number of units, 380–81
Expected time in queue, 380–81
Expected value, 385, 518–20
Expense, 331–32
Exponential distribution, 380–81, 516
Exponentially weighted moving average, 191
External failure cost, 220
Extruding, 60
Exxon Chemical Company, 212, 223

Fabrycky, W.J., 36, 338, 340, 363, 440
Facility layout, 99
  computerized layout planning, 113
Facility location, 80
  analytical techniques, 83
  mathematical programming, 89
  multiple objectives, 88
  public sector, 96
Factory of the future, 1, 278
Factory system, 276
Fatigue resistance, 58
Fear, 217
Feedback control in systems, 38–41
Fein, M., 271
Financial management, 329
Finishing, 67
Fishbone diagram (*See* Cause-and-effect diagram)
Fixed position layout, 102
Fixtures, 68
Flask, 58
Flexible manufacturing system (FMS), 278, 290
Flow diagram, 160
Flow process chart, 105, 154
Flowchart, 223
Forecasting, 184, 190–94
Forging, 60
Forward pass, 417–18
Francis, R.L., 84
Frequency function, 509
Freund, J.E., 192
From-to chart, 106
Functional groupings in organizations, 459
Fundamental data operations, 471

Gamma distribution, 380
Gantt chart, 14, 163, 412–13
Gantt, Henry L, 14
Gas welding, 67
Gate, 58
General-Purpose Simulation System (*See* GPSS), 408
Geometric programming, 373
Ghare, P.M., 363
Gilbreth, Frank, 13
Gilbreth, Lillian, 13
Global minimum, 358
Goal, 454
Goals, 216
GPSS, 408
Gradient, 359
Graduate education in industrial engineering, 19

Grinding, 62
Grinter report, 10
Groover, M.P., 206, 286
Group incentive plan, 253, 269
Group technology (GT), 278, 285
Grouping of organizational functions, 459
Grudnitski, G., 471

Hadley, T.M., 388
Halo effect, 488
Hammond, Ross W., 14
Hardness, 58
Hazardous materials management, 313, 324
Herzberg, 495
Hessian matrix, 359
Heuristic, 86, 138
Hierarchical nature of systems, 434
Hines, W.W., 192
Histogram, 230
Hoists, 127
Honing, 67
Human activity systems, 26, 40
Human engineering, 165
Human interface with information/ communication systems, 308
Human interface with machines, 307
Human interface with organizational/ supervisory structure, 308
Human interface with robots and intelligent machines, 309
Human interface with work environment, 306
Human interface with world of work, 305–10
Human resource development and management, 214
Hungarian method, 136, 367–68
Hyde, W.F., 285

***IIE Transactions*, 17**
Impact resistance, 58
Improshare, 271
In control, 218, 234
Incentives, 153
Income statements, 329, 331–33
Independent, 503
Indirect cost, 424, 427
Indirect labor, 335
Indirect material, 335
Industrial Engineering, 17
Industrial engineering, 350
Industrial engineering definition, 18
Industrial engineering education, 18–19
Industrial engineering historical developments, 2, 5, 12–22
Industrial engineering, organization of, 33–36
Industrial engineering organizations, 15–18
Industrial trucks, 127
Infinite population, 381
Infinite queue, 381
Information, 471
Information and analysis, 214
Information system design methodology, 473
Information systems, 470–72
Inspection, 217
Integer programming, 373
Interchangeable parts, 276
Interest factors, 337
Internal failure cost, 220
Interruptiple rate schedules, 319
Intersection, 503
Inventory carrying charge, 388
Inventory control, 384–89
Inventory planning and control, 187, 197–200
Involvement, 215
Ishikawa diagram (*See* Cause-and-effect diagram)

Jigs, 68
Job, 253
Job analysis, 253, 491
Job description, 252, 254, 460, 490
Job evaluation, 254, 352
  classification or grade description, 256
  factor comparison, 257
  point rating, 259
  ranking method, 256
Job lot manufacturing, 100
Job rotation, 490
Job specification, 252, 254
Journal of Industrial Enginering, 17
Juran, J.M., 217
Juran trilogy, 217
Just in time (JIT), 207–8, 212–13

Kafrissen, E., 293
Kalpakjian, Serope, 49–50
Kemeny, John G., 390
Kemper, J.D., 2
Kennedy, John F., 452
Key results areas, 452
Konz, S., 155

Labor relations, 490
Laplace transform, 444
Lapping, 67
Latest allowable activity finish time, 416, 418
Latest allowable activity start time, 416, 418
Latest allowable event occurrence time, 416, 417
Laws of chance, 501
Lead time, 384
Lead time distribution, 388
Leadership, 214, 217
Learning organizations, 466
Least cost assignment, 86
Left hand-right hand charts, 156
Liability, 330
Lifts, 127
Line function, 461

Linear programming, 362–63, 366–67
Local area networks, 480
Lohmann, M.R., 11
Long-term relationship, 217
Loss, 332
Lot size-reorder point model, 387
Luggen, W.W., 207, 291

Machine intelligence, 310
Macro-to-micro, 230
Mainframes, 345, 479
Maintenance, emergency, 71
Maintenance, preventive, 71
Maintenance systems design, 71
Malcolm Baldrige National Quality Award, 212–14
Malleability, 58
Management, 215
Management by objective, 217
Management control systems, 26, 40
Management controls, 460, 463
Management information systems (MIS), 470–72
Management of process quality, 214
Manufacturability, 48
Manufacturing engineering definition, 43
Manufacturing, history of, 275–76
Market success, 215
Markov chain, 389–93
Maslow, 494
Mass production, 4, 276
Material handling, 126
Material handling cost, 100
Material requirements planning (*See* MRP systems)
Mathematical programming problem, 262–64, 354
MBNQA (*See* Malcolm Baldrige National Quality Award)
McCormick, E.J., 302
Mean, 509, 511–13, 515–20
Measured daywork, 267
Mechanical engineering, 4, 22, 25
Median, 522
Meet target, 218, 233
Merit rating, 217
Metal cutting, 60
Metal forming, 59
Methods time measurement (MTM), 173
Military engineering, 3, 22
Miller, Irwin, 192
Milling, 61
Minicomputers, 479
Minimize variation, 218, 233–34
Minimum point, 357, 360
Mize, J.H., 183, 501
Mode, 522
Models, 28
Modeling, 378
Molds, 58
Molina, E.C., 528
Montgomery, D.C., 192
Moodie, C.L., 456
Moore, J.M., 82, 89
Morse, Samuel, 4
Most likely time, 420
Motivation (employee), 494
Moving average, 190
MRP systems, 203
Multi factor productivity measurement model, 176, 178
Multiple activity chart, 163
Multiple traveling salesman problem, 140
Multiplication theorem, 503
Multiplicative congruential random number generator, 406
Mundel, M.E., 163
Murphy's law, 420
Muther, R., 103
Mutually exclusive, 502
Myers, C.E., 14

Naehring, D.C.E., 12
National Institute of Standards and Technology, 212
Necessary condition, 355–56
Net worth, 330
Newsboy model, 385
Node, 414–15
Nonconformity, 245
Noncritical path, 415
Nonlinear programming, 363, 372
Normal cost, 425
Normal distribution, 514–15
Normal distribution tables, 531–32
Normal pace, 165
Normal time, 165, 425
Normalization of data base, 477
Normative group technique, 176
Normative productivity measurement, 176
N-step transition probability, 392
Numerical control (NC), 278, 283

Objective, 454
Off-line, 218
On-hand plus on-order, 388
On-line, 218
One-at-a-time, 361
Open-loop control, 442
Open vs. closed systems, 38
Operations planning, 186, 194–97
Operations process chart, 55, 105, 163
Operations research, 15, 19–20, 22, 349–50
Operations Research Society of America, 17
Operations scheduling, 188, 200–203
Optimistic time, 420
Optimum lot size, 388

Optimum order quantity, 386
Order, 385
Order quantity, 384
Organization chart, 461
Organization design, 461
Organization life cycle, 464–67
Organization structure, 461
Organizational functions, 459
Organizational manual, 462
Organizational performance, 152
Organizational redesign, 465
Organizational renewal, 465
Organizational structure, 34, 461
Out of control, 218, 234
Out of stock, 385
Outcome, 502, 505
Overhead, 70, 335

p chart, 235, 243
Packaging systems, 72
Paired block procedure, 361
Pareto analysis, 228
Path standard deviation, 423
Performance appraisal, 486
  checklist, 487
  comparison, 487
  rating scale, 486
Performance measures, 383
Periodic review model, 388
Personnel management, 483
PERT, 412–13, 415, 419–20, 501
Pessimistic time, 420
Physiological aspects of human performance, 302–4
Piecework incentive, 267
Plackett-Burman designs, 233
Plane, 366
Planning, 415
Planning sheet, 70–71
Plant location, 351
Point-to-point network topology, 480
Poisson distribution, 381, 511–13
Poisson distribution tables, 528–30
Population, 379, 520–21
Population parameter, 521–22
Position, 253
Pouring basin, 58
Power factor, 318
Precedence relationship, 413, 415
Predetermined times, 173
Present worth amount, 343–44
Pressure welding, 67
Price of nonconformance (PONC), 220
Price tag, 217
Pride of workmanship, 217
Principles of material handling, 131
Principles of motion economy, 163
Pritsker, A. Alan B., 408
Probabalistic, 352–54
Probabilistic inventory, 384
Probabilistic models, 378
Probability, 501–23
Probability density function, 507, 514–17
Probability distribution, 508–18
Probability function, 506–20
Probability theory, 502–5
Process, 234
Process capability analysis, 48
Process characteristics, 218
Process control points, 218
Process engineering, 45
Process layout flow, 100, 102
Process standard deviation, 242
Procurement cost, 198
Product control characteristics, 218
Product features, 215
Product layout flow, 100, 102
Product-production design interaction, 44
Product structure, 45–46
Product structure diagram, 204
Productivity, 23, 151, 175
Professional licensing, 8
Professional registration, 8–10
Profit, 332, 385–87
Profit sharing, 270
Program evaluations and review techniques (*See* PERT)
Project management, defined, 411
Project planning, 414
Psychological aspects of human activities, 304–5

Quadratic programming, 373
Quality and operational results, 214
Quality control, 217
Quality control, described, 212, 217
Quality cost systems, 220
Quality function deployment, 218
Quality improvement, 217
Quality planning, 217
Quality policy, 215
Quality values, 215
Queue, 379
Queueing notation, 380
Queueing theory, 378–84

R chart, 235–37
Racks, 127
Random number generation, 403–6
Random numbers, 398
Random variable, 384–85, 387, 505, 508–22
Range, 522
Ratchet clause (energy), 318
Ray, T.L., 91
Recognition system, 216
Rectangular distribution, 516–17
Rectilinear distances, 83, 97

Reddin, W.J., 455
Refining and alloying, 56–58
Regression analysis, 192
Regular Markov chain, 390–92
Relational data base, 477
Reorder point, 387
Repetition, 506–8
Resistance welding, 64
Resolution, 361
Response-time, 215
Retained earnings, 330
Revenue, 331–32
Reward system, 216
Rework, 220
Ring network topology, 480
Riser, 58
Ritchey, John A., 14, 456
Robbins, J.A., 145
Robot, 278, 291
Robot Society of America, 17
Robotic applications, 295
Robotics, 291–95
Rolling, 59
Route sheet, 56–57
Rucker, 271
Runner, 58

Saddle point, 359
Safety programs, 492
Safety stock, 387
Salvage, 384
Sample, 520–21
Sample mean, 521–22
Sample space, 502
Sample statistic, 521–22
Sample variance, 521
Sanders, M.S., 302
Sarchet, B.R., 498
Sawing-filing, 62
Scanlon plan, 270
Scatter plot, 232
Scheduling, 414–15
Scientific management, 15
Scientific model, 349–50, 352
Screening designs, 233
Search theory, 360
Sectioning, 361
Seehof, J.M., 113
Sensitivity checks for control charts, 241, 244
Service discipline, 379
Service facility, 379
Service industries, 21
Service rate, 379–84
Setup charge, 384
Shannon, R.E., 498
Shaping and planing, 61
Shearing, 60
Shewhart, W.A., 14, 235
Shortage cost, 388
Simplex method, 366–67
SIMSCRIPT II, 408
Simulated, 360
Simulation, 137, 396
Simulation experiment, 397
Simulation languages, 407
Single payment compound amount factor, 337–38
Single period model-no setup cost, 384
Single-service channel, 381
Sink, D.S., 151–52, 176, 178
Slack, 415
SLAM, 408
Smith, R.J., 2
Society for Decision Sciences, 17
Society of American Value Engineers, 17
Society of Manufacturing Engineers, 17
Soldering, 67
Solution point, 369
Span of control, 460
Special cause, 235
Spriegel, W.R., 14
Sprue, 58
Stable process, 218, 234
Staff function, 461
Standard data, 171, 174
Standard deviation, 509
Standard hour, 268
Standard normal distribution, 515
Standard operating procedures, 463
Standard time, 165
Star network topology, 480
State, 389–93
State of statistical control (SOSC), 234
Stationary, 390
Stationary point, 355
Statistical inference, 520
Statistical process control (SPC), 223
Statistics, 393, 501–23
Stephans, M., 293
Stochastic demand, 384
Stochastic process, 389
Stockout, 387
Strategic planning, 216
Strategic quality planning, 214
Subgroups, 236, 243
Subinterval, 508
Sufficient condition, 355–56
Sullivan, William G., 356
Superfund, 325
Supervision, 496
Supplier, 216
Swaim, J.C., 178
Swim, L.K., 151
Synergism, 434
System calibration, 463, 466
System classification, 38

System definition, 37, 437
System design process, 440
System integrator, 299
System representation, 442
Systematic layout planning, 102
Systems analysis, 472
Systems engineering, 15, 22, 440–42
Systems thinking, 433–36
Systems view of an organization, 449–56

Taft-Hartley Act, 491
Taguchi designs, 233
Task, 253
Taylor, Frederick W., 13, 53
Tensile strength, 58
The Institute for Management Sciences, 17
The learning organization, 466
Thermit Welding, 67
Theusen, G.J., 338, 340
Time-cost trade off, 424
Time diagram, 336
Time-flow mechanism for simulation, 407
Time of use, 319
Time-step incrementation, 407
Time study, 167
Time value, 337
Time value of money, 344
Tool design, 68
Torgersen, P.E., 363
Total activity slack, 416, 419
Total quality management (TQM), 213
Total system, 440
Trade-off of costs, 427–28
Training, 489
Transfer function, 443
Transition matrix, 389–93
Transition probability, 389–93
Transportation linear programming, 84, 134
Transportation problem, 369–72
Transportation routing problem, 141
Trauxal, John, 442
Traveling salesman problem, 138
Turner, W.C., 145, 415
Turning, 61
Turnover, 384
Two-step transition probability, 391

Unconstrained optimization, 355
Uniform distribution, 385, 513
Union, 503
Unstable process, 218, 234
Urwick, L., 14
U.S. Department of Commerce, 212
Useful many, 228

Variables, 236
Variance, 509, 511–13, 515–23
Vidosic, Joseph P., 53
Vision, 215
Vital few, 228
Voice of the customer, 219
Vollman, T.E., 203

Wage survey, 253, 262
Waiting line, 379–84
Warehouse location, 137
Water management, 313, 322
Welding, 64
Western management, 217
White, C.R., 183
White, J.A., 84
Whitin, T.M., 388
Whitson, C.W., 26, 448
Whybark, D.C., 203
Wide area networks, 480
Wire drawing, 60
Within-subgroup variation, 236
Work design, 152
Work distribution chart, 163
Work factor, 173
Work measurement, 152, 165
Work sampling, 174
Work simplification, 153
Work standards, 217
World class, 213
World War II, 213
Wright, J., 141
Wright, Robert, 434, 463

$\bar{X}$ chart, 235–37

Zero defects, 217
Zero-one programming, 373